atomic weights) of

1	Hydrogen	1.00794
2	Helium	4.002602
3	Lithium	6.941
4	Beryllium	9.012182
5	Boron	10.811
6	Carbon	12.011
7	Nitrogen	14.00674
8	Oxygen	15.9994
9	Fluorine	18.9984032
10	Neon	20.1797
11	Sodium	22.989768
12	Magnesium	24.3050
13	Aluminium	26.981539
14	Silicon	28.0855
15	Phosphorus	30.973762
16	Sulfur	32.066
17	Chlorine	35.4527
18	Argon	39.948
19	Potassium	39.0983
20	Calcium	40.078
21	Scandium	44.995910
22	Titanium	47.867
23	Vanadium	50.9415
24	Chromium	51.9961
25	Manganese	54.93805
26	Iron	55.845
27	Cobolt	58.93320
28	Nickel	58.6934
29	Copper	63.546
30	Zinc	65.39
31	Gallium	69.723
32	Germanium	72.61
33	Arsenic	74.92159
34	Selenium	78.96
35	Bromine	79.904
36	Krypton	83.80
37	Rubidium	85.4678
38	Strontium	87.62
39	Yttrium	88.90585
40	Zirconium	91.224
41	Niobium	92.90638
42	Molybdenum	95.94
43	Technetium-99	98.9063
44	Ruthenium	101.07
45	Rhodium	102.90550
46	Palladium	106.42
47	Silver	107.8682
48	Cadmium	112.411
49	Indium	114.818
50	Tin	118.710
51	Antimony	121.757
52	Tellurium	127.60
53	Iodine	126.90447
54	Xenon	131.29
55	Caesium	132.90543
[illegible]	Barium	[illegible]
[illegible]	Lanthanum	[illegible]
58	Cerium	140.115
59	Praseodymium	140.90765
60	Neodymium	144.24
61	Promethium-145	144.9127
62	Samarium	150.36
63	Europium	151.965
64	Gadolinium	157.25
65	Terbium	158.92534
66	Dysprosium	162.50
67	Holmium	164.93032
68	Erbium	167.26
69	Thulium	168.93421
70	Ytterbium	173.04
71	Lutetium	174.967
72	Hafnium	178.49
73	Tantalum	180.9479
74	Tungsten	183.84
75	Rhenium	186.207
76	Osmium	190.23
77	Iridium	192.217
78	Platinum	195.08
79	Gold	196.96654
80	Mercury	200.59
81	Thallium	204.3833
82	Lead	207.2
83	Bismuth	208.98037
84	Polonium	(209)
85	Astatine	(210)
86	Radon	(222)
87	Francium	(223)
88	Radium-226	226.0254
89	Actinium	(227)
90	Thorium	232.0381
91	Protactinium	231.03588
92	Uranium	238.0289
93	Neptunium	237.0482
94	Plutonium-244	244.0642
95	Americium-243	243.0614
96	Curium-247	247.0703
97	Berkelium-247	247.0703
98	Californium-251	251.0796
99	Einsteinium	(254)
100	Fermium	(257)
101	Mendelevium	(258)
102	Nobelium	(259)
103	Lawrencium	(260)
104	Rutherfordium	(261)
105	Dubnium	(262)
106	Seaborgium	(266)
107	Bohrium	(262)
108	Hassium	(265)
109	Meitnerium	(266)

* Where an isotope is indicated this is the most commonly used one and its relative atomic mass is given. Where all isotopes are short-lived then the relative atomic mass of the predominant isomer is given in parentheses.

THE ELEMENTS

THE ELEMENTS

THIRD EDITION

◆

Written and compiled by

JOHN EMSLEY

Department of Chemistry, University of Cambridge

OXFORD
UNIVERSITY PRESS
Great Clarendon Street, Oxford, OX2 6DP

Oxford University Press is a department of the University of Oxford.
It furthers the Universities objective of excellence in research, scholarship,
and educationby publishing worldwide in

Oxford New York

Athens Auckland Bangkok Bogotá Bombay Buenos Aires Calcutta
Cape Town Chennai Dar es Salaam Delhi Florence Hong Kong Istanbul
Karachi Kuala Lumpur Madrid Melbourne Mexico City Mumbai
Nairobi Paris São Paulo Singapore Taipei Tokyo Toronto Warsaw
with associated companies in Berlin Ibadan

Published in the United States
by Oxford University Press Inc., New York

First edition 1989
Second edition 1991
Third edition 1998
Reprinted 1998, 1999 (with corrections), 2000

British Library Cataloguing in Publication Data
Emsley, John
The elements.
1. Chemical elements, chemical compounds—
Technical data—For schools
I. Title 540'.212

Library of Congress Cataloging in Publication Data
Emsley, J. (John)
The elements/written and compiled by John Emsley.
Includes index.
1. Chemical elements—Handbooks, manuals, etc. I. Title
QD466.E48 1988 546—dc19 88-19011

ISBN 0 19 855819 8 (Hbk)
ISBN 0 19 855818 X (Pbk)

Printed in Great Britain by
Bookcraft (Bath) Ltd, Midsomer Norton, Avon

Preface to the third edition

THE second edition of *The elements* included a questionnaire, which hundreds of readers kindly filled in and returned. It asked which parts of the book they found most useful, and invited them to suggest extra data they would like to see in a third edition. If several respondents requested the same information, then this clearly indicated a demand that should be met. The topics which came into this category were binding energies, toxicity, health hazards, mineral sources, radioactive isotopes, and bulk metal properties. Several readers wanted more information about the periodic table of the elements.

A surprising number of people wanted to know more about the names of the elements, how they had been derived and how they should be pronounced. These requests were easy to accommodate, and at the top right hand page of each element there is now more information about the discovery, etymology, pronunciation, and foreign names.

It became clear that to span a wider range of data the third edition of *The elements* would require a complete revision. Hence the three new sections: biological data, crystal data and geological data. The first of these includes data on toxicity, health hazards, and the amount of each element in the average human being. The second section is in response to the growing importance of materials science and the investigation of solid phases. The third section on geological data contains a table of the most common minerals of a particular element, as well as data about its important ores and where they are mined.

Those sections of the earlier editions of *The elements* dealing with chemical data, physical data, nuclear data, and electron shell data have also been revised and expanded. The nuclear data now include most of the radioactive isotopes, given in an Appendix, and the electron shell data now have a table of electron binding energies.

In the second edition of *The elements*, I included elements 104 and 105 under the temporary names of unnilquadium and unnilpentium, given them by the International Union of Pure and Applied Chemistry (IUPAC) pending the outcome of an investigation into who had first made them. When IUPAC finally pronounced on this two years ago they then found themselves embroiled in a row over what all the elements should be called that are heavier than fermium (element 100).

In theory, once IUPAC had decided on the discoverers, those people then had the traditional right to name them, and such names would normally be validated by IUPAC. The American Chemical Society supported the continuing use of the names rutherfordium for element 104, hahnium for 105, and in 1994 they approved the name seaborgium for 106 as chosen by its discoverer Ghiorso. The names picked by the Germans who first made the heaviest elements were: nielsbohrium for 107, hassium for 108, and meitnerium for 109.

However, the issue was far from settled, and in 1994 IUPAC announced new names for the elements 104–109. These were dubnium for 104, joliotium for 105, rutherfordium for 106, bohrium for 107, hahnium for 108, and meitnerium for 109. Only the last was allowed to retain the name chosen by the discoverers.

These names did not meet with the general approval that usually follows a decision by IUPAC, and in the USA especially, there was much opposition to those suggested for elements 104, 105, and 106. As a result IUPAC was urged to reconsider, and in February 1997 a revised list was proposed that appeared to be acceptable to all and in August 1997 they were approved. These are the names I use in the third edition of *The elements*, and they are as follows:

- 104 rutherfordium
- 105 dubnium
- 106 seaborgium
- 107 bohrium
- 108 hassium
- 109 meitnerium

Elements 110 and 111 were reported towards the end of 1994 and element 112 in 1996, but as yet they are un-named.

Finally, and in response to many requests, *The elements* now contains a long chapter on the periodic table—how it was discovered, the changes leading up to its present form, and why it is as it is. This chapter also tells the full story of the controversy which surrounds the heaviest elements.

Acknowledgements

I should like to thank the following people in particular for not only sending me suggestions for new material, but also for supplying it or telling me where it could be found: James Catallo; John S. Coe; M. T. Deans; Fernando Dufour; Ian Gameson; Randall M. German; Chaim Horovitz, T. Pedersen; N. W. Larsen; David Pendlebury; Pekka Pyykkö; David J. Stabb; John Phillips; Jonathan Hare; R. W. Darlington; Tim Jones, M. Webster, Walter Saxon, John Ellis, and Simon Cotton.

Although this third edition does not include a questionnaire inviting comments, these would always be welcome and should be addressed to John Emsley, c/o Oxford University Press, Great Clarendon Street, Oxford OX2 6DP, United Kingdom.

London J. E.
May 1997

Contents

The key to *The elements*

THE curious thing about numerical information is the way it varies slightly from source to source. Usually these variations are of the order of ±2 per cent, small but irritating. The need to standardize data has been recognized by several organizations, such as the International Union of Pure and Applied Chemistry (IUPAC) and the National Bureau of Standards (NBS) in Washington DC. When such bodies set up committees to assess data and decide on the most reliable values, the job of an author like myself becomes much easier. For example, the table of thermodynamic data of the elements comes from an NBS book while the standard reduction potentials come from an IUPAC publication.

Rather than quote alternative values, or a range of values, for certain properties I have assumed that well established specialists in the collecting of data, such as those of the *CRC handbook of chemistry and physics* (reference 1) and *Lange's handbook of chemistry* (reference 2) provide reliable information, even though it may not be in SI units. Other compilations with extensive chemical data are references 3–6. In the sections which follow I give the sources of my information, explain the SI units used, and convert the data to other commonly used units.

All numerical data are given in the units of the Système Internationale (SI) the primary units of which are:

Quantity	Name of unit	Symbol
length	metre	m
mass	kilogram	kg
amount	mole	mol
time	second	s
current	ampere	A
temperature	kelvin	K

Some units are subdivisions of the primary units. The unit of length in chemistry is the picometre (pm), which is 10^{-12} m. To convert picometres to nanometres (nm) divide by 1000, to convert to Ångstroms (Å) divide by 100.

Some units are derived from the primary units. The ones encountered in *The elements* are:

Quantity	Name of unit	Symbol	Equivalent
electric potentials	volt	V	WA^{-1}
electric resistance	ohm	Ω	VA^{-1}
energy	joule	J	Nm
pressure	pascal	Pa	Nm^{-2}

Some units have become so common within a particular area of science that it would be counter–informative not to use them. One such is the electronvolt (eV), a unit of energy which can be converted to SI units thus: 1 eV = 0.160 219 x 10^{-18} J.

Note: n.a. in the tables means that the data are not available

The elements

Name and chemical formula

IUPAC is the body responsible for approving the names and formulae given to elements. Changes are reported in *Pure and Applied Chemistry*, which is IUPAC's official journal.

The US name for aluminium is aluminum, and for caesium is cesium, and whereas the former does not affect the element's position in an alphabetical listing, the latter is located differently. The British name for sulfur is sulphur, but this does not affects its location. Some names of the elements 104–108 have been disputed and this is discussed in the chapter on the periodic table, page 278. In this edition of *The elements* the following names have been used: 104, rutherfordium; 105, dubnium; 106, seaborgium; 107 bohrium; 108, hassium and 109, meitnerium.

Pronunciation

The pronunciation of the English names for the elements are based on the following:

symbol	sound	symbol	sound	symbol	sound	symbol	sound
a	cat	f	fat	o	dog	th	this
ah	part	g	go	oh	go	u	sun
ai	hair	h	hat	oo	soon	uh	your
aw	saw	i	pit	ow	cow	v	vat
ay	day	iy	lie	oy	joy	w	win
b	big	j	joy	p	pin	y	yet
ch	chin	k	kit	r	red	z	zip
d	dig	l	lid	s	set	zh	treasure
e	set	m	mat	sh	shop		
ee	see	n	not	t	top		
er	term	ng	sing	*th*	thin		

Relative atomic masses

These are authenticated regularly by IUPAC and published in their official journal. The values used in this edition of *The elements* are taken from *Pure and Applied Chemistry*, 1994, **66**, 2420.

Discovery

The most comprehensive volume on the discovery and history of the elements is M. E. Weeks and H. M. Leicester's *Discovery of the elements*, 7th edition, published by the *Journal of Chemical Education*, Easton, PA, 1968. An outline of the discovery of each element is also given in reference 1. Occasional articles in the *Journal of Chemical Education* give excellent histories of individual elements. The discovery of the transuranium elements is given by Glenn T. Seaborg, in Chapter 2, of *Transuranium elements: a half century*, edited by L. R. Morss and J. Fuger, American Chemical Society, Washington DC, 1992.

D. W. Ball, in the *Journal of Chemical Education*, 1985, **62**, 787 and J. G. Stark and H. G. Wallace, in *Education in Chemistry*, 1970, 152, explain how the names of the elements were chosen.

All chemicals have a Chemical Abstracts Service (CAS) registry number and this is given, in square brackets, below the name. If you are on–line to the Chemical Abstracts database then using the CAS number is the easiest way of gaining access to a particular chemical.

Chemical data

Descriptions

The brief descriptions of the elements and their reactivity towards air, water, acids, and alkalis have been taken from references 7–9. The *Encyclopedia of the chemical elements*, edited by C. A. Hampel (Reinhold Book Corporation, New York,1968) gives a comprehensive, if somewhat dated account, of the uses of each element. The uses to which the lanthanide elements are put are taken from J. T. Kilbourn *Metallurgical applications of the lanthanides and yttrium* (Molycorp Inc., White Plains, NY, 1987).

Radii

To convert radii, which are given in picometres (pm), to metres divide by 10^{12}. To convert to nanometres (nm) divide by 1000; and to convert to angstroms (Å) divide by 100.

The radius of an atom depends upon several factors: oxidation state, degree of ionization, and coordination number (for metals this is generally 12). When it is part of a molecule two radii are defined: the covalent radius, which refers to the role it plays in forming bonds, and the van der Waals radius, which refers to the radius it presents to the world beyond the molecule. Many textbooks quote some of these radii, but the best sources appear to be references 2 and 4.

Data for the actinides and radioactive elements beyond uranium have also been taken from Glenn T. Seaborg and Walter D. Loveland *The elements beyond uranium*, John Wiley & Sons, New York, 1990.

Electronegativity

This quantity is well understood but ill-defined. It refers to the potential of an atom to attract electrons to itself. The higher the electronegativity the stronger is this ability. Fluorine is the most electronegative of all elements.

Electronegativity was first calculated by Pauling, whose method is based on bond energies, and these values have been up-dated from time to time, e.g. by A. L. Allred, in *Journal of Inorganic and Nuclear Chemistry*, 1961, **17**, 215. Pauling's values are given in most reference works, such as reference 4. Allred, in collaboration with E. G. Rochow, proposed an alternative method of calculating electronegativity based on the effective nuclear charge and the covalent radius of an atom: see A. L. Allred and E. G. Rochow (*Journal of Inorganic and Nuclear Chemistry*, 1958, **5**, 261). Again these values, labelled (Allred), have been amended by later workers. The values quoted here are taken from *Inorganic chemistry, principles of structure and reactivity*, 4th edition, by J. E. Huheey, E. A. Keiter, and R. L. Keiter (Harper & Row, New York, 1993) pages 187–190. The units of electronegativity are rarely quoted for either Pauling's or Allred and Rochow's values.

R. G. Pearson (*Inorganic chemistry* 1988, **27**, 734) has proposed a scale of absolute electronegativity, which is defined as the average of the first ionization energy and the electron affinity of the neutral atom. Pearson uses units of electron volts (eV) for these quantities, and consequently the absolute electronegativity is in eV, and is given here as such. To convert from eV to kJ mol^{-1} multiply by 96.486. Established electronegativity scales generally lie in the range 0 to 4, and this is extended from 0 to 10.41 for absolute electronegativities. There is little to be gained by converting absolute electronegativities to SI units.

Calculating the absolute electronegativities of the f-block elements cannot be done with precision since individual electron affinities are known only to be $\leq$50 kJ mol^{-1}; the absolute negativities for these metals should therefore be quoted with caution. For the noble gases it is also possible to calculate absolute electronegativities with more precision, and for krypton (6.8 eV) and xenon (5.85 eV) the values have meaning in that compounds are known for these elements. For helium, neon, and argon it is also possible to give values, showing these to be the most electronegative of the elements, e.g. helium = 12.3 eV. Since these elements form no molecules the electronegativity values are enclosed in square brackets.

Effective nuclear charge, Z_e

Like electronegativity, this quantity is easier to understand than calculate. Although Z_e, can be defined for any electron within an atom, only the Z_e for the valence shell electrons is of interest to the chemist. Z_e is the charge due to the protons of the nucleus less a screening factor due to the other electrons of the atom. There are several slightly different ways of calculating this screening, and consequently there are several values for Z_e. Those quoted were computed by J. C. Slater (*Physical Review*, 1930, **36**, 57), E. Clementi and D. L. Raimondi (*Journal of Chemical Physics*, 1963, **38**, 2686), E. Clementi, D. L. Raimondi, and W. P. Reinhardt (*Journal of Chemical*

Physics, 1967, **47**, 1300), and C. Froese-Fischer (*Atomic Data*, 1972, **4**, 301, and *Atomic Data and Nuclear Data Tables*, 1973, **12**, 87).

Standard reduction potentials

In these diagrams the element is arranged with the highest oxidation state on the left. Potentials are given in volts. The higher the value of $E^\ominus$, the stronger is the oxidant as an oxidizing agent; the lower $E^\ominus$, the stronger is the reductant as a reducing agent. These diagrams are taken from *Standard potentials in aqueous solutions*, edited by A. J. Bard, R. Parsons, and J. Jordan. [Marcel Dekker (for IUPAC), New York, 1985.] Other compilations of $E^\ominus$ data are to be found in references 1–6.

Oxidation states

Although it is merely a formalism, the concept of oxidation number is much used in describing the changes that happen to an element in its chemical reactions. Consequently it is a useful way of classifying and explaining the compounds of that element. From the multitude of compounds known for elements in their various oxidation states I have chosen where possible to give the oxides, hydroxides or acids, hydrides, fluorides, and chlorides (after which 'etc.' means the corresponding bromides and iodides), and the species present in aqueous solutions of simple salts of the element (denoted 'aq'). Salts, complexes, and organometallic compounds are also given if these are special; otherwise I have merely indicated that such substances exist, and references 7–9 should be consulted for further details.

Bond lengths and energies

To convert bond lengths, *r*, given in picometres (pm), to metres, divide by 10^{12}; to convert them to nanometres (nm) divide by 1000; and to convert them to angstroms (Å) divide by 100. To convert bond energies, *E*, given in kJ mol^{-1} to kcal $mole^{-1}$, divide by 4.184. Bond energies are more correctly known as bond enthalpies.

Tables of bond lengths and bond energies are given only for the elements of the p–block and for beryllium. They are taken from a variety of sources, notably bond lengths from references 1 and 4 and bond enthalpies from references 1, 3, and 4 and from *Bond energies, ionization potentials and electron affinities* by V. I. Vedeneyev, L. V. Gurvich, V. N. Kondrat'yev, V. A. Mededev and Ye. L. Frankevich (Edward Arnold, London, 1966). *SI chemical data* by G. H. Aylward and T. J. V. Findley (Wiley, Sydney, 1971) also gives *r* and *E* for many bonds.

The most recent text to discuss these parameters, and list them for most bonds, is *Inorganic chemistry, principles of structure and reactivity*, 4/e, by J. E. Huheey, E. A. Keiter and R. L. Keiter (Harper & Row, New York, 1993) appendix E, pages A–21 to A–34. Where possible I have given preference to *r* and *E* values taken from this source, which gives them in SI units and in the older units of Å and kcal $mole^{-1}$.

Bond lengths to metals are to be found in a compilation from the Cambridge Structural Database: A. G. Orpen, L. Brammer, F. H. Allen, O. Kennard, D. G. Watson and R. Taylor, Tables of Bond Lengths Determined by X–Ray and Neutron Diffraction. Part 2. Organometallic compounds and coordination complexes of the d- and f-block metals, *J. Chem. Soc., Dalton Trans.*, 1989, supplement, which follows page 2462 of that journal.

Note: *r* and *E* should be treated with caution because they are generally the averages of values that vary widely and which depend upon the oxidation state of the element and the other bonds to which that element is attached.

Physical data

Melting points and boiling points

These are given in Kelvin (K). They can be converted to degrees Celsius (°C) by subtracting 273.15. The values quoted are based on reference 1, but are also given with slight variations in all major data books (references 2–9). Critical temperatures, pressures, and volumes are also given for the gaseous elements. Pressure is reported in kilopascals (kPa); to convert to bars divide by 100, to convert to Torr (mmHg) multiply by 7.500; to convert to atmospheres divide by 101.325. Values are taken from R. C. Reid, J. M. Prausnitz, and T. K. Sherwood, *The properties of gases and liquids* (3rd edition) (McGraw-Hill, New York,1977).

The critical temperature, above which a gas cannot be liquefied no matter how high the pressure, is given for hydrogen, helium, nitrogen, fluorine, neon, chlorine, argon, krypton, xenon and radon. These values are taken from reference 1 with the exceptions of chlorine and radon which are taken from reference 6. The corresponding critical pressures are also given for these elements.

Enthalpy of fusion, ΔH_{fusion} and enthalpy of vaporization, ΔH_{vap}

These, given in kJ mol $^{-1}$, can be converted to kcal mol $^{-1}$ by dividing by 4.184. The values are taken mainly from R. Loebel's compilation in reference I (where they are given in c.g.s. units), supplemented by references 4 and 6, where they are in SI units. The values for ΔH_{vap} are taken mainly from reference 6.

Thermodynamic properties

To convert kJ mol^{-1} to kcal mol^{-1} divide by 4.184. To convert entropies in J K^{-1} mol^{-1} to eu (entropy units, i.e. cal K^{-1} mol^{-1}) divide by 4.184. To convert specific heats, C_p, to cal g^{-1} K^{-1} divide first by 4.184 and then by the relative atomic mass of the element concerned.

The thermodynamic properties are taken from *The NBS tables of chemical thermodynamic properties* by D. D. Wagman, W. H. Evans, and V. B. Parker, which was published jointly by the American Chemical Society and the American Institute of Physics for the National Bureau of Standards, Washington DC, in 1982. Although thermodynamic data for the elements are to be found in references 1–5, the NBS compilation is preferred and is in SI units.

Density

Since the basic SI unit of weight is the kilogram (kg) and of length the metre (m), the preferred unit of density is kg m^{-3}. The more common unit, however, is g cm^{-3} and to convert from kg m^{-3} to g cm^{-3} divide by 1000.

Many sources list densities at only one temperature, such as reference 2 (293 K), others at various temperatures, e.g. reference 6, which reports them in SI units. The densities used here are based mainly on reference 1 for the solid elements. In the same work can be found the densities of the liquid elements at their melting points, compiled by G. Lang.

Thermal conductivity

The SI unit for this property is watts per metre per Kelvin (Ω m^{-1} K^{-1}). To convert to W cm^{-1} divide by 100.

The values are taken from C. Y. Ho, R. W. Powell, and P. E. Liley (*Journal of Physical Chemistry Reference Data*, 1974, **3**, suppl. 1), with the values for carbon being taken from reference 1 for the directions perpendicular and parallel to the graphite axis. Reference 2 also reports the thermal conductivity of the elements.

Electrical resistivity

The SI units are ohm metres (Ω m) and the electrical resistivity of metals are of the order of 10^{-8} Ω m. To convert to the more common units of $\mu\Omega$ cm multiply the Ω m values by 10^{8}.

The data are taken from reference 1. Reference 6, pp. 102–3, also gives the electrical resistivity in Ωm at temperatures of 78, 273, 373, 573, and 1473 K. Values in non-SI units are given in reference 3, pp. 580–684.

Mass magnetic susceptibility, χ

This is obtained from the volume magnetic susceptibility, κ, which is unitless, by dividing by the density (kg m^{-3} in SI units). To convert mass magnetic susceptibilities from the SI units of kg $^{-1}$ m^3 to c.g.s. units of g $^{-1}$ cm^3, multiply by $1000/4\pi$, i.e. 79.6. To convert to molar magnetic susceptibility multiply first by 79.6 then by the relative atomic mass of the element. The use of SI units for magnetic properties is discussed by T. I. Quickenden and R. C. Marshall, in the *Journal of Chemical Education*, 1972, **49**, 114.

The magnetic susceptibility data are taken from *Constantes sélectionnées. Diamagnétisme et paramagnétisme* by G. Foëx (Masson et Cie, Paris, 1957), and *Modern magnetism* (4th edition) by L. F. Bates (Cambridge University Press, Cambridge, 1963). Values for certain of the lanthanides were taken from J. M. Lock (*Proceedings of the Physical Society*, 1957, **B70**, 476 and 566). Reference 1 reports molar magnetic susceptibilities in c.g.s. units.

Coefficient of linear thermal expansion, α

This is the same in SI and c.g.s. units, i.e. K^{-1}. It is often reported as $10^6\alpha$. The data are taken from reference 3 (pp. 580–684).

Molar volume (also known as atomic volume)

This represents a slight problem for SI whose unit of volume is m^3. Traditionally atomic volume is obtained by dividing the relative atomic mass of an element in grams by the density in g cm^{-3}, and is consequently traditionally expressed in cm^3. To express the molar volume in m^3 divide the values given by 10^6. Expressing it in this way, however, implies a density measured in g m^{-3}, which is not a recognized way of expressing this quantity.

Because atomic volume for an element depends upon the density, it also depends upon the phase, the allotrope, and the temperature. The values reported here are based where possible on the solid state at room temperature (298 K or as indicated), and are taken mainly from C. N. Singman (*Journal of Chemical Education*, 1984, **61**, 137).

Elastic properties

The elastic constants (Young's modulus, rigidity modulus, bulk modulus and Poisson's ratio) of the polycrystalline metals at room temperature are taken mainly from *Smithells metals reference book*, 7th edition, E. A. Brandes, (ed.), Butterworths, London, 1991; the values for the lanthanides (elements with atomic numbers 58–71) are taken from reference 1 section B–214. The values for iron are taken from reference 6 although there is a much more comprehensive listing of the various grades of cast iron in *Smithells metals reference book*. Elastic moduli are given in pressures of gigapascals (GPa) equivalent to 10^9 Pa.

Young's modulus refers to longitudinal strain; the rigidity modulus, also known as the shear modulus, refers to the change of shape produced by a tangential stress; the bulk modulus is the ratio of pressure to the decrease in volume, which is the inverse of the compressibility; and Poisson's ratio is the lateral contraction per unit breadth divided by the longitudinal extension per unit length.

Physical property data for the f-block elements (lanthanides and actinides) are taken mainly from reference 12, and also from S. Cotton, *Lanthanides and actinides* (Macmillan, Basingstoke, UK, 1991) and L. R. Morss and J. Fuger, (ed.), *Transuranium elements* (American Chemical Society, Washington DC, 1992).

Biological data

Biological role

Information about essential elements is from reference 11 and R. J. Kutsky, *Handbook of vitamins, minerals and hormones* (2nd edition), Van Nostrand Reinhold, New York, 1981.

Toxicity

Toxicological information about the elements depends on individual compounds, as well as the elements themselves. Useful sources of information are *Metal toxicology in mammals (2* volumes) by T. D. Luckey and B. Venugopal (Plenum Press, New York, 1977), and the *Handbook of toxicity of inorganic compounds*, edited by H. G. Seiler and H. Sigel (Marcel Dekker, New York,1988), which covers a majority of elements in alphabetical order.

For ten toxic elements (antimony, arsenic, bismuth, cadmium, indium, lead, mercury, selenium, tellurium, thallium) the subject is covered in great detail in J. E. Fergusson, *The heavy elements: chemistry, environmental impact and health effects*, Pergamon Press, Oxford 1990.

The toxicity of an element depends very much of the form in which it is taken into the body. 'Toxic intake' is meant to give some guidance as to the elements ability to affect humans; TWA means time–weighted average and refers to a level above which there is a threat to health.

'Lethal intake' generally refers to a specific dose that will cause death and the data are based on animal studies in most cases. Human data in this section naturally refers to a few individual cases. LD_{50} (lethal dose 50 per cent) is the level of exposure given to a large group of animals in which half the animals died. The LD_{50} data are given for a typical compound of the element, with the method of exposure and the species on which it was tested. LC_{50} refers to the lethal concentration of a gas. The information is taken from reference 16.

'Health hazards' is a brief summary of the dangers to human health which the element represents. Where an element, or its compounds, is said to be carcinogenic this means that it is known to cause cancer in humans, whereas the term *experimental* carcinogen means that the effect has been observed only with laboratory animals. The term teratogenic means that it can cause a deformed fetus. The information is taken from reference 16.

Physical hazards associated with handling the element itself are given in the Geological Data section under the heading Specimen.

Level in humans

The level of an element in a human depends on many factors and generally a range of values is quoted for a particular organ. Reference 11 has tables of the elemental composition of various human tissue of which muscle, liver and bone are given here (p.p.m. refers to parts per million, which is milligrams per kilogram of dry weight). The elemental composition of human blood (whole blood) is given as milligrams per dm^3 (= litre), also taken from reference 11. The daily dietary intake is taken from the same source.

The total mass of an element in the human body has been taken from references 11, 10, and 15 and is based on the average person weighing 70 kg. For the rarer elements it has been calculated from the levels in the major organs, taking into account the relative mass of these in the human body. In some cases where the data are available from only one part of the body, such as the blood, then it has been assumed that this figure would apply throughout the body. This is a doubtful assumption and such data should be treated with caution.

Nuclear data

Isotopes

The tables of isotopes in the main entry includes all the stable isotopes for an element and any that are radioactive but occur naturally on Earth. The longer lived radioactive isotopes are given in Appendix 1. For elements beyond bismuth, where all isotopes are radioactive, I have included only those with half lives greater than a given time, indicated at the foot of the table, but for those beyond fermium I have included all the isotopes because this is about the only data we have for these elements.

The per cent natural abundance is given as 'trace' for certain shortlived nuclides that are part of a natural decay series. The half-life, $T_{1/2}$, is expressed in seconds (s), minutes (m), hours (h), days (d), or years (y). The decay mode is shown as β^- for electron emission, β^+ for positron emission, α for alpha decay, EC for electron capture, IT for isomeric transition, and SF for spontaneous fission. Some nuclei decay by two routes. The energy (in mega electronvolts, MeV) of the radiation is given in parentheses. γ indicates the emission of gamma radiation.

The nuclide data is taken from reference 1. Similar tables are given in references 2 and 3. The most comprehensive listing is to be found in the *Table of isotopes* (7th edition) by C. M. Lederer and V. S. Shirley (John Wiley & Sons, New York, 1978). This work lists the full data on 2600 known isotopes, and its compilation was supported by the US NBS Office of Standard Reference Data.

Nuclear spin (I) is reported in units of $h/2\pi$. The nuclear magnetic moment (μ) is reported in nuclear magnetons with diamagnetic correction. Both I and μ are taken from *Nuclear spins and moments* by G. H. Fuller (*Journal of Physical Chemistry Reference Data*, 1976, **5**, 835).

The abbreviations under the heading 'uses' refer to the following:

- **E** (experimental) means that isotopically enriched samples of stable isotopes are available for experimental purposes;
- **NMR** (nuclear magnetic resonance) means that this spectroscopic technique can be used to study compounds of the element;
- **R** (radioactive isotopes) means that these nuclides are available commercially;
- **D** (diagnostic) refers to the use of the nuclide for medical diagnosis;
- **T** (therapy) means that the nuclide is used in hospital treatment for disease.

E, R, D and T data are taken from reference 12, and from *Isotopes, products and services catalog* (Oak Ridge National Laboratory, PO Box X, Oak Ridge, Tennessee 37831, USA); and *Biochemicals* (The Radiochemical Centre, PO Box 16, Amersham, Bucks, HP7 9LL, UK). *Radionuclide tracers* by M. F. L'Annunziata (Academic Press, New York, London, 1987), also contains appendices of available isotopes and their radiation characteristics.

NMR

Where data for two nuclei are given, the one indicated in bold is more frequently used for NMR studies. Nuclear spins are given in the Key Isotopes table.

Relative sensitivity is at constant field for equal numbers of nuclei. Absolute sensitivity or receptivity is commonly quoted relative to $^{13}C = 1.00$.

The magnetogyric ratio is given in rad T^{-1} s^{-1}, but is often quoted as γ values which are in units of rad T^{-1} s^{-1} multiplied by 10^7. Thus for ^1H $\gamma = 26.75$, as opposed to 26.75×10^7 rad T^{-1} s^{-1}. γ is a constant of proportionality between frequency and field strength, and so the units are frequency (rad s^{-1})/field (T), i.e. rad T^{-1} s^{-1}.

Nuclear quadrupole moments are given in units of m^2. They can be converted to cm^2 by multiplying by 10^4, to barns by multiplying by 10^{28}, or to millibarns (mb) by multiplying by 10^{31}. The values were taken from the updated listing of nuclear quadrupole moments compiled by Pekka Pyykkö for IUPAC: *Quantities, units and symbols in physical chemistry* (2nd edition), Blackwells, Oxford, 1992. The quadrupole moments of the first twenty elements of the periodic table, hydrogen to calcium, are discussed in P. Pyykkö, *Zeitschrift für Naturforschung*, 1992, **47a**, 189. A complete periodic table of the NQRs of the elements by Pekka Pyykkö and Jian Li is available from the Department of Chemistry, University of Helsinki, Et. Hesperiankatu 4, SF–00100 Helsinki, Finland. Nuclei which do not have a quadrupole moment are indicated with a dash (–).

Frequency is quoted relative to the ^1H signal of $Si(CH_3)_4$, which is exactly 100

MHz in a magnetic field of 2.3488 T. For NMR spectrometers with ^{1}H at 60, 90, 200, 250, 360, or 400 MHz the frequency and field vary in direct proportion; in other words multiply the frequencies given in the tables by 0.6, 0.9, 2, 2.5, 3.6, or 4 respectively.

The data for the NMR tables were compiled from the *Handbook of high resolution multinuclear NMR* by C. Brevard and P. Granger (John Wiley & Sons, New York, 1981), *NMR and the periodic table* by R. K. Harris and B. E. Mann (Academic Press, London,1978), *Multinuclear NMR* by J. Mason (Plenum Press, New York and London, 1987), and from Bruker Scientific Instruments publications.

Electron shell data

Ground state electron configuration and term symbol

These are given in most inorganic textbooks, such as references 7 and 8. The data given here were taken from the *Handbook of atomic data* by S. Fraga, J. Karwowski, and K. M. S. Saxena (Elsevier, Amsterdam, 1976).

Electron affinity

To convert the values given in kJ mol^{-1} to electron volts (eV) divide by 96.486; to convert to MJ mol^{-1} divide by 1000.

Electron affinity is conventionally reported as positive if the addition of an electron to an atom releases energy, as it almost invariably does for the step $M \rightarrow M^-$, and negative if the process is energy absorbing, as it is for the addition of a second electron, i.e. $M^- \rightarrow M^{2-}$. This energy convention is the opposite of that used in reporting ionization energies and most other energy changes.

Although many chemistry textbooks report some electron affinities, and the major compilations, references 1 and 2, give extensive lists (in eV), the data reported here come from H. Hotop and W. C. Lineberger, *Journal of Physical Chemistry Reference Data*, 1985, **14**, 731, also reported in eV. An earlier compilation by R. J. Zollweg (*Journal of Chemical Physics*, 1969, **50**) also gives some values for the heavier elements. Their variation from element to element is discussed by E. C. M . Chen and W. E. Wentworth, *Journal of Chemical Education*, 1975, **52**, 486.

Ionization energies (ionization potentials)

To convert from kJ mol^{-1} to electron volts (eV) divide the values given by 96.486; to convert to MJ mol^{-1} divide by 1000.

Ionization energies are known to a high degree of accuracy for removal of the first, second, third, fourth, and fifth electrons for most elements, and for subsequent electron removal from the lighter elements. Values given in parentheses are less reliable. For some elements ionization energies beyond the tenth are available.

Many textbooks and compilations list ionization energies, some in SI units such as reference 4. The NBS has at different times published ionization data, beginning with E. C. Moore's *Atomic energy levels*, Volume III (NBS Circular 467, 1958). Reference 1 quotes *Analyses of optical spectra*, NSRDS-NBS 34 (Office of Standard Reference Data, NBS, Washington DC,1970). The values for the lanthanides and actinides were taken from W. C. Martin, L. Hagan, J. Reader, and J. Sugar (*Journal of Physical Chemistry Reference Data*, 1974, **3**, 771).

Electron binding energies

These are the energies required to remove electrons from the various shells and sub–shells (orbitals) of an atom and are given in units of electronvolts (eV), the commonly used unit for electron binding energies. To convert to kJ mol^{-1} multiply by 96.486, or to MJ mol^{-1} multiply by 0.096486. The electron binding energies are referred to the Fermi level (metals), valence band maxima (semiconductors) and vacuum level (noble gases). The values are taken from *Electron binding energies of the elements* version 2, January 1992, compiled by Gwyn P. Williams of the National Synchrotron Light Source, Brookhaven National Laboratory, Upton, New York,

11973, USA. The references used in this compilation were: J. C. Fuggle and N. Martensson *Journal of Electron Spectroscopy and Related Phenomena* 1980, **21**, 275; M. Cardona, and L. Ley, *Photoemission in solids*, Springer Verlag (1978); and J. A. Bearden and A. F. Burr, *Reviews of Modern Physics*, 1967, **39**, 125.

Reference 1 also gives the G. P. Williams's table of electron binding energies.

Principal lines in atomic spectrum

The wavelengths are given in nanometres (nm); to convert to angstrom units (Å) multiply by 10. The stronger lines in the spectrum are listed with the strongest shown in bold. Lines arising from the neutral atom are indicated by Species 1, and those arising from the singly charged ion M^+ are Species II. Application in atom absorption spectrometry is indicated by (AA) and is taken from *Pye Unicam atomic absorption data book* (3rd edition) by P. J. Whiteside (Pye Unicam, Cambridge, UK, 1979). For some transuranium elements, with many intense lines, only the first seven lines identified as arising from the neutral atom are given.

Data are taken from *Line spectra of the elements* by J. Reader and C. H. Corliss in reference l, and were prepared under the auspices of the Committee on Line Spectra of the Elements of the National Academy of Science, National Research Council, USA.

Crystal data

Crystal structure

To convert the cell parameters to angstroms (Å) divide by 100. To convert them to nm divide by 1000.

The abbreviation b.c.c. means body-centred cubic, f.c.c. means face-centred cubic, and h.c.p. means hexagonal close packed. The data are taken from *Landolt-Bornstein*, New Series, Group III, Vol. 6, edited by K. H. Hellwege and A. M. Hellwege. Crystal structure data are also to be found in references 3, 5, and 9; and *The structures of the elements* by J. Donohue (John Wiley & Sons, 1974).

X–Ray diffraction mass absorption coefficients

These are taken from *International tables for X-ray crystallography, Vol. III, Physical and Chemical Tables*, section 3.2, published for the International Union of Crystallographers by the Kynock Press (Birmingham, UK, 1962).

Neutron scattering lengths and thermal neutron capture cross–section

These are given in barns, a unit defined as 10^{-24} cm^2. To convert to m^2 multiply by 10^{-4}.

Given here are the neutron scattering length (b), more correctly referred to as the bound coherent scattering length, and the neutron capture cross–section (σ_a) for all naturally occurring isotopes. Values are taken from V. F. Sears *Neutron scattering lengths and cross sections*, published by AECL Research, Chalk River Laboratories, Ontario, Canada, 1992. This publication also lists the bound incoherent scattering lengths, and both coherent and incoherent scattering cross–sections.

Geological data

Minerals

In the main text the table of minerals for each element lists up to four or five of the most important ones, and gives for each its name, chemical composition, density, hardness and appearance. Where there are other important minerals the table is continued in Appendix 3 on page 259 and this is indicated.

Density is given in g cm^{-3} and to convert to the SI units of kg m^{-3} multiply by 1000.

The hardness is based on Mohs' scale and ranges from 1 to 10. This scale depends on the ability of one mineral to scratch another. It is not a linear scale, but is more akin to a logarithmic scale; in other words a mineral of hardness 10 is about ten times harder that one of 9, a hundred times harder than one of 8, a thousand times harder than one of 7, and so on. The minerals chosen by Mohs for his benchmarks are: 1 talc, 2 gypsum, 3 calcite, 4 fluorite, 5 apatite, 6 orthoclase, 7 quartz, 8 topaz, 9 corundum, 10 diamond. Under the heading 'appearance' I first give the crystal system, for which the abbreviations are as follows:

cub. = cubic	**tet.** = tetragonal
hex. = hexagonal	**rhom.** = rhombohedral
orth. = orthorhombic	**mon.** = monoclinic
tri. = triclinic (anorthic)	**amor.** = amorphous

then a brief description of the appearance of the crystal, and its colours; 'gem' indicates that the mineral has gem–like qualities. Other abbreviations used to describe the appearance of crystals are: adam. for adamantine (meaning hard); met. for metallic; res. for resinous; trans. for translucent; and vit. for vitreous (meaning glass like.

The mineral data is taken from W. L. Roberts, T. J. Campbell, and G. R. Rapp Jr. *Encylcopedia of minerals* (2nd edition), Van Nostrand Reinhold, New York, 1990. This compilation contains 3200 minerals listed alphabetically and includes for each mineral its crystal system, class, space group, lattice constants, three strongest diffraction lines, optical constants, cleavage, habit, colour–lustre, mode of occurrence and selected references. E. H. Nickel, and M. C. Nichols *Mineral reference manual*, Van Nostrand Reinhold, New York, 1991 is briefer, although it lists 3700 minerals and contains a literature reference for each one. A. Mottana, R. Crespi, and G. Liborio, *The Macdonald encyclopedia of rocks and minerals*, Macdonald, London, 1991 (English translation by C. Atthill, H. Young and S. Pleasance) lists only the 276 most common minerals but illustrates them all in colour.

Chief ores, main mining areas, world production and reserves

References 1 and 7 give the names of the chief ores of each element.

For certain key elements the data have been taken from W. Büchner, R. Schliebs, G. Winter, and K. H. Büchel, *Industrial inorganic chemistry*, translated by D. R. Terrell (VCH. Weinheim), 1989, which also gives the known reserves. The *Annual review of mining 1986*, published by Mining Journal Ltd, London, gives an account of 55 elements of commercial importance. Together these two publications give most of the data for the year 1985, which I have taken as representing a normal year, if there can be such a thing in the world economy. J. E. Ferguson, *Inorganic chemistry and the earth* (Pergamon Press, Oxford, 1982), also has some useful data, including Table 3.4 listing reserves for certain key metals.

The production figures refer to total world production of the element or its principal compound, although for some elements this should be treated with caution. The data for the rare earth elements (f block) was provided by Molycorp Inc., White Plains. NY, and for the platinum groups of metals by Johnson Matthey, London, UK. The production data for cadmium, cobalt, copper, lead, manganese and zinc are taken from C. R. German, *Chemistry in Britain*, September 1994, p. 720.

The data for carbon were taken from *BP statistical review of world energy* (London, June, 1997) recorded in units of tonnes of oil equivalent, which corresponds roughly to a chemical formula of CH_2. The figures for natural gas and oil are also expressed in units of oil equivalent.

Specimen

These data are for those who seek small samples of the elements, perhaps to make a collection for display purposes. Most elements are covered in S. Solomon and D. J. Bates *Journal of chemical education*, 1991, **68**, 991 who provide a list of companies which will supply them, together with their cost in $US, and a guide on safe handling. For *The elements* I have extended their list to all the elements, including the radioactive ones, although only officially registered users would be able to obtain these. The following precautions are indicated:

- ***safe*** means that no special precautions need be taken;
- ***care*** means the element is best handled with gloves;
- ***warning*** means the element should be handled with gloves in a fume hood;
- ***danger*** means this element is very dangerous and recommended guidelines must be followed when handling it;
- ***radioactive*** probably means that this element is unobtainable except for research purposes.

The names and addresses of suppliers are given below, and refer to the US divisions of these suppliers, many of whom have outlets in other countries.

- Johnson Matthey: Aesar Group, 892 Lafayette Road, PO Box 1087, Seabrook, NH 03874–1087, or Alfa Products, PO Box 8247, Ward Hill, MA 01835–0747.
- Aldrich Chemical Company, Inc., 1001 West Saint Paul Avenue, Milwaukee, WI 53233. They produce a catalogue *Aldrich Inorganics* which provides a comprehensive listing of all non–radioactive elements.
- Flinn Scientific, PO Box 219, 131 Flinn Street, Batavia, IL 60510.
- Fisher Scientific, Inc., 711 Forbes Avenue, Pittsburgh, PA 15219.
- Ward's National Science Establishment, PO Box 92912, Rochester, NY 14692–9012.

Suppliers of radioactive materials are:

- Oak Ridge National Laboratory, PO Box 2008, Oak Ridge, Tennessee 37831, USA;
- Amersham International plc, Lincoln Place, Green End, Aylesbury, Bucks HP20 2TP, UK.

Abundances

For atmosphere, Earth's crust, and seawater the units are parts per million (p.p.m.), defined in the case of the atmosphere as cubic centimetres per cubic metre, in the case of the Earth's crust as grams per metric tonne (1000 kg), which is the same as milligrams per kilogram. The relative abundances of elements in the Sun are taken from J. E. Ross and L. H. Aller, *Science*, 1976, **191**, 1223, where they are reported relative to hydrogen (taken as 1×10^{12}) and given as log(relative abundances). These logarithmic values are also reproduced in Appendix A of reference 10. Solar abundances for arsenic, selenium, tellurium, iodine, tantalum, krypton, and xenon are not available because their spectra lines are masked by lines of more abundant elements. Some elements, notably the heavier radioactive elements, are not reported because they are too rare to be detected.

The atmospheric abundances are taken from reference 6. The abundance of elements in the Earth's crust are an average taken from Table 3.3 of reference 11.

The concentrations of the elements in the oceans, their residence times, and oxidation states are taken chiefly from M. Whitfield and D. R. Turner, *Aquatic surface chemistry*, Chapter 17, pp. 457–93, edited by W. Stumm (John Wiley & Sons Inc., New York, 1987). The values are expressed as parts per million (p.p.m.) by weight, i.e. milligram of the element per kilogram of seawater. To convert to molarities, multiply by the relative density of seawater (e.g. average at 277 K is 1.028 g cm^{-3}), divide by 1000 and by the relative atomic mass of the element. The more restricted data for the rarer elements are taken from Appendix A of reference 10, which is based on K. W. Bruland, *Trace elements in the ocean*, in *Chemical oceanography*, Volume 8, edited by J. Riley and R. Chester (Academic Press, New York, 1983), P. Henderson, *Inorganic geochemistry* (Pergamon Press, Oxford, 1982), and

K. Schwochau, *Extraction of metals from seawater*, in *Topics in Current Chemistry*, 1984, **124**, 91.

Whitfield and Turner classify the elements as accumulating, recycled, and scavenged according to their oceanic profiles. 'Accumulating' means residence times in excess of 10^5 years with a concentration little changed with depth; 'recycled' refers to residence times of 10^3 to 10^5 years with a concentration that increases with depth; and scavenged elements have residence times of less than 1000 years and a concentration profile that decreases with depth. Readers requiring an approximate general value of ocean abundance should take an average of the Atlantic and Pacific deep values.

General references

1. Lide, D. R. (ed.) (1994) *CRC handbook of chemistry and physics*, (75th edn), CRC Press, Boca Raton, Fl.
2. Dean, J.A. (ed.) (1992) *Lange's handbook of chemistry* (14th edn), McGraw–Hill, New York.
3. Moses, A.J. (1978) *The practising scientist's handbook*, Van Nostrand Reinhold, New York.
4. Ball, M.C. and Norbury, A.H. (1974) *Physical data for inorganic chemists*, Longman, London.
5. Samsonov, G.V. (ed) (1968) *Handbook of the physicochemical properties of the elements*, IFI–Plenum, New York.
6. Kaye, G.W.C. and Laby, T.H. (1993) *Tables of physical and chemical constants* (15th edn), Longman, London.
7. Greenwood, N.N. and Earnshaw, A. (1984) *Chemistry of the elements*, Pergamon Press, Oxford.
8. Cotton, F.A. and Wilkinson, G. (1988) *Advanced inorganic chemistry* (5th edn), John Wiley & Sons, New York.
9. Bailar, J.C., Emeléus, H.J., Nyholm, R., and Trotman–Dickenson, A.F. (eds.) (1973) *Comprehensive inorganic chemistry* (5 vols), Pergamon Press, Oxford.
10. Cox, P.A. (1989) *The elements: their origin, abundance and distribution*, Oxford University Press, Oxford.
11. Bowen, H.J.M. (1979) *Environmental chemistry of the elements*, Academic Press, London.
12. Seaborg, G.T. and Loveland, W.D. (1990) *The elements beyond uranium*, Wiley Interscience, New York.
13. Merian, E. (ed.) (1991) *Metals and their compounds in the environment*, VCH, Weinheim, Germany.
14. Budavari, S. (ed.) (1989) *The Merck Index*, (11th edn), Merck & Co. Inc., Rahway, NJ.
15. James, A.M. and Lord, M.P. (1992) *Macmillan's chemical and physical data*, Macmillan Press Ltd., London.
16. Sax, N.I. and Lewis Sr., R.J. (1989) *Dangerous Properties of Industrial Materials* (7 edn), Van Nostrand Reinhold, New York, (3 volumes).

THE ELEMENTS

Ac

Atomic number: 89
Relative atomic mass ($^{12}C = 12.0000$): 227.0728 (Ac-227)

CAS: [7440-34-8]

• CHEMICAL DATA

Description: Actinium is a soft, silvery-white, radioactive metal which glows in the dark. Pure samples of the metal have been separated by reducing AcF_3 with Li at about 1200 °C. Actinium metal reacts with water to release hydrogen gas.

Radii/pm: Ac^{3+} 118; atomic 188
Electronegativity: 1.1 (Pauling); 1.00 (Allred); 5.3 eV (absolute)
Effective nuclear charge: 1.80 (Slater)

Standard reduction potentials $E^{\ominus}$/V

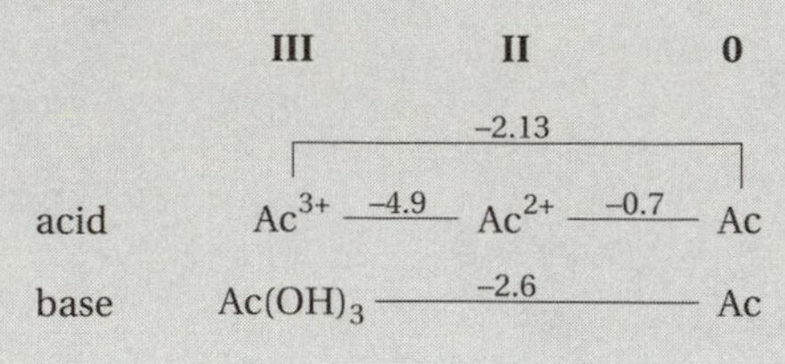

Oxidation states

Ac^{III} [Rn] Ac_2O_3, $Ac(OH)_3$ insoluble [AcH_2 and AcH_3 are known but are thought to be compounds of Ac in oxidation state III.]

• PHYSICAL DATA

Melting point/K: 1320 ± 50
Boiling point/K: 3470 ± 300

ΔH_{fusion}/kJ mol^{-1}: 14.2
ΔH_{vap}/kJ mol^{-1}: 418
ΔH_{subl}/kJ mol^{-1}: 51.9 ± 0.5

Thermodynamic properties (298.15 K, 0.1 MPa)

State	$\Delta_f H^{\ominus}$/kJ mol^{-1}	$\Delta_f G^{\ominus}$/kJ mol^{-1}	$S^{\ominus}$/J K^{-1} mol^{-1}	C_p/J K^{-1} mol^{-1}
Solid	0	0	56.5	27.2
Gas	406	366	188.1	20.84

Density/kg m^{-3}: 10 060 [293 K]
Molar volume/cm^3: 22.6
Thermal conductivity/W m^{-1} K^{-1}: 12 [300 K]
Coefficient of linear thermal expansion/K^{-1}: 14.9×10^{-6}
Electrical resistivity/Ω m: n.a.
Mass magnetic susceptibility/kg^{-1} m^3: n.a.

Young's modulus/GPa: n.a.
Rigidity modulus/GPa: n.a.
Bulk modulus/GPa: n.a.
Poisson's ratio/GPa: n.a.

• BIOLOGICAL DATA

Biological role

None.

Toxicity

Toxic intake: n.a.
Lethal intake: n.a.

Hazards

Actinium is never encountered normally. It is dangerous because it is a powerful source of α-radiation. This element is only to be found inside nuclear facilities or research laboratories.

Levels in humans

nil

Daily dietary intake: nil
Total mass of element in average (70 kg) person: nil

Discovered in 1899 by Andre Debierne at Paris, France.
[Greek, *aktinos* = ray]
French, *actinium*; German, *Actinium*; Italian, *attinio*; Spanish, *actinio*

Actinium

[ak-tin-iuhm]

• NUCLEAR DATA

Number of isotopes (including nuclear isomers): 26 **Isotope mass range:** 209 → 232

Key isotopes

Nuclide	Atomic mass	Half life ($T_{1/2}$)	Decay mode and energy (MeV)	Nuclear spin I	Nucl. mag. moment μ	Uses
^{224}Ac	224.021 685	2.9 h	EC (1.397) 90%; α (6.323) 10%; γ			
^{225}Ac	225.023 205	10.0 d	α (5.935); γ	3/2+		tracer
^{226}Ac	226.026 084	29 h	EC (0.635) 17%; β^- (1.117) 83%; α; γ	1–		
^{227}Ac*	227.027 750	21.77 y	β^- (0.0410) 99%; α (5.043) 1%; γ	3/2–	+1.1	NMR
^{228}Ac*	228.031 015	6.13 h	β^- (2.142); γ	3+		
^{229}Ac*	229.032 980	1.05 h	β^- (1.140); γ	3/2+		

*Traces of these isotopes occur naturally.

NMR [Reference: not recorded]	^{227}Ac
Relative sensitivity (^{1}H = 1.00)	–
Receptivity (^{13}C = 1.00)	–
Magnetogyric ratio/rad $T^{-1} s^{-1}$	3.5×10^7
Nuclear quadrupole moment/m^2	1.7×10^{-28}
Frequency (^{1}H = 100 Hz; 2.3488T)/MHz	13.1

• ELECTRON SHELL DATA

Ground state electron configuration: $[Rn]6d^1 7s^2$
Term symbol: $^2D_{3/2}$
Electron affinity ($M \rightarrow M^-$)/kJ mol^{-1}**:** n.a.

Ionization energies/kJ mol^{-1}**:**

1.	$M \rightarrow M^+$	499
2.	$M^+ \rightarrow M^{2+}$	1170
3.	$M^{2+} \rightarrow M^{3+}$	1900
4.	$M^{3+} \rightarrow M^{4+}$	(4700)
5.	$M^{4+} \rightarrow M^{5+}$	(6000)
6.	$M^{5+} \rightarrow M^{6+}$	(7300)
7.	$M^{6+} \rightarrow M^{7+}$	(9200)
8.	$M^{7+} \rightarrow M^{8+}$	(10 500)
9.	$M^{8+} \rightarrow M^{9+}$	(11 900)
10.	$M^{9+} \rightarrow M^{10+}$	(15 800)

Electron binding energies/eV

K	1s	106 755
L_I	2s	19 840
L_{II}	$2p_{1/2}$	19 083
L_{III}	$2p_{3/2}$	15 871
M_I	3s	5002
M_{II}	$3p_{1/2}$	4656
M_{III}	$3p_{3/2}$	3909
M_{IV}	$3d_{3/2}$	3370
M_V	$3d_{5/2}$	3219

continued in Appendix 2, p255

Main lines in atomic spectrum
[Wavelength/nm(species)]

386.312 (II)
408.844 (II)
416.840 (II)
438.641 (II)
450.720 (II)
591.085 (II)

• CRYSTAL DATA

Crystal structure (cell dimensions/pm), space group
f.c.c. (a = 531.1), Fm3m

X-ray diffraction: mass absorption coefficients (μ/ρ)/cm^2 g^{-1}**:** CuK_α n.a. MoK_α n.a.
Neutron scattering length, $b/10^{-12}$ cm**:** n.a.
Thermal neutron capture cross-section, σ_a/barns**:** 515 (^{227}Ac)

• GEOLOGICAL DATA

Minerals

None as such.

Chief source: actinium is a decay product of ^{235}U, and was first extracted from uranium ores, in which it is present to the extent of 0.2 p.p.m. The longest lived isotope is ^{227}Ac which decays to ^{227}Th, ^{223}Ra, and other short-lived products. Actinium has also been made by the bombardment of ^{236}Ra with neutrons.

World production: n.a. but probably less than a gram

Specimen: not commercially available.

Abundances

Sun (relative to H = 1×10^{12})**:** n.a.
Earth's crust/p.p.m.**:** minute traces
Seawater/p.p.m.**:** nil

Al

Atomic number: 13
Relative atomic mass ($^{12}C = 12.0000$)**:** 26.981539

CAS: [7429-90-5]

• CHEMICAL DATA

Description: Pure aluminium is soft and malleable, but can be toughened by alloying with small amounts of other metals like copper and magnesium. Aluminium objects are protected from reacting with air and water by an oxide film which forms rapidly on the surface. Aluminium is soluble in concentrated hydrochloric acid and in sodium hydroxide solution. The metal and its alloys have hundreds of uses in the vehicle, aircraft, and construction industries. It is used to make cans, kegs, wrapping foil, household utensils, etc.

Radii/pm: Al^{3+} 57; atomic 143; covalent 125; van der Waals 205
Electronegativity: 1.61 (Pauling); 1.47 (Allred); 3.23 eV (absolute)
Effective nuclear charge: 3.50 (Slater); 4.07 (Clementi); 3.64 (Froese-Fischer)

Standard reduction potentials $E^\ominus$/V

	III		0
acid	Al^{3+}	−1.676	Al
	$[AlF_6]_3^-$	−2.067	Al
base	$Al(OH)_3$	−2.300	Al
	$[Al(OH)_4]^-$	−2.310	Al

Oxidation states

Al^{I}	s^2	AlCl in gas phase
Al^{III}	[Ne]	Al_2O_3 (amphoteric), AlO(OH), $Al(OH)_3$, $[Al(OH_2)_6]^{3+}$ (aq), Al^{3+} salts, AlH_3, $LiAlH_4$, AlF_3, Na_3AlF_6, Al_2Cl_6

Covalent bonds

Bond	r/ pm	E/ kJ mol^{-1}
Al—H	*c.* 170	285
Al—C	224	255
Al—O	162	585
Al—F	163	583
Al—Cl	206	421
Al—Al	286	*c.* 200

• PHYSICAL DATA

Melting point/K: 933.52
Boiling point/K: 2740

ΔH_{fusion}/kJ mol^{-1}: 10.67
ΔH_{vap}/kJ mol^{-1}: 293.72

Thermodynamic properties (298.15 K, 0.1 MPa)

State	$\Delta_f H^\ominus$/kJ mol^{-1}	$\Delta_f G^\ominus$/kJ mol^{-1}	$S^\ominus$/J K^{-1} mol^{-1}	C_p/J K^{-1} mol^{-1}
Solid	0	0	28.33	24.35
Gas	326.4	285.7	165.54	21.38

Density/kg m^{-3}: 2698 [293 K]; 2390 [liquid at m.p.]
Molar volume/cm^3: 10.00
Thermal conductivity/W m^{-1} K^{-1}: 237 [300 K]
Coefficient of linear thermal expansion/K^{-1}: 23.03×10^{-6}
Electrical resistivity/Ω m: 2.6548×10^{-8} [293 K]
Mass magnetic susceptibility/kg^{-1} m^3: $+7.7 \times 10^{-9}$ (s)

Young's modulus/GPa: 70.6
Rigidity modulus/GPa: 26.2
Bulk modulus/GPa: 75.2
Poisson's ratio/GPa: 0.345

• BIOLOGICAL DATA

Biological role

Aluminium has no known biological role.

Toxicity

Toxic intake: 5 g

Lethal intake: n.a.

Hazards

Aluminium accumulates in the body from a daily intake. Its compounds are used as food additives and in indigestion tablets.

Levels in humans

Blood/mg dm^{-3}: 0.39
Bone/p.p.m.: 4 – 27
Liver/p.p.m.: 3 – 23
Muscle/p.p.m.: 0.7 – 28
Daily dietary intake: 2.45 mg
Total mass of element in average (70 kg) person: 60 mg

Discovered in 1825 by Oersted at Copenhagen, Denmark.
[Latin, *alumen* = alum] French, *aluminium*;
German, *Aluminium*; Italian, *alluminio*; Spanish, *aluminio*

Aluminium/Aluminum (US)

[al-yoo-min-iuhm] [US: al-oo-min-um]

• NUCLEAR DATA

Number of isotopes (including nuclear isomers): 11 **Isotope mass range:** $22 \rightarrow 31$

Key isotopes

Nuclide	Atomic mass	Natural abundance (%)	Nuclear spin I	Nuclear magnetic moment μ	Uses
^{27}Al	26.981 538 6	100	5/2+	+3.641 504	NMR

A table of radioactive isotopes is given in Appendix 1, on p235.

NMR [Reference: Al^{3+} (aq)]	^{27}Al
Relative sensitivity (^1H = 1.00)	0.21
Receptivity (^{13}C = 1.00)	1.17×10^3
Magnetogyric ratio/rad $T^{-1} s^{-1}$	6.9704×10^7
Nuclear quadrupole moment/m^2	0.1403×10^{-28}
Frequency (^1H = 100 Hz; 2.3488T)/MHz	26.057

• ELECTRON SHELL DATA

Ground state electron configuration: $[Ne]3s^23p^1$
Term symbol: $^2P_{1/2}$
Electron affinity ($M \rightarrow M^-$)/kJ mol^{-1}: 44

Ionization energies/kJ mol^{-1}:

1.	$M \rightarrow M^+$	577.4
2.	$M^+ \rightarrow M^{2+}$	1816.6
3.	$M^{2+} \rightarrow M^{3+}$	2744.6
4.	$M^{3+} \rightarrow M^{4+}$	11 575
5.	$M^{4+} \rightarrow M^{5+}$	14 839
6.	$M^{5+} \rightarrow M^{6+}$	18 376
7.	$M^{6+} \rightarrow M^{7+}$	23 293
8.	$M^{7+} \rightarrow M^{8+}$	27 457
9.	$M^{8+} \rightarrow M^{9+}$	31 857
10.	$M^{9+} \rightarrow M^{10+}$	38 459

Electron binding energies/eV

K	1s	1559.0
L_I	2s	117.8
L_{II}	$2p_{1/2}$	72.9
L_{III}	$2p_{3/2}$	72.5

Main lines in atomic spectrum
[Wavelength/nm(species)]
308.215 (I)
309.271 (I) (AA)
309.281 (I) (AA)
394.401 (I)
396.152 (I)

• CRYSTAL DATA

Crystal structure (cell dimensions/pm), space group
f.c.c. (a = 404.959), Fm3m

X-ray diffraction: mass absorption coefficients (μ/ρ)/$cm^2 g^{-1}$: CuK_α 48.6 MoK_α 5.16
Neutron scattering length, $b/10^{-12}$ cm: 0.3449
Thermal neutron capture cross-section, σ_a/barns: 0.231

• GEOLOGICAL DATA

Minerals

Many minerals are known, and aluminium is present in many other minerals.

Mineral	Formula	Density	Hardness	Crystal appearance
Bauxite*	AlO(OH)	2.3	1 – 3	rarely found as single crystals
Boehmite	AlO(OH)	3.07	3	orth., white/brown, microscopic
Diaspore	AlO(OH)	3.3 – 3.5	6.5 – 7	orth., white, thin plates
Gibbsite	$Al(OH)_3$	2.40	2.5 – 3.5	mon., pearly white, hexagonal

*Rock often found with clay.
Continued in Appendix 3 on page 259.

Chief ore: bauxite; diaspore is a constituent of bauxite.

World production/tonnes y^{-1}: 15×10^6

Main mining areas: Surinam, Jamaica, Ghana, Indonesia, Russia.

Reserves/tonnes: 6×10^9

Specimen: available as foil, granules, ingots, pellets, powder, rod, shot or wire. Safe. Aluminium powder can react dangerously with other materials.

Abundances

Sun (relative to H = 1×10^{12}): 3.3×10^6
Earth's crust/p.p.m.: 82 000
Seawater/p.p.m.:
Atlantic surface: 9.7×10^{-4}
Atlantic deep: 5.2×10^{-4}
Pacific surface: 1.3×10^{-4}
Pacific deep: 0.13×10^{-4}
Residence time/years: 150
Classification: scavenged
Oxidation state: III

Am

Atomic number: 95
Relative atomic mass ($^{12}C = 12.0000$)**:** 243.0614 (Am-243)

CAS: [7440-35-9]

• CHEMICAL DATA

Description: Americium is a radioactive, silvery metal which does not occur naturally. The metal itself has been isolated by reacting AmF_3 with barium metal at 1100 °C. It is attacked by air, steam and acids, but not alkalis. ^{241}Am has been used as a source of radiation for γ-radiography.

Radii/pm: Am^{6+} 80; Am^{5+} 86; Am^{4+} 92; Am^{3+} 107; atomic 173
Electronegativity: 1.3 (Pauling); 1.2 (est.) (Allred); n.a. (absolute)
Effective nuclear charge: 1.65 (Slater)

Standard reduction potentials $E^{\ominus}/V$

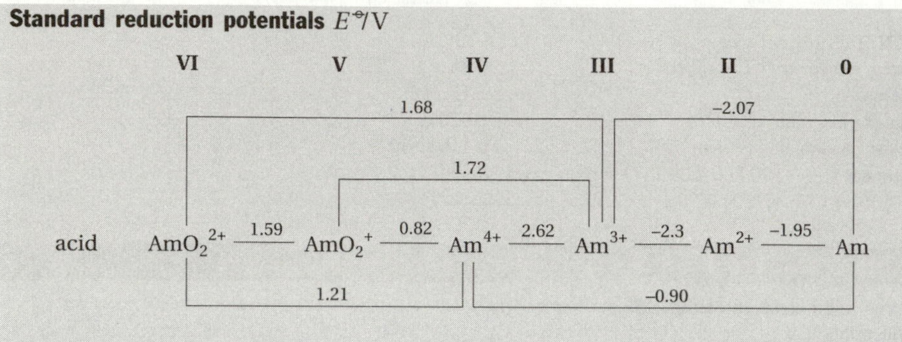

Oxidation states

Am^{II}	f^7	AmO, $AmCl_2$ etc.
Am^{III}	f^6	Am_2O_3, AmF_3, $AmCl_3$ etc., $[AmCl_6]^{3+}$, Am^{3+} (aq), $[Am(C_5H_5)_3]$
Am^{IV}	f^5	AmO_2, AmF_4
Am^{V}	f^4	AmO_2^{+} (aq) unstable due to reduction by radioactive decay products
Am^{VI}	f^3	AmO_2^{2+} (aq) unstable due to reduction by radioactive decay products

• PHYSICAL DATA

Melting point/K: 1445
Boiling point/K: 2880

ΔH_{fusion}/kJ mol^{-1}: 14.4
ΔH_{vap}/kJ mol^{-1}: 284
ΔH_{subl}/kJ mol^{-1}: 34.2

Thermodynamic properties (298.15 K, 0.1 MPa)

State	$\Delta_f H^{\ominus}$/kJ mol^{-1}	$\Delta_f G^{\ominus}$/kJ mol^{-1}	$S^{\ominus}$/J K^{-1} mol^{-1}	C_p/J K^{-1} mol^{-1}
Solid	0	0	n.a.	n.a.
Gas	n.a.	n.a.	n.a.	n.a.

Density/kg m^{-3}: 13 670 [293 K]
Molar volume/cm^3: 17.78
Thermal conductivity/W m^{-1} K^{-1}: 10 (est.) [300 K]
Coefficient of linear thermal expansion/K^{-1}: n.a.
Electrical resistivity/Ω m: 68×10^{-8}
Mass magnetic susceptibility/kg^{-1} m^3: $+5 \times 10^{-8}$

Young's modulus/GPa: n.a.
Rigidity modulus/GPa: n.a.
Bulk modulus/GPa: n.a.
Poisson's ratio/GPa: n.a.

• BIOLOGICAL DATA

Biological role

None.

Toxicity

Toxic intake: n.a.
Lethal intake: n.a.

Hazards

Americium is never encountered normally. It is highly dangerous because of its intense α-radiation. The maximum permissible body burden of ^{241}Am is 0.03 μCi - it targets bone. When americium compounds are handled in gram amounts the γ-radiation is also a problem.

Levels in humans

nil
Daily dietary intake: nil
Total mass of element in average (70 kg) person: nil

Americium

[amer-is-iuhm]

• NUCLEAR DATA

Discovery: Americium was discovered in 1944 by Glen T. Seaborg, R.A. James, L.O. Morgan, and A. Ghiorso at Chicago, Illinois, USA.

Number of isotopes (including nuclear isomers): 23 **Isotope mass range:** 227 → 247

Key isotopes

Nuclide	Atomic mass	Half life ($T_{1/2}$)	Decay mode and energy (MeV)	Nuclear spin I	Nucl. mag. moment μ	Uses
^{237}Am	237.050 050	1.22 h	EC (1.6) 99.98%; α (6.20) 0.02%; γ	5/2–		
^{238}Am	238.051 980	1.63 h	EC (2.26); α (6.04) <0.1%; γ	1+		
^{239}Am	239.053 016	11.9 h	EC (0.800) 99.99%; α (5.924) 0.01%; γ	5/2–		
^{240}Am	240.055 278	2.12 d	EC (1.38); α (5.592); γ	3–		
^{241}Am	241.056 823	432.2 y	α (5.637); γ	5/2–	+1.61	R, T, D
^{242}Am	242.056 541	16.02 h	β^- (0.663) 83%; EC (0.750) 17%; γ	1–	+0.388	
^{242m}Am		141 y	IT (0.048) 99.5%; α (5.62) 0.5%; γ	5–	+1.0	
^{243}Am	243.061 375	7370 y	α (5.438); γ	5/2–	+1.61	R, NMR
^{244}Am	244.064 279	10.1 h	β^- (1.427); γ			
^{245}Am	245.066 444	2.05 h	β^- (0.894); γ	5/2+		

Other isotopes of americium have half lives less than 1 h.

NMR [Reference: n.a.]

	^{243}Am
Relative sensitivity (^{1}H = 1.00)	–
Receptivity (^{13}C = 1.00)	–
Magnetogyric ratio/rad $T^{-1} s^{-1}$	1.54×10^7
Nuclear quadrupole moment/m^2	$+4.210 \times 10^{-28}$
Frequency (^{1}H = 100 Hz; 2.3488T)/MHz	5.76

• ELECTRON SHELL DATA

Ground state electron configuration: [Rn]$5f^7 7s^2$
Term symbol: $^8S_{7/2}$
Electron affinity ($M \rightarrow M^-$)/kJ mol^{-1}: n.a.

Ionization energies/kJ mol^{-1}:
1. $M \rightarrow M^+$ 578.2

Electron binding energies/eV
n.a.

Main lines in atomic spectrum
[Wavelength/nm(species)]
367.312 (I)
377.750 (II)
392.625 (II)
408.929 (II)
428.926 (I)
450.945 (II)
457.559 (II)
466.279 (II)
605.464 (I)

• CRYSTAL DATA

Crystal structure (cell dimensions/pm), space group
α-Am h.c.p. (a = 346.80, c = 1124.0), $P6_3/mmc$
β-Am f.c.c. (a = 489.4), Fm3m
$T(\alpha \rightarrow \beta)$ = 1347 K

X-ray diffraction: mass absorption coefficients (μ/ρ)/cm^2 g^{-1}: CuK_α n.a. MoK_α n.a.
Neutron scattering length, $b/10^{-12}$ cm: 0.83
Thermal neutron capture cross-section, σ_a/barns: 75.3 (^{243}Am)

• GEOLOGICAL DATA

Minerals
None.

Chief source: ^{243}Am is produced in 100 g quantities by neutron bombardment of ^{239}Pu, and has a relatively long half-life. Although ^{241}Am has a shorter half-life, it can be extracted from ^{241}Pu that has been subjected to neutron bombardment over a period of years.

World production: n.a. but probably a few kgs per year.

Specimen: ^{241}Am and ^{243}Am are commercially available, under licence - see Key.

Abundances
Sun (relative to H = 1×10^{12}): n.a.
Earth's crust/p.p.m.: nil
Seawater/p.p.m.: nil

Sb

Atomic number: 51
Relative atomic mass ($^{12}C = 12.0000$)**:** 112.760

CAS: [7440-36-0]

• CHEMICAL DATA

Description: Antimony is a metalloid element with three forms. The metallic form is the more stable and is bright, silvery, hard and brittle. It is stable in dry air, and is not attacked by dilute acids or alkalis. The addition of antimony will harden other metals, and it is used in storage batteries, bearings, etc.

Radii/pm**:** Sb^{5+} 62; Sb^{3+} 89; Sb^{2-} 245; atomic 182; covalent 141; van der Waals 220
Electronegativity: 2.05 (Pauling); 1.82 (Allred); 4.85 eV (absolute)
Effective nuclear charge: 6.30 (Slater); 9.99 (Clementi); 12.37 (Froese-Fischer)

Standard reduction potentials $E^{\ominus}$/V

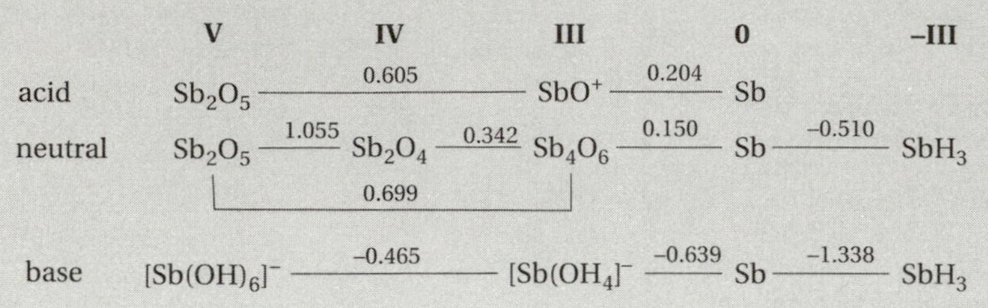

Oxidation states

Sb^{-III}	[Xe]	SbH_3
Sb^{III}	s^2	Sb_4O_6, SbO_3^{3-} (aq), SbF_3, $SbCl_3$ etc., $[SbF_5]^{2-}$, Sb_2S_3
Sb^{V}	d^{10}	Sb_4O_{10}, $[Sb(OH)_6]^-$ (aq), SbF_5, $SbCl_5$ etc., $[SbCl_6]^-$, $[SbBr_6]^-$

Covalent bonds

Bond	r/ pm	E/ kJ mol^{-1}
Sb—H	171	257
Sb—C	220	215
Sb—O	200	314
Sb—F	203	440
Sb—Cl	233	316
Sb—Sb	290	295

• PHYSICAL DATA

Melting point/K**:** 903.89
Boiling point/K**:** 1908

ΔH_{fusion}/kJ mol^{-1}**:** 20.9
ΔH_{vap}/kJ mol^{-1}**:** 67.91

Thermodynamic properties (298.15 K, 0.1 MPa)

State	$\Delta_f H^{\ominus}$/kJ mol^{-1}	$\Delta_f G^{\ominus}$/kJ mol^{-1}	$S^{\ominus}$/J K^{-1} mol^{-1}	C_p/J K^{-1} mol^{-1}
Solid	0	0	45.69	25.23
Gas	262.3	222.1	180.27	20.79

Density/kg m^{-3}**:** 6691 [293 K]; 6 483 [liquid at m.p.]
Molar volume/cm^3**:** 18.20
Thermal conductivity/W m^{-1} K^{-1}**:** 24.3 [300 K]
Coefficient of linear thermal expansion/K^{-1}**:** 8.5×10^{-6}
Electrical resistivity/Ω m**:** 39.0×10^{-8} [273 K]
Mass magnetic susceptibility/kg^{-1} m^3**:** -1.0×10^{-8} (s)

Young's modulus/GPa**:** 54.7
Rigidity modulus/GPa**:** 20.7
Bulk modulus/GPa**:** n.a.
Poisson's ratio/GPa**:** 0.25 - 0.33

• BIOLOGICAL DATA

Biological role

None.

Toxicity

Toxic intake: 100 mg

Lethal intake: antimony provokes vomiting, and was once prescribed for this purpose, but medical dose is near to toxic dose and antimony can kill. LD_{50} (oral) for antimony potassium tartrate is *c.* 140 mg.

Hazards

Small doses of antimony stimulate metabolism, large doses cause liver damage.

Levels in humans

Blood/mg dm^{-3}**:** 0.0033
Bone/p.p.m.**:** 0.01 – 0.6
Liver/p.p.m.**:** 0.011 – 0.42
Muscle/p.p.m.**:** 0.042 – 0.191
Daily dietary intake: 0.002 – 1.3 mg
Total mass of element in average (70 kg) person: 2 mg

Probably known to the ancients and certainly to the alchemists.
[Greek, *anti* + *monos* = not alone; Latin, *stibium*]
French, *antimoine*; German, *Antimon*; Italian, *antimonio*; Spanish, *antimonio*

Antimony

[anti-moni]

• NUCLEAR DATA

Number of isotopes (including nuclear isomers): 40 **Isotope mass range:** 108 → 136

Key isotopes

Nuclide	Atomic mass	Natural abundance (%)	Nuclear spin *I*	Nuclear magnetic moment μ	Uses
^{121}Sb	120.903 821 2	57.3	5/2+	+3.363 4	E, NMR
^{123}Sb	122.904 216 0	42.7	7/2+	+2.549 8	E, NMR

A table of radioactive isotopes is given in Appendix 1, on p235.

NMR [Reference: $[N(C_2H_5)_4][SbCl_6]$]	^{121}Sb	^{123}Sb
Relative sensitivity (^1H = 1.00)	0.16	4.57×10^{-2}
Receptivity (^{13}C = 1.00)	520	111
Magnetogyric ratio/rad $T^{-1}s^{-1}$	6.4016×10^7	3.4668×10^7
Nuclear quadrupole moment/m^2	-0.360×10^{-28}	-0.490×10^{-28}
Frequency (^1H = 100 Hz; 2.3488T)/MHz	23.930	12.959

• ELECTRON SHELL DATA

Ground state electron configuration: $[Kr]4d^{10}5s^25p^3$
Term symbol: $^4S_{3/2}$
Electron affinity ($M \rightarrow M^-$)/kJ mol^{-1}**:** 101

Ionization energies/kJ mol^{-1}**:**

1.	$M \rightarrow M^+$	833.7
2.	$M^+ \rightarrow M^{2+}$	1794
3.	$M^{2+} \rightarrow M^{3+}$	2443
4.	$M^{3+} \rightarrow M^{4+}$	4260
5.	$M^{4+} \rightarrow M^{5+}$	5400
6.	$M^{5+} \rightarrow M^{6+}$	10400
7.	$M^{6+} \rightarrow M^{7+}$	(12 700)
8.	$M^{7+} \rightarrow M^{8+}$	(15 200)
9.	$M^{8+} \rightarrow M^{9+}$	(17 800)
10.	$M^{9+} \rightarrow M^{10+}$	(20 400)

Electron binding energies/eV

K	1s	30 491
L_I	2s	4698
L_{II}	$2p_{1/2}$	4380
L_{III}	$2p_{3/2}$	4132
M_I	3s	946
M_{II}	$3p_{1/2}$	812.7
M_{III}	$3p_{3/2}$	766.4
M_{IV}	$3d_{3/2}$	537.5
M_V	$3d_{5/2}$	528.2

continued in Appendix 2, p255

Main lines in atomic spectrum
[Wavelength/nm(species)]
206.833 (I) (AA)
217.581 (I)
231.147 (I)
252.852 (I)
259.805 (I)

• CRYSTAL DATA

Crystal structure (cell dimensions/pm), space group
grey rhombohedral (a = 430.84, c = 1124.7), R$\bar{3}$m
(grey) cubic (a = 298.6), Pm3m
metal h.c.p. (a = 336.9, c = 533), $P6_3$/mmc

X-ray diffraction: mass absorption coefficients (μ/ρ)/cm^2 g^{-1}**:** CuK_α 270 MoK_α 33.1
Neutron scattering length, $b/10^{-12}$ cm**:** 0.557
Thermal neutron capture cross-section, σ_a/barns**:** 4.91

• GEOLOGICAL DATA

Minerals

A little native antimony occurs naturally as granular masses or nodules, generally in silver-bearing lodes, and has been found in Sweden, Germany, Italy, and the USA.

Mineral	Formula	Density	Hardness	Crystal appearance
Sibiconite	$Sb_3O_6(OH)$	5.58	4 – 5.5	hex., res./adam. black
Stibnite	Sb_2S_3	4.63	2	hex., yellow, tiny prisms
Tetrahedrite	$(Cu,Fe)_{12}Sb_4S_{13}$	4.97	3 – 4.5	rhom., adam. white/brown
Ullmannite	NiSbS	6.65	5 – 5.5	cub., metallic grey

Chief ores: stibnite. Tetrahedrite, although mainly a copper ore, yields antimony as a by-product.

World production/tonnes y^{-1}**:** 53 000

Main mining areas: China, Italy, Peru, Mexico, Bolivia, France.

Reserves/tonnes: 2.5×10^6

Specimen: available as pieces, powder or shot. *Care!*

Abundances

Sun (relative to H = 1×10^{12})**:** 10
Earth's crust/p.p.m.**:** 0.2
Seawater/p.p.m.**:** *c.* 3×10^{-4}
Residence time/years**:** *c.* 3.5×10^5
Classification: accumulating
Oxidation state: III

Atomic number: 18
Relative atomic mass ($^{12}C = 12.0000$)**:** 39.948

CAS: [7440-37-1]

• CHEMICAL DATA

Discovery: Argon was discovered in 1894 by Lord Rayleigh (London) and Sir William Ramsay (Bristol), England.
Description: Argon is a colourless, odourless gas comprising 1% of the atmosphere, from which it is extracted after liquefaction. Argon is inert towards all other elements and chemicals. It is used as an inert atmosphere in lamps and high temperature metallurgy.

Radii/pm: atomic 174; van der Waals 191
Electronegativity: n.a. (Pauling); 3.20 (Allred); [7.70 eV (absolute) - see Key]
Effective nuclear charge: 6.75 (Slater); 6.76 (Clementi); 7.52 (Froese-Fischer)

Oxidation states

Ar^0 [Ar] $Ar_8(H_2O)_{46}$ and $Ar(quinol)_3$. These are not true compounds but clathrates in which argon atoms are trapped inside a lattice of other molecules.

• PHYSICAL DATA

Melting point/K: 83.78
Boiling point/K: 87.29
Critical temperature/K: 150.87
Critical pressure/ kPa: 4862

ΔH_{fusion}/kJ mol^{-1}**:** 1.21
ΔH_{vap}/kJ mol^{-1}**:** 6.53

Thermodynamic properties (298.15 K, 0.1 MPa)

State	$\Delta_f H^\circ$/kJ mol^{-1}	$\Delta_f G^\circ$/kJ mol^{-1}	S°/J K^{-1} mol^{-1}	C_p/J K^{-1} mol^{-1}
Gas	0	0	154.843	20.786

Density/kg m^{-3}**:** 1656 [40 K]; 1380 [liquid b.p.]; 1.784 [273 K]
Molar volume/cm^3**:** 24.12 [40 K]
Thermal conductivity/W m^{-1} K^{-1}**:** 0.0177 [300 K] (g)
Mass magnetic susceptibility/kg^{-1} m^3**:** -6.16×10^{-9} (g)

• BIOLOGICAL DATA

Biological role

None.

Toxicity

Non-toxic.

Hazards

Argon is a harmless gas, although it could asphyxiate if it excluded oxygen from the lungs.

Levels in humans

Blood/mg dm^{-3}**:** trace
Bone/p.p.m.**:** nil
Liver/p.p.m.**:** nil
Muscle/p.p.m.**:** nil
Daily dietary intake: n.a. but low
Total mass of element in average (70 kg) person: n.a. but small

Discovery: see Chemical Data section.
[Greek, *argos* = inactive]
French, *argon*; German, *Argon*; Italian, *argo*; Spanish, *argón*

Argon

[ar-gon]

• NUCLEAR DATA

Number of isotopes (including nuclear isomers): 15 **Isotope mass range:** 32 → 46

Key isotopes

Nuclide	Atomic mass	Natural abundance (%)	Nuclear spin I	Nuclear magnetic moment μ	Uses
^{36}Ar	35.967 545 52	0.337	0+	0	
^{38}Ar	37.962 732 5	0.063	0+	0	
^{40}Ar	39.962 383 7	99.600	0+	0	

A table of radioactive isotopes is given in Appendix 1, on p235.

NMR [Reference: no known compounds] n.a.

• ELECTRON SHELL DATA

Ground state electron configuration: $[Ne]3s^2 3p^6$ = [Ar]
Term symbol: $^1S_0$
Electron affinity $(M \rightarrow M^-)/kJ\ mol^{-1}$**:** –35 (calc.)

Ionization energies/$kJ\ mol^{-1}$**:**

1.	$M \rightarrow M^+$	1520.4
2.	$M^+ \rightarrow M^{2+}$	2665.2
3.	$M^{2+} \rightarrow M^{3+}$	3928
4.	$M^{3+} \rightarrow M^{4+}$	5770
5.	$M^{4+} \rightarrow M^{5+}$	7238
6.	$M^{5+} \rightarrow M^{6+}$	8811
7.	$M^{6+} \rightarrow M^{7+}$	12 021
8.	$M^{7+} \rightarrow M^{8+}$	13 844
9.	$M^{8+} \rightarrow M^{9+}$	40 759
10.	$M^{9+} \rightarrow M^{10+}$	46 186

Electron binding energies/eV

K	1s	3205.9
L_I	2s	326.3
L_{II}	$2p_{1/2}$	250.6
L_{III}	$2p_{3/2}$	248.4
M_I	3s	29.3
M_{II}	$3p_{1/2}$	15.9
M_{III}	$3p_{3/2}$	15.7

Main lines in atomic spectrum
[Wavelength/nm(species)]

696.5431 (I)
706.7218 (I)
750.3869 (I)
801.4786 (I)
811.5311 (I)
912.2967 (I)
965.7786 (I)

• CRYSTAL DATA

Crystal structure (cell dimensions/pm), space group
f.c.c. (40 K) (a = 531.088), Fm3m

X-ray diffraction: mass absorption coefficients $(\mu/\rho)/cm^2\ g^{-1}$**:** CuK_α 123 MoK_α 13.5
Neutron scattering length, $b/10^{-12}$ cm**:** 0.1909
Thermal neutron capture cross-section, σ_a/barns**:** 0.675

• GEOLOGICAL DATA

Minerals

None as such.

Chief source: liquid air
World production/tonnes y^{-1}**:** 700 000
Reserves/tonnes**:** 6.6×10^{13} (atmosphere)
Specimen: available in small pressurized canisters. Safe.

Abundances

Sun (relative to H = 1×10^{12})**:** 1×10^6
Earth's crust/p.p.m.**:** 1.2
Atmosphere/p.p.m. (volume)**:** 9300
Seawater/p.p.m.**:** 0.45
Residence time/years**:** 28 000
Oxidation state: 0

As

Atomic number: 33
Relative atomic mass ($^{12}C = 12.0000$)**:** 74.92159

CAS: [7440-38-2]

• CHEMICAL DATA

Description: Arsenic is a metalloid element with two main forms. Grey α-arsenic is metallic and is brittle, tarnishes and burns in oxygen. Arsenic resists attack by water, ordinary acids and alkalis. However, it reacts with hot acids and molten NaOH. Arsenic is used in alloys, semiconductors, pesticides, wood preservatives and glass.

Radii/pm: As^{5+} 46; As^{3+} 69; atomic 125; covalent 121; van der Waals 200
Electronegativity: 2.18 (Pauling); 2.20 (Allred); 5.3 eV (absolute)
Effective nuclear charge: 6.30 (Slater); 7.45 (Clementi); 8.98 (Froese-Fischer)

Standard reduction potentials $E^{\ominus}$/V

	V		III		0		–III
acid	H_3AsO_4	0.560	$HAsO_2$	0.240	As	–0.225	AsH_3
base	AsO_4^{3-}	–0.67	AsO_2^-	–0.68	As	1.37	AsH_3

Oxidation states

As^{-III}	[Kr]	AsH_3
As^{III}	s^2	As_4O_6, H_3AsO_3, $H_2AsO_3^-$ (aq), AsF_3, $AsCl_3$ etc.
As^V	d^{10}	As_4O_{10}, H_3AsO_4, $H_2AsO_4^-$, etc. (aq), $NaAsO_3$, AsF_5

Covalent bonds

Bond	r/ pm	E/ kJ mol^{-1}
As—H	152	*c.* 245
As—C	198	200
As—O	178	301
As—F	171	484
As—Cl	216	322
As—As	243	146

• PHYSICAL DATA

Melting point/K: 1090 (α, under pressure)
Boiling point/K: 889 (sublimes)
Critical temperature/K: 1673

ΔH_{fusion}/kJ mol^{-1}: 27.7
ΔH_{vap}/kJ mol^{-1}: 31.9

Thermodynamic properties (298.15 K, 0.1 MPa)

State	$\Delta_f H^{\ominus}$/kJ mol^{-1}	$\Delta_f G^{\ominus}$/kJ mol^{-1}	$S^{\ominus}$/J K^{-1} mol^{-1}	C_p/J K^{-1} mol^{-1}
Solid	0	0	35.1	24.64
Gas	302.5	261.0	174.21	20.786

Density/kg m^{-3}: 5780 (α); 4700 (β) [293 K]
Molar volume/cm^3: 12.95 (α); 15.9 (β)
Thermal conductivity/W m^{-1} K^{-1}: 50.0 (α) [300 K]
Coefficient of linear thermal expansion/K^{-1}: 4.7×10^{-6}
Electrical resistivity/Ω m: 26×10^{-8} [273 K]
Mass magnetic susceptibility/kg^{-1} m^3: -9.17×10^{-10} (α); -3.97×10^{-9} (β)

• BIOLOGICAL DATA

Biological role

Essential to some species including humans.

Toxicity

Toxic intake: 5 – 50 mg
Lethal intake: > 50 – 300 mg; LD_{50} (oral) = 100 mg

Hazards

Arsenic salts and arsine gases are very poisonous. In small doses arsenic acts to stimulate metabolism, but it is carcinogenic and possibly teratogenic.

Levels in humans

Blood/mg dm^{-3}: 0.0017 – 0.09
Bone/p.p.m.: 0.08 – 1.6
Liver/p.p.m.: 0.023 – 1.61
Muscle/p.p.m.: 0.009 – 0.65
Daily dietary intake: 0.04 – 1.4 mg
Total mass of element in average (70 kg) person: 7 mg (fluctuates over range 0.5–15 mg depending upon diet)

Probably first isolated by Albertus Magnus (1193-1280).
[Greek, *arsenikon* = yellow orpiment]
French, *arsenic*; German, *Arsen*; Italian, *arsenico*; Spanish, *arsénico*

Arsenic

[ahrs-nik]

• NUCLEAR DATA

Number of isotopes (including nuclear isomers): 22 **Isotope mass range:** 67 → 87

Key isotopes

Nuclide	Atomic mass	Natural abundance (%)	Nuclear spin I	Nuclear magnetic moment μ	Uses
^{75}As	74.921 594 2	100	3/2–	+1.439 47	NMR

A table of radioactive isotopes is given in Appendix 1, on p236.

NMR [Reference: $KAsF_6$]	^{75}As
Relative sensitivity (^1H = 1.00)	0.0251
Receptivity (^{13}C = 1.00)	143
Magnetogyric ratio/rad $T^{-1}s^{-1}$	4.5804×10^7
Nuclear quadrupole moment/m^2	0.314×10^{-28}
Frequency (^1H = 100 Hz; 2.3488T)/MHz	17.126

• ELECTRON SHELL DATA

Ground state electron configuration: $[Ar]3d^{10}4s^24p^3$
Term symbol: $^4S_{3/2}$
Electron affinity ($M \rightarrow M^-$)/kJ mol^{-1}**:** 78

Ionization energies/kJ mol^{-1}**:**

1.	$M \rightarrow M^+$	947.0
2.	$M^+ \rightarrow M^{2+}$	1798
3.	$M^{2+} \rightarrow M^{3+}$	2735
4.	$M^{3+} \rightarrow M^{4+}$	4837
5.	$M^{4+} \rightarrow M^{5+}$	6042
6.	$M^{5+} \rightarrow M^{6+}$	12 305
7.	$M^{6+} \rightarrow M^{7+}$	(15 400)
8.	$M^{7+} \rightarrow M^{8+}$	(18 900)
9.	$M^{8+} \rightarrow M^{9+}$	(22 600)
10.	$M^{9+} \rightarrow M^{10+}$	(26 400)

Electron binding energies/eV

K	1s	11 867
L_I	2s	1527.0
L_{II}	$2p_{1/2}$	1359.1
L_{III}	$2p_{3/2}$	1323.6
M_I	3s	204.7
M_{II}	$3p_{1/2}$	146.2
M_{III}	$3p_{3/2}$	141.2
M_{IV}	$3d_{3/2}$	41.7
M_V	$3d_{5/2}$	41.7

Main lines in atomic spectrum
[Wavelength/nm(species)]

193.759 (I) (AA)
419.008 (II)
445.847 (II)
446.635 (II)
449.423 (II)
450.766 (II)
454.348 (II)

• CRYSTAL DATA

Crystal structure (cell dimensions/pm), space group

α-As rhombohedral (a = 413.18, α = 54° 10'), $R\bar{3}m$, metallic form
β-As hexagonal (a = 376.0, c = 10.548), yellow
grey amorphous
$T(\alpha \rightarrow \beta)$ = 501 K; $T(\beta \rightarrow$ grey) = room temperature

X-ray diffraction: mass absorption coefficients (μ/ρ)/$cm^2\ g^{-1}$**:** CuK_α 83.4 MoK_α 69.7
Neutron scattering length, $b/10^{-12}$ cm**:** 0.658
Thermal neutron capture cross-section, σ_a/barns**:** 4.30

• GEOLOGICAL DATA

Minerals

A little native arsenic occurs naturally as microcrystalline masses, found in Siberia, Germany, France, Italy, Romania and the USA.

Mineral	Formula	Density	Hardness	Crystal appearance
Arsenopyrite	FeAsS	6.07	5.5 – 6	tet., met. silvery-white
Conichalcite	$CaCu(AsO_4)(OH)$	4.33	4.5	orth., vitreous green
Enargite	Cu_3AsS_4	4.45	3	orth., met. grey-black
Löllingite	$FeAs_2$	7.40	5 – 5.5	orth., met. white
Olivenite	$Cu_2(AsO_4)(OH)$	4.46	3	mon., adam./vitreous green
Orpiment	As_2S_3	3.49	1.5 – 2	mon., trans. resinous golden
Realgar	β-As_4S_4	3.5	1.5 – 2	mon., res. red-orange

Chief ores: arsenopyrite, realgar, orpiment.
World production/tonnes y^{-1}**:** 47 000 (As_2O_3)
Main mining areas: not much mined as such because more than required is produced as a by-product of refining certain sulfide ores.
Reserves/tonnes: n.a.
Specimen: available as pieces or powder. *Danger!*

Abundances

Sun (relative to H = 1×10^{12})**:** n.a.
Earth's crust/p.p.m.**:** 1.5
Atmosphere/p.p.m. (volume)**:** trace
Seawater/p.p.m.**:**
Atlantic surface: 1.45×10^{-3}
Atlantic deep: 1.53×10^{-3}
Pacific surface: 1.45×10^{-3}
Pacific deep: 1.75×10^{-3}
Residence time/years**:** 90 000
Classification: As(III) scavenged, As(V) recycled
Oxidation state: III but mainly V

At

Atomic number: 85
Relative atomic mass ($^{12}C = 12.0000$)**:** 209.9871 (At-210)

CAS: [7440-68-8]

• CHEMICAL DATA

Description: Astatine is a reactive, radioactive non-metal element which resembles iodine. It has been little researched because all its isotopes have short half-lives.

Radii/pm: At^{5+} 57; At^{-} 227
Electronegativity: 2.2 (Pauling); 1.96 (Allred); 6.2 eV (absolute)
Effective nuclear charge: 7.60 (Slater); 15.16 (Clementi); 19.61 (Froese-Fischer)

Standard reduction potentials $E^{\ominus}$/V

	V		I		0		–I
acid	$HAtO_3$	1.4	HAtO	0.7	At_2	0.2	At^-
base	AtO_3^-	0.5	AtO^-	0.0	At_2	0.2	At^-

Oxidation states

At^{-I}	[Rn]	At^- (aq)
At^{I}	s^2p^4	$AtBr_2^-$
At^{V}	s^2	AtO_3^- (aq)

Covalent bonds

Bond	r/ pm	E/ kJ mol^{-1}
At—At	*c.* 290	116

• PHYSICAL DATA

Melting point/K: 575 (est.)
Boiling point/K: 610 (est.)
ΔH_{fusion}/kJ mol^{-1}**:** 23.8 (est.)
ΔH_{vap}/kJ mol^{-1}**:** n.a.

Thermodynamic properties (298.15 K, 0.1 MPa)

State	$\Delta_f H^{\ominus}$/kJ mol^{-1}	$\Delta_f G^{\ominus}$/kJ mol^{-1}	$S^{\ominus}$/J K^{-1} mol^{-1}	C_p/J K^{-1} mol^{-1}
Solid	0	0	n.a.	n.a.
Gas	n.a.	n.a.	n.a.	n.a.

Density/kg m^{-3}**:** n.a.
Molar volume/cm^3**:** n.a.
Thermal conductivity/W m^{-1} K^{-1}**:** 1.7 [300 K]
Coefficient of linear thermal expansion/K^{-1}**:** n.a.
Electrical resistivity/Ω m**:** n.a.
Mass magnetic susceptibility/kg^{-1} m^3**:** n.a.

• BIOLOGICAL DATA

Biological role

None.

Toxicity

N.a., but should be similar in toxicity to iodine.

Hazards

Astatine is never encountered normally. It is dangerous because it is a powerful source of radiation. This element is only to be found inside a nuclear facility or research institute.

Levels in humans

nil

Daily dietary intake: nil
Total mass of element in average (70 kg) person: nil

Discovery: see Nuclear Data section.
[Greek, *astatos* = unstable]
French, *astate*; German, *Astat*; Italian, *astato*; Spanish, *astato*

Astatine

[as-tat-een]

• NUCLEAR DATA

Discovery: Astatine was first produced in 1940 by D.R. Corson, K.R. Mackenzie, and E. Segré at the University of California, Berkeley, USA.

Number of isotopes (including nuclear isomers): 29   **Isotope mass range:** 196 → 219

Key isotopes

Nuclide	Atomic mass	Half life ($T_{1/2}$)	Decay mode and energy (MeV)	Nuclear spin I	Nucl. mag. moment μ	Uses
^{205}At	204.986 000	26.2 m	β^+, EC (4.5) 90%; α (6.020) 10%; γ	9/2–		
^{206}At	205.986 580	29.4 m	β^+, EC (5.7) 99%; α (5.881) 1%; γ	5+		
^{207}At	206.985 730	1.81 h	β^+, EC (3.88) 90%; α (5.873) 10%; γ	9/2–		
^{208}At	207.986 510	1.63 h	β^+, EC (4.93) 99%; α (5.752) 1%; γ	6+		
^{209}At	208.986 149	5.4 h	EC (4.93) 96%; α (5.757) 4%; γ	6+		
^{210}At	209.987 126	8.1 h	EC (3.98) 99.8%; α (5.63) 0.2%	5+		
^{211}At	210.987 469	7.21 h	EC (0.78) 58%; α (5.980) 42%	9/2–		

Other isotopes of astatine have half lives less than 10 m.

NMR [Not recorded]

• ELECTRON SHELL DATA

Ground state electron configuration: $[Xe]4f^{14}5d^{10}6s^26p^5$
Term symbol: $^3P_{3/2}$
Electron affinity ($M \rightarrow M^-$)/kJ mol^{-1}**:** 270

Ionization energies/kJ mol^{-1}**:**

1.	$M \rightarrow M^+$	930
2.	$M^+ \rightarrow M^{2+}$	1600
3.	$M^{2+} \rightarrow M^{3+}$	(2900)
4.	$M^{3+} \rightarrow M^{4+}$	(4000)
5.	$M^{4+} \rightarrow M^{5+}$	(4900)
6.	$M^{5+} \rightarrow M^{6+}$	(7500)
7.	$M^{6+} \rightarrow M^{7+}$	(8800)
8.	$M^{7+} \rightarrow M^{8+}$	(13 300)
9.	$M^{8+} \rightarrow M^{9+}$	(15 400)
10.	$M^{9+} \rightarrow M^{10+}$	(17 700)

Electron binding energies/eV

K	1s	95 730
L_I	2s	17 493
L_{II}	$2p_{1/2}$	16 785
L_{III}	$2p_{3/2}$	14 214
M_I	3s	4317
M_{II}	$3p_{1/2}$	4008
M_{III}	$3p_{3/2}$	3426
M_{IV}	$3d_{3/2}$	2909
M_V	$3d_{5/2}$	2787

continued in Appendix 2, p255

Main lines in atomic spectrum
[Wavelength/nm(species)]
216.225 (I)
224.401 (I)

• CRYSTAL DATA

Crystal structure (cell dimensions/pm), space group
n.a.

X-ray diffraction: mass absorption coefficients (μ/ρ)/cm^2 g^{-1}**:** CuK_α n.a. MoK_α n.a.
Neutron scattering length, $b/10^{-12}$ cm**:** n.a.
Thermal neutron capture cross-section, σ_a/barns**:** n.a.

• GEOLOGICAL DATA

Minerals

Some isotopes of astatine, namely ^{215}At, ^{218}At and ^{219}At, are present in uranium and thorium minerals as part of the natural decay series, but the total amount present in the Earth's crust is estimated to be less than 30 g. Astatine (^{210}At and ^{211}At) can be produced by bombarding bismuth with high energy α-particles, and it is easily separated because astatine is volatile.

World production: world production of astatine to date is estimated to be only about 50 nanograms (50×10^{-9} g)

Specimen: not commercially available.

Abundances

Sun (relative to $H = 1 \times 10^{12}$)**:** n.a.
Earth's crust/p.p.m.**:** traces in some minerals
Seawater/p.p.m.**:** nil

Ba

Atomic number: 56
Relative atomic mass ($^{12}C = 12.0000$): 137.327

CAS: [7440-39-3]

• CHEMICAL DATA

Description: Barium is a soft, silvery-white metal which oxidises in air and reacts with water. It is obtained from barium oxide (BaO) by heating with aluminium. Barium sulfate is used commercially in drilling fluids for oil and gas exploration, and small amounts of barium compounds are used in paints and glass.

Radii/pm: Ba^{2+} 143; atomic 217; covalent 198
Electronegativity: 0.89 (Pauling); 0.97 (Allred); 2.4 eV (absolute)
Effective nuclear charge: 2.85 (Slater); 7.58 (Clementi); 10.27 (Froese-Fischer)

Standard reduction potentials $E^{\ominus}$/V

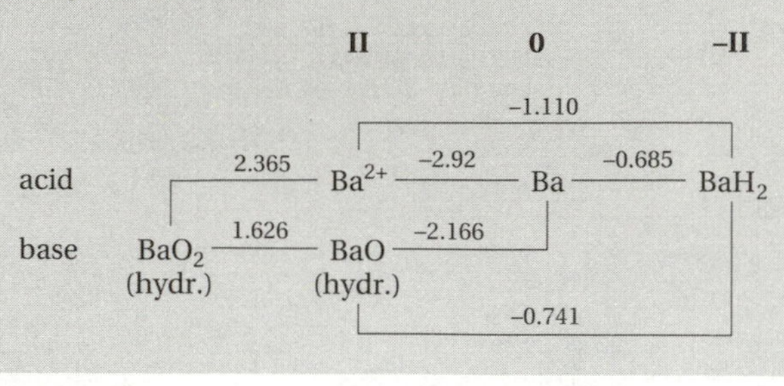

Oxidation states

Ba^{II}	[Xe]	BaH_2, BaO, $Ba(OH)_2$ (basic), BaO_2 (peroxide), Ba^{2+} (aq), BaF_2, $BaCl_2$ etc., $BaCO_3$, $BaSO_4$ insoluble, many other salts

• PHYSICAL DATA

Melting point/K: 1002
Boiling point/K: 1910

ΔH_{fusion}/kJ mol^{-1}: 7.66
ΔH_{vap}/kJ mol^{-1}: 150.9

Thermodynamic properties (298.15 K, 0.1 MPa)

State	$\Delta_f H^{\ominus}$/kJ mol^{-1}	$\Delta_f G^{\ominus}$/kJ mol^{-1}	$S^{\ominus}$/J K^{-1} mol^{-1}	C_p/J K^{-1} mol^{-1}
Solid	0	0	62.8	28.07
Gas	180	146	170.243	20.786

Density/kg m^{-3}: 3594 [293 K]; 3325 [liquid at m.p.]
Molar volume/cm^3: 38.21
Thermal conductivity/W m^{-1} K^{-1}: 18.4 [300 K]
Coefficient of linear thermal expansion/K^{-1}: $(18.1–21.0) \times 10^{-6}$
Electrical resistivity/Ω m: 50×10^{-8} [273 K]
Mass magnetic susceptibility/kg^{-1} m^3: $+1.9 \times 10^{-9}$ (s)

Young's modulus/GPa: 12.8
Rigidity modulus/GPa: 4.86
Bulk modulus/GPa: n.a.
Poisson's ratio/GPa: 0.28

• BIOLOGICAL DATA

Biological role

None, but barium is stimulatory.

Toxicity

Toxic intake: 100 – 200 mg
Lethal intake: LD_{50} ($BaCO_3$, oral, rat) = 418 mg kg^{-1}

Hazards

Barium sulfate is insoluble and used for body imaging, but soluble barium salts are highly toxic causing vomiting, colic, diarrhoea, tremors, paralysis, etc.

Levels in humans

Blood/mg dm^{-3}: 0.068
Bone/p.p.m.: 3 – 70
Liver/p.p.m.: 0.04 – 1.2
Muscle/p.p.m.: 0.09
Daily dietary intake: 0.6 – 1.7 mg
Total mass of element in average (70 kg) person: 22 mg

Discovered in 1808 by Sir Humphry Davy at London, England.
[Greek, *barys* = heavy]
French, *baryum*; German, *Barium*; Italian, *bario*; Spanish, *bario*

Barium

[bare-iuhm]

• NUCLEAR DATA

Number of isotopes (including nuclear isomers): 37 **Isotope mass range:** 117 → 148

Key isotopes

Nuclide	Atomic mass	Natural abundance (%)	Nuclear spin *I*	Nuclear magnetic moment *μ*	Uses
^{130}Ba	129.906 282	0.106	0+		E
^{132}Ba	131.905 042	0.101	0+		E
^{134}Ba	133.904 486	2.417	0+		E
^{135}Ba	134.905 665	6.592	3/2+	+0.837 943	E, NMR
^{136}Ba	135.904 553	7.854	0+		E
^{137}Ba	136.905 812	11.23	3/2+	+0.937 365	E, NMR
^{138}Ba	137.905 232	71.70	0+		E

A table of radioactive isotopes is given in Appendix 1, on p236.

NMR [Reference: $BaCl_2$ (aq)]	^{135}Ba	^{137}Ba
Relative sensitivity (^1H = 1.00)	4.99×10^{-3}	6.97×10^{-3}
Receptivity (^{13}C = 1.00)	1.83	4.41
Magnetogyric ratio/rad $T^{-1} s^{-1}$	2.6575×10^7	2.9728×10^7
Nuclear quadrupole moment/m^2	$+0.160 \times 10^{-28}$	$+0.245 \times 10^{-28}$
Frequency (^1H = 100 Hz; 2.3488T)/MHz	9.934	11.113

• ELECTRON SHELL DATA

Ground state electron configuration: [Xe]$6s^2$
Term symbol: $^1S_0$
Electron affinity ($M \to M^-$)/kJ mol^{-1}**:** –46

Ionization energies/kJ mol^{-1}**:**

1.	$M \to M^+$	502.8
2.	$M^+ \to M^{2+}$	965.1
3.	$M^{2+} \to M^{3+}$	(3600)
4.	$M^{3+} \to M^{4+}$	(4700)
5.	$M^{4+} \to M^{5+}$	(6000)
6.	$M^{5+} \to M^{6+}$	(7700)
7.	$M^{6+} \to M^{7+}$	(9000)
8.	$M^{7+} \to M^{8+}$	(10 200)
9.	$M^{8+} \to M^{9+}$	(13 500)
10.	$M^{9+} \to M^{10+}$	(15 100)

Electron binding energies/eV

K	1s	37 441
L_I	2s	5989
L_{II}	$2p_{1/2}$	5624
L_{III}	$2p_{3/2}$	5247
M_I	3s	1293
M_{II}	$3p_{1/2}$	1137
M_{III}	$3p_{3/2}$	1063
M_{IV}	$3d_{3/2}$	795.7
M_V	$3d_{5/2}$	780.5

continued in Appendix 2, p255

Main lines in atomic spectrum
[Wavelength/nm(species)]
350.111 (I)
455.403 (II)
493.409 (II)
553.548 (II) (AA)
614.172 (II)
649.690 (II)
705.994 (I)

• CRYSTAL DATA

Crystal structure (cell dimensions/pm), space group
b.c.c. (a = 502.5), Im3m
high pressure form: (a = 390.1, c = 615.5), $P6_3/mmc$

X-ray diffraction: mass absorption coefficients (μ/ρ)/$cm^2 g^{-1}$**:** CuK_α 330 MoK_α 43.5
Neutron scattering length, $b/10^{-12}$ cm: 0.507
Thermal neutron capture cross-section, σ_a/barns: 1.3

• GEOLOGICAL DATA

Minerals

Mineral	Formula	Density	Hardness	Crystal appearance
Barite	$BaSO_4$	4.50	3 – 3.5	orth., vitr./res. colourless-yellow
Benitoite*	$BaTiSi_3O_9$	3.6	6.3	hex. trans. blue
Witherite	$BaCO_3$	4.219	3–3.5	orth., vitr./res. colourless/grey

*gemstone

Chief ores: barite mainly, witherite occasionally
World production/tonnes y^{-1}**:** 6×10^6 (barium ores)
Main mining areas: UK, Italy, Czech Republic (Pribram), USA, Germany.
Reserves/tonnes: 450×10^6
Specimen: available as granules, rods or stick, *Warning!*

Abundances

Sun (relative to H = 1×10^{12})**:** 123
Earth's crust/p.p.m.**:** 500
Seawater/p.p.m.**:**
Atlantic surface: 4.7×10^{-3}
Atlantic deep: 9.3×10^{-3}
Pacific surface: 4.7×10^{-3}
Pacific deep: 20.0×10^{-3}
Residence time/years**:** 10 000
Classification: recycled
Oxidation state: II

Bk

Atomic number: 97
Relative atomic mass ($^{12}C = 12.0000$): 247.0703 (Bk-247)

CAS: [7440-40-6]

• CHEMICAL DATA

Description: Berkelium is a silvery, radioactive metal. It is attacked by oxygen, steam and acids, but not alkalis.

Radii/pm: Bk^{4+} 87; Bk^{3+} 98; Bk^{2+} 118; atomic 170
Electronegativity: 1.3 (Pauling); 1.2 (est.) (Allred); n.a. (absolute)
Effective nuclear charge: 1.65 (Slater)

Standard reduction potentials E°/V

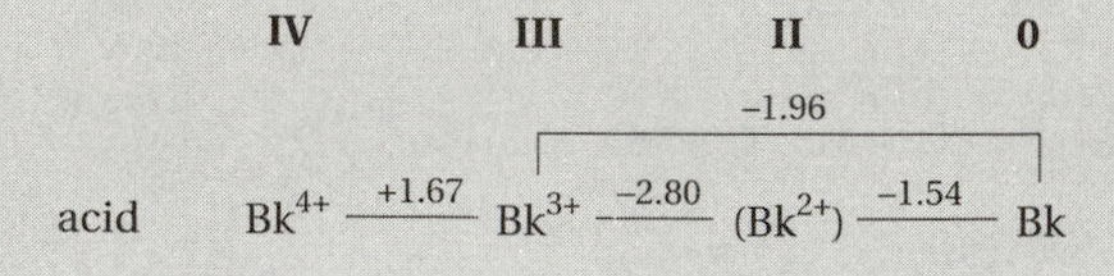

Oxidation states

Bk^{II}	f^9	BkO
Bk^{III}	f^8	Bk_2O_3, BkF_3, $BkCl_3$ etc., $[BkCl_6]^{3-}$, Bk^{3+} (aq), $[Bk(C_5H_5)_3]$
Bk^{IV}	f^7	BkO_2, BkF_4

• PHYSICAL DATA

Melting point/K: 1320
Boiling point/K: n.a.

ΔH_{fusion}/kJ mol^{-1}: n.a.
ΔH_{vap}/kJ mol^{-1}: 310

Thermodynamic properties (298.15 K, 0.1 MPa)

State	$\Delta_f H^{\circ}$/kJ mol^{-1}	$\Delta_f G^{\circ}$/kJ mol^{-1}	S°/J K^{-1} mol^{-1}	C_p/J K^{-1} mol^{-1}
Solid	0	0	n.a.	n.a.
Gas	n.a.	n.a.	n.a.	n.a.

Density/kg m^{-3}: 14 790 [293 K]
Molar volume/cm^3: 16.70
Thermal conductivity/W m^{-1} K^{-1}: 10 (est.) [300 K]
Coefficient of linear thermal expansion/K^{-1}: n.a.
Electrical resistivity/Ω m: n.a.
Mass magnetic susceptibility/kg^{-1} m^3: n.a.

Young's modulus/GPa: n.a.
Rigidity modulus/GPa: n.a.
Bulk modulus/GPa: n.a.
Poisson's ratio/GPa: n.a.

• BIOLOGICAL DATA

Biological role
None.

Toxicity
Toxic intake: n.a.
Lethal intake: n.a.

Hazards
Berkelium is never encountered normally. It is dangerous because it is a powerful source of radiation and the maximum body burden is 0.0004 μCi. This element is only to be found inside nuclear facilities or research laboratories.

Levels in humans
nil
Daily dietary intake: nil
Total mass of element in average (70 kg) person: nil

• NUCLEAR DATA

Discovery: Berkelium was first produced in 1949 by S.G. Thompson, A. Ghiorso, and G.T. Seaborg at Berkeley, California, USA.

Number of isotopes (including nuclear isomers): 15 **Isotope mass range:** 240 → 251

Key isotopes

Nuclide	Atomic mass	Half life ($T_{1/2}$)	Decay mode and energy (MeV)	Nuclear spin *I*	Nucl. mag. moment *μ*	Uses
^{243}Bk	243.062 997	4.5 h	EC (1.505) 99.8%; α (6.871) 0.2%; γ	3/2–		
^{244}Bk	244.065 160	4.4 h	EC (2.25) 99.99%; α (6.778) 0.01%; γ	4–		
^{245}Bk	245.066 357	4.94 d	EC (0.812) 99.9%; α (6.453) 0.1%; γ	3/2–		
^{246}Bk	246.068 720	1.80 d	EC (1.5); γ	2–		
^{247}Bk	247.070 300	1400 y	α (5.889); γ	3/2–		
^{248}Bk	248.073 106	23.7 h	$β^-$ (0.87); 70%; EC (0.72) 30%; γ	1–		
^{249}Bk	249.074 980	320 d	$β^-$ (0.125); α (5.525) 0.001%	7/2+	2.0	R
^{250}Bk	250.078 312	3.217 h	$β^-$ (1.781); γ	2–		

Other isotopes of berkelium have half lives less than 1 h.

NMR [not recorded]

• ELECTRON SHELL DATA

Ground state electron configuration: [Rn]$5f^9 7s^2$
Term symbol: $^6H_{15/2}$
Electron affinity ($M \rightarrow M^-$)/kJ mol^{-1}**:** n.a.

Ionization energies/kJ mol^{-1}**:**
1. $M \rightarrow M^+$ 601

Electron binding energies/eV
n.a.

Main lines in atomic spectrum*
[Wavelength/nm(species)]
323.972 (I)
325.219 (I)
328.875 (I)
328.935 (I)
333.526 (I)
340.828 (I)
342.695 (I)
*first seven lines associated with the neutral atom

• CRYSTAL DATA

Crystal structure (cell dimensions/pm), space group

α-Bk h.c.p.
β-Bk f.c.c.
$T(\alpha \rightarrow \beta)$ = 1203 K

X-ray diffraction: mass absorption coefficients (μ/ρ)/cm^2 g^{-1}**:** CuK_α n.a. MoK_α n.a.
Neutron scattering length, $b/10^{-12}$ cm**:** n.a.
Thermal neutron capture cross-section, σ_a/barns**:** 710 (^{249}Bk)

• GEOLOGICAL DATA

Minerals
None.

Chief source: berkelium-249 can be obtained in 100 mg quantities by the bombardment of ^{239}Pu with neutrons.

World production: n.a. but probably less than 1 g per year.

Specimen: commercially available, under licence - see Key.

Abundances
Sun (relative to H = 1×10^{12})**:** n.a.
Earth's crust/p.p.m.**:** nil
Seawater/p.p.m.**:** nil

Be

Atomic number: 4
Relative atomic mass ($^{12}C = 12.0000$): **9.012182**

CAS: [7440-41-7]

• CHEMICAL DATA

Discovery: Beryllium was discovered in 1797 by Nicholas Louis Vauquelin at Paris, France. Isolated in 1828 by F. Wöhler at Berlin, Germany, and independently by A.A.B. Bussy at Paris, France.
Description: Beryllium is a silvery-white, lustrous, relatively soft metal, which is obtained e.g. by the electrolysis of fused $BeCl_2$. It is unaffected by air or water, even at red heat. Beryllium us used in alloys with copper and nickel, and imparts excellent electrical and thermal conductivities. The copper alloy is used to make spark-proof tools.

Radii/pm: Be^{2+} 34; atomic 113; covalent 89
Electronegativity: 1.57 (Pauling); 1.47 (Allred); 4.9 eV (absolute)
Effective nuclear charge: 1.95 (Slater); 1.91 (Clementi); 2.27 (Froese-Fischer)

Standard reduction potentials $E^{\ominus}$/V

II 0

$Be^{2+} \xrightarrow{-1.97} Be$

Oxidation states

Be^{II} [He] BeO, $Be(OH)_2$, BeH_2, $[Be(OH_2)_4]^{2+}$ (aq), BeF_2, $BeCl_2$, etc, $BeCO_3$, salts

Covalent bonds

Bond	r/ pm	E/ kJ mol^{-1}
Be—H	163	226
Be—C	193	
Be—O	133	523
Be—F	143	632
Be—Cl	177	461
Be—Be	223	

• PHYSICAL DATA

Melting point/K: 1551 ± 5
Boiling point/K: 3243 (under pressure)
ΔH_{fusion}/kJ mol^{-1}: 9.80
ΔH_{vap}/kJ mol^{-1}: 308.8

Thermodynamic properties (298.15 K, 0.1 MPa)

State	$\Delta_f H^{\ominus}$/kJ mol^{-1}	$\Delta_f G^{\ominus}$/kJ mol^{-1}	$S^{\ominus}$/J K^{-1} mol^{-1}	C_p/J K^{-1} mol^{-1}
Solid	0	0	9.50	16.44
Gas	324.6	286.6	136.269	20.786

Density/kg m^{-3}: 1847.7 [293 K]
Molar volume/cm^3: 4.88
Thermal conductivity/W m^{-1} K^{-1}: 200 [300 K]
Coefficient of linear thermal expansion/K^{-1}: 11.5×10^{-6}
Electrical resistivity/Ω m: 4.0×10^{-8} [293 K]
Mass magnetic susceptibility/kg^{-1} m^3: -1.3×10^{-8} (s)

Young's modulus/GPa: 318
Rigidity modulus/GPa: 156
Bulk modulus/GPa: 110
Poisson's ratio/GPa: 0.02

• BIOLOGICAL DATA

Biological role

None.

Toxicity

Toxic intake: 13 mg kg^{-1} (rat)
Lethal intake: LD_{50} (acetate, intraperitoneal, rat) = 317 mg kg^{-1}

Hazards

Beryllium is a deadly poison. It is also carcinogenic for laboratory animals and maybe for humans. Inhalation of beryllium dust causes severe and irreparable lung damage.

Levels in humans

Blood/mg dm^{-3}: $< 1 \times 10^{-5}$
Bone/p.p.m.: 0.003
Liver/p.p.m.: 0.0016
Muscle/p.p.m.: 0.00075
Daily dietary intake: 0.01 mg
Total mass of element in average (70 kg) person: 0.036 mg

Discovery: see Chemical Data section.
[Greek, *beryllos* = beryl]
French, *béryllium*; German, *Beryllium*; Italian, *berillio*; Spanish, *berilio*

Beryllium

[be-ril-iuhm]

• NUCLEAR DATA

Number of isotopes (including nuclear isomers): 6 **Isotope mass range:** $6 \rightarrow 11$

Key isotopes

Nuclide	Atomic mass	Natural abundance (%)	Nuclear spin I	Nuclear magnetic moment μ	Uses
^{9}Be	9.012 182 2	100	3/2–	–1.177 9	NMR

A table of radioactive isotopes is given in Appendix 1, on p236.

NMR [Reference: $Be(NO_3)_2$ (aq)]	^{9}Be
Relative sensitivity (^{1}H = 1.00)	1.39×10^{-2}
Receptivity (^{13}C = 1.00)	78.8
Magnetogyric ratio/rad $T^{-1} s^{-1}$	3.7589×10^{7}
Nuclear quadrupole moment/m^2	0.05288×10^{-28}
Frequency (^{1}H = 100 Hz; 2.3488T)/MHz	14.053

• ELECTRON SHELL DATA

Ground state electron configuration: $[He]2s^2$
Term symbol: $^{1}S_0$
Electron affinity ($M \rightarrow M^-$)/kJ mol^{-1}**:** –18

Ionization energies/kJ mol^{-1}**:**

1.	$M \rightarrow M^{+}$	899.4
2.	$M^{+} \rightarrow M^{2+}$	1757.1
3.	$M^{2+} \rightarrow M^{3+}$	14 848
4.	$M^{3+} \rightarrow M^{4+}$	21 006

Electron binding energies/eV

K	1s	111.5

Main lines in atomic spectrum
[Wavelength/nm(species)]

234.861 (I) (AA)
381.345 (I)
436.099 (II)
467.333 (II)
467.342 (II)
527.081 (II)

• CRYSTAL DATA

Crystal structure (cell dimensions/pm), space group

α-Be h.c.p. (a = 228.55, c = 358.32), $P6_3/mmc$
β-Be b.c.c. (a = 255.15), Im3m
$T(\alpha \rightarrow \beta)$ = 1523 K

X-ray diffraction: mass absorption coefficients $(\mu/\rho)/cm^2\ g^{-1}$**:** CuK_α 1.50 MoK_α 0.298
Neutron scattering length, $b/10^{-12}$ cm**:** 0.779
Thermal neutron capture cross-section, σ_a/barns**:** 0.0092

• GEOLOGICAL DATA

Minerals

Mineral	Formula	Density	Hardness	Crystal appearance
Beryl*	$Be_3Al_2Si_6O_{18}$	2.8	7.5 – 8	hex., vit. green/blue
Bertrandite	$Be_2Si_2O_7(OH)_2$	2.60	6 – 7	orth., vit./pearly colourless
Chrysoberyl**	$BeAl_2O_4$	3.75	8.5	orth., vit. green, yellow, brown
Gadolinite	$Be_2FeY_2Si_2O_{10}$	4.4	6.5 – 7	mon., vit./greasy greenish-black
Herderite	$CaBe(PO_4)(F,OH)$	3.01	5 – 5.5	mon., vit. colourless/pale yellow

*gem quality = emerald; ** gem quality = Alexandrite, also known as "cat's eyes"

Chief ores: beryl, bertrandite
World production/tonnes y^{-1}**:** 364
Main mining areas: Brazil, USA, Madagascar, Germany, Czech Republic, Russia, India.
Reserves/tonnes**:** 400 000
Specimen: available as lumps or powder. *Danger!*

Abundances

Sun (relative to H = 1×10^{12})**:** 14
Earth's crust/p.p.m.**:** 2.6
Seawater/p.p.m.**:**
Atlantic surface: 8.8×10^{-8}
Atlantic deep: 17.5×10^{-8}
Pacific surface: 3.5×10^{-8}
Pacific deep: 22×10^{-8}
Residence time/years**:** 4000
Classification: recycled
Oxidation state: II

Atomic number: 83
Relative atomic mass (^{12}C = 12.0000)**:** 208.98037

CAS: [7440-69-9]

• CHEMICAL DATA

Description: Bismuth is a brittle metal with a silvery lustre and a pink tinge. It is stable to oxygen and water, but dissolves in concentrated HNO_3. Bismuth is used in alloys, pharmaceuticals, electronics, catalysts, cosmetics and pigments. The metal *expands* on solidification.

Radii/pm: Bi^{5+} 74; Bi^{3+} 96; atomic 155; covalent 152; van der Waals 240
Electronegativity: 2.02 (Pauling); 1.67 (Allred); 4.69 eV (absolute)
Effective nuclear charge: 6.30 (Slater); 13.34 (Clementi); 16.90 (Froese-Fischer)

Standard reduction potentials $E^{\ominus}$/V

V		III		0		–III
Bi^{5+}	*c.* 2	Bi^{3+}	0.317	Bi	–0.97	BiH_3

Oxidation states

Bi^{-III}	[Rn]	BiH_3
Bi^{I}	s^2p^2	Bi^+
Bi^{III}	s^2	Bi_2O_3, $Bi(OH)_3$, Bi^{3+} (aq), BiOCl, BiF_3, $BiCl_3$ etc., $[BiBr_6]^{3-}$, salts
Bi^{V}	d^{10}	Bi_2O_5 unstable, $[Bi(OH)_6]^-$ (aq), $NaBiO_3$, BiF_5, $KBiF_6$

Covalent bonds

Bond	r/ pm	E/ kJ mol^{-1}
Bi—H	n.a.	194
Bi—C	230	143
Bi—O	232	339
Bi^{V}—F	235	297
Bi^{III}—Cl	248	274
Bi—Bi	309	192

• PHYSICAL DATA

Melting point/K: 544.5
Boiling point/K: 1883 ± 5

ΔH_{fusion}/kJ mol^{-1}: 10.48
ΔH_{vap}/kJ mol^{-1}: 179.1

Thermodynamic properties (298.15 K, 0.1 MPa)

State	$\Delta_f H^{\ominus}$/kJ mol^{-1}	$\Delta_f G^{\ominus}$/kJ mol^{-1}	$S^{\ominus}$/J K^{-1} mol^{-1}	C_p/J K^{-1} mol^{-1}
Solid	0	0	56.74	25.52
Gas	207.1	168.2	187.005	20.786

Density/kg m^{-3}: 9747 [293 K]; 10 050 [liquid at m.p.]
Molar volume/cm^3: 21.44
Thermal conductivity/W m^{-1} K^{-1}: 7.87 [300 K]
Coefficient of linear thermal expansion/K^{-1}: 13.4 × 10^{-6}
Electrical resistivity/Ω m: 106.8 × 10^{-8} [273 K]
Mass magnetic susceptibility/kg^{-1} m^3: –1.684 × 10^{-8} (s)

Young's modulus/GPa: 34.0
Rigidity modulus/GPa: 12.8
Bulk modulus/GPa: n.a.
Poisson's ratio/GPa: 0.33

• BIOLOGICAL DATA

Biological role

None.

Toxicity

Toxic intake: n.a.

Lethal intake: *c.* 15 g (only one case reported)

Hazards

Bismuth is regarded as one of the less toxic heavy metals and it is commonly used as a medicine for stomach upsets. Excess bismuth can cause mild kidney damage.

Levels in humans

Blood/mg dm^{-3}: *c.* 0.016
Bone/p.p.m.: < 0.2
Liver/p.p.m.: 0.015 – 0.33
Muscle/p.p.m.: 0.032
Daily dietary intake: 0.005 – 0.02 mg
Total mass of element in average (70 kg) person: < 0.5 mg

Known in the fifteenth century, discoverer unknown.
[German, *bisemutum*]
French, *bismuth*; German, *Bismut*; Italian, *bismuto*; Spanish, *bismuto*

Bismuth

[biz-muth]

• NUCLEAR DATA

Number of isotopes (including nuclear isomers): 37 **Isotope mass range:** 189 → 215

Key isotopes

Nuclide	Atomic mass	Natural abundance (%)	Nuclear spin I	Nuclear magnetic moment μ	Uses
^{209}Bi	208.980 374	100	9/2–	+4.110 6	NMR

A table of radioactive isotopes is given in Appendix 1, on p236.

NMR [Reference: $KBiF_6$]	^{209}Bi
Relative sensitivity (^1H = 1.00)	0.13
Receptivity (^{13}C = 1.00)	777
Magnetogyric ratio/rad $T^{-1} s^{-1}$	4.2986×10^7
Nuclear quadrupole moment/m^2	-0.500×10^{-28}
Frequency (^1H = 100 Hz; 2.3488T)/MHz	16.069

• ELECTRON SHELL DATA

Ground state electron configuration: $[Xe]4f^{14}5d^{10}6s^26p^3$
Term symbol: $^4S_{3/2}$
Electron affinity ($M \rightarrow M^-$)/kJ mol^{-1}**:** 91.3

Ionization energies/kJ mol^{-1}**:**

1.	$M \rightarrow M^+$	703.2
2.	$M^+ \rightarrow M^{2+}$	1610
3.	$M^{2+} \rightarrow M^{3+}$	2466
4.	$M^{3+} \rightarrow M^{4+}$	4372
5.	$M^{4+} \rightarrow M^{5+}$	5400
6.	$M^{5+} \rightarrow M^{6+}$	8520
7.	$M^{6+} \rightarrow M^{7+}$	(10 300)
8.	$M^{7+} \rightarrow M^{8+}$	(12 300)
9.	$M^{8+} \rightarrow M^{9+}$	(14 300)
10.	$M^{9+} \rightarrow M^{10+}$	(16 300)

Electron binding energies/eV

K	1s	90 526
L_I	2s	16 388
L_{II}	$2p_{1/2}$	15 711
L_{III}	$2p_{3/2}$	13 419
M_I	3s	3999
M_{II}	$3p_{1/2}$	3696
M_{III}	$3p_{3/2}$	3177
M_{IV}	$3d_{3/2}$	2688
M_V	$3d_{5/2}$	2580

continued in Appendix 2, p255

Main lines in atomic spectrum
[Wavelength/nm(species)]
202.121 (I)
206.170 (I)
211.026 (I)
223.061 (I) (AA)
289.798 (I)
306.772 (I)

• CRYSTAL DATA

Crystal structure (cell dimensions/pm), space group
rhombohedral (a = 454.950, c = 1186.225), R3m

X-ray diffraction: mass absorption coefficients (μ/ρ)/$cm^2 g^{-1}$**:** CuK_α 240 MoK_α 120
Neutron scattering length, $b/10^{-12}$ cm**:** 0.8533
Thermal neutron capture cross-section, σ_a/barns**:** 0.034

• GEOLOGICAL DATA

Minerals

Native bismuth occurs naturally as metallic crystals associated with nickel, cobalt, silver, tin and uranium sulfide ores; found in Brazil, England, Norway and Canada.

Mineral	Formula	Density	Hardness	Crystal appearance
Bismite	α-Bi_2O_3	8.64	4.4	mon., sub-adam. yellow
Bismuthinite	Bi_2S_3	6.78	2	orth., met. grey
Bismutite	$Bi_2O_2(CO_3)$	8.15	2.5 – 3.5	tet., vit. yellow

Chief ores: native bismuth and bismuthinite; mainly produced as a by-product from lead and copper smelters, especially in the USA

World production/tonnes y^{-1}**:** 3000

Main mining areas: Bolivia, Peru, Japan, Mexico, Canada, Australia.

Reserves/tonnes: n.a.

Specimen: available as ingot, pieces, powder and shot. Safe.

Abundances

Sun (relative to H = 1×10^{12})**:** < 80
Earth's crust/p.p.m.**:** 0.048
Seawater/p.p.m.**:**
Atlantic surface: 5.1×10^{-8}
Atlantic deep: n.a.
Pacific surface: 4×10^{-8}
Pacific deep: 0.4×10^{-8}
Residence time/years**:** n.a.
Classification: scavenged
Oxidation state: III

Bh

Atomic number: 107

Relative atomic mass ($^{12}C = 12.0000$)**:** 262.12 (Bh-262)

CAS: [54037-14-8]

• CHEMICAL DATA

Description: Bohrium is a radioactive metal which does not occur naturally, and is of research interest only.

Radii/pm: Bh^{5+} 83 (est.); atomic 128 (est.)

Electronegativity: n.a.

Effective nuclear charge: n.a.

Standard reduction potentials $E^{\ominus}$/V

V　　　　0

$$Bh^{5+} \xrightarrow{+0.1 \text{ (est.)}} Bh$$

Oxidation states

Bh^{III}	d^4	predicted
Bh^{IV}	d^3	predicted
Bh^{V}	d^2	predicted
Bh^{VI}	d^1	predicted
Bh^{VII}	[f^{14}]	predicted, most stable?

• PHYSICAL DATA

Melting point/K: n.a.

Boiling point/K: n.a.

ΔH_{fusion}/kJ mol^{-1}: n.a.

ΔH_{vap}/kJ mol^{-1}: n.a.

ΔH_{subl}/kJ mol^{-1}: 753 (est.)

Thermodynamic properties (298.15 K, 0.1 MPa)

State	$\Delta_f H^{\ominus}$/kJ mol^{-1}	$\Delta_f G^{\ominus}$/kJ mol^{-1}	$S^{\ominus}$/J K^{-1} mol^{-1}	C_p/J K^{-1} mol^{-1}
Solid	0	0	n.a.	n.a.
Gas	n.a.	n.a.	n.a.	n.a.

Density/kg m^{-3}: 37 000 (est.)

Molar volume/cm^3: n.a.

Thermal conductivity/W m^{-1} K^{-1}: n.a.

Coefficient of linear thermal expansion/K^{-1}: n.a.

Electrical resistivity/Ω m: n.a.

Mass magnetic susceptibility/kg^{-1} m^3: n.a.

• BIOLOGICAL DATA

Biological role

None.

Toxicity

Toxic intake: n.a.

Lethal intake: n.a.

Hazards

Bohrium is never encountered normally, and only a few atoms have ever been made. It would be dangerous because of its intense radioactivity.

Levels in humans

nil

Daily dietary intake: nil

Total mass of element in average (70 kg) person: nil

• NUCLEAR DATA

Discovery: Bohrium was first made in 1981 by Peter Armbruster, Gottfried Münzenberg and their co-workers at Gesellschaft für Schwerionenforschung in Darmstadt, Germany.

Number of isotopes (including nuclear isomers): 2 **Isotope mass range:** 261 → 262

Key isotopes

Nuclide	Atomic mass	Half life ($T_{1/2}$)	Decay mode and energy (MeV)	Nuclear spin I	Nucl. mag. moment μ	Uses
^{261}Bh	261.127	0.012 s	α; SF			
^{262}Bh	262.1231	0.1 s	α			
^{262m}Bh		8×10^{-3} s	α			

NMR [Not recorded]

• ELECTRON SHELL DATA

Ground state electron configuration: $[Rn]5f^{14}6d^{5}7s^{2}$
Term symbol: $^{6}S_{5/2}$
Electron affinity ($M \rightarrow M^{-}$)/kJ mol^{-1}: n.a.

Ionization energies/kJ mol^{-1}:
1. $M \rightarrow M^{+}$ 660 (est.)

Electron binding energies/eV
n.a.

Main lines in atomic spectrum
[Wavelength/nm(species)]
n.a.

• CRYSTAL DATA

Crystal structure (cell dimensions/pm), space group
n.a.

X-ray diffraction: mass absorption coefficients (μ/ρ)/cm^{2} g^{-1}: CuK_{α} n.a. MoK_{α} n.a.
Neutron scattering length, $b/10^{-12}$ cm: n.a.
Thermal neutron capture cross-section, σ_{a}/barns: n.a.

• GEOLOGICAL DATA

Minerals
Not found on Earth.

Chief source: bohrium was made by the so-called cold fusion method in which a target of bismuth was bombarded with atoms of chromium having just the right energy to cause fusion. An atom of bohrium was detected:

$^{209}Bi + ^{54}Cr \rightarrow ^{262}Bh + n$

Specimen: not available commercially.

Abundances
Sun (relative to $H = 1 \times 10^{12}$): n.a.
Earth's crust/p.p.m.: nil
Seawater/p.p.m.: nil

B

Atomic number: 5
Relative atomic mass ($^{12}C = 12.0000$): **10.811**

CAS: [7440-42-8]

• CHEMICAL DATA

Discovery: Boron was discovered in 1808 by L.J. Lussac and L.J. Thenard in Paris, France and Sir Humphry Davy in London, England.
Description: Boron is a non-metal element with several forms - the most common is amorphous boron, which is a dark powder, unreactive to oxygen, water, acids and alkalis. It reacts with metals to form borides. Boron compounds are used in borosilicate glass, detergents and fire-retardants.

Radii/pm: B^{3+} 23; atomic 83; covalent 88; van der Waals 208
Electronegativity: 2.04 (Pauling); 2.01 (Allred); 4.29 eV (absolute)
Effective nuclear charge: 2.60 (Slater); 2.42 (Clementi); 2.27 (Froese-Fischer)

Standard reduction potentials $E^{\ominus}$/V

III		0
$B(OH)_3$	−0.890	B
BF_4^-	−1.284	B

Oxidation states

B^{III} [He] B_2O_3, H_3BO_3 (= $B(OH)_3$), borates e.g. borax $Na_2[B_4O_5(OH)_4].8H_2O$, B_2H_6, B_4H_{10} etc., $NaBH_4$, BF_3, BCl_3 etc.

Covalent bonds

Bond	r/ pm	E/ kJ mol^{-1}
B—H	119	381
B—H—B	132	439
B—C	156	372
B—O	136	536
B—F	129	613
B—Cl	175	456
B—Br	187	410
B—B	175	293

• PHYSICAL DATA

Melting point/K: 2573
Boiling point/K: 3931
ΔH_{fusion}/kJ mol^{-1}: 22.2
ΔH_{vap}/kJ mol^{-1}: 538.9

Thermodynamic properties (298.15 K, 0.1 MPa)

State	$\Delta_f H^{\ominus}$/kJ mol^{-1}	$\Delta_f G^{\ominus}$/kJ mol^{-1}	$S^{\ominus}$/J K^{-1} mol^{-1}	C_p/J K^{-1} mol^{-1}
Solid (α)	0	0	5.86	11.09
Gas	562.7	518.8	153.45	20.799

Density/kg m^{-3}: 2340 (β-rhomb.) [293 K]
Molar volume/cm^3: 4.62
Thermal conductivity/W m^{-1} K^{-1}: 27.0 [300 K]
Coefficient of linear thermal expansion/K^{-1}: 5×10^{-6}
Electrical resistivity/Ω m: 18 000 [273 K]
Mass magnetic susceptibility/kg^{-1} m^3: -7.8×10^{-9} (s)

• BIOLOGICAL DATA

Biological role

Essential to plants; toxic in excess.

Toxicity

Toxic intake: 5 g (boric acid)
Lethal intake: 10 – 20 g (boric acid). LD_{50} (boric acid, oral, rat) = 2.66 g kg^{-1}.

Hazards

Boric acid and borates are a human poison in excess, although once used in medicines.

Levels in humans

Blood/mg dm^{-3}: 0.13
Bone/p.p.m.: 1.1 – 3.3
Liver/p.p.m.: 0.4 – 3.3
Muscle/p.p.m.: 0.33 – 1
Daily dietary intake: 1 – 3 mg
Total mass of element in average (70 kg) person: 18 mg

Discovery: see Chemical Data section.
[Arabic, *buraq*]
French, *bore*; German, *Bor*; Italian, *boro*; Spanish, *boro*

Boron

[bohr-on]

• NUCLEAR DATA

Number of isotopes (including nuclear isomers): 6 **Isotope mass range:** 8 → 13

Key isotopes

Nuclide	Atomic mass	Natural abundance (%)	Nuclear spin *I*	Nuclear magnetic moment μ	Uses
^{10}B	10.012 936 9	19.9	3+	+1.800 65	NMR
^{11}B	11.009 305 4	80.1	3/2–	+2.688 637	NMR

A table of radioactive isotopes is given in Appendix 1, on p237.

NMR [Reference: $(C_2H_5)_2O/BF_3$]	^{10}B	^{11}B
Relative sensitivity (^1H = 1.00)	0.0199	0.17
Receptivity (^{13}C = 1.00)	22.1	754
Magnetogyric ratio/rad $T^{-1}s^{-1}$	2.8740×10^7	8.5794×10^7
Nuclear quadrupole moment/m^2	0.08459×10^{-28}	0.04059×10^{-28}
Frequency (^1H = 100 Hz; 2.3488T)/MHz	10.746	32.084

• ELECTRON SHELL DATA

Ground state electron configuration: $[He]2s^22p^1$
Term symbol: $^2P_{1/2}$
Electron affinity ($M \rightarrow M^-$)/kJ mol^{-1}**:** 26.7

Ionization energies/kJ mol^{-1}**:**

1.	$M \rightarrow M^+$	800.6
2.	$M^+ \rightarrow M^{2+}$	2427
3.	$M^{2+} \rightarrow M^{3+}$	3660
4.	$M^{3+} \rightarrow M^{4+}$	25 025
5.	$M^{4+} \rightarrow M^{5+}$	32 822

Electron binding energies/eV

K	1s	188

Main lines in atomic spectrum
[Wavelength/nm(species)]

208.891 (I)
208.957 (I)
249.667 (I)
249.773 (I) (AA)
345.129 (I)
1166.004 (I)
1166.247 (I)

• CRYSTAL DATA

Crystal structure (cell dimensions/pm), space group

Tetragonal (a = 874.0; c = 506), $P4_2/nnm$
α-B rhombohedral (a = 506.7, α = 58° 4'), R3m
β-B rhombohedral (a = 1014.5, α = 65° 12'), R3m, R32, R3m
Orthorhombic (a = 1015, b = 895, c = 1790)
Monoclinic (a = 1013, b = 893, c = 1786, $\alpha \approx 90°$, $\beta \approx 90°$, $\gamma \approx 90°$) or triclinic
Hexagonal (a = 1198, c = 954)

X-ray diffraction: mass absorption coefficients $(\mu/\rho)/cm^2\ g^{-1}$**:** CuK_α 2.39 MoK_α 0.392
Neutron scattering length, $b/10^{-12}$ cm**:** 0.535
Thermal neutron capture cross-section, σ_a/barns**:** 767

• GEOLOGICAL DATA

Minerals

Mineral	Formula	Density	Hardness	Crystal appearance
Borax	$Na_2B_4O_5(OH)_4.8H_2O$	1.715	2 – 2.5	mon., vit./res./earthy colourless
Colemanite	$CaB_3O_4(OH)_3.H_2O$	2.423	4.5	mon., vit./adam. colourless
Datolite	$CaBSiO_4(OH)$	3.0	5 – 5.5	mon., vit. white
Kernite	$Na_2B_4O_6.(OH)_2.3H_2O$	1.908	2.5	mon., vit. colourless
Ulexite	$NaCaB_5O_6(OH)_6.5H_2O$	1.955	2.5	tric., silky/vit. colourless

Chief ores: kernite, borax, ulexite, colemanite
World production/tonnes y^{-1}**:** 1×10^6 (B_2O_3)
Main mining areas: ulexite in USA, Tibet, Chile; colemanite in USA, Turkey
Reserves/tonnes: 270×10^6 as B_2O_3
Specimen: available as crystals, pieces or powder. Safe.

Abundances

Sun (relative to H = 1×10^{12})**:** 2.63×10^5
Earth's crust/p.p.m.**:** 950
Seawater/p.p.m.**:** 4.41
Residence time/years**:** 1×10^7
Classification: accumulating
Oxidation state: III

Br

Atomic number: 35
Relative atomic mass ($^{12}C = 12.0000$): **79.904**

CAS: [7726-95-6]

• CHEMICAL DATA

Discovery: Bromine was discovered in 1826 by A.J.Balard at Montpellier, France, and C. Löwig at Heidelberg, Germany.
Description: Bromine is a red, dense, sharp-smelling liquid (Br_2) that is extracted industrially from sea water. Bromine compounds are used in fuel additives, pesticides, flame-retardants and photography.

Radii/pm: Br^- 196; covalent 114; van der Waals 195
Electronegativity: 2.96 (Pauling); 2.74 (Allred); 7.59 eV (absolute)
Effective nuclear charge: 7.60 (Slater); 9.03 (Clementi); 10.89 (Froese-Fischer)

Standard reduction potentials $E^⊖$/V

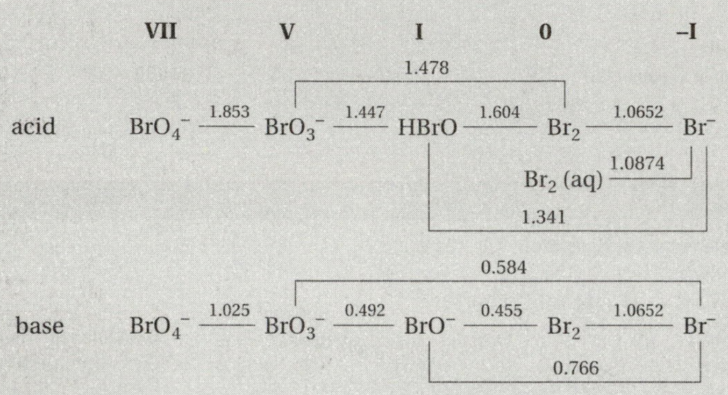

Oxidation states

Br^{-I}	[Kr]	Br^- (aq), HBr, KBr etc.
Br^0	s^2p^5	Br_2
Br^I	s^2p^4	Br_2O, $BrCl_2^-$
Br^{III}	s^2p^2	BrF_3, BrF_4^-
Br^{IV}	s^2p^1	BrO_2
Br^V	s^2	BrO_3^- (aq), BrF_5, BrF_6^-
Br^{VII}	d^{10}	$KBrO_4$, BrF_6^+

Covalent bonds

Bond	r/ pm	E/ kJ mol^{-1}
Br—H	141	366
Br—C	194	285
Br—O	160	201
Br—F	176	249
Br—Cl	214	216
Br—Br	228	191

Other bonds to bromine: see other elements

• PHYSICAL DATA

Melting point/K: 265.9
Boiling point/K: 331.93
Critical temperature/K: 588

ΔH_{fusion}/kJ mol^{-1}: 10.8
ΔH_{vap}/kJ mol^{-1}: 30.0

Thermodynamic properties (298.15 K, 0.1 MPa)

State	$\Delta_f H^⊖$/kJ mol^{-1}	$\Delta_f G^⊖$/kJ mol^{-1}	$S^⊖$/J K^{-1} mol^{-1}	C_p/J K^{-1} mol^{-1}
Liquid	0	0	152.231	75.689
Gas (atom)	111.884	82.396	175.022	20.786

Density/kg m^{-3}: 4050 [123 K]; 3122.6 [293 K]; 7. 59 (gas)
Molar volume/cm^3: 19.73 [123 K]
Thermal conductivity/W m^{-1} K^{-1}: 0.122 [300 K] (l)
Mass magnetic susceptibility/kg^{-1} m^3: -4.44×10^{-9} (l)

• BIOLOGICAL DATA

Biological role
None proved.

Toxicity
Br_2 is very toxic; bromide is slightly toxic.

Toxic intake: 3 g (Br^-)

Lethal intake: LD_{50} (Br_2, oral, human) = *c.* 1 g; bromide > 35 g

Hazards
Bromine is corrosive and its vapour attacks the eyes and lungs. Bromide intake leads to depression and loss of weight.

Levels in humans
Blood/mg dm^{-3}: 4.7
Bone/p.p.m.: 6.7
Liver/p.p.m.: 0.2 – 7
Muscle/p.p.m.: 7.7
Daily dietary intake: 0.8 – 24 mg
Total mass of element in average (70 kg) person: 260 mg

Discovery: see Chemical Data section.
[Greek, *bromos* = stench]
French, *brome*; German, *Brom*; Italian, *bromo*; Spanish, *bromo*

Bromine
[broh-meen]

• NUCLEAR DATA

Number of isotopes (including nuclear isomers): 28 **Isotope mass range:** 72 → 92

Key isotopes

Nuclide	Atomic mass	Natural abundance (%)	Nuclear spin *I*	Nuclear magnetic moment μ	Uses
^{79}Br	78.918 336 1	50.69	3/2–	+2.106 399	E, NMR
^{81}Br	80.916 289	49.31	3/2–	+2.270 560	E, NMR

A table of radioactive isotopes is given in Appendix 1, on p237.

NMR [Reference: NaBr (aq)]	^{79}Br	^{81}Br
Relative sensitivity (^1H = 1.00)	0.0786	0.0985
Receptivity (^{13}C = 1.00)	226	277
Magnetogyric ratio/rad $T^{-1}s^{-1}$	6.7023×10^7	7.2246×10^7
Nuclear quadrupole moment/m^2	$+0.331 \times 10^{-28}$	$+0.276 \times 10^{-28}$
Frequency (^1H = 100 Hz; 2.3488T)/MHz	25.053	27.006

• ELECTRON SHELL DATA

Ground state electron configuration: $[Ar]3d^{10}4s^24p^5$
Term symbol: $^2P_{3/2}$
Electron affinity ($M \rightarrow M^-$)/kJ mol^{-1}**:** 324.7

Ionization energies/kJ mol^{-1}**:**

1.	$M \rightarrow M^+$	1139.9
2.	$M^+ \rightarrow M^{2+}$	2104
3.	$M^{2+} \rightarrow M^{3+}$	3500
4.	$M^{3+} \rightarrow M^{4+}$	4560
5.	$M^{4+} \rightarrow M^{5+}$	5760
6.	$M^{5+} \rightarrow M^{6+}$	8550
7.	$M^{6+} \rightarrow M^{7+}$	9940
8.	$M^{7+} \rightarrow M^{8+}$	18 600
9.	$M^{8+} \rightarrow M^{9+}$	(23 900)
10.	$M^{9+} \rightarrow M^{10+}$	(28 100)

Electron binding energies/eV

K	1s	13 474
L_I	2s	1782
L_{II}	$2p_{1/2}$	1596
L_{III}	$2p_{3/2}$	1550
M_I	3s	257
M_{II}	$3p_{1/2}$	189
M_{III}	$3p_{3/2}$	182
M_{IV}	$3d_{3/2}$	70
M_V	$3d_{5/2}$	69

Main lines in atomic spectrum
[Wavelength/nm(species)]

614.860 (I)
635.073 (I)
655.980 (I)
663.162 (I)
751.296 (I)
827.244 (I)
844.655 (I)
926.542 (I)

• CRYSTAL DATA

Crystal structure (cell dimensions/pm), space group
orthorhombic (120 K) (*a* = 673.7, *b* = 454.8, *c* = 876.1), Cmca

X-ray diffraction: mass absorption coefficients (μ/ρ)/$cm^2\ g^{-1}$**:** CuK_α 99.6 MoK_α 79.8
Neutron scattering length, $b/10^{-12}$ cm: 0.679
Thermal neutron capture cross-section, σ_a/barns**:** 6.8

• GEOLOGICAL DATA

Minerals

Mineral	Formula	Density	Hardness	Crystal appearance
Bromargyrite	AgBr	6.474	2.5	res./adam. colourless

Chief source: : sea water, Dead Sea and natural brines; salt-lake evaporates

World production/tonnes y^{-1}**:** 330 000

Main mining areas: USA, Israel, UK, Russia, France, Japan.

Reserves/tonnes: almost unlimited

Specimen: available as the liquid in sealed ampoules. *Danger!*

Abundances

Sun (relative to H = 1×10^{12})**:** n.a.
Earth's crust/p.p.m.**:** 0.37
Seawater/p.p.m.**:** 65
Residence time/years**:** 1×10^8
Classification: accumulating
Oxidation state: –I

Cd

Atomic number: 48
Relative atomic mass ($^{12}C = 12.0000$)**:** 112.411

CAS: [7440-43-9]

• CHEMICAL DATA

Description: Cadmium is a silvery metal that tarnishes in air, and is soluble in acids but not alkalis. It is used in rechargeable batteries, alloys and pigments, but because of its toxicity these uses are being phased out wherever possible.

Radii/pm: Cd^{2+} 103; Cd^{+} 114; atomic 149; covalent 141
Electronegativity: 1.69 (Pauling); 1.46 (Allred); 4.33 eV (absolute)
Effective nuclear charge: 4.35 (Slater); 8.19 (Clementi); 11.58 (Froese-Fischer)

Standard reduction potentials $E^{\ominus}$/V

	II		0
acid	Cd^{2+}	−0.4025	Cd
base	$Cd(OH)_2$	−0.824	Cd
	$[Cd(NH_3)_4]^{2+}$	−0.622	Cd
	$[Cd(CN)_4]^{2-}$	−1.09	Cd

Oxidation states

Cd^{I}	$d^{10}s^{1}$	rare e.g. $Cd_2[AlCl_4]_2$
Cd^{II}	d^{10}	CdO (basic), CdS, $Cd(OH)_2$, CdF_2, $CdCl_2$ etc., many salts, $[Cd(OH_2)_6]^{2+}$ (aq), many complexes, e.g. $[Cd(SCN)_4]^{2-}$

• PHYSICAL DATA

Melting point/K: 594.1
Boiling point/K: 1038
ΔH_{fusion}/kJ mol^{-1}: 6.11
ΔH_{vap}/kJ mol^{-1}: 99.87

Thermodynamic properties (298.15 K, 0.1 MPa)

State	$\Delta_f H^{\ominus}$/kJ mol^{-1}	$\Delta_f G^{\ominus}$/kJ mol^{-1}	$S^{\ominus}$/J K^{-1} mol^{-1}	C_p/J K^{-1} mol^{-1}
Solid	0	0	51.76	25.98
Gas	112.01	77.41	167.746	20.786

Density/kg m^{-3}: 8650 [293 K]; 7996 [liquid at m.p.]
Molar volume/cm^3: 13.00
Thermal conductivity/W m^{-1} K^{-1}: 96.8 [300 K]
Coefficient of linear thermal expansion/K^{-1}: 29.8×10^{-6}
Electrical resistivity/Ω m: 6.83×10^{-8} [273 K]
Mass magnetic susceptibility/kg^{-1} m^3: -2.21×10^{-9} (s)

Young's modulus/GPa: 62.6
Rigidity modulus/GPa: 24.0
Bulk modulus/GPa: 51.0
Poisson's ratio/GPa: 0.30

• BIOLOGICAL DATA

Biological role

None has been proved, although suspected. It is stimulatory.

Toxicity

Toxic intake: 17 mg kg^{-1} (chloride, oral, rat)
Lethal intake: LD_{50} (chloride, oral, guinea pig) = 63 mg kg^{-1}

Hazards

Cadmium is toxic but its emetic action means that little is absorbed, so fatal poisoning rarely occurs. Cadmium is carcinogenic and teratogenic.

Levels in humans

Blood/mg dm^{-3}: 0.0052
Bone/p.p.m.: 1.8
Liver/p.p.m.: 2 – 22
Muscle/p.p.m.: 0.14 – 3.2
Daily dietary intake: 0.007 – 3 mg
Total mass of element in average (70 kg) person: 50 mg

Discovered in 1817 by F. Stromeyer at Göttingen, Germany.
[Latin, *cadmia* = calamine]
French, *cadmium*; German, *Kadmium*; Italian, *cadmio*; Spanish, *cadmio*

Cadmium

[cad-mium]

• NUCLEAR DATA

Number of isotopes (including nuclear isomers): 31 **Isotope mass range:** 99 → 124

Key isotopes

Nuclide	Atomic mass	Natural abundance (%)	Nuclear spin I	Nuclear magnetic moment μ	Uses
^{106}Cd	105.906 461	1.25	0+		E
^{108}Cd	107.904 176	0.89	0+		E
^{110}Cd	109.903 005	12.49	0+		E
^{111}Cd	110.904 182	12.80	1/2+	–0.594 885 7	E, NMR
^{112}Cd	111.902 757	24.13	0+		E
^{113}Cd	112.904 400	12.22	1/2+	–0.622 300 5	E, NMR
^{114}Cd	113.903 357	28.73	0+		E
^{116}Cd	115.904 755	7.49	0+		E

A table of radioactive isotopes is given in Appendix 1, on p237.

NMR [Reference: $Cd(ClO_4)_2$ (aq); $Cd(CH_3)_2$]	^{111}Cd	^{113}Cd
Relative sensitivity (^1H = 1.00)	9.54×10^{-3}	0.0109
Receptivity (^{13}C = 1.00)	6.93	7.6
Magnetogyric ratio/rad $T^{-1} s^{-1}$	-5.6714×10^7	-5.9328×10^7
Frequency (^1H = 100 Hz; 2.3488T)/MHz	21.205	22.182

• ELECTRON SHELL DATA

Ground state electron configuration: $[Kr]4d^{10}5s^2$
Term symbol: $^1S_0$
Electron affinity ($M \rightarrow M^-$)/kJ mol^{-1}**:** –26

Ionization energies/kJ mol^{-1}**:**

1.	$M \rightarrow M^+$	867.6
2.	$M^+ \rightarrow M^{2+}$	1631
3.	$M^{2+} \rightarrow M^{3+}$	3616
4.	$M^{3+} \rightarrow M^{4+}$	(5300)
5.	$M^{4+} \rightarrow M^{5+}$	(7000)
6.	$M^{5+} \rightarrow M^{6+}$	(9100)
7.	$M^{6+} \rightarrow M^{7+}$	(11 100)
8.	$M^{7+} \rightarrow M^{8+}$	(14 100)
9.	$M^{8+} \rightarrow M^{9+}$	(16 400)
10.	$M^{9+} \rightarrow M^{10+}$	(18 800)

Electron binding energies/eV

K	1s	26 711
L_I	2s	4018
L_{II}	$2p_{1/2}$	3727
L_{III}	$2p_{3/2}$	3538
M_I	3s	772.0
M_{II}	$3p_{1/2}$	652.6
M_{III}	$3p_{3/2}$	618.4
M_{IV}	$3d_{3/2}$	411.9
M_V	$3d_{5/2}$	405.2

continued in Appendix 2, p255

Main lines in atomic spectrum
[Wavelength/nm(species)]
214.441 (II)
226.502 (II)
228.802 (I) (AA)
326.106 (I)
643.847 (I)

• CRYSTAL DATA

Crystal structure (cell dimensions/pm), space group
h.c.p. (a = 297.94, c = 561.86), $P6_3/mmc$

X-ray diffraction: mass absorption coefficients (μ/ρ)/$cm^2 g^{-1}$**:** CuK_α 231 MoK_α 27.5
Neutron scattering length, $b/10^{-12}$ cm: 0.51
Thermal neutron capture cross-section, σ_a/barns**:** 2450

• GEOLOGICAL DATA

Minerals

Very rare.

Mineral	Formula	Density	Hardness	Crystal appearance
Cadmoselite	β-CdSe	5.6	n.a.	hex., res./adam. black
Greenockite	CdS	4.9	3 – 3.5	hex., yellow, tiny prisms
Otavite	$CdCO_3$	5.03	n.a.	rhom., adam. white/brown

Chief ores: none as such, most cadmium is produced as a by-product of the smelting of zinc from its ore ZnS, in which CdS is a significant impurity.

World production/tonnes y^{-1}**:** 13 900

Main mining areas: see zinc

Reserves/tonnes**:** see zinc

Specimen: available as foil, granules, powder, rod, shot, stick or wire. *Care!*

Abundances

Sun (relative to H = 1×10^{12})**:** 71
Earth's crust/p.p.m.**:** 0.11
Seawater/p.p.m.**:**
Atlantic surface: 1.1×10^{-6}
Atlantic deep: 38×10^{-6}
Pacific surface: 1.1×10^{-6}
Pacific deep: 100×10^{-6}
Residence time/years**:** 30
Classification: recycled
Oxidation state: II

Cs

Atomic number: 55
Relative atomic mass ($^{12}C = 12.0000$)**:** 132.90543

CAS: [7440-46-2]

• CHEMICAL DATA

Description: Caesium is a soft, shiny, gold-coloured metal which oxidises rapidly in air and reacts explosively with water. It is obtained by the electrolysis of molten caesium cyanide, and by other methods. Caesium and its salts are used commercially as catalyst promoters, in special glasses, and in radiation monitoring equipment.

Radii/pm: Cs^+ 165; atomic 265.4; covalent 235; van der Waals 262
Electronegativity: 0.79 (Pauling); 0.86 (Allred); 2.18 eV (absolute)
Effective nuclear charge: 2.20 (Slater); 6.36 (Clementi); 8.56 (Froese-Fischer)

Standard reduction potentials $E^\ominus$/V

I		0
Cs^+	$\xrightarrow{-2.923}$	Cs

Oxidation states

Cs^{-I}	s^2	caesium metal in liquid ammonia
Cs^{I}	[Xe]	Cs_2O, Cs_2O_2, CsO_2, CsOH, CsH, CsF, CsCl etc., $[Cs(OH_2)_x]^{3+}$ (aq), Cs_2CO_3, many salts and salt hydrates, some complexes with crown ethers etc.

• PHYSICAL DATA

Melting point/K: 301.55
Boiling point/K: 951.6

ΔH_{fusion}/kJ mol^{-1}: 2.09
ΔH_{vap}/kJ mol^{-1}: 65.90

Thermodynamic properties (298.15 K, 0.1 MPa)

State	$\Delta_f H^\ominus$/kJ mol^{-1}	$\Delta_f G^\ominus$/kJ mol^{-1}	$S^\ominus$/J K^{-1} mol^{-1}	C_p/J K^{-1} mol^{-1}
Solid	0	0	85.23	32.17
Gas	76.065	49.121	175.595	20.786

Density/kg m^{-3}: 1873 [293 K]; 1843 [liquid at m.p.]
Molar volume/cm^3: 70.96
Thermal conductivity/W m^{-1} K^{-1}: 35.9 [300 K]
Coefficient of linear thermal expansion/K^{-1}: 97×10^{-6}
Electrical resistivity/Ω m: 20.0×10^{-8} [293 K]
Mass magnetic susceptibility/kg^{-1} m^3: $+2.8 \times 10^{-9}$ (s)

Young's modulus/GPa: 1.7
Rigidity modulus/GPa: 0.65
Bulk modulus/GPa: n.a.
Poisson's ratio/GPa: 0.295

• BIOLOGICAL DATA

Biological role

No known biological role, but it may partly replace potassium.

Toxicity

Toxic intake: n.a. but regarded as fairly toxic
Lethal intake: LD_{50} (Cs_2CO_3, oral, rat) = 2333 mg kg^{-1}

Hazards

Although similar to potassium, caesium can have serious effects on the body if taken in excess. Rats fed Cs in place of K died after two weeks. ^{134}Cs and ^{137}Cs are dangerous radioactive pollutants which have escaped from nuclear reactors.

Levels in humans

Blood/mg dm^{-3}: 0.0038
Bone/p.p.m.: 0.013 – 0.052
Liver/p.p.m.: 0.04 – 0.05
Muscle/p.p.m.: 0.07 – 1.6
Daily dietary intake: 0.004 – 0.03 mg
Total mass of element in average (70 kg) person: *c.* 6 mg

Discovered in 1860 by R. Bunsen and G.R. Kirchhoff at Heidelberg, Germany. [Latin, *caesius* = sky blue]
French, *césium*; German, *Caesium*; Italian, *cesio*; Spanish, *cesio*

Caesium/Cesium (US)

[seez-iuhm]

• NUCLEAR DATA

Number of isotopes (including nuclear isomers): 40 **Isotope mass range:** 114 → 145

Key isotopes

Nuclide	Atomic mass	Natural abundance (%)	Nuclear spin I	Nuclear magnetic moment μ	Uses
^{133}Cs	133.905 429	100	7/2+	+2.582 024	NMR

A table of radioactive isotopes is given in Appendix 1, on p237.

NMR [Reference: 0.5M CsBr(aq)]	^{133}Cs
Relative sensitivity (^1H = 1.00)	0.0474
Receptivity (^{13}C = 1.00)	269
Magnetogyric ratio/rad $T^{-1} s^{-1}$	3.5087×10^7
Nuclear quadrupole moment/m^2	-0.0037×10^{-28}
Frequency (^1H = 100 Hz; 2.3488T)/MHz	13.117

• ELECTRON SHELL DATA

Ground state electron configuration: $[Xe]6s^1$
Term symbol: $^2S_{1/2}$
Electron affinity ($M \rightarrow M^-$)/kJ mol^{-1}**:** 45.5

Ionization energies/kJ mol^{-1}**:**

1.	$M \rightarrow M^+$	375.7
2.	$M^+ \rightarrow M^{2+}$	2420
3.	$M^{2+} \rightarrow M^{3+}$	(3400)
4.	$M^{3+} \rightarrow M^{4+}$	(4400)
5.	$M^{4+} \rightarrow M^{5+}$	(6000)
6.	$M^{5+} \rightarrow M^{6+}$	(7100)
7.	$M^{6+} \rightarrow M^{7+}$	(8300)
8.	$M^{7+} \rightarrow M^{8+}$	(11 300)
9.	$M^{8+} \rightarrow M^{9+}$	(12 700)
10.	$M^{9+} \rightarrow M^{10+}$	(23 700)

Electron binding energies/eV

K	1s	35 985
L_I	2s	5714
L_{II}	$2p_{1/2}$	5359
L_{III}	$2p_{3/2}$	5012
M_I	3s	1211
M_{II}	$3p_{1/2}$	1071
M_{III}	$3p_{3/2}$	1003
M_{IV}	$3d_{3/2}$	740.5
M_V	$3d_{5/2}$	726.6

continued in Appendix 2, p255

Main lines in atomic spectrum
[Wavelength/nm(species)]

455.528 (I)
460.376 (II)
522.704 (II)
592.563 (II)
852.113 (AA) (I)
895.347 (I)

• CRYSTAL DATA

Crystal structure (cell dimensions/pm), space group
b.c.c. (78 K) (a = 614), Im3m
High pressure forms: (a = 598.4), Fm3m; (a = 580.0), Fm3m
X-ray diffraction: mass absorption coefficients (μ/ρ)/$cm^2 g^{-1}$**:** CuK_α 318 MoK_α 41.3
Neutron scattering length, $b/10^{-12}$ cm**:** 0.542
Thermal neutron capture cross-section, σ_a/barns**:** 29

• GEOLOGICAL DATA

Minerals

Few are known.

Mineral	Formula	Density	Hardness	Crystal appearance
Cesium kupleskite	$Cs_3(Mn,Fe)_7(Ti,Nb)_2Si_8O_{24}.(OH,F)_7$	3.68	4	tric., dull gold-brown
Pollucite	$(Cs,Na)_2Al_2Si_4O_{12}.nH_2O$	2.94	6.5	cub., col., vit.

Chief ores: pollucite; caesium is also found in lepidolite (see lithium)

World production/tonnes y^{-1}**:** *c.* 20 (caesium compounds)

Main mining areas: Bernic Lake (Manitoba, Canada), Bikita (Zimbabwe) and South-West Africa.

Reserves/tonnes**:** *c.* 100 000 (60 000 at Bernic Lake)

Specimen: available as small ingots in sealed ampoules. *Danger!*

Abundances

Sun (relative to H = 1×10^{12})**:** < 80
Earth's crust/p.p.m.**:** 3
Seawater/p.p.m.**:** 3.0×10^{-4}
Atlanticsurface: n.a.
Atlanticdeep: n.a.
Pacific surface: n.a.
Pacific deep: n.a.
Residence time/years**:** 600 000
Classification: accumulating
Oxidation state: I

Ca

Atomic number: 20
Relative atomic mass ($^{12}C = 12.0000$)**:** 40.078

CAS: [7440-70-2]

• CHEMICAL DATA

Description: Calcium is a silvery-white, relatively soft metal that is obtained by heating calcium oxide (CaO) with aluminium metal in a vacuum. Although calcium metal is attacked by oxygen and water, the bulk metal is protected by an oxide-nitride film and can be worked as a metal. It is used in alloys and in the manufacture of zirconium, thorium, uranium and the rare earth metals. Calcium oxide (lime) is used in metallurgy, water treatment, chemicals industry, cement, etc.

Radii/pm: Ca^{2+} 106; atomic 197 (α-form); covalent 174
Electronegativity: 1.00 (Pauling); 1.04 (Allred); 2.2 eV (absolute)
Effective nuclear charge: 2.85 (Slater); 4.40 (Clementi); 5.69 (Froese-Fischer)

Standard reduction potentials $E^{\ominus}$/V

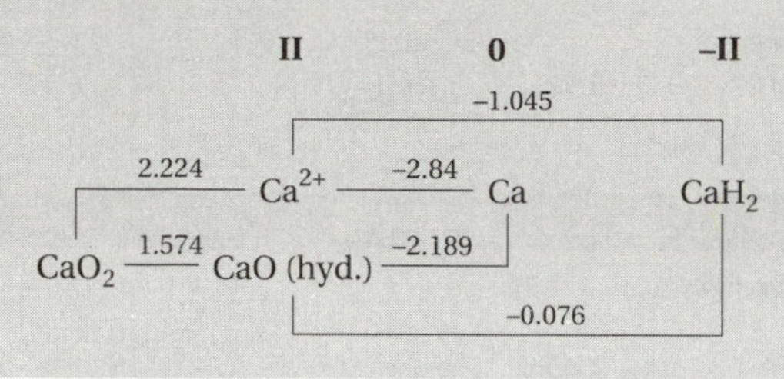

Oxidation states

Ca^{II} [Ar] CaO, CaO_2 (peroxide), $Ca(OH)_2$, CaH_2, CaF_2, $CaCl_2$ etc., Ca^{2+} (aq), $CaCO_3$, $CaSO_4.2H_2O$ (gypsum) $CaSO_4.{}^1/_2H_2O$ (plaster of Paris), CaC_2 (calcium carbide), many salts, few complexes

• PHYSICAL DATA

Melting point/K: 1112
Boiling point/K: 1757

ΔH_{fusion}/kJ mol^{-1}: 9.33
ΔH_{vap}/kJ mol^{-1}: 149.95

Thermodynamic properties (298.15 K, 0.1 MPa)

State	$\Delta_f H^{\ominus}$/kJ mol^{-1}	$\Delta_f G^{\ominus}$/kJ mol^{-1}	$S^{\ominus}$/J K^{-1} mol^{-1}	C_p/J K^{-1} mol^{-1}
Solid (α)	0	0	41.42	25.31
Gas	178.2	144.3	154.884	20.786

Density/kg m^{-3}: 1550 [293 K]; 1365 [liquid at m.p.]
Molar volume/cm^3: 25.86
Thermal conductivity/W m^{-1} K^{-1}: 200 [300 K]
Coefficient of linear thermal expansion/K^{-1}: 22×10^{-6}
Electrical resistivity/Ω m: 3.43×10^{-8} [293 K]
Mass magnetic susceptibility/kg^{-1} m^3: $+1.4 \times 10^{-8}$ (s)

Young's modulus/GPa: 19.6
Rigidity modulus/GPa: 7.9
Bulk modulus/GPa: 17.2
Poisson's ratio/GPa: 0.31

• BIOLOGICAL DATA

Biological role
Essential to all species.

Toxicity
Non-toxic.
Toxic intake: n.a.
Lethal intake: LD_{50} (carbonate, oral, rat) = 6450 mg kg^{-1}

Hazards
Calcium compounds are only toxic via their other components.

Levels in humans
Blood/mg dm^{-3}: 60.5
Bone/p.p.m.: 170 000
Liver/p.p.m.: 100 – 360
Muscle/p.p.m.: 140 – 700
Daily dietary intake: 600 – 1400 mg
Total mass of element in average (70 kg) person: 1.00 kg

Isolated in 1808 by Sir Humphry Davy at London, England.
[Latin, *calx* = lime]
French, *calcium*; German, *Kalzium*; Italian, *calcio*; Spanish, *calcio*

[kal-sium]

• NUCLEAR DATA

Number of isotopes (including nuclear isomers): 16 **Isotope mass range:** 36 → 51

Key isotopes

Nuclide	Atomic mass	Natural abundance (%)	Nuclear spin *I*	Nuclear magnetic moment μ	Uses
^{40}Ca	39.962 590 6	96.941	0+	0	E
^{42}Ca	41.958 617 6	0.647	0+	0	E
^{43}Ca	42.958 766 2	0.135	7/2–	–1.317 27	E, NMR
^{44}Ca	43.955 480 6	2.086	0+	0	E
^{46}Ca	45.953 689	0.004	0+	0	E
^{48}Ca	47.952 533	0.187	0+	0	E

A table of radioactive isotopes is given in Appendix 1, on p238.

NMR [Reference: $CaCl_2$ (aq)]	^{43}Ca
Relative sensitivity (^1H = 1.00)	6.40×10^{-3}
Receptivity (^{13}C = 1.00)	0.0527
Magnetogyric ratio/rad $T^{-1} s^{-1}$	-1.8001×10^7
Nuclear quadrupole moment/m^2	-0.0408×10^{-28}
Frequency (^1H = 100 Hz; 2.3488T)/MHz	6.728

• ELECTRON SHELL DATA

Ground state electron configuration: $[Ar]4s^2$
Term symbol: $^1S_0$
Electron affinity ($M \rightarrow M^-$)/kJ mol^{-1}**:** –186

Ionization energies/kJ mol^{-1}**:**

		kJ mol^{-1}
1.	$M \rightarrow M^+$	589.7
2.	$M^+ \rightarrow M^{2+}$	1145
3.	$M^{2+} \rightarrow M^{3+}$	4910
4.	$M^{3+} \rightarrow M^{4+}$	6474
5.	$M^{4+} \rightarrow M^{5+}$	8144
6.	$M^{5+} \rightarrow M^{6+}$	10 496
7.	$M^{6+} \rightarrow M^{7+}$	12 320
8.	$M^{7+} \rightarrow M^{8+}$	14 207
9.	$M^{8+} \rightarrow M^{9+}$	18 191
10.	$M^{9+} \rightarrow M^{10+}$	20 385

Electron binding energies/eV

		eV
K	1s	4038.5
L_I	2s	438.4
L_{II}	$2p_{1/2}$	349.7
L_{III}	$2p_{3/2}$	346.2
M_I	3s	44.3
M_{II}	$3p_{1/2}$	25.4
M_{III}	$3p_{3/2}$	25.4

Main lines in atomic spectrum
[Wavelength/nm(species)]

239.856 (I)
317.933 (II)
373.690 (II)
393.366 (II)
393.847 (II)
422.673 (I) (AA)

• CRYSTAL DATA

Crystal structure (cell dimensions/pm), space group

α-Ca f.c.c. (*a* = 558.84), Fm3m
β-Ca b.c.c. (*a* = 448.0), Im3m
γ-Ca h.c.p. (*a* = 397, *a* = 649), $P6_3mmc$ [may contain H]
$T(\alpha \rightarrow \beta)$ = 573 K; $T(\beta \rightarrow \gamma)$ = 723 K

X-ray diffraction: mass absorption coefficients $(\mu/\rho)/cm^2\,g^{-1}$**:** CuK_α 162 MoK_α 18.3
Neutron scattering length, $b/10^{-12}$ cm: 0.476
Thermal neutron capture cross-section, σ_a/barns**:** 0.43

• GEOLOGICAL DATA

Minerals

Calcium occurs in many minerals.

Mineral	Formula	Density	Hardness	Crystal appearance
Anhydrite	$CaSO_4$	2.98	3.5	orth., vit./pearly col.
Aragonite	$CaCO_3$	2.947	3.5 – 4	orth., vit. col.-white
Calcite	$CaCO_3$	2.710	3	rhom., vit. col. (gem, onyx)
Dolomite	$CaMg(CO_3)_2$	2.85	3.4 – 4	rhom., vit. col.

This table is continued in Appendix 3, p259.

Chief ores: calcite, dolomite, gypsum (used in cement and plaster) anhydrite (used to make H_2SO_4)

World production/tonnes y^{-1}**:** 2000 (calcium metal); 112×10^6 (lime, CaO)

Main mining areas: common everywhere.

Reserves/tonnes**:** almost unlimited

Specimen: available as granules, pieces or turnings. *Care!*

Abundances

Sun (relative to H = 1×10^{12})**:** 2.24×10^6
Earth's crust/p.p.m.**:** 41 000
Seawater/p.p.m.**:**
Atlantic surface: 390
Atlantic deep: 430
Pacific surface: 390
Pacific deep: 440
Residence time/years**:** 1×10^6
Classification: recycled
Oxidation state: II

Cf

Atomic number: 98
Relative atomic mass ($^{12}C = 12.0000$)**:** 251.0796 (Cf-251)

CAS: [7440-71-3]

• CHEMICAL DATA

Description: Californium is a silvery, radioactive metal which does not occur naturally. It is attacked by oxygen, steam and acids, but not alkalis. ^{252}Cf is a strong neutron emitter, 1 μg releases 170×10^6 neutrons per minute, and it is used in portable neutron sources used in moisture gauges, in core analysis in drilling oil wells, and in on-the-spot activation analysis in gold prospecting. It is also used in cancer therapy.

Radii/pm: Cf^{4+} 86; Cf^{3+} 98; Cf^{2+} 117; atomic 169; van der Waals n.a.
Electronegativity: 1.3 (Pauling); 1.2 (est.) (Allred); n.a. (absolute)
Effective nuclear charge: 1.65 (Slater); n.a. (Clementi); n.a. (Froese-Fischer)

Standard reduction potentials $E^{\ominus}$/V

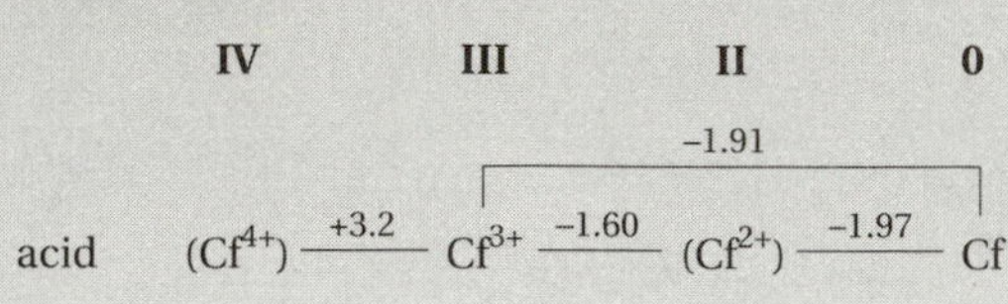

Oxidation states

Cf^{II}	f^{10}	CfO?, $CfBr_2$, CfI_2
Cf^{III}	f^9	Cf_2O_3, CfF_3, $CfCl_3$ etc., Cf^{3+} (aq), $[Cf(C_5H_5)_3]$
Cf^{IV}	f^8	CfO_2, CfF_4
Cf^{V}	f^7	suspected

• PHYSICAL DATA

Melting point/K: 1170
Boiling point/K: n.a.

ΔH_{fusion}/kJ mol^{-1}: n.a.
ΔH_{vap}/kJ mol^{-1}: 196

Thermodynamic properties (298.15 K, 0.1 MPa)

State	$\Delta_f H^{\ominus}$/kJ mol^{-1}	$\Delta_f G^{\ominus}$/kJ mol^{-1}	$S^{\ominus}$/J K^{-1} mol^{-1}	C_p/J K^{-1} mol^{-1}
Solid	0	0	n.a.	25.98
Gas	n.a.	n.a.	n.a.	n.a.

Density/kg m^{-3}: n.a.
Molar volume/cm^3: n.a.
Thermal conductivity/W m^{-1} K^{-1}: 10 (est.) [300 K]
Coefficient of linear thermal expansion/K^{-1}: n.a.
Electrical resistivity/Ω m: n.a.
Mass magnetic susceptibility/kg^{-1} m^3: n.a.

Young's modulus/GPa: n.a.
Rigidity modulus/GPa: n.a.
Bulk modulus/GPa: n.a.
Poisson's ratio/GPa: n.a.

• BIOLOGICAL DATA

Biological role

None.

Toxicity

Toxic intake: n.a.
Lethal intake: n.a.

Hazards

Californium is rarely encountered outside certain prospecting and diagnostic uses. Special precautions are needed because it is not only a powerful source of radiation but also a dangerous neutron emitter.

Levels in humans

nil
Daily dietary intake: nil
Total mass of element in average (70 kg) person: nil

Discovery: see Nuclear Data section.
[Named after California]
French, *californium*; German, *Californium*; Italian, *californio*; Spanish, *californio*

Californium

[kali-forn-iuhm]

• NUCLEAR DATA

Discovery: Produced in 1950 by S.G. Thompson, K. Street Jr., A. Ghiorso and G.T. Seaborg at Berkeley, California, USA .

Number of isotopes (including nuclear isomers): 18 **Isotope mass range:** 239 → 256

Key isotopes

Nuclide	Atomic mass	Half life ($T_{1/2}$)	Decay mode and energy (MeV)	Nuclear spin *I*	Nucl. mag. moment μ	Uses
^{246}Cf	246.068 800	1.49 h	α (6.869); γ	0+		
^{247}Cf	247.071 020	3.11 h	EC (0.65) 99.96%; α (6.55) 0.04%; γ	7/2+		
^{248}Cf	248.072 183	334 d	α (6.369); no γ	0+		
^{249}Cf	249.074 844	351 y	α (6.295); γ	9/2–		
^{250}Cf	250.076 400	13.1 y	α (6.129); no γ	0+		
^{251}Cf	251.079 580	900 y	α (6.172); γ	1/2+		
^{252}Cf	252.081 621	2.64 y	α (6.217); 96.9%; SF 3.1%; γ	0+		R, D, T
^{253}Cf	253.085 127	17.8 d	β^- (0.29) 99.7%; α (6.126) 0.3%	7/2+		
^{254}Cf	254.087 318	60.5 d	SF 99.7%; α (5.930) 0.3%	0+		
^{255}Cf		1.4 h	β^-			

Other isotopes of californium have half-lives shorter than 1 hour.

NMR [Reference: n.a.]

	^{252}Cf
Relative sensitivity (^{1}H = 1.00)	n.a.
Receptivity (^{13}C = 1.00)	n.a.
Magnetogyric ratio/rad $T^{-1}s^{-1}$	n.a.
Nuclear quadrupole moment/m^2	n.a.
Frequency (^{1}H = 100 Hz; 2.3488T)/MHz	n.a.

• ELECTRON SHELL DATA

Ground state electron configuration: [Rn]$5f^{10}7s^2$
Term symbol: $^5I_8$
Electron affinity ($M \rightarrow M^-$)/kJ mol^{-1}**:** n.a.

Ionization energies/kJ mol^{-1}**:**

1. $M \rightarrow M^+$ 608

Electron binding energies/eV

n.a.

Main lines in atomic spectrum
[Wavelength/nm(species)]

339.222 (I)
353.149 (I)
354.098 (I)
359.877 (I)
360.532 (I)
361.211 (II)
362.676 (II)

• CRYSTAL DATA

Crystal structure (cell dimensions/pm), space group

α-Cf h.c.p.
β-Cf f.c.c.
$T(\alpha \rightarrow \beta)$ = 1173 K

X-ray diffraction: mass absorption coefficients (μ/ρ)/$cm^2 g^{-1}$**:** CuK_α n.a. MoK_α n.a.
Neutron scattering length, $b/10^{-12}$ cm**:** n.a.
Thermal neutron capture cross-section, σ_a/barns**:** 2900 (^{251}Cf)

• GEOLOGICAL DATA

Minerals

Not found on Earth.

Chief source: californium-249 and 252 are obtained in gram quantities by the bombardment of ^{239}Pu with neutrons.
World production: in excess of 100 g has been produced, and a few grams are made each year.
Specimen: commercially available under licence - see Key.

Abundances

Sun (relative to H = 1×10^{12})**:** n.a.
Earth's crust/p.p.m.**:** nil
Seawater/p.p.m.**:** nil

C

Atomic number: 6
Relative atomic mass ($^{12}C = 12.0000$)**:** 12.011

CAS: [7440-44-0]

• CHEMICAL DATA

Description: Carbon occurs in three forms: graphite, diamond and buckminsterfullerene C_{60}. It is mainly used in its amorphous forms: as coke in steel making, as carbon black in printing, and as a filler, and as activated charcoal in sugar refining, water treatment and in respirators.

Radii/pm: C^{4-} 260; atomic 77; covalent C–C 77; C=C 67; C≡C 60; van der Waals 185
Electronegativity: 2.55 (Pauling); 2.50 (Allred); 6.27 eV (absolute)
Effective nuclear charge: 3.25 (Slater); 3.14 (Clementi); 2.87 (Froese-Fischer)

Standard reduction potentials $E^{\ominus}/V$

	VI		II		0		–II		–IV
acid	CO_2	–0.106	CO	0.517	C	0.132	CH_4		
	CO_2	–0.20	HCO_2H	0.034	HCHO	0.232	CH_3OH	0.59	CH_4
base	CO_2	–1.01	HCO_2^-	–1.07	HCHO	0.59	CH_3OH	–0.2	CH_4

Oxidation states

This concept is rarely used in discussing carbon and its compounds because of their subtleties of bonding. However for simple compounds with a single carbon we can use it:

C^{-IV}	[Ne]	CH_4
C^{-II}	s^2p^4	CO
C^{IV}	[He]	CO_2, CO_3^{2-}, CF_4, etc.

Covalent bonds

Bond	r/ pm	E/ kJ mol^{-1}
C–H	190	411
C–C	154	346
C=C	134	602
C≡C	120	835
C–N	147	305
C=N	130	615
C≡N	116	887
C–O	143	358
C=O	120	799
C≡O	113	1072

Other bonds to carbon: see other elements

• PHYSICAL DATA

Melting point/K: *c.* 3820 (diam.); 3800 (graph.); 800 (C_{60}, sublimes)
Boiling point/K: 5100 (sublimes)
ΔH_{fusion}/kJ mol^{-1}**:** 105.1
ΔH_{vap}/kJ mol^{-1}**:** 710.9

Thermodynamic properties (298.15 K, 0.1 MPa)

State	$\Delta_f H^{\ominus}$/kJ mol^{-1}	$\Delta_f G^{\ominus}$/kJ mol^{-1}	$S^{\ominus}$/J K^{-1} mol^{-1}	C_p/J K^{-1} mol^{-1}
Solid (graphite)	0	0	5.740	8.527
Solid (diamond)	1.895	2.900	2.377	6.113
Gas	716.682	671.257	158.096	20.838

Density/kg m^{-3}: 3513 (diam); 2260 (graph.); 1650 (C_{60}) [293 K]
Molar volume/cm^3: 3.42 (diam.)
Thermal conductivity/W m^{-1} K^{-1}: 990–2320 (diam.); $5.7^{\perp}$; $1960^{\parallel}$ (graph.) [298 K]
Coefficient of linear thermal expansion/K^{-1}: 1.19×10^{-6} (diam.)
Electrical resistivity/Ω m: 1×10^{11} (diam.); 1.375×10^{-5} (graph.); $1 \times 10_{14}$ (C_{60}) [293 K]
Mass magnetic susceptibility/kg^{-1} m^3: -6.3×10^{-9} (graph.); -6.2×10^{-9} (diam.)

• BIOLOGICAL DATA

Biological role

Constituent element of DNA.

Toxicity

Non-toxic as the element, but some simple compounds can be very toxic, such as CO or cyanide CN^-.

Lethal intake: n.a.

Hazards

Carbon black can be a nuisance dust but is not itself dangerous, although soot may harbour carcinogenic materials.

Levels in humans

Blood/mg dm^{-3}: 0.0016 – 0.075
Bone/p.p.m.: 300 000
Liver/p.p.m.: 670 000
Muscle/p.p.m.: 670 000
Daily dietary intake: 300 g
Total mass of element in average (70 kg) person: 16 kg

Occurs naturally as graphite and diamond; known to prehistoric humans.
[Latin, *carbo* = charcoal]
French, *carbone*; German, *Kohlenstoff*; Italian, *carbonio*; Spanish, *carbono*

Carbon

[kar-bon]

• NUCLEAR DATA

Number of isotopes (including nuclear isomers): 8 **Isotope mass range:** 9 → 16

Key isotopes

Nuclide	Atomic mass	Natural abundance (%)	Nuclear spin *I*	Nuclear magnetic moment μ	Uses
^{12}C	12.000 000 000*	98.90	0+	0	
^{13}C	13.003 354 826	1.10	1/2–	+0.702 411	NMR

* by definition.
A table of radioactive isotopes is given in Appendix 1, p238.

NMR [Reference: $Si(CH_3)_4$]	^{13}C
Relative sensitivity (^1H = 1.00)	0.0159
Receptivity (^{13}C = 1.00)	1.00 (by defn.)
Magnetogyric ratio/rad $T^{-1}s^{-1}$	6.7263×10^7
Nuclear quadrupole moment/m^2	–
Frequency (^1H = 100 Hz; 2.3488T)/MHz	25.144

• ELECTRON SHELL DATA

Ground state electron configuration: $[He]2s^22p^2$
Term symbol: $^3P_0$
Electron affinity ($M \rightarrow M^-$)/kJ mol^{-1}**:** 121.9

Ionization energies/kJ mol^{-1}**:**

1.	$M \rightarrow M^+$	1086.2
2.	$M^+ \rightarrow M^{2+}$	2352
3.	$M^{2+} \rightarrow M^{3+}$	4620
4.	$M^{3+} \rightarrow M^{4+}$	6222
5.	$M^{4+} \rightarrow M^{5+}$	37 827
6.	$M^{5+} \rightarrow M^{6+}$	47 270

Electron binding energies/eV

K	1s	284.2

Main lines in atomic spectrum
[Wavelength/nm(species)]

247.856 (I)
283.671 (II)
426.726 (II)
723.642 (II)

• CRYSTAL DATA

Crystal structure (cell dimensions/pm), space group

Cubic diamond (*a* = 356.703), Fd3m
Hexagonal diamond (*a* = 252, *c* = 412), $P6_3$/mmc
Hexagonal graphite (*a* = 246.12, *c* = 670.78), $P6_3$mc
Rhombohedral graphite (*a* = 364.2, α = 39° 30'), R3m
Hexagonal carbon [chaoite] (*a* = 894.8, *c* = 1408)
F.c.c. buckminsterfullerene C_{60} (a = 1414)

X-ray diffraction: mass absorption coefficients (μ/ρ)/$cm^2 g^{-1}$**:** CuK_α 4.60 MoK_α 0.625
Neutron scattering length, $b/10^{-12}$ cm: 0.66460
Thermal neutron capture cross-section, σ_a/barns**:** 0.0035

• GEOLOGICAL DATA

Minerals

Carbon is commonly found as graphite, very rarely as diamond, and only in minute traces as C_{60}. Carbon is also found as fossil fuel deposits - see below - and as carbonates, in particular calcium/magnesium carbonates - see limestone, dolomite, etc.

Mineral	Formula	Density	Hardness	Crystal appearance
Diamond	C	3.51	10	cub., crystalline, col./pale tints
Graphite	C	2.2	1 – 2	hex., met. black sheets, sometimes crystals

Chief ores: graphite

Main mining areas: graphite deposits: Sri Lanka, Madagascar, Russia, South Korea, Mexico, Czech Republic, Italy. Diamonds: South Africa, USA, Russia, Brazil, Zaire, Sierra Leone, Ghana.

World production/tonnes y^{-1}**:** 8.6×10^9 (fossil carbon, 1996). Fossil fuel production: natural gas, 2.0×10^9; oil, 3.3×10^9; coal, 2.3×10^9.

Reserves/tonnes**:** natural gas, 127×10^9; oil, 140×10^9; coal, 1000×10^9; tar sands, n.a. but large.

Specimen: available as amorphous, fullerenes, bucky tubes, diamond, graphite and soot. Safe.

Abundances

Sun (relative to H = 1×10^{12})**:** 4.17×10^8
Earth's crust/p.p.m.**:** 480
Atmosphere/p.p.m. (volume)**:** *c.* 350 (CO_2)
Seawater/p.p.m.**:**
Atlantic surface: 23
Atlantic deep: 26
Pacific surface: 23
Pacific deep: 28
Residence time/years**:** 800 000
Classification: recycled
Oxidation state: IV

Ce

Atomic number: 58
Relative atomic mass ($^{12}C = 12.0000$): **140.115**

CAS: [7440-45-1]

• CHEMICAL DATA

Discovery: Cerium was discovered in 1803 by J.J. Berzelius and W. Hisinger at Vestmanland, Sweden. First isolated by W.F. Hillebrand and T.H. Norton in 1875 at Washington DC, USA.
Description: Cerium is a reactive, grey metal and is the most abundant of the so-called rare earth metals (more correctly termed the lanthanides). It tarnishes in air, burns easily if ignited, reacts rapidly with water, and dissolves in acids. Cerium is used in glass, flints, ceramics and alloys.

Radii/pm: Ce^{4+} 94; Ce^{3+} 107; atomic 182.5; covalent 165
Electronegativity: 1.12 (Pauling); 1.06 (Allred); ≤ 3.0 eV (absolute)
Effective nuclear charge: 2.85 (Slater); 10.80 (Clementi); 10.57 (Froese-Fischer)

Standard reduction potentials $E^{\ominus}/V$

	IV		III		0
acid	Ce^{4+}	1.72	Ce^{3+}	−2.34	Ce
base	CeO_2	−0.7	$Ce(OH)_3$	−2.78	Ce

Oxidation states

Ce^{III}	f^1	Ce_2O_3, $Ce(OH)_3$, $[Ce(OH_2)_6]^{3+}$ (aq), Ce^{3+} salts, CeF_3, $CeCl_3$ etc., complexes
Ce^{IV}	[Xe]	CeO_2, CeF_4, $[CeCl_6]^{2-}$, $[Ce(NO_3)_6]^{2-}$ (aq)

• PHYSICAL DATA

Melting point/K: 1072
Boiling point/K: 3699
ΔH_{fusion}/kJ mol^{-1}: 8.87
ΔH_{vap}/kJ mol^{-1}: 313.8

Thermodynamic properties (298.15 K, 0.1 MPa)

State	$\Delta_f H^{\ominus}$/kJ mol^{-1}	$\Delta_f G^{\ominus}$/kJ mol^{-1}	$S^{\ominus}$/J K^{-1} mol^{-1}	C_p/J K^{-1} mol^{-1}
Solid	0	0	72.0	26.94
Gas	423	385	191.776	23.075

Density/kg m^{-3}: 8240 (α); 6749 (β); 6773 (γ); 6700 (δ) [298 K]
Molar volume/cm^3: 17.00
Thermal conductivity/W m^{-1} K^{-1}: 11.4 [300 K]
Coefficient of linear thermal expansion/K^{-1}: 8.5×10^{-6}
Electrical resistivity/Ω m: 73×10^{-8} [273 K]
Mass magnetic susceptibility/kg^{-1} m^3: $+2.17 \times 10^{-7}$ (s)

Young's modulus/GPa: 33.5
Rigidity modulus/GPa: 13.5
Bulk modulus/GPa: n.a.
Poisson's ratio/GPa: 0.248

• BIOLOGICAL DATA

Biological role

None, but acts to stimulate metabolism.

Toxicity

Toxic intake: n.a.
Lethal intake: LD_{50} (chloride, oral, mouse) = 2100 mg kg^{-1}

Hazards

Cerium is mildly toxic by ingestion, but insoluble salts, such as the oxalate, are non-toxic and doses of up to 500 mg were once prescribed to prevent travel sickness and morning sickness.

Levels in humans

Blood/mg dm^{-3}: < 0.002
Bone/p.p.m.: 2.7
Liver/p.p.m.: 0.29
Muscle/p.p.m.: n.a.
Daily dietary intake: n.a. but very low
Total mass of element in average (70 kg) person: 40 mg

Discovery: see Chemical Data section.
[Named after the asteroid Ceres discovered in 1801]
French, *cérium*; German, *Cer*; Italian, *cerio*; Spanish, *cerio*

Cerium

[seer-iuhm]

• NUCLEAR DATA

Number of isotopes (including nuclear isomers): 28 **Isotope mass range:** 129 → 151

Key isotopes

Nuclide	Atomic mass	Natural abundance (%)	Nuclear spin I	Nuclear magnetic moment μ	Uses
^{136}Ce	135.907 140	0.19	0+		E
^{138}Ce	137.905 985	0.25	0+		E
^{140}Ce	139.905 433	88.48	0+		E
^{142}Ce	141.909 24	11.08	0+		E

A table of radioactive isotopes is given in Appendix 1, on p238.

NMR [Reference: none] none

• ELECTRON SHELL DATA

Ground state electron configuration: [Xe]$4f5d6s^2$
Term symbol: $^3H_4$
Electron affinity ($M \rightarrow M^-$)/kJ mol^{-1}**:** ≤ 50

Ionization energies/kJ mol^{-1}**:**

1.	$M \rightarrow M^+$	527.4
2.	$M^+ \rightarrow M^{2+}$	1047
3.	$M^{2+} \rightarrow M^{3+}$	1949
4.	$M^{3+} \rightarrow M^{4+}$	3547
5.	$M^{4+} \rightarrow M^{5+}$	(6800)
6.	$M^{5+} \rightarrow M^{6+}$	(8200)
7.	$M^{6+} \rightarrow M^{7+}$	(9700)
8.	$M^{7+} \rightarrow M^{8+}$	(11 800)
9.	$M^{8+} \rightarrow M^{9+}$	(13 200)
10.	$M^{9+} \rightarrow M^{10+}$	(14 700)

Electron binding energies/eV

K	1s	40 443
L_I	2s	6548
L_{II}	$2p_{1/2}$	6164
L_{III}	$2p_{3/2}$	5723
M_I	3s	1436
M_{II}	$3p_{1/2}$	1274
M_{III}	$3p_{3/2}$	1187
M_{IV}	$3d_{3/2}$	902.4
M_V	$3d_{5/2}$	883.8

continued in Appendix 2, p255

Main lines in atomic spectrum
[Wavelength/nm(species)]

349.275 (II)
395.254 (II)
399.924 (II)
401.239 (II)
413.380 (II)
418.660 (II)

• CRYSTAL DATA

Crystal structure (cell dimensions/pm), space group

α-Ce f.c.c. (a = 485), Fm3m
β-Ce hexagonal (a = 367.3, c = 1180.2), $P6_3/mmc$
γ-Ce f.c.c. (a = 516.01), Fm3m
δ-Ce f.c.c. (a = 412), Im3m
$T(\beta \rightarrow \gamma)$ = 441 K

X-ray diffraction: mass absorption coefficients $(\mu/\rho)/cm^2\ g^{-1}$**:** CuK_α 352 MoK_α 48.2
Neutron scattering length, $b/10^{-12}$ cm**:** 0.484
Thermal neutron capture cross-section, σ_a/barns**:** 0.6

• GEOLOGICAL DATA

Minerals

Mineral	Formula	Density	Hardness	Crystal appearance
Bastnäsite-Ce*	(Ce,La, etc,) CO_3F	4.9	4 – 4.5	hex., vit./greasy yellow
Monazite-Ce*	(Ce, La, Nd, Th, etc.)PO_4	5.20	5 – 5.5	mon., waxy/vit. yellow-brown

*Varieties of these minerals that are particularly rich in cerium.

Chief ores: monazite, bastnäsite. Perovskite (Ti mineral) can also be rich in cerium
World production/tonnes y^{-1}**:** 24 000
Main mining areas: USA, Brazil, India, Sri Lanka, Australia, China.
Reserves/tonnes: *c.* 15×10^6
Specimen: available as chips, ingots or powder. Safe.

Abundances

Sun (relative to H = 1×10^{12})**:** 35.5
Earth's crust/p.p.m.**:** 68
Seawater/p.p.m.**:**
Atlantic surface: 9.0×10^{-6}
Atlantic deep: 2.6×10^{-6}
Pacific surface: 1.5×10^{-6}
Pacific deep: 0.5×10^{-6}
Residence time/years**:** 100
Classification: scavenged
Oxidation state: III

Cl	Atomic number: 17 Relative atomic mass ($^{12}C = 12.0000$): 35.4527	CAS: [7782-50-5]

• CHEMICAL DATA

Description: Chlorine is a yellow-green, dense, sharp-smelling gas (Cl_2) produced on the million tonne scale by the electrolysis of sodium chloride solution. It is used as a bleaching agent and as a sterilising agent for water supplies. It is a key industrial chemical and used for the manufacture of organochlorine solvents and PVC.

Radii/pm: Cl^- 181; covalent 99; van der Waals 181

Electronegativity: 3.16 (Pauling); 2.83 (Allred); 8.30 eV (absolute)

Effective nuclear charge: 6.10 (Slater); 6.12 (Clementi); 6.79 (Froese-Fischer)

Standard reduction potentials $E^{\ominus}$/V

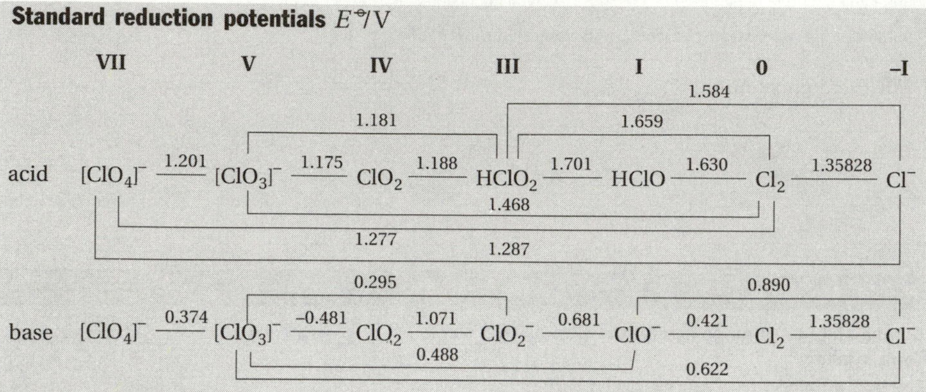

Oxidation states

Cl^{-I}	[Ar]	Cl^- (aq), HCl, NaCl etc.
Cl^0	s^2p^5	Cl_2
Cl^I	s^2p^4	Cl_2O, HOCl, salts, ClO^- (aq), ClF
Cl^{III}	s^2p^2	$NaClO_2$, ClF_3
Cl^{IV}	s^2p^1	ClO_2
Cl^V	s^2	$HClO_3$, salts, ClO_3^- (aq), ClF_5, ClF_3O
Cl^{VI}	s^1	Cl_2O_6
Cl^{VII}	[Ne]	Cl_2O_7, $HClO_4$, salts, ClO_4^- (aq), $ClFO_3$

Covalent bonds

Bond	r/ pm	E/ kJ mol^{-1}
Cl—O	170	218
Cl—F	163	249
Cl—Cl	199	240

For other bonds to chlorine: see other elements

• PHYSICAL DATA

Melting point/K: 172.17

Boiling point/K: 239.18

Critical temperature/K: 417

Critical pressure/ kPa: 7700

ΔH_{fusion}/kJ mol^{-1}: 6.41

ΔH_{vap}/kJ mol^{-1}: 20.4033

Thermodynamic properties (298.15 K, 0.1 MPa)

State	$\Delta_f H^{\ominus}$/kJ mol^{-1}	$\Delta_f G^{\ominus}$/kJ mol^{-1}	$S^{\ominus}$/J K^{-1} mol^{-1}	C_p/J K^{-1} mol^{-1}
Gas (Cl_2)	0	0	223.066	33.907
Gas (atoms)	121.679	105.680	165.198	21.840

Density/kg m^{-3}: 2030 [113 K]; 1507 [239 K]; 3.214 [273 K]

Molar volume/cm^3: 17.46 [113 K]

Thermal conductivity/W m^{-1} K^{-1}: 0.0089 [300 K] (g)

Mass magnetic susceptibility/kg^{-1} m^3: -7.2×10^{-9} (g)

• BIOLOGICAL DATA

Biological role

Chloride, Cl^-, is essential to many species, including humans.

Toxicity

Cl_2 is very toxic; chloride is non-toxic.

Toxic intake: affects eyes and lungs at 3 p.p.m. in air

Lethal intake: LC_{50} (Cl_2, inhalation, human) = 500 ppm for 5 minutes

Hazards

Chlorine is corrosive; its vapour attacks the eyes and lungs. In air, 15 ppm Cl_2 produces throat irritation, 50 ppm is dangerous even in short doses. TWA = 0.5 ppm.

Levels in humans

Blood/mg dm^{-3}: 2890 (chloride)

Bone/p.p.m.: 900 (chloride)

Liver/p.p.m.: 3000 – 7200 (chloride)

Muscle/p.p.m.: 2000 – 5200 (chloride)

Daily dietary intake: 3.00 – 6.50 g

Total mass of element in average (70 kg) person: 95 g

Discovered in 1774 by C.W. Scheele at Uppsala, Sweden.
[Greek, *chloros* = pale green]
French, *chlore*; German, *Chlor*; Italian, *cloro*; Spanish, *cloro*

Chlorine

[klor-een]

• NUCLEAR DATA

Number of isotopes (including nuclear isomers): 13 **Isotope mass range:** 31 → 41

Key isotopes

Nuclide	Atomic mass	Natural abundance (%)	Nuclear spin I	Nuclear magnetic moment μ	Uses
^{35}Cl	34.968 852 721	75.77	3/2+	+0.821 873 6	E, NMR
^{37}Cl	36.965 902 62	24.23	3/2+	+0.684 123 0	E, NMR

A table of radioactive isotopes is given in Appendix 1, on p238.

NMR [Reference: NaCl (aq)]	^{35}Cl	^{37}Cl
Relative sensitivity (^1H = 1.00)	4.70×10^{-3}	2.71×10^{-3}
Receptivity (^{13}C = 1.00)	20.2	3.8
Magnetogyric ratio/rad $T^{-1}s^{-1}$	2.6210×10^7	2.1718×10^7
Nuclear quadrupole moment/m^2	-0.08165×10^{-28}	-0.06435×10^{-28}
Frequency (^1H = 100 Hz; 2.3488T)/MHz	9.798	8.156

• ELECTRON SHELL DATA

Ground state electron configuration: $[Ne]3s^23p^5$
Term symbol: $^2P_{3/2}$
Electron affinity ($M \rightarrow M^-$)/kJ mol^{-1}**:** 349.0

Ionization energies/kJ mol^{-1}**:**

1.	$M \rightarrow M^+$	1251.1
2.	$M^+ \rightarrow M^{2+}$	2297
3.	$M^{2+} \rightarrow M^{3+}$	3826
4.	$M^{3+} \rightarrow M^{4+}$	5158
5.	$M^{4+} \rightarrow M^{5+}$	6540
6.	$M^{5+} \rightarrow M^{6+}$	9362
7.	$M^{6+} \rightarrow M^{7+}$	11 020
8.	$M^{7+} \rightarrow M^{8+}$	33 610
9.	$M^{8+} \rightarrow M^{9+}$	38 600
10.	$M^{9+} \rightarrow M^{10+}$	43 960

Electron binding energies/eV

K	1s	2822
L_I	2s	270
L_{II}	$2p_{1/2}$	202
L_{III}	$2p_{3/2}$	200

Main lines in atomic spectrum
[Wavelength/nm(species)]

479.455 (II)
489.677 (II)
542.323 (II)
837.574 (I)
858.597 (I)

• CRYSTAL DATA

Crystal structure (cell dimensions/pm), space group

Tetragonal (a = 856; c = 612), P4/ncm
Orthorhombic (a = 624; b = 448; c = 826), Cmca
T(tetragonal → orthorhombic) = 100 K

X-ray diffraction: mass absorption coefficients (μ/ρ)/cm^2 g^{-1}**:** CuK_α 106 MoK_α 11.4
Neutron scattering length, $b/10^{-12}$ cm**:** 0.95770
Thermal neutron capture cross-section, σ_a/barns**:** 35.5

• GEOLOGICAL DATA

Minerals

Mineral	Formula	Density	Hardness	Crystal appearance
Halite	NaCl	2.168	2	cub., vit. colourless

See also carnallite and sylvite under potassium.

Chief ore: halite (rock salt)

World production/tonnes y^{-1}**:** 168×10^6

Main mining areas: vast deposits in USA, Poland, Russia, Germany, China, India, Australia.

Reserves/tonnes: $> 1 \times 10^{13}$

Specimen: chlorine gas is available in small pressurized canisters. *Danger!*

Abundances

Sun (relative to H = 1×10^{12})**:** 3.2×10^5
Earth's crust/p.p.m.: 130
Atmosphere/p.p.m. (volume)**:** traces as organochlorine compounds
Seawater/p.p.m.: 18 000
Residence time/years: 4×10^8
Classification: accumulating
Oxidation state: –I

Cr

Atomic number: 24
Relative atomic mass ($^{12}C = 12.0000$)**: 51.9961**

CAS: [7440-47-3]

• CHEMICAL DATA

Description: Chromium is a hard, blue-white metal. It will dissolve in HCl and H_2SO_4, but not in HNO_3, H_3PO_4 or $HClO_4$ due to reaction and formation of a protective layer on the surface. Chromium can be polished to a high shine and resists oxidation in air. It main uses are in alloys, chrome plating and metal ceramics.

Radii/pm: Cr^{4+} 56; Cr^{3+} 64; Cr^{2+} 84; atomic 125
Electronegativity: 1.66 (Pauling); 1.56 (Allred); 3.72 eV (absolute)
Effective nuclear charge: 3.45 (Slater); 5.13 (Clementi); 6.92 (Froese-Fischer)

Standard reduction potentials $E^{\ominus}$/V

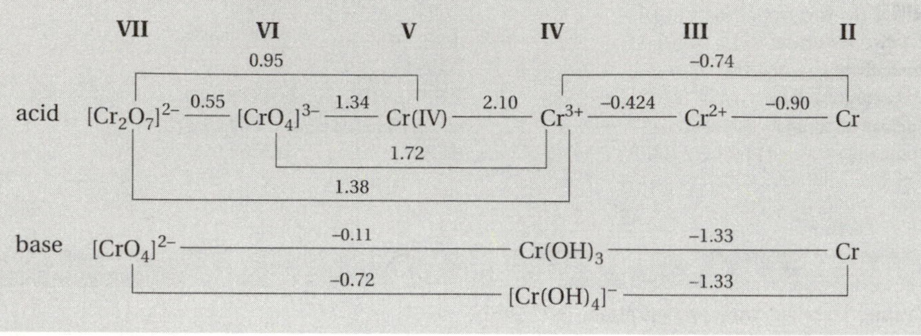

Oxidation states

Cr^{-II}	d^8	$Na_2[Cr(CO)_5]$
Cr^{-I}	d^7	$Na_2[Cr_2(CO)_{10}]$
Cr^0	d^6	$[Cr(CO)_6]$
Cr^I	d^5	$[Cr(bipyridyl)_3]^+$
Cr^{II}	d^4	CrO, CrF_2, $CrCl_2$ etc., CrS
Cr^{III}	d^3	Cr_2O_3, CrF_3, $CrCl_3$ etc., $[Cr(OH_2)_6]^{3+}$ (aq), $Cr(OH)_3$, salts, complexes
Cr^{IV}	d^2	CrO_2, CrF_4
Cr^V	d^1	CrF_5
Cr^{VI}	d^0 [Ar]	CrO_3, $Na_2Cr_2O_7$, $[CrO_4]^{2-}$, CrF_4O

• PHYSICAL DATA

Melting point/K: 2130 ± 20
Boiling point/K: 2945
ΔH_{fusion}/kJ mol^{-1}: 15.3
ΔH_{vap}/kJ mol^{-1}: 348.78

Thermodynamic properties (298.15 K, 0.1 MPa)

State	$\Delta_f H^{\ominus}$/kJ mol^{-1}	$\Delta_f G^{\ominus}$/kJ mol^{-1}	$S^{\ominus}$/J K^{-1} mol^{-1}	C_p/J K^{-1} mol^{-1}
Solid	0	0	23.47	23.35
Gas	396.6	351.8	174.50	20.79

Density/kg m^{-3}: 7190 [293 K]; 6460 [liquid at m.p.]
Molar volume/cm^3: 7.23
Thermal conductivity/W m^{-1} K^{-1}: 93.7 [300 K]
Coefficient of linear thermal expansion/K^{-1}: 6.2×10^{-6}
Electrical resistivity/Ω m: 12.7×10^{-8} [273 K]
Mass magnetic susceptibility/kg^{-1} m^3: $+4.45 \times 10^{-8}$ (s)

Young's modulus/GPa: 279
Rigidity modulus/GPa: 115.3
Bulk modulus/GPa: 160.2
Poisson's ratio/GPa: 0.21

• BIOLOGICAL DATA

Biological role

Essential to some species, including humans; it is also stimulatory.

Toxicity

Toxic intake: 200 mg
Lethal intake: metal, oral, human = 70 mg kg^{-1}. LD_{50} (acetate, oral, rat) = 11 000 mg kg^{-1}

Hazards

Chromium is a human poison by ingestion, it is also a suspected carcinogen. Chromates have a corrosive action on skin and tissue.

Levels in humans

Blood/mg dm^{-3}: 0.006 – 0.11
Bone/p.p.m.: 0.1 – 0.33
Liver/p.p.m.: 0.02 – 3.3
Muscle/p.p.m.: 0.024 – 0.84
Daily dietary intake: 0.01 – 1.2 mg
Total mass of element in average (70 kg) person: 14 mg

Discovered and isolated in 1780 by Nicholas Louis Vauquelin at Paris, France.
[Greek, *chroma* = colour]
French, *chrome*; German, *Chrom*; Italian, *cromo*; Spanish, *cromo*

[kroh-mi-uhm]

• NUCLEAR DATA

Number of isotopes (including nuclear isomers): 13 **Isotope mass range:** 45 → 57

Key isotopes

Nuclide	Atomic mass	Natural abundance (%)	Nuclear spin *I*	Nuclear magnetic moment μ	Uses
^{50}Cr	49.946 046 4	4.345	0+		E
^{52}Cr	51.940 509 8	83.789	0+		E
^{53}Cr	52.940 651 3	9.501	3/2–	–0.474 54	E, NMR
^{54}Cr	53.938 882 5	2.365	0+		E

A table of radioactive isotopes is given in Appendix 1, on p238.

NMR [Reference: $[CrO_4]^{2-}$]	^{53}Cr
Relative sensitivity (^1H = 1.00)	9.03×10^{-4}
Receptivity (^{13}C = 1.00)	0.49
Magnetogyric ratio/rad $T^{-1} s^{-1}$	-1.5120×10^7
Nuclear quadrupole moment/m^2	-0.150×10^{-28}
Frequency (^1H = 100 Hz; 2.3488T)/MHz	5.652

• ELECTRON SHELL DATA

Ground state electron configuration: $[Ar]3d^5 4s^1$
Term symbol: $^7S_3$
Electron affinity ($M \to M^-$)/kJ mol^{-1}**:** 64.3

Ionization energies/kJ mol^{-1}**:**

1.	$M \to M^+$	652.7
2.	$M^+ \to M^{2+}$	1592
3.	$M^{2+} \to M^{3+}$	2987
4.	$M^{3+} \to M^{4+}$	4740
5.	$M^{4+} \to M^{5+}$	6690
6.	$M^{5+} \to M^{6+}$	8738
7.	$M^{6+} \to M^{7+}$	15 550
8.	$M^{7+} \to M^{8+}$	17 830
9.	$M^{8+} \to M^{9+}$	20 220
10.	$M^{9+} \to M^{10+}$	23 580

Electron binding energies/eV

K	1s	5989
L_I	2s	696.0
L_{II}	$2p_{1/2}$	583.8
L_{III}	$2p_{3/2}$	574.1
M_I	3s	74.1
M_{II}	$3p_{1/2}$	42.2
M_{III}	$3p_{3/2}$	42.2

Main lines in atomic spectrum
[Wavelength/nm(species)]

357.869 (I) (AA)
359.349 (I)
360.533 (I)
425.435 (I)
427.480 (I)
428.972 (I)
520.844 (I)

• CRYSTAL DATA

Crystal structure (cell dimensions/pm), space group
b.c.c. (a = 288.46), Im3m

X-ray diffraction: mass absorption coefficients (μ/ρ)/$cm^2 g^{-1}$**:** CuK_α 260 MoK_α 31.1
Neutron scattering length, $b/10^{-12}$ cm**:** 0.3635
Thermal neutron capture cross-section, σ_a/barns**:** 3.1

• GEOLOGICAL DATA

Minerals

Mineral	Formula	Density	Hardness	Crystal appearance
Chromite	$FeCr_2O_4$	4.7	5.5	cub., compacted, met. black
Crocoite	$PbCrO_4$	6.0	2.5 – 3	mon., adam./vit. red-orange

Chief ore: chromite

World production/tonnes y^{-1}**:** *c.* 20 000 (chromium metal); 9.6×10^6 (chromite ore)

Main mining areas: Turkey, South Africa, Zimbabwe, Russia, Philippines.

Reserves/tonnes**:** 1×10^9

Specimen: available as chips, chunks, crystallites or powder. Safe.

Abundances

Sun (relative to H = 1×10^{12})**:** 5.13×10^5
Earth's crust/p.p.m.**:** *c.* 100
Seawater/p.p.m.**:**
Atlantic surface: 1.8×10^{-4}
Atlantic deep: 2.3×10^{-4}
Pacific surface: 1.5×10^{-4}
Pacific deep: 2.5×10^{-4}
Residence time/years**:** 10 000
Classification: recycled
Oxidation state: VI

Co

Atomic number: 27
Relative atomic mass ($^{12}C = 12.0000$)**:** 58.93320

CAS: [7440-48-4]

• CHEMICAL DATA

Description: Cobalt is a lustrous, silvery-blue, hard metal which is also ferromagnetic. It is stable in air, unaffected by water, but slowly attacked by dilute acids. ^{60}Co is a useful radioisotope. Cobalt is used in alloys for magnets, in ceramics, in catalysts and in paints.

Radii/pm: Co^{3+} 64; Co^{2+} 82; atomic 125; covalent 116
Electronegativity: 1.88 (Pauling); 1.70 (Allred); 4.3 eV (absolute)
Effective nuclear charge: 3.90 (Slater); 5.58 (Clementi); 7.63 (Froese-Fischer)

Standard reduction potentials $E^\ominus$/V

	IV		III		II		0
acid	CoO_2	1.416	Co^{3+}	1.92	Co^{2+}	−0.277	Co
base	CoO_2	0.7	$Co(OH)_3$	0.17	$Co(OH)_2$	−0.733	Co

Oxidation states

Co^{-I}	d^{10}	rare $[Co(CO)_4]^-$
Co^{0}	d^{9}	rare $[Co_2(CO)_8]$, $[Co_4(CO)_{12}]$, $[Co_6(CO)_{16}]$
Co^{I}	d^{8}	rare $[Co(NCCH_2)_5]^+$
Co^{II}	d^{7}	CoO, Co_3O_4 (= $Co^{II}Co^{III}{}_2O_4$), $Co(OH)_2$, CoF_2, $CoCl_2$ etc., $[Co(OH_2)_6]^{2+}$ (aq), many salts and complexes, $[Co(\eta\text{-}C_5H_5)_2]$
Co^{III}	d^{6}	$Co(OH)_3$, CoF_3, $[Co(OH_2)_6]^{3+}$ (aq), $[Co(NH_3)_6]^{3+}$, many complexes
Co^{IV}	d^{5}	CoO_2 ?, CoS_2, $[CoF_6]^{2-}$
Co^{V}	d^{4}	K_3CoO_4

• PHYSICAL DATA

Melting point/K: 1768
Boiling point/K: 3143
ΔH_{fusion}/kJ mol^{-1}: 15.2
ΔH_{vap}/kJ mol^{-1}: 382.4

Thermodynamic properties (298.15 K, 0.1 MPa)

State	$\Delta_f H^\ominus$/kJ mol^{-1}	$\Delta_f G^\ominus$/kJ mol^{-1}	$S^\ominus$/J K^{-1} mol^{-1}	C_p/J K^{-1} mol^{-1}
Solid	0	0	30.04	24.81
Gas	424.7	380.3	179.515	23.020

Density/kg m^{-3}: 8900 [293 K]; 7670 [liquid at m.p.]
Molar volume/cm^3: 6.62
Thermal conductivity/W m^{-1} K^{-1}: 100 [300 K]
Coefficient of linear thermal expansion/K^{-1}: 13.36×10^{-6}
Electrical resistivity/Ω m: 6.24×10^{-8} [293 K]
Mass magnetic susceptibility/kg^{-1} m^3: ferromagnetic
Young's modulus/GPa: 211
Rigidity modulus/GPa: 82
Bulk modulus/GPa: 181.5
Poisson's ratio/GPa: 0.32

• BIOLOGICAL DATA

Biological role

Essential to most species, including humans.

Toxicity

Toxic intake: 500 mg
Lethal intake: LD_{50} (chloride, oral, rat) = 80 mg kg^{-1}

Hazards

For humans, cobalt compounds generally have low toxicity by ingestion, but produce vomiting. Cobalt is a suspected carcinogen.

Levels in humans

Blood/mg dm^{-3}: 0.0002 – 0.04
Bone/p.p.m.: 0.01 – 0.04
Liver/p.p.m.: 0.06 – 1.1
Muscle/p.p.m.: 0.028 – 0.65
Daily dietary intake: 0.005 – 1.8 mg
Total mass of element in average (70 kg) person: 3 mg

Discovered in 1735 by Georg Brandt at Stockholm, Sweden.
[German, *kobald* = goblin]
French, *cobalt*; German, *Kobalt*; Italian, *cobalto*; Spanish, *cobalto*

Cobalt

[koh-bolt]

• NUCLEAR DATA

Number of isotopes (including nuclear isomers): 17 **Isotope mass range:** 35m → 64

Key isotopes

Nuclide	Atomic mass	Natural abundance (%)	Nuclear spin *I*	Nuclear magnetic moment μ	Uses
^{59}Co	58.933 197 6	100	7/2–	+4.627	NMR

A table of radioactive isotopes is given in Appendix 1, on p239.

NMR [Reference: $K_3[Co(CN)_6]$]	^{59}Co
Relative sensitivity (^1H = 1.00)	0.28
Receptivity (^{13}C = 1.00)	1570
Magnetogyric ratio/rad $T^{-1}s^{-1}$	6.3472×10^7
Nuclear quadrupole moment/m^2	$+0.420 \times 10^{-28}$
Frequency (^1H = 100 Hz; 2.3488T)/MHz	23.614

• ELECTRON SHELL DATA

Ground state electron configuration: $[Ar]3d^74s^2$
Term symbol: $^4F_{9/2}$
Electron affinity ($M \rightarrow M^-$)/kJ mol^{-1}**:** 63.8

Ionization energies/kJ mol^{-1}**:**

1.	$M \rightarrow M^+$	760.0
2.	$M^+ \rightarrow M^{2+}$	1646
3.	$M^{2+} \rightarrow M^{3+}$	3232
4.	$M^{3+} \rightarrow M^{4+}$	4950
5.	$M^{4+} \rightarrow M^{5+}$	7670
6.	$M^{5+} \rightarrow M^{6+}$	9840
7.	$M^{6+} \rightarrow M^{7+}$	12 400
8.	$M^{7+} \rightarrow M^{8+}$	15 100
9.	$M^{8+} \rightarrow M^{9+}$	17 900
10.	$M^{9+} \rightarrow M^{10+}$	26 600

Electron binding energies/eV

K	1s	7709
L_I	2s	925.1
L_{II}	$2p_{1/2}$	793.2
L_{III}	$2p_{3/2}$	778.1
M_I	3s	101.0
M_{II}	$3p_{1/2}$	58.9
M_{III}	$3p_{3/2}$	58.9

Main lines in atomic spectrum
[Wavelength/nm(species)]

240.725 (I) (AA)
242.493 (I)
340.512 (I)
344.364 (I)
345.350 (I)
350.228 (I)
356.938 (I)

• CRYSTAL DATA

Crystal structure (cell dimensions/pm), space group

α-Co f.c.c. (a = 354.41), Fm3m
ε-Co h.c.p. (a = 250.7, c = 406.9), $P6_3$/mmc
$T(\alpha \rightarrow \varepsilon)$ = 690 K

X-ray diffraction: mass absorption coefficients $(\mu/\rho)/cm^2\ g^{-1}$**:** CuK_α 313 MoK_α 42.5
Neutron scattering length, $b/10^{-12}$ cm**:** 0.278
Thermal neutron capture cross-section, σ_a/barns**:** 37.2

• GEOLOGICAL DATA

Minerals

Mineral	Formula	Density	Hardness	Crystal appearance
Cobaltite	CoAsS	6.33	5.5	orth., met. steel grey
Erythrite	$Co_3(AsO_4)_2.8H_2O$	3.09	1.5 – 2.5	mon., adam./pearly crimson red
Glaucodot	(Co,Fe)AsS	6.04	5	orth., met. white
Linnaeite	Co_3S_4	4.6	4.5 – 5.5	cub., met. grey/white in reflected light
Skutterudite*	$CoAs_3$	6.5	5.5 – 6	cub., met. white/grey

*Also known as smaltite.

Chief ores: cobaltite, skutterudite
World production/tonnes y^{-1}**:** 17 000
Main mining areas: Zaire, Morocco, Sweden, Canada.
Reserves/tonnes**:** n.a.
Specimen: available as foil, pieces, powder, rod and wire. *Care!*

Abundances

Sun (relative to H = 1×10^{12}): 7.94×10^4
Earth's crust/p.p.m.**:** 20
Seawater/p.p.m.**:**
Atlantic surface: n.a.
Atlantic deep: n.a.
Pacific surface: 6.9×10^{-6}
Pacific deep: 1.1×10^{-6}
Residence time/years**:** 40
Classification: scavenged
Oxidation state: II

Cu

Atomic number: 29
Relative atomic mass ($^{12}C = 12.0000$)**:** 63.546

CAS: [7440-50-8]

• CHEMICAL DATA

Description: Copper is a reddish metal, malleable and ductile, with high electrical conductivity, making it ideal for electrical wiring, probably its most important commercial use. Copper is resistant to air and water, and is used as roofing material for public buildings, where it slowly weathers to an attractive green surface patina of copper carbonate. Historically copper has been important as one of the first worked metals, especially as the alloy bronze. It is still used in coins.

Radii/pm: Cu^{2+} 72; Cu^{+} 96; atomic 128; covalent 117
Electronegativity: 1.90 (Pauling); 1.75 (Allred); 4.48 eV (absolute)
Effective nuclear charge: 4.20 (Slater); 5.84 (Clementi); 8.07 (Froese-Fischer)

Standard reduction potentials $E^{\ominus}$/V

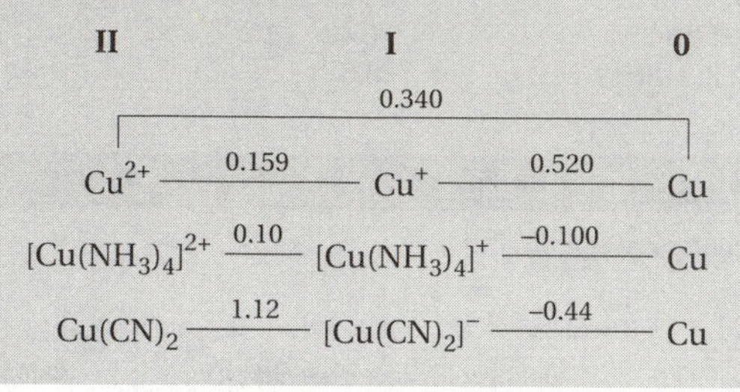

Oxidation states

Cu^{0}	$d^{10}s^{1}$	rare, $[Cu(CO)_3]$ at 10 K
Cu^{I}	d^{10}	Cu_2O, CuCl, $K[Cu(CN)_2]$
Cu^{II}	d^{9}	CuO, $CuCl_2$, Cu^{2+}(aq), Cu^{2+} salts
Cu^{III}	d^{8}	$K_3[CuF_6]$
Cu^{IV}	d^{7}	rare, $Cs_2[CuF_6]$

• PHYSICAL DATA

Melting point/K: 1356.6
Boiling point/K: 2840
ΔH_{fusion}/kJ mol^{-1}: 13.0
ΔH_{vap}/kJ mol^{-1}: 304.6

Thermodynamic properties (298.15 K, 0.1 MPa)

State	$\Delta_f H^{\ominus}$/kJ mol^{-1}	$\Delta_f G^{\ominus}$/kJ mol^{-1}	$S^{\ominus}$/J K^{-1} mol^{-1}	C_p/J K^{-1} mol^{-1}
Solid	0	0	33.150	24.435
Gas	338.32	298.58	166.38	20.786

Density/kg m^{-3}: 8960 [293 K]; 7940 [liquid at m.p.]
Molar volume/cm^3: 7.09
Thermal conductivity/W m^{-1} K^{-1}: 401 [300 K]
Coefficient of linear thermal expansion/K^{-1}: 16.5×10^{-6}
Electrical resistivity/Ω m: 1.6730×10^{-8} [293 K]
Mass magnetic susceptibility/kg^{-1} m^3: -1.081×10^{-9} (s)

Young's modulus/GPa: 129.8
Rigidity modulus/GPa: 48.3
Bulk modulus/GPa: 137.8
Poisson's ratio/GPa: 0.343

• BIOLOGICAL DATA

Biological role

Essential to all species.

Toxicity

Toxic intake: 85 g of metal; 20 g of $CuSO_4$
Lethal intake: 60 g of $CuSO_4$; LD_{50} (sulfate, oral, rat) = 300 mg kg^{-1}

Hazards

As little as 30 g of $CuSO_4$ has been known to be fatal when eaten.

Levels in humans

Blood/mg dm^{-3}: 1.01
Bone/p.p.m.: 1 – 26
Liver/p.p.m.: 30
Muscle/p.p.m.: 10
Daily dietary intake: 0.50 – 6 mg
Total mass of element in average (70 kg) person: 72 mg

Known to ancient civilizations.
[Latin, *cuprum* = Cyprus]
French, *cuivre*; German, *Kupfer*; Italian, *rame*; Spanish, *cobre*

Copper

[kop-er]

• NUCLEAR DATA

Number of isotopes (including nuclear isomers): 18 **Isotope mass range:** 58 → 73

Key isotopes

Nuclide	Atomic mass	Natural abundance (%)	Nuclear spin *I*	Nuclear magnetic moment μ	Uses
^{63}Cu	62.929 598 9	69.17	3/2–	+2.223 3	E, NMR
^{65}Cu	64.927 792 9	30.83	3/2–	+2.381 7	E, NMR

A table of radioactive isotopes is given in Appendix 1, on p239.

NMR [Reference: $Cu(CH_3CN)_4BF_4/CH_3CN$]	^{63}Cu	^{65}Cu
Relative sensitivity (^1H = 1.00)	0.0931	0.11
Receptivity (^{13}C = 1.00)	365	201
Magnetogyric ratio/rad $T^{-1}s^{-1}$	7.0965×10^7	7.6108×10^7
Nuclear quadrupole moment/m^2	-0.220×10^{-28}	-0.204×10^{-28}
Frequency (^1H = 100 Hz; 2.3488T)/MHz	26.505	28.394

• ELECTRON SHELL DATA

Ground state electron configuration: $[Ar]3d^{10}4s^1$
Term symbol: $^2S_{1/2}$
Electron affinity ($M \rightarrow M^-$)/kJ mol^{-1}**:** 118.5

Ionization energies/kJ mol^{-1}**:**

1.	$M \rightarrow M^+$	745.4
2.	$M^+ \rightarrow M^{2+}$	1958
3.	$M^{2+} \rightarrow M^{3+}$	3554
4.	$M^{3+} \rightarrow M^{4+}$	5326
5.	$M^{4+} \rightarrow M^{5+}$	7709
6.	$M^{5+} \rightarrow M^{6+}$	(9940)
7.	$M^{6+} \rightarrow M^{7+}$	(13 400)
8.	$M^{7+} \rightarrow M^{8+}$	(16 000)
9.	$M^{8+} \rightarrow M^{9+}$	(19 200)
10.	$M^{9+} \rightarrow M^{10+}$	(22 400)

Electron binding energies/eV

K	1s	8979
L_I	2s	1096.7
L_{II}	$2p_{1/2}$	952.3
L_{III}	$2p_{3/2}$	932.5
M_I	3s	122.5
M_{II}	$3p_{1/2}$	77.3
M_{III}	$3p_{3/2}$	75.1

Main lines in atomic spectrum
[Wavelength/nm(species)]
216.509 (I)
217.894 (I)
324.754 (I) (AA)
327.396 (I)
521.820 (I)

• CRYSTAL DATA

Crystal structure (cell dimensions/pm), space group
f.c.c. (a = 361.47), Fm3m

X-ray diffraction: mass absorption coefficients (μ/ρ)/$cm^2 g^{-1}$**:** CuK_α 52.9 MoK_α 50.9
Neutron scattering length, $b/10^{-12}$ cm**:** 0.7718
Thermal neutron capture cross-section, σ_a/barns**:** 3.78

• GEOLOGICAL DATA

Minerals

Crystals of native copper occur naturally and there are small deposits in the USA, Germany, Zambia, Chile and Italy.

Mineral	Formula	Density	Hardness	Crystal appearance
Atacamite	$Cu_2Cl(OH)_3$	3.77	3 – 3.5	orth., adam. vit. green
Azurite	$Cu(CO_3)_2(OH)_2$	3.773	3.5 – 4	mon., vit. blue (ornamental)
Bornite	Cu_5FeS_4	5.07	3	tet., met. copper-red brown
Brochantite	$Cu_4(SO_4)(OH)_4$	3.97	3.5 – 4	mon., vit. green
Chalcanthite	$CuSO_4.5H_2O$	2.286	2.5	tric., vit. blue
Chalcocite	Cu_2S	5.7	2.5 – 3	hex., met. blackish-grey
Chalcopyrite	$CuFeS_2$	4.2	3.5 – 4	tet., met. yellow

This table is continued in Appendix 3 on page 259.

Chief ores: chalcopyrite accounts for *c.* 80% of world's copper (with silver and gold as by-products), chalcanthite, brochantite. Malachite is used for polished slabs, tables and columns.

World production/tonnes y^{-1}**:** 6.54×10^6

Main mining areas: chalcopyrite in USA, Zaire, Zambia, Canada, Chile, Cyprus, Russia; malachite in Russia, Zaire, Zambia, Chile, Australia; chalcanthite in Chile.

Reserves/tonnes**:** 310×10^6

Specimen: available as bars, foil, powder, shot, turnings or wire. Safe.

Abundances

Sun (relative to H = 1×10^{12})**:** 1.15×10^4
Earth's crust/p.p.m.**:** 50
Seawater/p.p.m.**:**
Atlantic surface: 8.0×10^{-5}
Atlantic deep: 12×10^{-5}
Pacific surface: 8.0×10^{-5}
Pacific deep: 28×10^{-5}
Residence time/years**:** 3000
Classification: recycled
Oxidation state: II

Cm

Atomic number: 96
Relative atomic mass ($^{12}C = 12.0000$)**:** 247.0703 (Cm-247)

CAS: [7440-51-9]

• CHEMICAL DATA

Description: Curium is a silvery, radioactive metal. It is attacked by oxygen, steam and acids, but not alkalis. The metal itself was produced by the reduction of CmF_3 with barium vapour at 1200 °C. It is a potential isotope power source because it gives off three watts of heat energy per gram.

Radii/pm: Cm^{4+} 88; Cm^{3+} 99; Cm^{2+} 119; atomic 174
Electronegativity: 1.3 (Pauling); 1.2 (est.) (Allred); n.a. (absolute)
Effective nuclear charge: 1.80 (Slater); n.a. (Clementi); n.a. (Froese-Fischer)

Standard reduction potentials $E^{\ominus}$/V

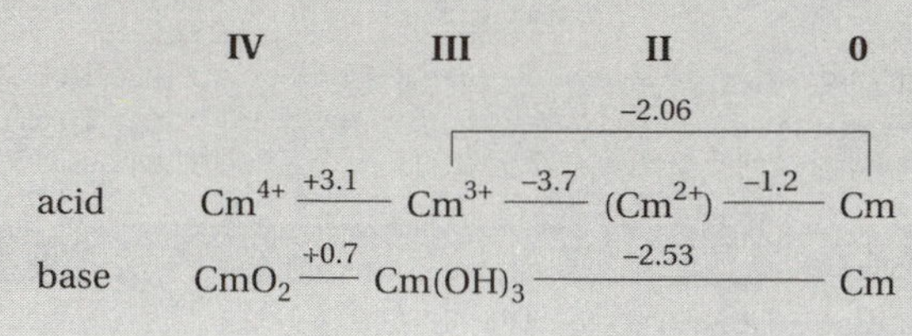

Oxidation states

Cm^{II} f^7d^1	CmO
Cm^{III} f^7	Cm_2O_3, $Cm(OH)_3$, CmF_3, $CmCl_3$ etc., $[CmCl_6]^{3+}$, Cm^{3+} (aq), $[Cm(C_5H_5)_3]$
Cm^{IV} f^6	CmO_2, CmF_4, Cm^{4+} (aq) very unstable

• PHYSICAL DATA

Melting point/K: 1610 ± 40
Boiling point/K: n.a.

ΔH_{fusion}/kJ mol^{-1}**:** n.a.
ΔH_{vap}/kJ mol^{-1}**:** 387

Thermodynamic properties (298.15 K, 0.1 MPa)

State	$\Delta_f H^{\ominus}$/kJ mol^{-1}	$\Delta_f G^{\ominus}$/kJ mol^{-1}	$S^{\ominus}$/J K^{-1} mol^{-1}	C_p/J K^{-1} mol^{-1}
Solid	0	0	n.a.	25.98
Gas	n.a.	n.a.	n.a.	n.a.

Density/kg m^{-3}: 13 300 [293 K]
Molar volume/cm^3: 18.6
Thermal conductivity/W m^{-1} K^{-1}: 10 (est.) [300 K]
Coefficient of linear thermal expansion/K^{-1}: n.a.
Electrical resistivity/Ω m: n.a.
Mass magnetic susceptibility/kg^{-1} m^3: approaches that of gadolinium [*c.* 1×10^{-5}]

Young's modulus/GPa: n.a.
Rigidity modulus/GPa: n.a.
Bulk modulus/GPa: n.a.
Poisson's ratio/GPa: n.a.

• BIOLOGICAL DATA

Biological role

None.

Toxicity

Toxic intake: n.a.
Lethal intake: n.a.

Hazards

Curium is never encountered normally. It is dangerous because it accumulates in bone marrow and its intense radiation destroys red cells. This element is only to be found inside nuclear facilities or research laboratories.

Levels in humans

nil
Daily dietary intake: nil
Total mass of element in average (70 kg) person: nil

Discovery: see Nuclear Data section.
[Named after Pierre and Marie Curie]
French, *curium*; German, *Curium*; Italian, *curio*; Spanish, *curio*

Curium

[kyuhr-iuhm]

• NUCLEAR DATA

Discovery: Curium was prepared in 1944 by G.T. Seaborg, R.A. James and A. Ghiorso at Berkeley, California, USA.

Number of isotopes (including nuclear isomers): 14

Isotope mass range: 238 → 251

Key isotopes

Nuclide	Atomic mass	Half life ($T_{1/2}$)	Decay mode and energy (MeV)	Nuclear spin I	Nucl. mag. moment μ	Uses
^{240}Cm	240.055 503	27 h	α (6.397)	0+		
^{241}Cm	241.057 645	32.8 d	EC (0.77) 99%; α (6.184) 1%; γ	1/2+		
^{242}Cm	242.058 830	162.8 d	α (6.126); γ	0+		
^{243}Cm	243.061 381	28.5 y	α (6.167); γ	5/2+	0.41	
^{244}Cm	244.062 747	18.11 y	α (5.902); γ	0+		R
^{245}Cm	245.065 483	8500 y	α (5.623); γ	7/2+	0.5	
^{246}Cm	246.067 218	4780 y	α (5.476); γ	0+		
^{247}Cm	247.070 347	1.56×10^7 y	α (5.362); γ	9/2–	+0.37	
^{248}Cm	248.072 343	3.4×10^5 y	α (5.162); 92%; SF 8%; no γ	0+		R
^{250}Cm	250.078 352	*c.* 9000 y	SF; α (5.27)	0+		

Other radioisotopes of curium have half-lives shorter than 5 hours.

NMR n.a.

• ELECTRON SHELL DATA

Ground state electron configuration: [Rn]$5f^7 6d^1 7s^2$
Term symbol: $^9D_2$
Electron affinity $(M \rightarrow M^-)$/kJ mol^{-1}**:** n.a.

Ionization energies/kJ mol^{-1}**:**
1. $M \rightarrow M^+$ 581

Electron binding energies/eV
n.a.

Main lines in atomic spectrum*
[Wavelength/nm(species)]
299.939 (I)
310.969 (I)
311.641 (I)
313.716 (I)
314.733 (I)
315.510 (I)
315.860 (I)
*first seven lines associated with the neutral atom

• CRYSTAL DATA

Crystal structure (cell dimensions/pm), space group
α-Cm h.c.p.
β-Cm f.c.c.
$T(\alpha \rightarrow \beta)$ = 1550 K

X-ray diffraction: mass absorption coefficients (μ/ρ)/cm^2 g^{-1}**:** CuK_α n.a. MoK_α n.a.
Neutron scattering length, $b/10^{-12}$ cm**:** 0.95
Thermal neutron capture cross-section, σ_a/barns**:** 79

• GEOLOGICAL DATA

Minerals
None.

Chief source: curium-242 and 244 are obtained in kilogram quantities by the bombardment of ^{239}Pu with neutrons.
World production: several kilograms are produced each year.
Reserves: attempts are being made to convert plutonium into curium so potential stocks may be quite large.
Specimen: commercially available, under licence - see Key.

Abundances
Sun (relative to H = 1×10^{12})**:** n.a.
Earth's crust/p.p.m.**:** nil
Seawater/p.p.m.**:** nil

Db

Atomic number: 105
Relative atomic mass ($^{12}C = 12.0000$)**: 262.114 (Db-262)**

CAS: [3850-35-4]

• CHEMICAL DATA

Description: Dubnium is a radioactive metal which does not occur naturally, and is of research interest only.

Radii/pm: Db^{5+} 68 (est.); atomic 139 (est.)
Electronegativity: n.a.
Effective nuclear charge: n.a.

Standard reduction potentials $E^{\ominus}$/V

V		0
Db^{5+}	$\xrightarrow{-0.8\ (est.)}$	Db

Oxidation states

Db^{IV}	d^1	?
Db^{V}	$[f^{14}]$	most stable?

• PHYSICAL DATA

Melting point/K: n.a.
Boiling point/K: n.a.

ΔH_{fusion}/kJ mol^{-1}: n.a.
ΔH_{vap}/kJ mol^{-1}: n.a.
ΔH_{subl}/kJ mol^{-1}: 795 (est.)

Thermodynamic properties (298.15 K, 0.1 MPa)

State	$\Delta_f H^{\ominus}$/kJ mol^{-1}	$\Delta_f G^{\ominus}$/kJ mol^{-1}	$S^{\ominus}$/J K^{-1} mol^{-1}	C_p/J K^{-1} mol^{-1}
Solid	0	0	n.a.	n.a.
Gas	n.a.	n.a.	n.a.	n.a.

Density/kg m^{-3}: 29 000
Molar volume/cm^3: n.a.
Thermal conductivity/W m^{-1} K^{-1}: n.a.
Coefficient of linear thermal expansion/K^{-1}: n.a.
Electrical resistivity/Ω m: n.a.
Mass magnetic susceptibility/kg^{-1} m^3: n.a.

• BIOLOGICAL DATA

Biological role

None.

Toxicity

Toxic intake: n.a.
Lethal intake: n.a.

Hazards

Dubnium is never encountered normally, and only a few atoms have ever been made. It would be dangerous because of its intense radioactivity.

Levels in humans

nil

Daily dietary intake: nil
Total mass of element in average (70 kg) person: nil

Discovery: see Nuclear Data section.
[Named after Dubna]
French, *dubnium*; German, *Dubnium*; Italian, *dubnio*; Spanish, *dubnio*

[dub-ni-uhm]

• NUCLEAR DATA

Discovery: Isotopes 260 and 261 were tentatively reported in 1967 by a group of scientists at Dubna, near Moscow, Russia. Isotope 260 was confirmed at both Berkeley, California, USA and Dubna in 1970. IUPAC concluded in 1992 that credit for the discovery should be shared between both groups.

Number of isotopes (including nuclear isomers): 7

Isotope mass range: 255 → 262

Key isotopes

Nuclide	Atomic mass	Half life ($T_{1/2}$)	Decay mode and energy (MeV)	Nuclear spin I	Nucl. mag. moment μ	Uses
^{255}Db		*c.* 1.5 s	SF			
^{257}Db	257.107 770	1.3 s	α; SF			
^{258}Db	258.109 020	4.4 s	α; EC (5.3)			
^{259}Db	259.109 580	*c.* 1.2s	SF			
^{260}Db	260.111 040	1.5 s	α; SF			
^{261}Db	261.111 820	1.8 s	α; SF			
^{262}Db	262.113 760	34 s	EC; α			
^{263}Db		27 s	SF; α			

NMR [Not recorded]

• ELECTRON SHELL DATA

Ground state electron configuration: $[Rn]5f^{14}6d^{3}7s^{2}$
Term symbol: $^{4}F_{3/2}$
Electron affinity ($M \rightarrow M^{-}$)/kJ mol^{-1}**:** n.a.

Ionization energies/kJ mol^{-1}**:**
1. $M \rightarrow M^{+}$ 640 (est.)

Electron binding energies/eV
n.a.

Main lines in atomic spectrum
[Wavelength/nm(species)]
n.a.

• CRYSTAL DATA

Crystal structure (cell dimensions/pm), space group
n.a.

X-ray diffraction: mass absorption coefficients (μ/ρ)/cm^2 g^{-1}**:** CuK_{α} n.a. MoK_{α} n.a.
Neutron scattering length, $b/10^{-12}$ cm**:** n.a.
Thermal neutron capture cross-section, σ_a/barns**:** n.a.

• GEOLOGICAL DATA

Minerals
Not found on Earth.

Chief source: several atoms of dubnium have been made from ^{249}Cf by bombarding with ^{15}N nuclei ($^{249}Cf + ^{15}N \rightarrow ^{260}Db + 4n$), or from ^{249}Bk by bombarding with ^{18}O nuclei.

Specimen: not available commercially.

Abundances
Sun (relative to $H = 1 \times 10^{12}$)**:** n.a.
Earth's crust/p.p.m.**:** nil
Seawater/p.p.m.**:** nil

Dy

Atomic number: 66
Relative atomic mass (^{12}C = 12.0000): 162.50

CAS: [7429-91-6]

• CHEMICAL DATA

Description: Dysprosium is a hard, silvery metal of the so-called rare earth group (more correctly termed the lanthanides). It is oxidised by oxygen, reacts rapidly with cold water, and dissolves in acids. Dysprosium is used in alloys for making magnets.

Radii/pm: Dy^{3+} 91; atomic 177; covalent 159
Electronegativity: 1.22 (Pauling); 1.10 (Allred); n.a. (absolute)
Effective nuclear charge: 2.85 (Slater); 8.34 (Clementi); 11.49 (Froese-Fischer)

Standard reduction potentials $E^{\ominus}$/V

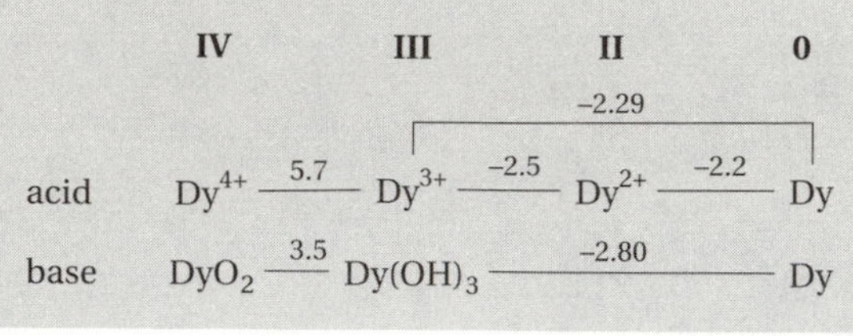

Oxidation states

Dy^{II}	f^{10}	$DyCl_2$, DyI_2
Dy^{III}	f^9	Dy_2O_3, $Dy(OH)_3$, $[Dy(OH_2)_x]^{3+}$(aq), Dy^{3+} salts, DyF_3, $DyCl_3$ etc., salts, $[DyCl_6]^{3-}$
Dy^{IV}	f^8	Cs_3DyF_7

• PHYSICAL DATA

Melting point/K: 1685
Boiling point/K: 2835

ΔH_{fusion}/kJ mol^{-1}: 17.2
ΔH_{vap}/kJ mol^{-1}: 293

Thermodynamic properties (298.15 K, 0.1 MPa)

State	$\Delta_f H^{\ominus}$/kJ mol^{-1}	$\Delta_f G^{\ominus}$/kJ mol^{-1}	$S^{\ominus}$/J K^{-1} mol^{-1}	C_p/J K^{-1} mol^{-1}
Solid	0	0	74.77	28.16
Gas	290.4	254.4	196.63	20.79

Density/kg m^{-3}: 8550 [293 K]
Molar volume/cm^3: 19.00
Thermal conductivity/W m^{-1} K^{-1}: 10.7 [300 K]
Coefficient of linear thermal expansion/K^{-1}: 10.0×10^{-6}
Electrical resistivity/Ω m: 57.0×10^{-8} [273 K]
Mass magnetic susceptibility/kg^{-1} m^3: $+8.00 \times 10^{-6}$ (s)

Young's modulus/GPa: 61.4
Rigidity modulus/GPa: 24.7
Bulk modulus/GPa: 40.5
Poisson's ratio/GPa: 0.247

• BIOLOGICAL DATA

Biological role

None.

Toxicity

Low.

Toxic intake: n.a.
Lethal intake: LD_{50} (chloride, oral, mouse) = 7650 mg kg^{-1}

Hazards

Dysprosium is mildly toxic by ingestion.

Levels in humans

Organs: n.a., but low
Daily dietary intake: n.a.
Total mass of element in average (70 kg) person: n.a.

Discovered in 1886 by P.É. Lecoq de Boisbaudran at Paris, France.
[Greek, *dysprositos* = hard to obtain]
French, *dysprosium*; German, *Dysprosium*; Italian, *disprosio*; Spanish, *disprosio*

Dysprosium

[dis-pro-zee-uhm]

• NUCLEAR DATA

Number of isotopes (including nuclear isomers): 24

Isotope mass range: 147m → 168

Key isotopes

Nuclide	Atomic mass	Natural abundance (%)	Nuclear spin I	Nuclear magnetic moment μ	Uses
^{156}Dy	155.924 277	0.06	0+		E
^{158}Dy	157.924 403	0.10	0+		E
^{160}Dy	159.925 193	2.34	0+		E
^{161}Dy	160.926 930	18.9	5/2+	−0.480 6	E, NMR
^{162}Dy	161.926 795	25.5	0+		E
^{163}Dy	162.928 728	24.9	5/2−	+0.672 6	E, NMR
^{164}Dy	163.929 171	28.2	0+		E

A table of radioactive isotopes is given in Appendix 1, on p239.

NMR [Not recorded]

	^{161}Dy	^{163}Dy
Relative sensitivity (^1H = 1.00)	4.17×10^{-4}	1.12×10^{-3}
Receptivity (^{13}C = 1.00)	0.509	1.79
Magnetogyric ratio/rad $T^{-1} s^{-1}$	-0.9206×10^7	1.2750×10^7
Nuclear quadrupole moment/m^2	$+2.468 \times 10^{-28}$	$+2.648 \times 10^{-28}$
Frequency (^1H = 100 Hz; 2.3488T)/MHz	3.294	4.583

• ELECTRON SHELL DATA

Ground state electron configuration: [Xe]$4f^{10}6s^2$
Term symbol: $^5I_8$
Electron affinity ($M \to M^-$)/kJ mol^{-1}**:** n.a.

Ionization energies/kJ mol^{-1}**:**

1.	$M \to M^+$	571.9
2.	$M^+ \to M^{2+}$	1126
3.	$M^{2+} \to M^{3+}$	2200
4.	$M^{3+} \to M^{4+}$	4001
5.	$M^{4+} \to M^{5+}$	5990

Electron binding energies/eV

K	1s	53 789
L_I	2s	9046
L_{II}	$2p_{1/2}$	8581
L_{III}	$2p_{3/2}$	7790
M_I	3s	2047
M_{II}	$3p_{1/2}$	1842
M_{III}	$3p_{3/2}$	1676
M_{IV}	$3d_{3/2}$	1333
M_V	$3d_{5/2}$	1292

continued in Appendix 2, p255

Main lines in atomic spectrum
[Wavelength/nm(species)]

353.170 (II)
364.540 (II)
394.468 (II)
396.839 (II)
404.597 (I)
418.682 (I)
421.172 (I) (AA)

• CRYSTAL DATA

Crystal structure (cell dimensions/pm), space group
Orthorhombic (a = 359.5, b = 618.3, c = 567.7), Cmcm
h.c.p. (a = 359.03, c = 564.75), $P6_3/mmc$
b.c.c. (a = 398), Im3m
T(orthorhombic → h.c.p.) = 86 K
high pressure form: (a = 334, c = 245), $R\bar{3}m$

X-ray diffraction: mass absorption coefficients (μ/ρ)/$cm^2 g^{-1}$**:** CuK_α 286 MoK_α 70.6
Neutron scattering length, $b/10^{-12}$ cm**:** 1.69
Thermal neutron capture cross-section, σ_a/barns**:** 920

• GEOLOGICAL DATA

Minerals

Mineral	Formula	Density	Hardness	Crystal appearance
Bastnäsite*	(Ce,La, etc.)CO_3F	4.9	4 – 4.5	hex., vit./greasy yellow
Monazite*	(Ce, La, Nd, Th, etc.)PO_4	5.20	5 – 5.5	mon., waxy/vit. yellow-brown

*Although not a major constituent, dysprosium is present in extractable amounts.

Chief ores: monazite, bastnäsite
World production/tonnes y^{-1}**:** *c.* 100
Main mining areas: USA, Brazil, India, Sri Lanka, Australia.
Reserves/tonnes: *c.* 1.5×10^5
Specimen: available as chips, foil, ingots or powder. Safe.

Abundances

Sun (relative to H = 1×10^{12})**:** 11.5
Earth's crust/p.p.m.**:** 6
Seawater/p.p.m.**:**
Atlantic surface: 8×10^{-7}
Atlantic deep: 9.6×10^{-7}
Pacific surface: n.a.
Pacific deep: n.a.
Residence time/years**:** 300
Classification: recycled
Oxidation state: III

Es

Atomic number: 99
Relative atomic mass ($^{12}C = 12.0000$)**:** 252.083 (Es-252)

CAS: [7429-92-7]

• CHEMICAL DATA

Description: Einsteinium is a radioactive metal element, which has yet to be isolated in the metallic form. It is expected to be attacked by oxygen, steam and acids, but not alkalis. There are, as yet, no known applications.

Radii/pm: Es^{4+} 85; Es^{3+} 98; Es^{2+} 117; atomic 203
Electronegativity: 1.3 (Pauling); 1.2 (est.) (Allred); ≤ 3.5 eV (absolute)
Effective nuclear charge: 1.65 (Slater)

Standard reduction potentials $E^{\ominus}$/V

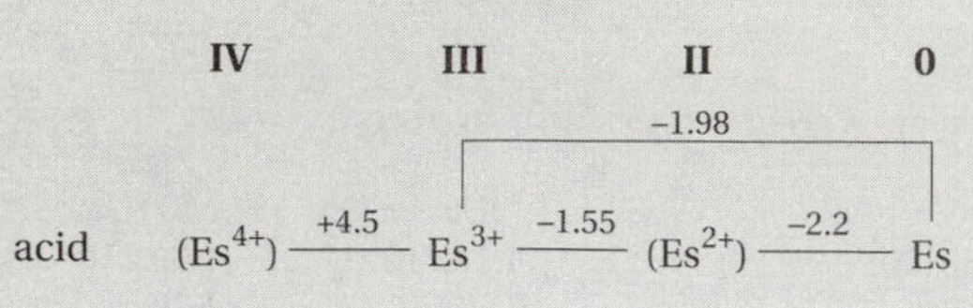

Oxidation states

Es^{II}	f^{11}	transient
Es^{III}	f^{10}	Es_2O_3, $Es(OH)_3$, $EsCl_3$, $EsBr_3$ etc., Es^{3+}(aq), EsOCl
Es^{IV}	f^{9}	suspected

• PHYSICAL DATA

Melting point/K: 1130 ± 30
Boiling point/K: n.a.

ΔH_{fusion}/kJ mol^{-1}: n.a.
ΔH_{vap}/kJ mol^{-1}: 133

Thermodynamic properties (298.15 K, 0.1 MPa)

State	$\Delta_f H^{\ominus}$/kJ mol^{-1}	$\Delta_f G^{\ominus}$/kJ mol^{-1}	$S^{\ominus}$/J K^{-1} mol^{-1}	C_p/J K^{-1} mol^{-1}
Solid	0	0	n.a.	25.98
Gas	n.a.	n.a.	n.a.	n.a.

Density/kg m^{-3}: n.a.
Molar volume/cm^3: n.a.
Thermal conductivity/W m^{-1} K^{-1}: 10 (est.) [300 K]
Coefficient of linear thermal expansion/K^{-1}: n.a.
Electrical resistivity/Ω m: n.a.
Mass magnetic susceptibility/kg^{-1} m^3: n.a.

Young's modulus/GPa: n.a.
Rigidity modulus/GPa: n.a.
Bulk modulus/GPa: n.a.
Poisson's ratio/GPa: n.a.

• BIOLOGICAL DATA

Biological role

None.

Toxicity

Toxic intake: n.a.
Lethal intake: n.a.

Hazards

Einsteinium is never encountered outside the laboratory or nuclear industry. It would be highly dangerous because of its intense radioactivity if it were to be produced in bulk.

Levels in humans

nil

Daily dietary intake: nil
Total mass of element in average (70 kg) person: nil

Discovery: see Nuclear Data section.
[Named after Albert Einstein]
French, *einsteinium*; German, *Einsteinium*; Italian, *einsteinio*; Spanish, *einsteinio*

Einsteinium

[iyn-stiyn-iuhm]

• NUCLEAR DATA

Discovery: Einsteinium was discovered in the debris of the 1952 thermonuclear explosion in the Pacific by G.R. Choppin, S.G. Thompson, A. Ghiorso and B.G. Harvey.

Number of isotopes (including nuclear isomers): 17 **Isotope mass range:** 243 → 256

Key isotopes

Nuclide	Atomic mass	Half life ($T_{1/2}$)	Decay mode and energy (MeV)	Nuclear spin *I*	Nucl. mag. moment *μ*	Uses
^{249}Es	249.076 340	1.70 h	EC (1.39) 99.4%; α 0.6%; no γ	7/2+		
^{250}Es	250.078 660	8.6 h	EC (2.1); γ	6+		
^{250m}Es		2.2 h	EC; β^+; γ			
^{251}Es	251.079 986	1.38 d	EC (0.38) 99.5%; α 0.5%	3/2–		
^{252}Es	252.082 944	1.29 y	EC (1.26) 24%; α 76%	5–		
^{253}Es	253.084 818	20.47 d	α; γ	7/2+	4.10	available
^{254}Es	254.088 019	276 d	α; γ	7+		
^{254m}Es		1.64 d	β^- 99.6%; α (6.617) 0.4%; γ	2+	2.9	
^{255}Es	255.090 270	40 d	β^- (0.30) 92%; α 8%; SF	7/2+		
^{256m}Es	256.093 560	7.6 h	β^- (1.7) 24%; γ	8+		

Other isotopes of einsteinium have half-lives shorter than 1 hour.

NMR [Not recorded]

• ELECTRON SHELL DATA

Ground state electron configuration: $[Rn]5f^{11}7s^2$
Term symbol: $^5I_{15/2}$
Electron affinity $(M \rightarrow M^-)$/kJ mol^{-1}**:** ≤ 50

Ionization energies/kJ mol^{-1}**:**
1. M → M^+ 619

Electron binding energies/eV
n.a.

Main lines in atomic spectrum
[Wavelength/nm(species)]
270.866 (II)
342.848 (I)
349.811 (I)
351.433 (I)
352.138 (I)
352.349 (I)
354.775 (II)

• CRYSTAL DATA

Crystal structure (cell dimensions/pm), space group
α-Es h.c.p.
β-Es f.c.c.
$T(\alpha \rightarrow \beta) = 1133$ K

X-ray diffraction: mass absorption coefficients (μ/ρ)/cm^2 g^{-1}**:** CuK_α n.a. MoK_α n.a.
Neutron scattering length, $b/10^{-12}$ cm**:** n.a.
Thermal neutron capture cross-section, σ_a/barns**:** 160 (^{253}Es)

• GEOLOGICAL DATA

Minerals
Not found on Earth.

Chief source: einsteinium-253 can be obtained in mg quantities by the bombardment of ^{239}Pu with neutrons.
World production: n.a. but probably less than 1 g per year.
Specimen: commercially available, under licence - see Key.

Abundances
Sun (relative to H = 1×10^{12})**:** n.a.
Earth's crust/p.p.m.**:** nil
Seawater/p.p.m.**:** nil

Er

Atomic number: 68
Relative atomic mass ($^{12}C = 12.0000$)**:** 167.26

CAS: [7440-52-0]

• CHEMICAL DATA

Description: Erbium is a silvery metal of the so-called rare earth group (more correctly termed the lanthanides). It slowly tarnishes in air, reacts slowly with water, and dissolves in acids. Erbium is used in infrared-absorbing glass and in alloys with titanium.

Radii/pm: Er^{3+} 89; atomic 176; covalent 157
Electronegativity: 1.24 (Pauling); 1.14 (Allred); ≤ 3.3 eV (absolute)
Effective nuclear charge: 2.85 (Slater); 8.48 (Clementi); 11.70 (Froese-Fischer)

Standard reduction potentials $E^\ominus$/V

	III		0
acid	Er^{3+}	−2.32	Er
base	$Er(OH)_3$	−2.84	Er

Oxidation states

Er^{III} f^{11} Er_2O_3, $Er(OH)_3$, ErF_3, $ErCl_3$ etc., $[Er(OH_2)_x]^{3+}$(aq), Er^{3+} salts, $[ErCl_6]^{3-}$, complexes, etc.

• PHYSICAL DATA

Melting point/K: 1802
Boiling point/K: 3136
ΔH_{fusion}/kJ mol^{-1}: 17.2
ΔH_{vap}/kJ mol^{-1}: 292.9

Thermodynamic properties (298.15 K, 0.1 MPa)

State	$\Delta_f H^\ominus$/kJ mol^{-1}	$\Delta_f G^\ominus$/kJ mol^{-1}	$S^\ominus$/J K^{-1} mol^{-1}	C_p/J K^{-1} mol^{-1}
Solid	0	0	73.18	28.12
Gas	317.1	280.7	195.59	20.79

Density/kg m^{-3}: 9066 [298 K]
Molar volume/cm^3: 18.44
Thermal conductivity/W m^{-1} K^{-1}: 14.3 [300 K]
Coefficient of linear thermal expansion/K^{-1}: 9.2×10^{-6}
Electrical resistivity/Ω m: 87×10^{-8} [298 K]
Mass magnetic susceptibility/kg^{-1} m^3: $+3.33 \times 10^{-6}$ (s)

Young's modulus/GPa: 69.9
Rigidity modulus/GPa: 28.3
Bulk modulus/GPa: 44.4
Poisson's ratio/GPa: 0.237

• BIOLOGICAL DATA

Biological role
None, but acts to stimulate metabolism.

Toxicity
Toxic intake: n.a.
Lethal intake: LD_{50} (chloride, oral, mouse) = 6200 mg kg^{-1}

Hazards
Erbium is mildly toxic by ingestion.

Levels in humans
Organs: n.a., but low
Daily dietary intake: n.a.
Total mass of element in average (70 kg) person: n.a.

Discovered in 1842 by C.G. Mosander at Stockholm, Sweden.
[Named after Ytterby, Sweden]
French, *erbium*; German, *Erbium*; Italian, *erbio*; Spanish, *erbio*

Erbium

[erb-iuhm]

• NUCLEAR DATA

Number of isotopes (including nuclear isomers): 25 **Isotope mass range:** 150 → 173

Key isotopes

Nuclide	Atomic mass	Natural abundance (%)	Nuclear spin I	Nuclear magnetic moment μ	Uses
^{162}Er	161.928 775	0.14	0+		E
^{164}Er	163.929 198	1.61	0+		E
^{166}Er	165.930 290	33.6	0+		E
^{167}Er	166.932 046	22.95	7/2+	–0.566 5	E, NMR
^{168}Er	167.932 368	26.8	0+		E
^{170}Er	169.935 46	14.9	0+		E

A table of radioactive isotopes is given in Appendix 1, on p239.

NMR [Reference: not reported]

	^{167}Er
Relative sensitivity (^1H = 1.00)	5.07×10^{-4}
Receptivity (^{13}C = 1.00)	0.665
Magnetogyric ratio/rad $T^{-1} s^{-1}$	-0.7752×10^{7}
Nuclear quadrupole moment/m^2	$+3.565 \times 10^{-28}$
Frequency (^1H = 100 Hz; 2.3488T)/MHz	2.890

• ELECTRON SHELL DATA

Ground state electron configuration: $[Xe]4f^{12}6s_2$
Term symbol: $^3H_6$
Electron affinity ($M \rightarrow M^-$)/kJ mol^{-1}**:** ≤ 50

Ionization energies/kJ mol^{-1}**:**

1.	$M \rightarrow M^+$	588.7
2.	$M^+ \rightarrow M^{2+}$	1151
3.	$M^{2+} \rightarrow M^{3+}$	2194
4.	$M^{3+} \rightarrow M^{4+}$	4115
5.	$M^{4+} \rightarrow M^{5+}$	6282

Electron binding energies/eV

K	1s	57 486
L_I	2s	9751
L_{II}	$2p_{1/2}$	9264
L_{III}	$2p_{3/2}$	8358
M_I	3s	2206
M_{II}	$3p_{1/2}$	2006
M_{III}	$3p_{3/2}$	1812
M_{IV}	$3d_{3/2}$	1453
M_V	$3d_{5/2}$	1409

continued in Appendix 2, p255

Main lines in atomic spectrum
[Wavelength/nm(species)]
369.265 (II)
386.285 (I)
389.268 (I)
390.631 (II)
400.796 (I) (AA)
415.111 (I)

• CRYSTAL DATA

Crystal structure (cell dimensions/pm), space group
α-Er h.c.p. (a = 355.88, c = 558.74), $P6_3/mmc$
β-Er b.c.c. (a = 394), Im3m
$T(\alpha \rightarrow \beta)$ = 1640 K

X-ray diffraction: mass absorption coefficients (μ/ρ)/$cm^2 g^{-1}$**:** CuK_α 134 MoK_α 77.3
Neutron scattering length, $b/10^{-12}$ cm**:** 0.816
Thermal neutron capture cross-section, σ_a/barns**:** 160

• GEOLOGICAL DATA

Minerals

Mineral	Formula	Density	Hardness	Crystal appearance
Bastnäsite*	(Ce,La, etc.)CO_3F	4.9	4 – 4.5	mon., waxy/vit. yellow
Monazite*	(Ce, La, Nd, Th, etc.)PO_4	5.2	5 – 5.5	mon., waxy/vit. yellow-brown

*Although not a major constituent, erbium is present in extractable amounts.

Chief ores: monazite, bastnäsite
World production/tonnes y^{-1}**:** *c.* 500
Main mining areas: USA, Brazil, India, Sri Lanka, Australia.
Reserves/tonnes: *c.* 1×10^6
Specimen: available as chips, ingots or powder. Safe.

Abundances

Sun (relative to H = 1×10^{12})**:** 5.8
Earth's crust/p.p.m.**:** 3.8
Seawater/p.p.m.**:**
Atlantic surface: 5.9×10^{-7}
Atlantic deep: 8.6×10^{-7}
Pacific surface: n.a.
Pacific deep: n.a.
Residence time/years**:** 400
Classification: recycled
Oxidation state: III

Eu

Atomic number: 63
Relative atomic mass ($^{12}C = 12.0000$)**:** 151.965

CAS: [7440-53-1]

• CHEMICAL DATA

Description: Europium is a soft, silvery metal which is one of the rarest of the so-called rare earth group (more correctly termed the lanthanides). It is the most reactive of these metals, reacting quickly with oxygen and water. It is little used, but some is employed in thin-film superconductor alloys.

Radii/pm: Eu^{3+} 98; Eu^{2+} 112; atomic 204; covalent 185
Electronegativity: n.a. (Pauling); 1.01 (Allred); ≤ 3.1 eV (absolute)
Effective nuclear charge: 2.85 (Slater); 8.11 (Clementi); 11.17 (Froese-Fischer)

Standard reduction potentials $E^{\ominus}$/V

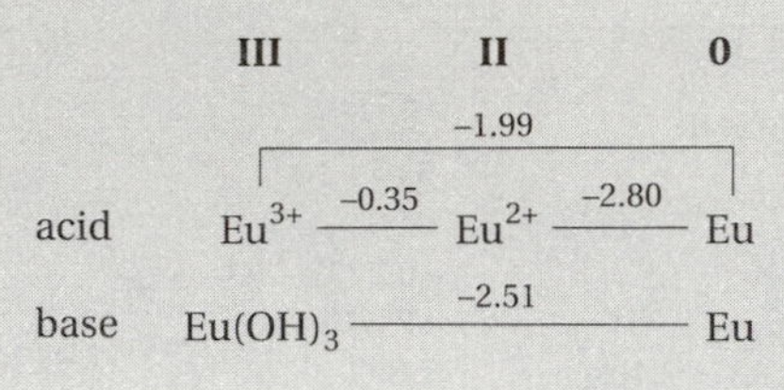

Oxidation states

Eu^{II}	f^7	EuO, EuS, EuF_2, $EuCl_2$ etc.
Eu^{III}	f^6	Eu_2O_3, $Eu(OH)_3$, $[Eu(OH_2)_x]^{3+}$ (aq), Eu^{3+} salts, EuF_3, $EuCl_3$ etc., complexes

• PHYSICAL DATA

Melting point/K: 1095
Boiling point/K: 1870
ΔH_{fusion}/kJ mol^{-1}: 10.5
ΔH_{vap}/kJ mol^{-1}: 175.7

Thermodynamic properties (298.15 K, 0.1 MPa)

State	$\Delta_f H^{\ominus}$/kJ mol^{-1}	$\Delta_f G^{\ominus}$/kJ mol^{-1}	$S^{\ominus}$/J K^{-1} mol^{-1}	C_p/J K^{-1} mol^{-1}
Solid	0	0	77.78	27.66
Gas	175.3	142.2	188.795	20.786

Density/kg m^{-3}: 5243 [293 K]
Molar volume/cm^3: 28.98
Thermal conductivity/W m^{-1} K^{-1}: 13.9 [300 K]
Coefficient of linear thermal expansion/K^{-1}: 32×10^{-6}
Electrical resistivity/Ω m: 990.0×10^{-8} [298 K]
Mass magnetic susceptibility/kg^{-1} m^3: $+2.81 \times 10^{-6}$ (s)

Young's modulus/GPa: 18.2
Rigidity modulus/GPa: 7.9
Bulk modulus/GPa: 8.3
Poisson's ratio/GPa: 0.152

• BIOLOGICAL DATA

Biological role

None.

Toxicity

Toxic intake: n.a.
Lethal intake: LD_{50} (nitrate, oral, mouse) = > 5000 mg kg^{-1}

Hazards

Europium is mildly toxic by ingestion.

Levels in humans

Organs: n.a., but very low
Daily dietary intake: n.a.
Total mass of element in average (70 kg) person: n.a., but very low

Discovered in 1901 by E.A. Demarçay at Paris, France.
[Named after Europe]
French, *europium*; German, *Europium*; Italian, *europio*; Spanish, *europio*

Europium

[yoo-roh-pi-uhm]

• NUCLEAR DATA

Number of isotopes (including nuclear isomers): 26 **Isotope mass range:** 141m → 160

Key isotopes

Nuclide	Atomic mass	Natural abundance (%)	Nuclear spin I	Nuclear magnetic moment μ	Uses
^{151}Eu	150.919 702	47.8	5/2+	+3.471 8	E, NMR
^{153}Eu	152.921 225	52.2	5/2+	+1.533 1	E, NMR

A table of radioactive isotopes is given in Appendix 1, on p240.

NMR [Reference: not given]	^{151}Eu	^{153}Eu
Relative sensitivity (^{1}H = 1.00)	0.18	0.0152
Receptivity (^{13}C = 1.00)	464	45.7
Magnetogyric ratio/rad $T^{-1}s^{-1}$	6.5477×10^7	2.9371×10^7
Nuclear quadrupole moment/m^2	0.903×10^{-28}	2.412×10^{-28}
Frequency (^{1}H = 100 Hz; 2.3488T)/MHz	24.801	10.951

• ELECTRON SHELL DATA

Ground state electron configuration: [Xe]$4f^7 6s^2$
Term symbol: $^8S_{7/2}$
Electron affinity ($M \rightarrow M^-$)/kJ mol^{-1}: ≤ 50

Ionization energies/kJ mol^{-1}:

1.	$M \rightarrow M^+$	546.7
2.	$M^+ \rightarrow M^{2+}$	1085
3.	$M^{2+} \rightarrow M^{3+}$	2404
4.	$M^{3+} \rightarrow M^{4+}$	4110
5.	$M^{4+} \rightarrow M^{5+}$	6101

Electron binding energies/eV

K	1s	48 519
L_I	2s	8052
L_{II}	$2p_{1/2}$	7617
L_{III}	$2p_{3/2}$	6977
M_I	3s	1800
M_{II}	$3p_{1/2}$	1614
M_{III}	$3p_{3/2}$	1481
M_{IV}	$3d_{3/2}$	1158.6
M_V	$3d_{5/2}$	1127.5

continued in Appendix 2, p255

Main lines in atomic spectrum
[Wavelength/nm(species)]

318.967 (II)
412.974 (II)
420.505 (II)
459.402 (I) (AA)
462.722 (I)
466.187 (I)

• CRYSTAL DATA

Crystal structure (cell dimensions/pm), space group
b.c.c. (a = 458.20), Im3m

X-ray diffraction: mass absorption coefficients (μ/ρ)/cm^2 g^{-1}: CuK_α 425 MoK_α 61.5
Neutron scattering length, b/10^{-12} cm: 0.722
Thermal neutron capture cross-section, σ_a/barns: 4600

• GEOLOGICAL DATA

Minerals

Mineral	Formula	Density	Hardness	Crystal appearance
Bastnäsite*	(Ce,La, etc.)CO_3F	4.9	4 – 4.5	hex., vit./greasy yellow
Monazite*	(Ce, La, Nd, Th, etc.)PO_4	5.20	5 – 5.5	mon., waxy/vit. yellow-brown

*Although not a major constituent, europium is present in extractable amounts.

Chief ores: monazite, bastnäsite
World production/tonnes y^{-1}: *c.* 400
Main mining areas: USA, Brazil, India, Sri Lanka, Australia, China.
Reserves/tonnes: *c.* 1.5×10^6
Specimen: available as ingots. Safe.

Abundances

Sun (relative to H = 1×10^{12}): 5
Earth's crust/p.p.m.: 2.1
Seawater/p.p.m.:
Atlantic surface: 0.9×10^{-7}
Atlantic deep: 1.5×10^{-7}
Pacific surface: 1.0×10^{-7}
Pacific deep: 2.7×10^{-7}
Residence time/years: 500
Classification: recycled
Oxidation state: III

Fm

Atomic number: 100
Relative atomic mass ($^{12}C = 12.0000$)**:** 257.0951 (Fm-257)
CAS: [7440-72-4]

• CHEMICAL DATA

Description: Fermium is a radioactive metal element which does not occur naturally. There are, as yet, no known applications.

Radii/pm: Fm^{4+} 84; Fm^{3+} 91; Fm^{2+} 115; atomic n.a.
Electronegativity: 1.3 (Pauling); 1.2 (est.) (Allred); n.a. (absolute)
Effective nuclear charge: 1.65 (Slater); n.a. (Clementi); n.a. (Froese-Fisher)

Standard reduction potentials $E^⊖$/V

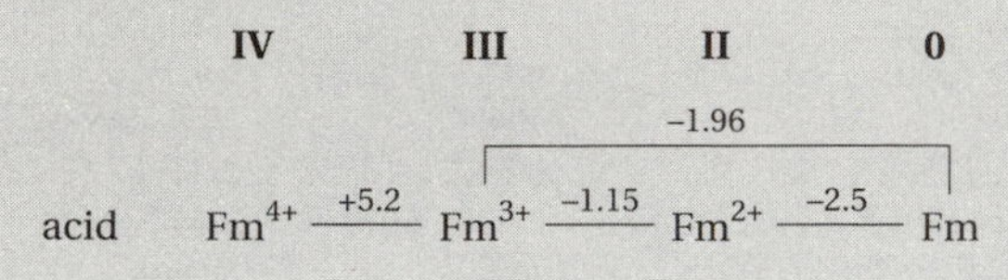

Oxidation states

Fm^{II}	f^{12}	?
Fm^{III}	f^{11}	$[Fm(OH_2)_x]^{3+}$ (aq)
Fm^{IV}	f^{10}	?

• PHYSICAL DATA

Melting point/K: n.a.
Boiling point/K: n.a.
ΔH_{fusion}/kJ mol^{-1}: n.a.
ΔH_{vap}/kJ mol^{-1}: n.a.

Thermodynamic properties (298.15 K, 0.1 MPa)

State	$\Delta_f H^⊖$/kJ mol^{-1}	$\Delta_f G^⊖$/kJ mol^{-1}	$S^⊖$/J K^{-1} mol^{-1}	C_p/J K^{-1} mol^{-1}
Solid	0	0	n.a.	n.a.
Gas	n.a.	n.a.	n.a.	n.a.

Density/kg m^{-3}**:** n.a.
Molar volume/cm^3**:** n.a.
Thermal conductivity/W m^{-1} K^{-1}**:** 10 (est.) [300 K]
Coefficient of linear thermal expansion/K^{-1}**:** n.a.
Electrical resistivity/Ω m**:** n.a.
Mass magnetic susceptibility/kg^{-1} m^3**:** n.a.

• BIOLOGICAL DATA

Biological role

None.

Toxicity

Toxic intake: n.a.
Lethal intake: n.a.

Hazards

Fermium is never encountered normally, and relatively few atoms have ever been made. It would be dangerous because of its intense radioactivity.

Levels in humans

nil
Daily dietary intake: nil
Total mass of element in average (70 kg) person: nil

Discovery: see Nuclear Data section.
[Named after Enrico Fermi]
French, *fermium*; German, *Fermium*; Italian, *fermio*; Spanish, *fermio*

Fermium

[ferm-iuhm]

• NUCLEAR DATA

Discovery: Fermium was discovered in the debris of the 1952 thermonuclear explosion in the Pacific by G.R. Choppin, S.G. Thompson, A. Ghiorso and B.G. Harvey.

Number of isotopes (including nuclear isomers): 20 **Isotope mass range:** 243 → 259

Key isotopes

Nuclide	Atomic mass	Half life ($T_{1/2}$)	Decay mode and energy (MeV)	Nuclear spin I	Nucl. mag. moment μ	Uses
^{251}Fm	251.081 590	5.3 h	EC (1.49) 98%; α (7.424) 2%; γ	0+		
^{252}Fm	252.082 466	1.058 d	α (7.154)	0+		
^{253}Fm	253.085 173	3 d	EC (0.334) 88%; α (7.200) 12%; γ	1/2+		
^{254}Fm	254.086 846	3.2 h	α (7.303)	0+		
^{255}Fm	255.089 948	20.1 h	α (7.240)	7/2+		
^{256}Fm	256.091 767	2.63 h	SF 92%; α (7.240) 18%	0+		
^{257}Fm	257.075 099	100.5 d	α (6.871); γ	9/2+		

Other isotopes of fermium have half-lives shorter than 1 hour.

NMR [Not recorded]

• ELECTRON SHELL DATA

Ground state electron configuration: [Rn]$5f^{12}7s^2$
Term symbol: $^3H_6$
Electron affinity (M → M$^-$)/kJ mol^{-1}: n.a.

Ionization energies/kJ mol^{-1}:
1. M → M$^+$ 627

Electron binding energies/eV
n.a.

Main lines in atomic spectrum
[Wavelength/nm(species)]
n.a.

• CRYSTAL DATA

Crystal structure (cell dimensions/pm), space group
n.a.

X-ray diffraction: mass absorption coefficients (μ/ρ)/cm^2 g^{-1}: CuK$_\alpha$ n.a. MoK$_\alpha$ n.a.
Neutron scattering length, b/10^{-12} cm: n.a.
Thermal neutron capture cross-section, σ_a/barns: 5800 (^{257}Fm)

• GEOLOGICAL DATA

Minerals

Not found on Earth.

Chief source: fermium-253 can be obtained in nanogram (10^{-9} g) quantities by the bombardment of ^{239}Pu with neutrons

World production: about 3×10^{-6} g in total.

Specimen: commercially available, under licence - see Key.

Abundances

Sun (relative to H = 1×10^{12}): n.a.
Earth's crust/p.p.m.: nil
Seawater/p.p.m.: nil

F

Atomic number: 9
Relative atomic mass ($^{12}C = 12.0000$)**: 18.998 4032**

CAS: [7782-41-4]

• CHEMICAL DATA

Description: Fluorine is a pale yellow gas (F_2) and is the most reactive of all the elements. It is also the strongest available oxidising agent. Fluorine is produced by the electrolysis of molten KF.2HF and is used to make UF_6, SF_6 and fluorinating agents such as ClF_3. A wide range of fluorinated materials are now in common use including polymers, pesticides and antibiotics. Some fluoride salts are also used, such as CaF_2 as a flux in metallurgy and AlF_3 in the production of aluminium.

Radii/pm: F^- 133; atomic 70.9; covalent 58; van der Waals 135
Electronegativity: 3.98 (Pauling); 4.10 (Allred); 10.41 eV (absolute)
Effective nuclear charge: 5.20 (Slater); 5.10 (Clementi); 4.61 (Froese-Fischer)

Standard reduction potentials $E^\ominus$/V

0		–I
F_2	2.866	F^-
F_2	2.979	HF_2^-
F_2	3.053	HF (aq)

Oxidation states

F^{-I}	[Ne]	F^- (aq), HF, KHF_2, CaF_2, many salts and derivatives of other elements
F^0	s^2p^5	F_2

Covalent bonds

Bond	r/ pm	E/ kJ mol^{-1}
F—O	147	190
F—N	137	272
F—F	142	159

For other bonds to fluorine, see other elements

• PHYSICAL DATA

Melting point/K: 53.53
Boiling point/K: 85.01
Critical temperature/K: 144.3
Critical pressure/ kPa: 5573

ΔH_{fusion}/kJ mol^{-1}: 5.10
ΔH_{vap}/kJ mol^{-1}: 6.548

Thermodynamic properties (298.15 K, 0.1 MPa)

State	$\Delta_f H^\ominus$/kJ mol^{-1}	$\Delta_f G^\ominus$/kJ mol^{-1}	$S^\ominus$/J K^{-1} mol^{-1}	C_p/J K^{-1} mol^{-1}
Gas (F_2)	0	0	202.78	31.30
Gas (atoms)	78.99	61.91	158.754	22.744

Density/kg m^{-3}: n.a. [solid]; 1516 [liquid, 85 K]; 1.696 [gas, 273 K]
Molar volume/cm^3: 18.05 [85 K]
Thermal conductivity/W m^{-1} K^{-1}: 0.0279 [300 K]; 0.0248 [273 K]
Mass magnetic susceptibility/kg^{-1} m^3: n.a.

• BIOLOGICAL DATA

Biological role

Essential trace element for mammals, including humans, in the form of fluoride (F^-).

Toxicity

Toxic intake: 250 mg NaF
Lethal intake: 5 – 25 g NaF; LD_{50} (NaF, oral, rat) = 52 mg kg^{-1}; LC_{50} (F_2 inhalation, rat) = 185 p.p.m. for 1 hour

Hazards

Fluorine gas is highly corrosive and toxic, and even exposure to low concentrations causes eye and lung irritation. Metal fluorides are also very toxic. Organic fluorides are generally much less toxic and are often quite harmless.

Levels in humans

Blood/mg dm^{-3}: 0.5
Bone/p.p.m.: 2000 – 12 000
Liver/p.p.m.: 0.22 – 7
Muscle/p.p.m.: 0.05
Daily dietary intake: 0.3 – 0.5 mg
Total mass of element in average (70 kg) person: 2.6 g

First isolated in 1886 by H. Moissan at Paris, France.
[Latin, *fluere* = to flow]
French, *fluor*; German, *Fluor*; Italian, *fluoro*; Spanish, *flúor*

Fluorine

[floor-een]

• NUCLEAR DATA

Number of isotopes (including nuclear isomers): 7 **Isotope mass range:** 17 → 23

Key isotopes

Nuclide	Atomic mass	Natural abundance (%)	Nuclear spin *I*	Nuclear magnetic moment μ	Uses
^{19}F	18.998 403 22	100	1/2+	+2.628 867	NMR

A table of radioactive isotopes is given in Appendix 1, on p240.

NMR [Reference: $CFCl_3$]	^{19}F
Relative sensitivity (^1H = 1.00)	0.83
Receptivity (^{13}C = 1.00)	4730
Magnetogyric ratio/rad $T^{-1}s^{-1}$	25.1665×10^7
Nuclear quadrupole moment/m^2	–
Frequency (^1H = 100 Hz; 2.3488T)/MHz	94.077

• ELECTRON SHELL DATA

Ground state electron configuration: $[He]2s^22p^5$
Term symbol: $^2P_{3/2}$
Electron affinity ($M \rightarrow M^-$)/kJ mol^{-1}**:** 328

Ionization energies/kJ mol^{-1}**:**

1.	$M \rightarrow M^+$	1681
2.	$M^+ \rightarrow M^{2+}$	3374
3.	$M^{2+} \rightarrow M^{3+}$	6050
4.	$M^{3+} \rightarrow M^{4+}$	8408
5.	$M^{4+} \rightarrow M^{5+}$	11 023
6.	$M^{5+} \rightarrow M^{6+}$	15 164
7.	$M^{6+} \rightarrow M^{7+}$	17 867
8.	$M^{7+} \rightarrow M^{8+}$	92 036
9.	$M^{8+} \rightarrow M^{9+}$	106 423

Electron binding energies/eV

K	1s	696.7

Main lines in atomic spectrum
[Wavelength/nm(species)]

685.603 (I)
690.248 (I)
703.747 (I)
712.789 (I)
775.470 (I)

• CRYSTAL DATA

Crystal structure (cell dimensions/pm), space group
α-F_2 monoclinic (*a* = 550, *b* = 328, *c* = 728, β = 102.17°), C2/m
β-F_2 cubic (*a* = 667), Pm3n
$T(\alpha \rightarrow \beta)$ = 45.6 K

X-ray diffraction: mass absorption coefficients (μ/ρ)/cm^2 g^{-1}**:** CuK_α 16.4 MoK_α 1.80
Neutron scattering length, $b/10^{-12}$ cm**:** 0.5654
Thermal neutron capture cross-section, σ_a/barns**:** 0.0096

• GEOLOGICAL DATA

Minerals

Mineral	Formula	Density	Hardness	Crystal appearance
Apatite*	$Ca_5(PO_4)_3(F, OH)$	3.3	5	hex., vit., varied colours
Cryolite	α-Na_3AlF_6	2.97	2.5	mon., vit., greasy white
Fluorite**	CaF_2	3.180	4	cub., vit. colourless to blue/green

*Apatite is known as hydroxyapatite or fluoroapatite according to its level of fluoride.
**Fluorite is also known as fluorspar.

Chief ores: fluorite
World production/tonnes y^{-1}**:** 2400 (fluorine gas, F_2); 4.7×10^6 (fluorite)
Main mining areas: Canada, USA, UK, Russia, Mexico, Italy.
Reserves/tonnes**:** 123×10^6 (fluorite)
Specimen: not available for sale as pure gas because it is too reactive and dangerous.

Abundances

Sun (relative to H = 1×10^{12})**:** 3.63×10^4
Earth's crust/p.p.m.**:** 950
Atmosphere/p.p.m. (volume)**:** n.a.
Seawater/p.p.m.**:** 1.3
Atlantic surface: 1.0×10^{-4}
Atlantic deep: 0.96×10^{-4}
Pacific surface: 1.0×10^{-4}
Pacific deep: 0.4×10^{-4}
Residence time/years**:** 400 000
Classification: accumulating
Oxidation state: –I

Atomic number: 87 **CAS:**
Relative atomic mass (^{12}C = 12.0000): 223.0197 (Fr-223) [7440-73-5]

• CHEMICAL DATA

Description: Francium is an intensely radioactive metal element, whose isotopes are all short-lived, and as a consequence it is little studied, although being an alkali metal its chemistry resembles that of caesium.

Radii/pm: Fr^+ 180; atomic *c.* 270
Electronegativity: 0.7 (Pauling); 0.86 (Allred); n.a. (absolute)
Effective nuclear charge: 2.20 (Slater); n.a. (Clementi); n.a. (Froese-Fischer)

Standard reduction potentials $E^\ominus$/V

I 0

$$Fr^+ \xrightarrow{-3.09} Fr$$

Oxidation states

Fr^I [Rn] little studied, Fr^+ (aq), $FrClO_4$ insoluble

• PHYSICAL DATA

Melting point/K: 300
Boiling point/K: 950
ΔH_{fusion}/kJ mol^{-1}: n.a.
ΔH_{vap}/kJ mol^{-1}: n.a.

Thermodynamic properties (298.15 K, 0.1 MPa)

State	$\Delta_f H^\ominus$/kJ mol^{-1}	$\Delta_f G^\ominus$/kJ mol^{-1}	$S^\ominus$/J K^{-1} mol^{-1}	C_p/J K^{-1} mol^{-1}
Solid	0	0	95.4	n.a.
Gas	72.8	n.a.	n.a.	n.a.

Density/kg m^{-3}: n.a.
Molar volume/cm^3: n.a.
Thermal conductivity/W m^{-1} K^{-1}: 15 (est.) [300 K]
Coefficient of linear thermal expansion/K^{-1}: n.a.
Electrical resistivity/Ω m: n.a.
Mass magnetic susceptibility/kg^{-1} m^3: n.a.

• BIOLOGICAL DATA

Biological role

None.

Toxicity

Toxic intake: n.a. (its toxicity should be like that of caesium)
Lethal intake: n.a.

Hazards

Francium is never encountered normally. It is dangerous because it is a powerful source of α-radiation. This element is only to be found inside nuclear facilities or research laboratories.

Levels in humans

Organs: nil
Daily dietary intake: nil
Total mass of element in average (70 kg) person: nil

Discovered in 1939 by Marguerite Perey at Paris, France .
[Named after France]
French, *francium*; German, *Francium*; Italian, *francio*; Spanish, *francio*

Francium

[fran-see-uhm]

• NUCLEAR DATA

Number of isotopes (including nuclear isomers): 30 **Isotope mass range:** 201 → 229

Key isotopes

Nuclide	Atomic mass	Half life ($T_{1/2}$)	Decay mode and energy (MeV)	Nuclear spin *I*	Nucl. mag. moment μ	Uses
^{210}Fr	209.996 340	3.2 m	α (6.670); EC (6.2); γ	6+	+4.4	
^{211}Fr	210.995 490	3.10 m	α (6.660); EC (4.58); γ	9/2–	+4.0	
^{212}Fr	211.996 130	20 m	α (6.529) 43%; EC (5.08), 57%; γ	5+	+4.6	
^{221}Fr	221.014 230	4.8 m	α (6.457); γ	5/2–	+1.58	
^{222}Fr		14.3 m	β^- (2.08); γ	2–	0.63	
^{223}Fr*	223.019 733	21.8 m	β^- (1.149); γ	3/2+	+1.17	
^{224}Fr	224.023 220	2.7 m	β^- (2.82); γ	1–	+0.40	
^{225}Fr	225.025 590	3.9 m	β^- (1.850)	3/2	+1.07	
^{227}Fr	227.031 770	2.48 m	β^- (2.4)	1/2	+1.50	

*Traces of this isotope occur naturally.

NMR [Not recorded]	^{223}Fr
Nuclear quadrupole moment/m^2	1.17×10^{-28}

• ELECTRON SHELL DATA

Ground state electron configuration: [Rn]$7s^1$
Term symbol: $^2S_{1/2}$
Electron affinity ($M \rightarrow M^-$)/kJ mol^{-1}**:** 44 (calc.)

Ionization energies/kJ mol^{-1}**:**

1.	$M \rightarrow M^+$	400
2.	$M^+ \rightarrow M^{2+}$	(2100)
3.	$M^{2+} \rightarrow M^{3+}$	(3100)
4.	$M^{3+} \rightarrow M^{4+}$	(4100)
5.	$M^{4+} \rightarrow M^{5+}$	(5700)
6.	$M^{5+} \rightarrow M^{6+}$	(6900)
7.	$M^{6+} \rightarrow M^{7+}$	(8100)
8.	$M^{7+} \rightarrow M^{8+}$	(12 300)
9.	$M^{8+} \rightarrow M^{9+}$	(12 800)
10.	$M^{9+} \rightarrow M^{10+}$	(29 300)

Electron binding energies/eV

K	1s	101 137
L_I	2s	18 639
L_{II}	$2p_{1/2}$	17 907
L_{III}	$2p_{3/2}$	15 031
M_I	3s	4652
M_{II}	$3p_{1/2}$	4327
M_{III}	$3p_{3/2}$	3663
M_{IV}	$3d_{3/2}$	3136
M_V	$3d_{5/2}$	3000

continued in Appendix 2, p255

Main lines in atomic spectrum
[Wavelength/nm(species)]
717.7 (I)

• CRYSTAL DATA

Crystal structure (cell dimensions/pm), space group
n.a.

X-ray diffraction: mass absorption coefficients (μ/ρ)/cm^2 g^{-1}**:** CuK_α n.a. MoK_α n.a.
Neutron scattering length, $b/10^{-12}$ cm**:** n.a.
Thermal neutron capture cross-section, σ_a/barns**:** n.a.

• GEOLOGICAL DATA

Minerals

Francium occurs naturally in uranium minerals, where ^{223}Fr occurs as the radioactive decay product of actinium isotope ^{227}Ac. There is probably less than 30 g of francium in the Earth's crust at any one time. Francium has been made by the neutron bombardment of radium, or the proton bombardment of thorium. No weighable quantities of francium have ever been made.

Specimen: not commercially available.

Abundances

Sun (relative to H = 1×10^{12})**:** n.a.
Earth's crust/p.p.m.**:** traces
Seawater/p.p.m.**:** nil

Gd

Atomic number: 64
Relative atomic mass ($^{12}C = 12.0000$)**:** 157.25

CAS: [7440-54-2]

• CHEMICAL DATA

Discovery: Gadolinium was discovered in 1880 by J.C. Galissard de Marignac at Geneva, Switzerland. Isolated in 1886 by P.E. Lecoq de Boisbaudren at Paris, France.
Description: Gadolinium is a soft, silvery metal of the so-called rare earth group (more correctly termed the lanthanides). It reacts slowly with oxygen and water, and dissolves in acids. Gadolinium is used in magnets, electronics, refractories, neutron radiography, and alloyed with iron, for magneto-optic recording devices.

Radii/pm: Gd^{3+} 97; atomic 180; covalent 161
Electronegativity: 1.20 (Pauling); 1.11 (Allred); ≤ 3.3 eV (absolute)
Effective nuclear charge: 2.85 (Slater); 8.22 (Clementi); 11.28 (Froese-Fischer)

Standard reduction potentials $E^{\ominus}$/V

	III		0
acid	Gd^{3+}	−2.28	Gd
base	$Gd(OH)_3$	−2.82	Gd

Oxidation states

Gd^{II}	f^8	GdI_2
Gd^{III}	f^7	Gd_2O_3, $Gd(OH)_3$, $[Gd(OH_2)_x]^{3+}$ (aq), Gd^{3+} salts, GdF_3, $GdCl_3$ etc., complexes

• PHYSICAL DATA

Melting point/K: 1586
Boiling point/K: 3539

ΔH_{fusion}/kJ mol^{-1}: 15.5
ΔH_{vap}/kJ mol^{-1}: 311.7

Thermodynamic properties (298.15 K, 0.1 MPa)

State	$\Delta_f H^{\ominus}$/kJ mol^{-1}	$\Delta_f G^{\ominus}$/kJ mol^{-1}	$S^{\ominus}$/J K^{-1} mol^{-1}	C_p/J K^{-1} mol^{-1}
Solid	0	0	68.07	37.03
Gas	397.5	359.8	193.314	27.547

Density/kg m^{-3}: 7900.4 [298 K]
Molar volume/cm^3: 19.90
Thermal conductivity/W m^{-1} K^{-1}: 10.6 [300 K]
Coefficient of linear thermal expansion/K^{-1}: 8.6×10^{-6}
Electrical resistivity/Ω m: 134.0×10^{-8} [298 K]
Mass magnetic susceptibility/kg^{-1} m^3: $+6.030 \times 10^{-5}$ (s)

Young's modulus/GPa: 54.8
Rigidity modulus/GPa: 21.8
Bulk modulus/GPa: 37.9
Poisson's ratio/GPa: 0.259

• BIOLOGICAL DATA

Biological role

None, but acts to stimulate metabolism.

Toxicity

Toxic intake: n.a.
Lethal intake: LD_{50} (chloride, oral, mouse) = > 2000 mg kg^{-1}

Hazards

Gadolinium is mildly toxic by ingestion, but is a skin and eye irritant and a suspected tumorigen.

Levels in humans

Organs: n.a., but very low
Daily dietary intake: n.a.
Total mass of element in average (70 kg) person: n.a., but very low

Discovery: see Chemical Data section.
[Named after J. Gadolin, a Finnish minerologist]
French, *gadolinium*; German, *Gadolinium*; Italian, *gadolinio*; Spanish, *gadolinio*

Gadolinium

[gad-oh-lin-iuhm]

• NUCLEAR DATA

Number of isotopes (including nuclear isomers): 23

Isotope mass range: 143m → 163

Key isotopes

Nuclide	Atomic mass	Natural abundance (%)	Nuclear spin I	Nuclear magnetic moment μ	Uses
^{152}Gd*	151.919 786	0.20	0+		E
^{154}Gd	153.920 861	2.18	0+		E
^{155}Gd	154.922 618	14.80	3/2–	–0.259 1	E, NMR
^{156}Gd	155.922 118	20.47	0+		E
^{157}Gd	156.923 956	15.65	3/2–	–0.339 9	E, NMR
^{158}Gd	157.924 019	24.84	0+		E
^{160}Gd	159.927 049	21.86	0+		E

* ^{152}Gd is radioactive with a half-life of 1.1×10^{14} y and decay mode α (2.24 MeV).
A table of radioactive isotopes is given in Appendix 1, on p240.

NMR [Reference: not recorded]	^{155}Gd	^{157}Gd
Relative sensitivity (^{1}H = 1.00)	2.79×10^{-4}	5.44×10^{-4}
Receptivity (^{13}C = 1.00)	0.124	0.292
Magnetogyric ratio/rad $T^{-1}s^{-1}$	-0.8273×10^7	-1.0792×10^7
Nuclear quadrupole moment/m^2	$+1.270 \times 10^{-28}$	$+1.350 \times 10^{-28}$
Frequency (^{1}H = 100 Hz; 2.3488T)/MHz	3.819	4.774

• ELECTRON SHELL DATA

Ground state electron configuration: [Xe]$4f^7 5d^1 6s^2$
Term symbol: $^9D_2$
Electron affinity ($M \rightarrow M^-$)/kJ mol^{-1}**:** ≤ 50

Ionization energies/kJ mol^{-1}**:**

1.	$M \rightarrow M^+$	592.5
2.	$M^+ \rightarrow M^{2+}$	1167
3.	$M^{2+} \rightarrow M^{3+}$	1990
4.	$M^{3+} \rightarrow M^{4+}$	4250
5.	$M^{4+} \rightarrow M^{5+}$	6249

Electron binding energies/eV

K	1s	50 239
L_I	2s	8376
L_{II}	$2p_{1/2}$	7930
L_{III}	$2p_{3/2}$	7243
M_I	3s	1881
M_{II}	$3p_{1/2}$	1688
M_{III}	$3p_{3/2}$	1544
M_{IV}	$3d_{3/2}$	1221.9
M_V	$3d_{5/2}$	1189.6

continued in Appendix 2, p255

Main lines in atomic spectrum
[Wavelength/nm(species)]
342.247 (II)
364.619 (II)
368.413 (I)
376.839 (II)
368.305 (I)
407.870 (I) (AA)

• CRYSTAL DATA

Crystal structure (cell dimensions/pm), space group
α-Gd h.c.p. (a = 363.60, c = 578.26), $P6_3/mmc$
β-Gd b.c.c. (a = 405), Im3m
$T(\alpha \rightarrow \beta)$ = 1535 K
High pressure form: (a = 361, c = 2603), $R\bar{3}m$
X-ray diffraction: mass absorption coefficients (μ/ρ)/cm^2 g^{-1}**:** CuK_α 439 MoK_α 64.4
Neutron scattering length, $b/10^{-12}$ cm**:** 0.65
Thermal neutron capture cross-section, σ_a/barns**:** 49 000

• GEOLOGICAL DATA

Minerals

Mineral	Formula	Density	Hardness	Crystal appearance
Bastnäsite*	(Ce,La, etc.)CO_3F	4.9	4 – 4.5	hex., vit/greasy yellow
Monazite*	(Ce, La, Nd, Th, etc.)PO_4	5.20	5 – 5.5	mon., waxy/vit. yellow-brown

*Although not a major constituent, gadolinium is present in extractable amounts.

Chief ores: monazite, bastnäsite
World production/tonnes y^{-1}**:** *c.* 400
Main mining areas: USA, Brazil, India, Sri Lanka, Australia, China.
Reserves/tonnes: *c.* 2×10^6
Specimen: available as chips, foil or ingots. Safe.

Abundances

Sun (relative to H = 1×10^{12})**:** 13.2
Earth's crust/p.p.m.**:** 7.7
Seawater/p.p.m.**:**
Atlantic surface: 5.2×10^{-7}
Atlantic deep: 9.3×10^{-7}
Pacific surface: 6.0×10^{-7}
Pacific deep: 15×10^{-7}
Residence time/years**:** 300
Classification: recycled
Oxidation state: III

Ga

Atomic number: 31
Relative atomic mass ($^{12}C = 12.0000$)**:** 69.723

CAS: [7440-55-3]

• CHEMICAL DATA

Description: Gallium is a soft, silvery-white metal, and has the longest liquid range of all the elements. It is stable in air and with water; it dissolves in acids and alkalis. Gallium has semiconductor properties, especially as gallium arsenide. It is used in light-emitting diodes and microwave equipment.

Radii/pm: Ga^{3+} 62; Ga^{+} 113; atomic 122; covalent 125
Electronegativity: 1.81 (Pauling); 1.82 (Allred); 3.2 eV (absolute)
Effective nuclear charge: 5.00 (Slater); 6.22 (Clementi); 6.72 (Froese-Fischer)

Standard reduction potentials $E^{\ominus}$/V

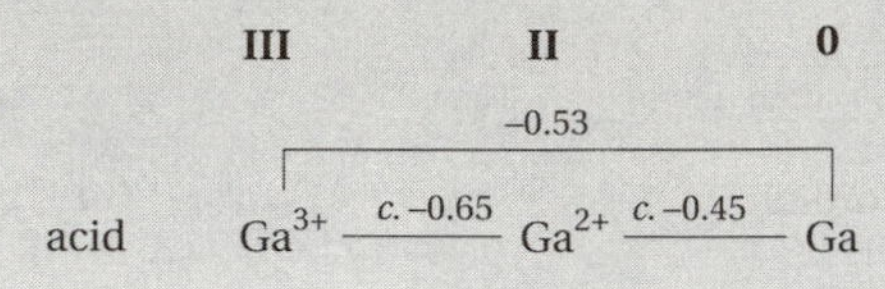

Oxidation states

Ga^{I}	s^2	Ga_2O, $GaCl_2$ etc., (Ga_2Cl_4 is $Ga^{I}[Ga^{III}Cl_4]$)
Ga^{II}	s^1	$[Ga_2Cl_6]^{2-}$
Ga^{III}	d^{10}	Ga_2O_3, $Ga(OH)_3$, $[Ga(OH_2)_6]^{3+}$ (aq), GaF_3, Ga_2Cl_6, $[GaCl_6]^{3-}$

Covalent bonds

Bond*	r/ pm	E/ kJ mol^{-1}
Ga—F	188	469
Ga—Cl	220	354
Ga—Br	235	302
Ga—I	257	327
Ga—Ga	244	113

*Ga^{I} except for GaF_3

• PHYSICAL DATA

Melting point/K: 302.93
Boiling point/K: 2676
ΔH_{fusion}/kJ mol^{-1}: 5.59
ΔH_{vap}/kJ mol^{-1}: 256.1

Thermodynamic properties (298.15 K, 0.1 MPa)

State	$\Delta_f H^{\ominus}$/kJ mol^{-1}	$\Delta_f G^{\ominus}$/kJ mol^{-1}	$S^{\ominus}$/J K^{-1} mol^{-1}	C_p/J K^{-1} mol^{-1}
Solid	0	0	40.88	25.86
Gas	277.0	238.9	169.06	25.36

Density/kg m^{-3}: 5907 [293 K]; 6113.6 [liquid at m.p.]
Molar volume/cm^3: 11.81
Thermal conductivity/W m^{-1} K^{-1}: 40.6 [300 K]
Coefficient of linear thermal expansion/K^{-1}: 11.5×10^{-6} (*a* axis); 31.5×10^{-6} (*b* axis); 16.5×10^{-6} (*c* axis)
Electrical resistivity/Ω m: 27×10^{-8} [273 K] varies with axis
Mass magnetic susceptibility/kg^{-1} m^3: -3.9×10^{-8} (s)

Young's modulus/GPa: 9.81
Rigidity modulus/GPa: 6.67
Bulk modulus/GPa: n.a.
Poisson's ratio/GPa: 0.47

• BIOLOGICAL DATA

Biological role

None, but gallium acts to stimulate metabolism.

Toxicity

Toxic intake: < 15 mg kg^{-1} tolerated without apparent harm

Lethal intake: LD_{50} (chloride, injection, rat) = 47 mg kg^{-1}

Hazards

Gallium salts generally have low toxicity.

Levels in humans

Blood/mg dm^{-3}: < 0.08
Bone/p.p.m.: n.a.
Liver/p.p.m.: 0.0025
Muscle/p.p.m.: 0.0014
Daily dietary intake: n.a. but low
Total mass of element in average (70 kg) person: < 0.7 mg

Discovered in 1875 by Paul-Émile Lecoq de Boisbaudran at Paris, France.
[Latin, *Gallia* = France]
French, *gallium*; German, *Gallium*; Italian, *gallio*; Spanish, *galio*

Gallium

[gal-iuhm]

• NUCLEAR DATA

Number of isotopes (including nuclear isomers): 23 **Isotope mass range:** 62 → 83

Key isotopes

Nuclide	Atomic mass	Natural abundance (%)	Nuclear spin *I*	Nuclear magnetic moment μ	Uses
^{69}Ga	68.925 580	60.1	3/2–	+2.016 59	E, NMR
^{71}Ga	70.924 700 5	39.9	3/2–	+2.562 27	E, NMR

A table of radioactive isotopes is given in Appendix 1, on p240.

NMR [Reference: Ga^{3+} (aq)]	^{69}Ga	^{71}Ga
Relative sensitivity (^1H = 1.00)	0.0691	0.14
Receptivity (^{13}C = 1.00)	237	319
Magnetogyric ratio/rad $T^{-1}s^{-1}$	6.420×10^7	8.158×10^7
Nuclear quadrupole moment/m^2	0.170×10^{-28}	0.100×10^{-28}
Frequency (^1H = 100 Hz; 2.3488T)/MHz	24.003	30.495

• ELECTRON SHELL DATA

Ground state electron configuration: [Ar]$3d^{10}4s^24p^1$
Term symbol: $^2P_{1/2}$
Electron affinity ($M \rightarrow M^-$)/kJ mol^{-1}**:** *c.* 30

Ionization energies/kJ mol^{-1}**:**

1.	$M \rightarrow M^+$	578.8
2.	$M^+ \rightarrow M^{2+}$	1979
3.	$M^{2+} \rightarrow M^{3+}$	2963
4.	$M^{3+} \rightarrow M^{4+}$	6200
5.	$M^{4+} \rightarrow M^{5+}$	(8700)
6.	$M^{5+} \rightarrow M^{6+}$	(11 400)
7.	$M^{6+} \rightarrow M^{7+}$	(14 400)
8.	$M^{7+} \rightarrow M^{8+}$	(17 700)
9.	$M^{8+} \rightarrow M^{9+}$	(22 300)
10.	$M^{9+} \rightarrow M^{10+}$	(26 100)

Electron binding energies/eV

K	1s	10 367
L_I	2s	1299.0
L_{II}	$2p_{1/2}$	1143.2
L_{III}	$2p_{3/2}$	1116.4
M_I	3s	159.5
M_{II}	$3p_{1/2}$	103.5
M_{III}	$3p_{3/2}$	100.0
M_{IV}	$3d_{3/2}$	18.7
M_V	$3d_{5/2}$	18.7

Main lines in atomic spectrum
[Wavelength/nm(species)]

287.424 (I) (AA)
294.364 (I)
403.299 (I)
417.204 (I)
639.656 (I)
641.344 (I)

• CRYSTAL DATA

Crystal structure (cell dimensions/pm), space group
α-Ga orthorhombic (*a* = 451.86, *b* = 765.70, *c* = 452.58), Cmca
β-Ga orthorhombic (*a* = 290, *b* = 813, *c* = 317), Cmcm (metastable form)
γ-Ga orthorhombic (*a* = 1060, *b* = 1356, *c* = 519), $Cmc2_1$
$T(\gamma \rightarrow \alpha)$ = 238 K
High pressure form: (*a* = 279, *c* = 438), I4/mmm

X-ray diffraction: mass absorption coefficients (μ/ρ)/cm^2 g^{-1}**:** CuK_α 67.9 MoK_α 60.1
Neutron scattering length, $b/10^{-12}$ cm**:** 0.7288
Thermal neutron capture cross-section, σ_a/barns**:** 2.9

• GEOLOGICAL DATA

Minerals

Gallium minerals are rare, but gallium occurs in other ores to the extent of 1%.

Mineral	Formula	Density	Hardness	Crystal appearance
Gallite	$CuGaS_2$	4.40 (calc.)	3 – 3.5	tet., met. grey

The ores diaspore, sphalerite, germanite and bauxite contain traces of gallium. Coal can also have a high gallium content.

Chief ores: gallium is recovered as a by-product of zinc and copper refining.
World production/tonnes y^{-1}**:** 30
Main mining areas: see copper and zinc.
Reserves/tonnes**:** n.a.
Specimen: available as ingot, and as ultrapure gallium. Safe.

Abundances

Sun (relative to H = 1×10^{12})**:** 631
Earth's crust/p.p.m.**:** 18
Seawater/p.p.m.**:** 3×10^{-5}
Residence time/years**:** 10 000
Classification: n.a.
Oxidation state: III

Ge

Atomic number: 32
Relative atomic mass ($^{12}C = 12.0000$)**:** 72.61

CAS: [7440-56-4]

• CHEMICAL DATA

Description: Ultrapure germanium is a silvery-white brittle metalloid element. It is stable in air and water, is unaffected by acids, except HNO_3, and alkalis. It is used in semiconductors, alloys and special glasses for infrared devices.

Radii/pm: Ge^{2+} 90; Ge^{4-} 272; atomic 123; covalent 122
Electronegativity: 2.01 (Pauling); 2.02 (Allred); 4.6 eV (absolute)
Effective nuclear charge: 5.65 (Slater); 6.78 (Clementi); 7.92 (Froese-Fischer)

Standard reduction potentials $E^\ominus$/V

	IV		II		0		–IV
acid	GeO_2	–0.370	GeO	0.255	Ge	–0.29	GeH_4
	Ge^{4+}	0.00	Ge^{2+}	–0.247 (to Ge)			

[basic solutions contain many different forms]

Oxidation states

Ge^{II}	s^2	GeO, GeS, GeF_2, $GeCl_2$ etc.
Ge^{IV}	d^{10}	GeO_2, GeH_4 etc., GeF_4, $GeCl_4$ etc., $[GeF_6]^{2-}$, $[GeCl_6]^{2-}$, GeS_2, $[Ge(OH)_3O]^-$ (aq)

Covalent bonds

Bond	r/ pm	E/ kJ mol^{-1}
Ge—H	153	288
Ge—C	194	237
Ge—O	165	363
Ge—F	168	452
Ge—Cl	210	188
Ge—Br	230	276
Ge—Ge	241	188

• PHYSICAL DATA

Melting point/K: 1210.6
Boiling point/K: 3103
ΔH_{fusion}/kJ mol^{-1}: 34.7
ΔH_{vap}/kJ mol^{-1}: 334.3

Thermodynamic properties (298.15 K, 0.1 MPa)

State	$\Delta_f H^\ominus$/kJ mol^{-1}	$\Delta_f G^\ominus$/kJ mol^{-1}	$S^\ominus$/J K^{-1} mol^{-1}	C_p/J K^{-1} mol^{-1}
Solid	0	0	31.09	23.347
Gas	376.6	335.9	167.900	30.731

Density/kg m^{-3}: 5323 [293 K]; 5490 [liquid at m.p.]
Molar volume/cm^3: 13.64
Thermal conductivity/W m^{-1} K^{-1}: 59.9 [300 K]
Coefficient of linear thermal expansion/K^{-1}: 5.57×10^{-6}
Electrical resistivity/Ω m: 0.46 [295 K]
Mass magnetic susceptibility/kg^{-1} m^3: -1.328×10^{-9} (s)

Young's modulus/GPa: 79.9
Rigidity modulus/GPa: 29.6
Bulk modulus/GPa: n.a.
Poisson's ratio/GPa: 0.32

• BIOLOGICAL DATA

Biological role

None, but germanium acts to stimulate metabolism.

Toxicity

Toxic intake: germanium salts generally have low toxicity

Lethal intake: LD_{50} (various, ingestion, rats etc.) = 500 – 5000 mg kg^{-1}

Hazards

The fumes of $GeCl_4$ liquid can irritate the eyes and lungs.

Levels in humans

Blood/mg dm^{-3}: *c.* 0.44
Bone/p.p.m.: n.a.
Liver/p.p.m.: 0.15
Muscle/p.p.m.: 0.14
Daily dietary intake: 0.4 – 1.5 mg
Total mass of element in average (70 kg) person: 5 mg

Discovered in 1886 by C.A. Winkler at Freiberg, Germany.
[Latin, *Germania* = Germany]
French, *germanium*; German, *Germanium*; Italian, *germanio*; Spanish, *germanio*

Germanium

[jer-may-niuhm]

• NUCLEAR DATA

Number of isotopes (including nuclear isomers): 24

Isotope mass range: 64 → 83

Key isotopes

Nuclide	Atomic mass	Natural abundance (%)	Nuclear spin *I*	Nuclear magnetic moment μ	Uses
^{70}Ge	69.924 249 7	20.5	0+		E
^{72}Ge	71.922 078 9	27.4	0+		E
^{73}Ge	72.923 462 6	7.8	9/2+	–0.879 466 9	E, NMR
^{74}Ge	73.921 177 4	36.5	0+		E
^{76}Ge	75.921 401 6	7.8	0+		E

A table of radioactive isotopes is given in Appendix 1, on p241.

NMR [Reference: $Ge(CH_3)_4$]

	^{73}Ge
Relative sensitivity (^{1}H = 1.00)	1.4×10^{-3}
Receptivity (^{13}C = 1.00)	0.617
Magnetogyric ratio/rad $T^{-1}s^{-1}$	-0.9331×10^{7}
Nuclear quadrupole moment/m^2	-0.173×10^{-28}
Frequency (^{1}H = 100 Hz; 2.3488T)/MHz	3.488

• ELECTRON SHELL DATA

Ground state electron configuration: $[Ar]3d^{10}4s^24p^2$
Term symbol: $^3P_0$
Electron affinity ($M \rightarrow M^-$)/kJ mol^{-1}**:** 116

Ionization energies/kJ mol^{-1}**:**

1.	$M \rightarrow M^+$	762.1
2.	$M^+ \rightarrow M^{2+}$	1537
3.	$M^{2+} \rightarrow M^{3+}$	3302
4.	$M^{3+} \rightarrow M^{4+}$	4410
5.	$M^{4+} \rightarrow M^{5+}$	9020
6.	$M^{5+} \rightarrow M^{6+}$	(11 900)
7.	$M^{6+} \rightarrow M^{7+}$	(15 000)
8.	$M^{7+} \rightarrow M^{8+}$	(18 200)
9.	$M^{8+} \rightarrow M^{9+}$	(21 800)
10.	$M^{9+} \rightarrow M^{10+}$	(27 000)

Electron binding energies/eV

K	1s	11103
L_I	2s	1414.6
L_{II}	$2p_{1/2}$	1248.1
L_{III}	$2p_{3/2}$	1217.0
M_I	3s	180.1
M_{II}	$3p_{1/2}$	124.9
M_{III}	$3p_{3/2}$	120.8
M_{IV}	$3d_{3/2}$	29.8
M_V	$3d_{5/2}$	29.2

Main lines in atomic spectrum
[Wavelength/nm(species)]

204.171 (I)
206.866 (I)
209.426 (I)
259.253 (I)
265.117 (I)
265.157 (I) (AA)

• CRYSTAL DATA

Crystal structure (cell dimensions/pm), space group
Cubic (*a* = 565.754), Fd3m, diamond structure
High pressure forms: (*a* = 488.4, *c* = 269.2), $I4_1$/amd; (*a* = 593, *c* = 698), $P4_32_12$; (*a* = 692), b.c.c.

X-ray diffraction: mass absorption coefficients (μ/ρ)/cm^2 g^{-1}**:** CuK_α 75.6 MoK_α 64.8
Neutron scattering length, $b/10^{-12}$ cm**:** 0.8193
Thermal neutron capture cross-section, σ_a/barns**:** 2.2

• GEOLOGICAL DATA

Minerals

Mineral	Formula	Density	Hardness	Crystal appearance
Germanite	$Cu_{26}Fe_4Ge_4S_{32}$	4.46	4	cub, met. pale greyish-pink

Chief ores: not mined as such; widely distributed in other minerals and Ge is recovered as a by-product of zinc and copper refining.

World production/tonnes y^{-1}**:** 80

Main mining areas: see zinc and copper.

Reserves/tonnes**:** n.a.

Specimen: available as chips, pieces or powder. Safe.

Abundances

Sun (relative to H = 1×10^{12})**:** 3160
Earth's crust/p.p.m.**:** 1.8
Seawater/p.p.m.**:**
Atlantic surface: 0.07×10^{-6}
Atlantic deep: 0.14×10^{-6}
Pacific surface: 0.35×10^{-6}
Pacific deep: 7.00×10^{-6}
Residence time/years**:** 20 000
Classification: recycled
Oxidation state: IV

Au

Atomic number: 79
Relative atomic mass ($^{12}C = 12.0000$)**:** **196.96654**

CAS: [7440-57-5]

• CHEMICAL DATA

Description: Gold is a soft metal with a characteristic shiny yellow colour. It has the highest malleability and ductility of any element, and can be beaten into a film only microns thick. Gold is unaffected by air, water, acids (except aqua regia, HNO_3-HCl) and alkalis. It is used as bullion, in jewellery, electronics and glass, to colour it and as a heat reflector.

Radii/pm: Au^{3+} 91; Au^{+} 137; atomic 144; covalent 134
Electronegativity: 2.54 (Pauling); 1.42 (Allred); 5.77 eV (absolute)
Effective nuclear charge: 4.20 (Slater); 10.94 (Clementi); 15.94 (Froese-Fischer)

Standard reduction potentials $E^{\ominus}$/V

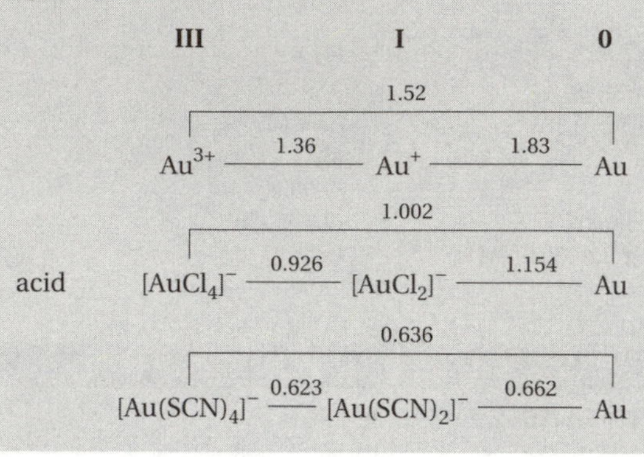

Oxidation states

Au^{-I}	$d^{10}s^2$	$[Au(NH_3)_n]^-$ in liquid ammonia
Au^{0}	$d^{10}s^1$	Gold clusters e.g. $[Au_8(PPh_3)_8]^{2+}$
Au^{I}	d^{10}	Au_2S, $[Au(CN)_2]^-$ and other complexes
Au^{II}	d^9	Rare but some complexes known
Au^{III}	d^8	Au_2O_3, $[Au(OH)_4]^-$ (aq), $[AuCl_4]^-$ (aq), $[AuCl_3(OH)]^-$ (aq), Au_2S_3, AuF_3, Au_2Cl_6, $AuBr_3$, complexes
Au^{V}	d^6	AuF_5
Au^{VII}	d^4	AuF_7

• PHYSICAL DATA

Melting point/K: 1337.58
Boiling point/K: 3080
ΔH_{fusion}/kJ mol^{-1}: 12.7
ΔH_{vap}/kJ mol^{-1}: 324.4

Thermodynamic properties (298.15 K, 0.1 MPa)

State	$\Delta_f H^{\ominus}$/kJ mol^{-1}	$\Delta_f G^{\ominus}$/kJ mol^{-1}	$S^{\ominus}$/J K^{-1} mol^{-1}	C_p/J K^{-1} mol^{-1}
Solid	0	0	47.40	25.418
Gas	336.1	326.3	180.503	20.786

Density/kg m^{-3}: 19 320 [293 K]; 17 280 [liquid at m.p.]
Molar volume/cm^3: 10.19
Thermal conductivity/W m^{-1} K^{-1}: 317 [300 K]
Coefficient of linear thermal expansion/K^{-1}: 14.16×10^{-6}
Electrical resistivity/Ω m: 2.35×10^{-8} [293 K]
Mass magnetic susceptibility/kg^{-1} m^3: -1.78×10^{-9} (s)

Young's modulus/GPa: 78.5
Rigidity modulus/GPa: 26.0
Bulk modulus/GPa: 171
Poisson's ratio/GPa: 0.42

• BIOLOGICAL DATA

Biological role

None, but acts to stimulate metabolism.

Toxicity

Toxic intake: gold metal and gold salts generally have low toxicity
Lethal intake: n.a.

Hazards

Gold is poorly absorbed by the body and poisoning by gold compounds is very rare. Gold-based anti-arthritics can cause liver damage and kidney damage.

Levels in humans

Blood/mg dm^{-3}: $(0.1-4.2) \times 10^{-4}$
Bone/p.p.m.: 0.016
Liver/p.p.m.: 0.0004
Muscle/p.p.m.: n.a.
Daily dietary intake: n.a., but very low
Total mass of element in average (70 kg) person: 0.2 mg

Known to ancient civilizations.
[Anglo-Saxon, *gold*; Latin, *aurum*]
French, *or*; German, *Gold*; Italian, *oro*; Spanish, *oro*

Gold

[gowld]

• NUCLEAR DATA

Number of isotopes (including nuclear isomers): 39 **Isotope mass range:** 176 → 204

Key isotopes

Nuclide	Atomic mass	Natural abundance (%)	Nuclear spin I	Nuclear magnetic moment μ	Uses
^{197}Au	196.966 543	100	3/2+	+0.148 159	NMR

A table of radioactive isotopes is given in Appendix 1, on p241.

NMR [Difficult to detect]	^{197}Au
Relative sensitivity (^{1}H = 1.00)	2.51×10^{-5}
Receptivity (^{13}C = 1.00)	0.06
Magnetogyric ratio/rad $T^{-1}s^{-1}$	0.357×10^{7}
Nuclear quadrupole moment/m^2	0.547×10^{-28}
Frequency (^{1}H = 100 Hz; 2.3488T)/MHz	1.712

• ELECTRON SHELL DATA

Ground state electron configuration: [Xe]$4f^{14}5d^{10}6s^1$
Term symbol: $^2S_{1/2}$
Electron affinity ($M \rightarrow M^-$)/kJ mol^{-1}**:** 222.8

Ionization energies/kJ mol^{-1}**:**

1.	$M \rightarrow M^+$	890.0
2.	$M^+ \rightarrow M^{2+}$	1980
3.	$M^{2+} \rightarrow M^{3+}$	(2900)
4.	$M^{3+} \rightarrow M^{4+}$	(4200)
5.	$M^{4+} \rightarrow M^{5+}$	(5600)
6.	$M^{5+} \rightarrow M^{6+}$	(7000)
7.	$M^{6+} \rightarrow M^{7+}$	(9300)
8.	$M^{7+} \rightarrow M^{8+}$	(11 000)
9.	$M^{8+} \rightarrow M^{9+}$	(12 800)
10.	$M^{9+} \rightarrow M^{10+}$	(14 800)

Electron binding energies/eV

K	1s	80 725
L_I	2s	14 353
L_{II}	$2p_{1/2}$	13 734
L_{III}	$2p_{3/2}$	11 919
M_I	3s	3425
M_{II}	$3p_{1/2}$	3148
M_{III}	$3p_{3/2}$	2743
M_{IV}	$3d_{3/2}$	2291
M_V	$3d_{5/2}$	2206

continued in Appendix 2, p256

Main lines in atomic spectrum
[Wavelength/nm(species)]
201.200 (I)
202.138 (I)
242.795 (I) (AA)
267.595 (I)
274.825 (I)
312.278 (I)

• CRYSTAL DATA

Crystal structure (cell dimensions/pm), space group
f.c.c. (a = 407.833), Fm3m

X-ray diffraction: mass absorption coefficients (μ/ρ)/cm^2 g^{-1}**:** CuK$_\alpha$ 208 MoK$_\alpha$ 115
Neutron scattering length, $b/10^{-12}$ cm**:** 0.763
Thermal neutron capture cross-section, σ_a/barns**:** 98.7

• GEOLOGICAL DATA

Minerals

Gold occurs mainly as the metal, occasionally as crystals, but more generally as grains, sheets and flakes in other rocks.

Mineral	Formula	Density	Hardness	Crystal appearance
Gold	Au	19.3	2.5 – 3	cub., met. white
Sylvanite	$AgAuTe_4$	8.16	1.5 – 2	hex., met. white

Chief ores: quartz veins in extrusive rocks
World production/tonnes y^{-1}**:** *c.* 1400
Main mining areas: South Africa, USA, Canada, Russia.
Reserves/tonnes**:** 15 000
Specimen: available as foil, powder, rod, shot, sponge or wire. Safe.

Abundances

Sun (relative to H = 1×10^{12})**:** 5.6
Earth's crust/p.p.m.**:** 0.0011
Seawater/p.p.m.**:** 1×10^{-5}
Residence time/years**:** n.a.
Classification: n.a.
Oxidation state: I

Hf

Atomic number: 72
Relative atomic mass ($^{12}C = 12.0000$)**:** 178.49

CAS: [7440-58-6]

• CHEMICAL DATA

Description: Hafnium is a lustrous, silvery, ductile metal that resists corrosion due to an oxide film on its surface. However, powdered hafnium will burn in air. The metal is unaffected by acids (except HF) and alkalis. It is used in control rods for nuclear reactors, and in high temperature alloys and ceramics.

Radii/pm: Hf^{3+} 84; atomic 156; covalent 144
Electronegativity: 1.3 (Pauling); 1.23 (Allred); 3.8 eV (absolute)
Effective nuclear charge: 3.15 (Slater); 9.16 (Clementi); 13.27 (Froese-Fischer)

Standard reduction potentials $E^{\ominus}$/V

IV		0
Hf^{4+}	−1.70	Hf
HfO_2	−1.57	Hf

Oxidation states

Hf^{I}	d^3	HfCl ?
Hf^{II}	d^2	$HfCl_2$?
Hf^{III}	d^1	$HfCl_3$, $HfBr_3$, HfI_3, Hf^{3+} reduces water
Hf^{IV}	$d^0[f^{14}]$	HfO_2, $Hf(OH)^{3+}$ (aq), HfF_4, $HfCl_4$ etc., $[HfF_6]^{2-}$, $[HfF_7]^{3-}$, $[HfF_8]^{4-}$

• PHYSICAL DATA

Melting point/K: 2503
Boiling point/K: 5470
ΔH_{fusion}/kJ mol^{-1}: 25.5
ΔH_{vap}/kJ mol^{-1}: 661.1

Thermodynamic properties (298.15 K, 0.1 MPa)

State	$\Delta_f H^{\ominus}$/kJ mol^{-1}	$\Delta_f G^{\ominus}$/kJ mol^{-1}	$S^{\ominus}$/J K^{-1} mol^{-1}	C_p/J K^{-1} mol^{-1}
Solid	0	0	43.56	25.73
Gas	619.2	576.5	186.892	20.803

Density/kg m^{-3}: 13 310 [293 K]; 12 000 [liquid at m.p.]
Molar volume/cm^3: 13.41
Thermal conductivity/W m^{-1} K^{-1}: 23.0 [300 K]
Coefficient of linear thermal expansion/K^{-1}: 5.9×10^{-6}
Electrical resistivity/Ω m: 35.1×10^{-8} [293 K]
Mass magnetic susceptibility/kg^{-1} m^3: $+5.3 \times 10^{-9}$ (s)
Young's modulus/GPa: 141
Rigidity modulus/GPa: 56
Bulk modulus/GPa: 109
Poisson's ratio/GPa: 0.26

• BIOLOGICAL DATA

Biological role

None.

Toxicity

Toxic intake: hafnium and hafnium salts generally have low toxicity
Lethal intake: LD_{50} (chloride, oral, rat) = 2400 mg kg^{-1}

Hazards

Hafnium is poorly absorbed by the body and poisoning by hafnium compounds is very rare.

Levels in humans

Organs: n.a.
Daily dietary intake: n.a.
Total mass of element in average (70 kg) person: n.a.

Discovered in 1923 by D. Coster and G.C. von Hevesey at Copenhagen, Denmark.
[Latin, *Hafnia* = Copenhagen]
French, *hafnium*; German, *Hafnium*; Italian, *afnio*; Spanish, *hafnio*

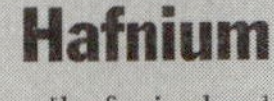

[haf-ni-uhm]

• NUCLEAR DATA

Number of isotopes (including nuclear isomers): 33 **Isotope mass range:** 158 → 184

Key isotopes

Nuclide	Atomic mass	Natural abundance (%)	Nuclear spin I	Nuclear magnetic moment μ	Uses
^{174}Hf	173.940 044	0.162	0+		E
^{176}Hf	175.941 406	5.206	0+		E
^{177}Hf	176.943 217	18.606	7/2+	+0.793 6	E
^{178}Hf	177.943 696	27.297	0+		E
^{179}Hf	178.945 812	13.629	9/2+	–0.6409	E, NMR
^{180}Hf	179.946 545	35.100	0+		E

A table of radioactive isotopes is given in Appendix 1, on p241.

NMR [Difficult to observe]	^{177}Hf	^{179}Hf
Relative sensitivity (^1H = 1.00)	6.38×10^{-4}	2.16×10^{-4}
Receptivity (^{13}C = 1.00)	0.88	0.27
Magnetogyric ratio/rad $T^{-1}s^{-1}$	$+0.945 \times 10^7$	-0.609×10^7
Nuclear quadrupole moment/m^2	$+3.365 \times 10^{-31}$	$+3.793 \times 10^{-31}$
Frequency (^1H = 100 Hz; 2.3488T)/MHz	3.120	1.869

• ELECTRON SHELL DATA

Ground state electron configuration: $[Xe]4f^{14}5d^26s^2$
Term symbol: $^3F_2$
Electron affinity ($M \rightarrow M^-$)/kJ mol^{-1}**:** *c.* 0

Ionization energies/kJ mol^{-1}**:**

1.	$M \rightarrow M^+$	642
2.	$M^+ \rightarrow M^{2+}$	1440
3.	$M^{2+} \rightarrow M^{3+}$	2250
4.	$M^{3+} \rightarrow M^{4+}$	3216
5.	$M^{4+} \rightarrow M^{5+}$	6596

Electron binding energies/eV

K	1s	65 351
L_I	2s	11 271
L_{II}	$2p_{1/2}$	10 739
L_{III}	$2p_{3/2}$	9561
M_I	3s	2601
M_{II}	$3p_{1/2}$	2365
M_{III}	$3p_{3/2}$	2107
M_{IV}	$3d_{3/2}$	1716
M_V	$3d_{5/2}$	1662

continued in Appendix 2, p256

Main lines in atomic spectrum
[Wavelength/nm(species)]

201.278 (II)
202.818 (II)
286.637 (I)
289.826 (I)
307.288 (I) (AA)
329.980 (II)
368.224 (I)

• CRYSTAL DATA

Crystal structure (cell dimensions/pm), space group
α-Hf h.c.p. (a = 319.46, c = 505.10), $P6_3/mmc$
β-Hf cubic (a = 362)
$T(\alpha \rightarrow \beta)$ = 2033 K

X-ray diffraction: mass absorption coefficients (μ/ρ)/cm^2 g^{-1}**:** CuK_α 159 MoK_α 91.7
Neutron scattering length, $b/10^{-12}$ cm**:** 0.777
Thermal neutron capture cross-section, σ_a/barns**:** 104

• GEOLOGICAL DATA

Minerals

Extremely rare. Hafnium generally occurs as a 1-5% impurity in zirconium minerals.

Mineral	Formula	Density	Hardness	Crystal appearance
Hafnon	$HfSiO_4$	6.97	n.a.	tet.

Chief source: hafnium is obtained as a by-product of zirconium refining
World production/tonnes y^{-1}**:** *c.* 50
Main mining areas: see zirconium.
Reserves/tonnes**:** n.a.
Specimen: available as foil, pieces, powder, sponge or wire. Safe.

Abundances

Sun (relative to H = 1×10^{12})**:** 6
Earth's crust/p.p.m.**:** 5.3
Seawater/p.p.m.**:** 7×10^{-6}
Residence time/years**:** n.a.
Classification: n.a.
Oxidation state: IV

Hs

Atomic number: 108
Relative atomic mass ($^{12}C = 12.0000$)**:** (265)

CAS: [54037-57-9]

• CHEMICAL DATA

Description: Hassium is a radioactive metal which does not occur naturally, and is of research interest only.

Radii/pm: Hs^{4+} 80 (est.); atomic 126 (est.)
Electronegativity: n.a.
Effective nuclear charge: n.a.

Standard reduction potentials $E^{\ominus}/V$

IV		0
Hs^{4+}	$\xrightarrow{+0.4\ (est.)}$	Hs

Oxidation states

Hs^{I}	d^6	predicted
Hs^{II}	d^5	predicted
Hs^{III}	d^4	predicted, most stable?
Hs^{IV}	d^3	predicted
Hs^{V}	d^2	predicted
Hs^{VI}	d^1	predicted
Hs^{VII}	$[f^{14}]$	predicted

• PHYSICAL DATA

Melting point/K: n.a.
Boiling point/K: n.a.

ΔH_{fusion}/kJ mol^{-1}: n.a.
ΔH_{vap}/kJ mol^{-1}: n.a.
ΔH_{subl}/kJ mol^{-1}: 628 (est.)

Thermodynamic properties (298.15 K, 0.1 MPa)

State	$\Delta_f H^{\ominus}$/kJ mol^{-1}	$\Delta_f G^{\ominus}$/kJ mol^{-1}	$S^{\ominus}$/J K^{-1} mol^{-1}	C_p/J K^{-1} mol^{-1}
Solid	0	0	n.a.	n.a.
Gas	n.a.	n.a.	n.a.	n.a.

Density/kg m^{-3}: 41 000 (est.)
Molar volume/cm^3: n.a.
Thermal conductivity/W m^{-1} K^{-1}: n.a.
Coefficient of linear thermal expansion/K^{-1}: n.a.
Electrical resistivity/Ω m: n.a.
Mass magnetic susceptibility/kg^{-1} m^3: n.a.

• BIOLOGICAL DATA

Biological role

None.

Toxicity

Toxic intake: n.a.
Lethal intake: n.a.

Hazards

Hassium is never encountered normally, and only a few atoms have ever been made. It would be dangerous because of its intense radioactivity.

Levels in humans

nil
Daily dietary intake: nil
Total mass of element in average (70 kg) person: nil

Discovery: see Nuclear Data section.
[Latin *Hassias* = Hesse, the German state]
French, *hassium*; German, *Hassium*; Italian, *hassio*; Spanish, *hassio*

Hassium

[hass-iuhm]

• NUCLEAR DATA

Discovery: Hassium was first made in 1984 by Peter Armbruster, Gottfried Münzenberg and their co-workers at Gesellschaft für Schwerionenforschung in Darmstadt, Germany.

Number of isotopes (including nuclear isomers): 3 **Isotope mass range:** 263 → 265

Key isotopes

Nuclide	Atomic mass	Half life ($T_{1/2}$)	Decay mode and energy (MeV)	Nuclear spin I	Nucl. mag. moment μ	Uses
^{264}Hs	264.129	*c.* 8×10^{-5} s	α, SF			
^{265}Hs	264.129	*c.* 2×10^{-3} s	α			

NMR [Not recorded]

• ELECTRON SHELL DATA

Ground state electron configuration: $[Rn]5f^{14}6d^{6}7s^{2}$
Term symbol: $^{5}D_{4}$
Electron affinity ($M \rightarrow M^{-}$)/kJ mol^{-1}**:** n.a.

Ionization energies/kJ mol^{-1}**:**
1. M → M^{+} 750 (est.)

Electron binding energies/eV
n.a.

Main lines in atomic spectrum
[Wavelength/nm(species)]
n.a.

• CRYSTAL DATA

Crystal structure (cell dimensions/pm), space group
n.a.

X-ray diffraction: mass absorption coefficients (μ/ρ)/cm^2 g^{-1}**:** CuK_{α} n.a. MoK_{α} n.a.
Neutron scattering length, $b/10^{-12}$ cm**:** n.a.
Thermal neutron capture cross-section, σ_a/barns**:** n.a.

• GEOLOGICAL DATA

Minerals
Not found on Earth.

Chief source: hassium has been made by the so-called cold fusion method, in which a target of lead was bombarded with atoms of iron to give an atom of hassium: $^{208}Pb + ^{58}Fe \rightarrow ^{265}Hs + n$
Specimen: not available commercially.

Abundances
Sun (relative to H = 1×10^{12})**:** n.a.
Earth's crust/p.p.m.**:** nil
Seawater/p.p.m.**:** nil

He

Atomic number: 2
Relative atomic mass ($^{12}C = 12.0000$)**:** 4.002602

CAS: [7440-59-7]

• CHEMICAL DATA

Discovery: Helium was observed in the sun's spectrum during the eclipse of 1868 by Norman Lockyer and Edward Frankland. Isolated in 1895 by Sir William Ramsey at London, England and independently by P.T. Cleve and N.A. Langlet at Uppsala, Sweden.
Description: Helium is a colourless, odourless gas obtained mainly from gas wells. It is inert towards all other elements and chemicals. Helium is used in deep-sea diving, weather balloons, and as a liquid for low-temperature research instruments.

Radii/pm: atomic 128; van der Waals 122
Electronegativity: n.a. (Pauling); 5.50 (Allred); [12.3 eV (absolute) - see Key]
Effective nuclear charge: 1.70 (Slater); 1.69 (Clementi); 1.62 (Froese-Fischer)

Oxidation states

He^0 [He] only He^1 as a gas

Covalent bonds

none

• PHYSICAL DATA

Melting point/K: 0.95 (under pressure)
Boiling point/K: 4.216
Critical temperature/K: 5.25
Critical pressure/ kPa: 229

ΔH_{fusion}/kJ mol^{-1}: 0.021
ΔH_{vap}/kJ mol^{-1}: 0.082

Thermodynamic properties (298.15 K, 0.1 MPa)

State	$\Delta_f H^{\ominus}$/kJ mol^{-1}	$\Delta_f G^{\ominus}$/kJ mol^{-1}	$S^{\ominus}$/J K^{-1} mol^{-1}	C_p/J K^{-1} mol^{-1}
Gas	0	0	126.150	20.786

Density/kg m^{-3}: n.a. [s.]; 124.8 [liq. at b.p.]; 0.178 5 [gas, 273 K]
Molar volume/cm^3: 32.07 [4 K]
Thermal conductivity/W m^{-1} K^{-1}: 0.152 [300 K] (g)
Mass magnetic susceptibility/kg^{-1} m^3: -5.9×10^{-9} (g)

• BIOLOGICAL DATA

Biological role

None.

Toxicity

Non-toxic.

Hazards

Helium is a harmless gas, although it could asphyxiate if it excluded oxygen from the lungs.

Levels in humans

Blood/mg dm^{-3}: trace
Bone/p.p.m.: nil
Liver/p.p.m.: nil
Muscle/p.p.m.: nil
Daily dietary intake: n.a. but low
Total mass of element in average (70 kg) person: n.a. but very small

Discovery: see Chemical Data section.
[Greek, *helios* = sun]
French, *hélium*; German, *Helium*; Italian, *elio*; Spanish, *helio*

[heel-iuhm]

• NUCLEAR DATA

Number of isotopes (including nuclear isomers): 5 **Isotope mass range:** $3 \rightarrow 8$ except 7

Key isotopes

Nuclide	Atomic mass	Natural abundance (%)	Nuclear spin I	Nuclear magnetic moment μ	Uses
^{3}He	3.016 029 31	0.000 137	1/2	–2.127 624	NMR, tracer
^{4}He	4.002 603 24	99.999 863	0+	0	

A table of radioactive isotopes is given in Appendix 1, on p241.

NMR [No known compounds]	^{3}He
Relative sensitivity (^{1}H = 1.00)	0.44
Receptivity (^{13}C = 1.00)	0.00326
Magnetogyric ratio/rad $T^{-1}s^{-1}$	-20.378×10^7
Nuclear quadrupole moment/m^2	n.a.
Frequency (^{1}H = 100 Hz; 2.3488T)/MHz	76.178

• ELECTRON SHELL DATA

Ground state electron configuration: $1s^2$ = [He]
Term symbol: $^1S_0$
Electron affinity ($M \rightarrow M^-$)/kJ mol^{-1}**:** 0.0

Ionization energies/kJ mol^{-1}**:**

		Energy
1.	$M \rightarrow M^+$	2372.3
2.	$M^+ \rightarrow M^{2+}$	5250.4

Electron binding energies/eV

K	1s	24.6

Main lines in atomic spectrum
[Wavelength/nm(species)]

388.865 (I)
587.562 (I)
1083.025 (I)
1083.034 (I)
1868.534 (I)
2058.130 (I)

• CRYSTAL DATA

Crystal structure (cell dimensions/pm), space group
α-He h.c.p. (a = 353.1, c = 569.3), $P6_3/mmc$. [T = 1.15 K, p = 6.69 MPa]
β-He f.c.c. (a = 424.0), Fm3m. [T = 16 K, p = 127 MPa]
γ-He b.c.c. (a = 411), Im3m. [T = 1.73 K, p = 2.94 MPa]

X-ray diffraction: mass absorption coefficients (μ/ρ)/cm^2 g^{-1}**:** CuK_α 0.383 MoK_α 0.207
Neutron scattering length, $b/10^{-12}$ cm**:** 0.326
Thermal neutron capture cross-section, σ_a/barns**:** *c.* 0.007

• GEOLOGICAL DATA

Minerals

None as such, although it is present in some minerals.

Chief sources: natural gas may contain up to 7% helium
World production/tonnes y^{-1}**:** *c.* 4500
Reserves/tonnes**:** atmosphere contains 3.7×10^9
Specimen: available in small pressurized canisters. Safe.

Abundances

Sun (relative to H = 1×10^{12})**:** 6.31×10^{10}
Earth's crust/p.p.m.**:** 0.008
Atmosphere/p.p.m. (volume)**:** 5.2
Seawater/p.p.m.**:** 4×10^{-6}
Residence time/years**:** n.a.
Oxidation state: 0

Ho

Atomic number: 67
Relative atomic mass ($^{12}C = 12.0000$)**:** 164.93032

CAS: [7440-60-0]

• CHEMICAL DATA

Discovery: Holmium was discovered in 1878 by P.T. Cleve at Uppsala, Sweden, and independently by M. Delafontaine and J.L. Soret at Geneva, Switzerland.
Description: Holmium is a silvery metal of the so-called rare earth group (more correctly termed the lanthanides). It is slowly attacked by oxygen and water, and dissolves in acids. Holmium is used as a flux concentrator for high magnetic fields.

Radii/pm: Ho^{3+} 89; atomic 177; covalent 158
Electronegativity: 1.23 (Pauling); 1.10 (Allred); ≤ 3.3 eV (absolute)
Effective nuclear charge: 2.85 (Slater); 8.44 (Clementi); 11.60 (Froese-Fischer)

Standard reduction potentials $E^{\ominus}$/V

	III		0
acid	Ho^{3+}	—— −2.33 ——	Ho
base	$Ho(OH)_3$	—— −2.85 ——	Ho

Oxidation states

Ho^{III} f^{10} — Ho_2O_3, $Ho(OH)_3$, $[Ho(OH_2)_x]^{3+}$ (aq), Ho^{3+} salts, HoF_3, $HoCl_3$ etc., $[HoCl_6]^{3-}$, complexes

• PHYSICAL DATA

Melting point/K: 1747
Boiling point/K: 2968
ΔH_{fusion}/kJ mol^{-1}: 17.2
ΔH_{vap}/kJ mol^{-1}: 251.0

Thermodynamic properties (298.15 K, 0.1 MPa)

State	$\Delta_f H^{\ominus}$/kJ mol^{-1}	$\Delta_f G^{\ominus}$/kJ mol^{-1}	$S^{\ominus}$/J K^{-1} mol^{-1}	C_p/J K^{-1} mol^{-1}
Solid	0	0	75.3	27.15
Gas	300.8	264.8	195.59	20.79

Density/kg m^{-3}: 8795 [298 K]
Molar volume/cm^3: 18.75
Thermal conductivity/W m^{-1} K^{-1}: 16.2 [300 K]
Coefficient of linear thermal expansion/K^{-1}: 9.5×10^{-6}
Electrical resistivity/Ω m: 87.0×10^{-8} [298 K]
Mass magnetic susceptibility/kg^{-1} m^3: $+5.49 \times 10^{-6}$ (s)

Young's modulus/GPa: 64.8
Rigidity modulus/GPa: 26.3
Bulk modulus/GPa: 40.2
Poisson's ratio/GPa: 0.231

• BIOLOGICAL DATA

Biological role
None, but acts to stimulate metabolism.

Toxicity
Toxic intake: n.a.
Lethal intake: LD_{50} (chloride, oral, mouse) = 7200 mg kg^{-1}

Hazards
Holmium is mildly toxic by ingestion.

Levels in humans
Organs: n.a., but low
Daily dietary intake: n.a.
Total mass of element in average (70 kg) person: n.a.

Discovery: see Chemical Data section.
[Greek, *Holmia* = Sweden]
French, *holmium*; German, *Holmium*; Italian, *olmio*; Spanish, *holmio*

Holmium

[hol-mi-uhm]

• NUCLEAR DATA

Number of isotopes (including nuclear isomers): 39 **Isotope mass range:** 148 → 170

Key isotopes

Nuclide	Atomic mass	Natural abundance (%)	Nuclear spin I	Nuclear magnetic moment μ	Uses
^{165}Ho	164.930 319	100	7/2–	+4.173	NMR

A table of radioactive isotopes is given in Appendix 1, on p241.

NMR [Reference: not recorded]	^{165}Ho
Relative sensitivity (^{1}H = 1.00)	0.18
Receptivity (^{13}C = 1.00)	1160
Magnetogyric ratio/rad $T^{-1}s^{-1}$	5.710×10^7
Nuclear quadrupole moment/m^2	$+3.580 \times 10^{-28}$
Frequency (^{1}H = 100 Hz; 2.3488T)/MHz	20.513

• ELECTRON SHELL DATA

Ground state electron configuration: [Xe]$4f^{11}6s^2$
Term symbol: $^4I_{15/2}$
Electron affinity ($M \rightarrow M^-$)/kJ mol^{-1}**:** ≤ 50

Ionization energies/kJ mol^{-1}**:**

1.	$M \rightarrow M^+$	580.7
2.	$M^+ \rightarrow M^{2+}$	1139
3.	$M^{2+} \rightarrow M^{3+}$	2204
4.	$M^{3+} \rightarrow M^{4+}$	4100
5.	$M^{4+} \rightarrow M^{5+}$	6169

Electron binding energies/eV

K	1s	55 618
L_I	2s	9394
L_{II}	$2p_{1/2}$	8918
L_{III}	$2p_{3/2}$	8071
M_I	3s	2128
M_{II}	$3p_{1/2}$	1923
M_{III}	$3p_{3/2}$	1741
M_{IV}	$3d_{3/2}$	1392
M_V	$3d_{5/2}$	1351

continued in Appendix 2, p256

Main lines in atomic spectrum
[Wavelength/nm(species)]

345.600 (II)
379.675 (II)
381.073 (II)
389.102 (II)
405.393 (I)
410.384 (I) (AA)
416.303 (I)

• CRYSTAL DATA

Crystal structure (cell dimensions/pm), space group
α-Ho h.c.p. (a = 357.73, c = 561.58), $P6_3/mmc$
β-Ho b.c.c. (a = 396), Im3m
$T(\alpha \rightarrow \beta)$ = just below melting point
High pressure form: (a = 334, c = 2 410), $R\bar{3}m$

X-ray diffraction: mass absorption coefficients (μ/ρ)/cm^2 g^{-1}**:** CuK_α 128 MoK_α 73.9
Neutron scattering length, $b/10^{-12}$ cm: 0.808
Thermal neutron capture cross-section, σ_a/barns**:** 65

• GEOLOGICAL DATA

Minerals

Mineral	Formula	Density	Hardness	Crystal appearance
Bastnäsite*	(Ce, La, etc.)CO_3F	4.9	4–4.5	hex., vit./greasy yellow
Monazite*	(Ce, La, Nd, Th, etc.)PO_4	5.20	5–5.5	mon., waxy/vit. yellow-brown

*Although not a major constituent, holmium is present in extractable amounts.

Chief ores: monazite, bastnäsite
World production/tonnes y^{-1}**:** *c.* 10
Main mining areas: USA, Brazil, India, Sri Lanka, Australia, China.
Reserves/tonnes: *c.* 4×10^5
Specimen: available as ingots or granules. Safe.

Abundances

Sun (relative to H = 1×10^{12})**:** n.a.
Earth's crust/p.p.m.**:** 1.4
Seawater/p.p.m.**:**
Atlantic surface: 2.4×10^{-7}
Atlantic deep: 2.9×10^{-7}
Pacific surface: 1.6×10^{-7}
Pacific deep: 5.8×10^{-7}
Residence time/years**:** n.a.
Classification: recycled
Oxidation state: III

Atomic number: 1
Relative atomic mass ($^{12}C = 12.0000$)**:** 1.00794

CAS: [1333-74-0]

• CHEMICAL DATA

Description: Hydrogen is a colourless, odourless gas, insoluble in water. Burns in air, and forms explosive mixtures with air. Used for making ammonia, cyclohexane, methanol, etc.

Radii/pm: H^+ 0.00066; H^- 154; atomic 78; covalent *c.* 30; van der Waals 120
Electronegativity: 2.20 (Pauling); 2.20 (Allred); 7.18 eV (absolute)
Effective nuclear charge: 1.00 (Slater); 1.00 (Clementi); 1.00 (Froese-Fischer)

Standard reduction potentials E°/V

	I		0		–I
acid	H_3O^+	0.00	H_2	−2.25	H^-
base	H_2O	0.828	H_2	−2.25	H^-

Oxidation states

H^{-I}	[He]	NaH, CaH_2, etc.
H^0	s^1	H_2
H^I	s^0	H_2O, H_3O^+, etc., OH^-, HF, HCl, etc., other acids, NH_3, etc., CH_4 etc.

Hydrogen is the most versatile of elements in its range of chemical bonds.

Covalent bonds

Bond	r/ pm	E/ kJ mol^{-1}
H—H	74.2	432
H—F	91.8	565
H—Cl	127.4	428
H—Br	140.8	362
H—I	160.8	295
^{2}H—^{2}H	74	447

For other bonds to hydrogen see other elements

• PHYSICAL DATA

Melting point/K: 14.01
Boiling point/K: 20.28
Critical temperature/K: 33.35
Critical pressure/ kPa: 1297

ΔH_{fusion}/kJ mol^{-1}**:** 0.12
ΔH_{vap}/kJ mol^{-1}**:** 0.46
Triple point/K: 13.96 K (7.2 kPa)

Thermodynamic properties (298.15 K, 0.1 MPa)

State	$\Delta_f H^\circ$/kJ mol^{-1}	$\Delta_f G^\circ$/kJ mol^{-1}	S°/J K^{-1} mol^{-1}	C_p/J K^{-1} mol^{-1}
Gas (H_2)	0	0	130.684	28.824
Gas (atoms)	217.965	203.247	114.713	20.784

Density/kg m^{-3}**:** 76.0 [s, 11 K]; 70.8 [liq, b.p.]; 0.089 88 [gas, 273 K]
Molar volume/cm^3**:** 13.26 [11 K]
Thermal conductivity/W m^{-1} K^{-1}**:** 0.1815 [300 K] (g)
Mass magnetic susceptibility/kg^{-1} m^3**:** -2.50×10^{-8} (g)

• BIOLOGICAL DATA

Biological role

Hydrogen is a constituent element of DNA. Equally important, it is also a component of water, which regulates life at all levels, from the living cell to the global environment.

Toxicity

Non-toxic.

Hazards

Hydrogen is flammable and explosive when mixed with air; it could asphyxiate if it excluded oxygen from the lungs.

Levels in humans

Blood/mg dm^{-3}**:** constituent of water
Bone/p.p.m.: 52 000
Liver/p.p.m.: 93 000
Muscle/p.p.m.: 93 000
Daily dietary intake: mainly as water
Total mass of element in average (70 kg) person: 7 kg

Recognized as an element in 1766 by H. Cavendish at London, England.
[Greek, *hydro genes* = water forming]
French, *hydrogène*; German, *Wasserstoff*; Italian, *idrogeno*; Spanish, *hidrógeno*

Hydrogen

[hy-dro-jen]

• NUCLEAR DATA

Number of isotopes (including nuclear isomers): 3

Isotope mass range: $1 \rightarrow 3$

Key isotopes

Nuclide	Atomic mass	Natural abundance (%)	Nuclear spin *I*	Nuclear magnetic moment μ	Uses
^{1}H	1.007 825 035	99.985	1/2	+2.792 845 6	NMR
^{2}H	2.014 101 779	0.015	1	+0.857 437 6	NMR
^{3}H*	3.016 05	0	1/2	+2.978 96	R, D

*^{3}H is radioactive with a half-life of 12.26 y and decay mode β^- (0.01861 MeV); no γ.

NMR [Reference: $Si(CH_3)_4$]	^{1}H	^{2}H	^{3}H
Relative sensitivity (^{1}H = 1.00)	1.00 (by def.)	9.65×10^{-3}	1.21
Receptivity (^{13}C = 1.00)	5680	8.2×10^{-3}	–
Magnetogyric ratio/rad $T^{-1} s^{-1}$	26.7510×10^{7}	4.1064×10^{7}	28.5335×10^{7}
Nuclear quadrupole moment/m^2	–	2.860×10^{-31}	–
Frequency (^{1}H = 100 Hz; 2.3488T)/MHz	100.000	15.351	106.663

• ELECTRON SHELL DATA

Ground state electron configuration: $1s^1$
Term symbol: $^{2}S_{1/2}$
Electron affinity ($M \rightarrow M^-$)/kJ mol^{-1}**:** 72.8

Ionization energies/kJ mol^{-1}**:**

1. $M \rightarrow M^+$ 1312.0

Electron binding energies/eV

K 1s 13.6

Main lines in atomic spectrum
[Wavelength/nm(species)]

434.047 (I)
486.133 (I)
656.272 (I)
656.285 (I)
1875.10 (I)

• CRYSTAL DATA

Crystal structure (cell dimensions/pm), space group

H_2 h.c.p. (a = 377.6, c = 616.2), $P6_3/mmc$
$^{2}H_2$ h.c.p. (a = 360.0, c = 585.8), $P6_3/mmc$
H_2 cubic (a = 533.8), Fm3m
$^{2}H_2$ cubic (a = 509.2), Fm3m
H_2 tetragonal (a = 450, c = 368), I4
$^{2}H_2$ tetragonal (a = 338, c = 560), I4
T (h.c.p. $\rightarrow$ tetragaonal) = 4.5 K

X-ray diffraction: mass absorption coefficients $(\mu/\rho)/cm^2\ g^{-1}$**:** CuK_α 0.435 MoK_α 0.380
Neutron scattering length, $b/10^{-12}$ cm**:** –0.37390
Thermal neutron capture cross-section, σ_a/barns**:** 0.3326

• GEOLOGICAL DATA

Minerals

None as such, although hydrogen is present in many as water molecules

Sources: Hydrogen gas is produced mainly from natural methane gas ($CH_4 + 2H_2O = 3H_2 + CO$). Some is also produced from the electrolysis of brine using a mercury amalgam cell, or by the action of steam on red hot coke which gives a mixture of H_2 and CO.

World production/$m^3\ y^{-1}$**:** 350×10^9 (hydrogen gas)
Reserves/tonnes: almost limitless
Specimen: available in small pressurized canisters. *Warning!*

Abundances

Sun (relative to H = 1×10^{12})**:** most abundant element, and taken as standard against which others are measured.

Earth's crust/p.p.m.**:** 1520
Atmosphere/p.p.m. (volume)**:** 0.5
Seawater/p.p.m.**:** constituent of water; some dissolved H_2 gas.
Residence time/years**:** n.a.
Oxidation state: I

In

Atomic number: 49
Relative atomic mass ($^{12}C = 12.0000$)**:** 114.818

CAS: [7440-74-6]

• CHEMICAL DATA

Discovery: Indium was discovered in 1863 by Ferdinand Reich and Hieronymous Richter at Freiberg, Germany.
Description: Indium is a soft, silvery-white metal, and has one of the longest liquid range of all the elements. It is stable in air and with water; it dissolves in acids. Indium is used in low-melting alloys in safety devices. Indium arsenide and indium antimonide have uses in transistors and thermistors.

Radii/pm: In^{3+} 92; In^{+} 132; atomic 163; covalent 150
Electronegativity: 1.78 (Pauling); 1.49 (Allred); 3.1 eV (absolute)
Effective nuclear charge: 5.00 (Slater); 8.47 (Clementi); 9.66 (Froese-Fischer)

Standard reduction potentials $E^\ominus$/V

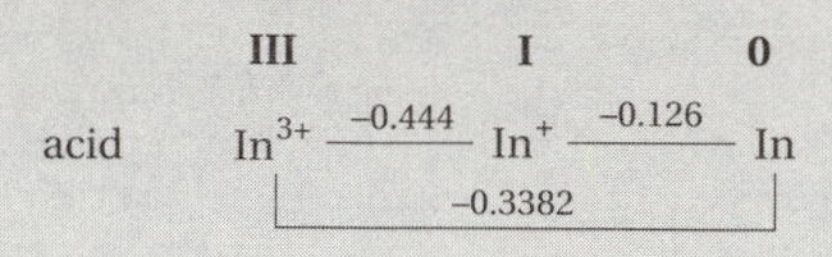

Oxidation states

In^{I}	s^2	InCl, InBr, InI
In^{II}	s^1	$[In_2Cl_6]^{2-}$, $[In_2Br_6]^{2-}$, $[In_2I_6]^2$
In^{III}	d^{10}	In_2O_3, $In(OH)_3$, $[In(OH_2)_6]^{3+}$ (aq), InF_3, $InCl_3$ etc., $[InCl_5]^{2-}$, $[InCl_6]^{3-}$, complexes

Covalent bonds

Bond	r/ pm	E/ kJ mol⁻¹
In—H	185	243
In—C	216	165
In—O	213	109
In^{I}—F	199	523
In^{I}—Cl	240	435
In^{I}—Br	254	406
In—In	325	100

• PHYSICAL DATA

Melting point/K: 429.32
Boiling point/K: 2353
ΔH_{fusion}/kJ mol^{-1}: 3.27
ΔH_{vap}/kJ mol^{-1}: 226.4

Thermodynamic properties (298.15 K, 0.1 MPa)

State	$\Delta_f H^\ominus$/kJ mol^{-1}	$\Delta_f G^\ominus$/kJ mol^{-1}	$S^\ominus$/J K^{-1} mol^{-1}	C_p/J K^{-1} mol^{-1}
Solid	0	0	57.82	26.74
Gas	243.30	208.71	173.79	20.84

Density/kg m^{-3}: 7310 [298 K]; 7032 [liquid at m.p.]
Molar volume/cm^3: 15.71
Thermal conductivity/W m^{-1} K^{-1}: 81.6 [300 K]
Coefficient of linear thermal expansion/K^{-1}: 33×10^{-6}
Electrical resistivity/Ω m: 8.37×10^{-8} [293 K]
Mass magnetic susceptibility/kg^{-1} m^3: -7.0×10^{-9} (s)

Young's modulus/GPa: 10.6
Rigidity modulus/GPa: 3.68
Bulk modulus/GPa: n.a.
Poisson's ratio/GPa: 0.45

• BIOLOGICAL DATA

Biological role
None, but acts to stimulate metabolism.

Toxicity
Toxic intake: 30 mg
Lethal intake: LD_{50} (sulfate, oral, rat) = *c.* 1200 mg kg^{-1}

Hazards
Indium is moderately toxic by ingestion and affects the liver, heart and kidneys. It may have teratogenic effects.

Levels in humans
Blood/mg dm^{-3}: n.a. but low
Bone/p.p.m.: n.a.
Liver/p.p.m.: n.a
Muscle/p.p.m.: *c.* 0.015
Daily dietary intake: n.a. but low
Total mass of element in average (70 kg) person: *c.* 0.4 mg

Discovery: see Chemical Data section.
[Named after the indigo line in its spectrum]
French, *indium*; German, *Indium*; Italian, *indio*; Spanish, *indio*

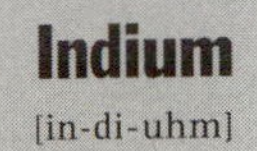
[in-di-uhm]

• NUCLEAR DATA

Number of isotopes (including nuclear isomers): 59 **Isotope mass range:** 102 → 132

Key isotopes

Nuclide	Atomic mass	Natural abundance (%)	Nuclear spin *I*	Nuclear magnetic moment μ	Uses
^{113}In	112.904 061	4.3	9/2+	+5.528 9	E, NMR
^{115}In*	114.903 882	95.7	9/2+	+5.540 8	E, NMR

* ^{115}In is radioactive with a half-life of 6×10^{14} y and decay mode β^- (0.496 MeV); no γ.
A table of radioactive isotopes is given in Appendix 1, on p242.

NMR [Reference: In^{3+} (aq)]	^{113}In	^{115}In
Relative sensitivity (^1H = 1.00)	0.345	0.348
Receptivity (^{13}C = 1.00)	83.8	1890
Magnetogyric ratio/rad $T^{-1}s^{-1}$	5.8493×10^7	5.8618×10^7
Nuclear quadrupole moment/m^2	$+0.799 \times 10^{-28}$	$+0.810 \times 10^{-28}$
Frequency (^1H = 100 Hz; 2.3488T)/MHz	21.866	21.914

• ELECTRON SHELL DATA

Ground state electron configuration: $[Kr]4d^{10}5s^25p^1$
Term symbol: $^2P_{1/2}$
Electron affinity ($M \rightarrow M^-$)/kJ mol^{-1}: *c.* 30

Ionization energies/kJ mol^{-1}:

1.	$M \rightarrow M^+$	558.3
2.	$M^+ \rightarrow M^{2+}$	1820.6
3.	$M^{2+} \rightarrow M^{3+}$	2704
4.	$M^{3+} \rightarrow M^{4+}$	5200
5.	$M^{4+} \rightarrow M^{5+}$	(7400)
6.	$M^{5+} \rightarrow M^{6+}$	(9500)
7.	$M^{6+} \rightarrow M^{7+}$	(11 700)
8.	$M^{7+} \rightarrow M^{8+}$	(13 900)
9.	$M^{8+} \rightarrow M^{9+}$	(17 200)
10.	$M^{9+} \rightarrow M^{10+}$	(19 700)

Electron binding energies/eV

K	1s	27 940
L_I	2s	4238
L_{II}	$2p_{1/2}$	3938
L_{III}	$2p_{3/2}$	3730
M_I	3s	827.2
M_{II}	$3p_{1/2}$	703.2
M_{III}	$3p_{3/2}$	665.3
M_{IV}	$3d_{3/2}$	451.4
M_V	$3d_{5/2}$	443.9

continued in Appendix 2, p256

Main lines in atomic spectrum
[Wavelength/nm(species)]
303.936 (I) (AA)
325.609 (I)
325.856 (I)
410.176 (I)
451.131 (I)

• CRYSTAL DATA

Crystal structure (cell dimensions/pm), space group
face centred tetragonal (a = 325.30, c = 494.55), I4/mmm

X-ray diffraction: mass absorption coefficients (μ/ρ)/$cm^2 g^{-1}$: CuK_α 243 MoK_α 29.3
Neutron scattering length, $b/10^{-12}$ cm: 0.4065
Thermal neutron capture cross-section, σ_a/barns: 194

• GEOLOGICAL DATA

Minerals

Indium has been reported to occur as the native element, but it mainly occurs in zinc sulfide and lead sulfide ores to the extent of 1%.

Mineral	Formula	Density	Hardness	Crystal appearance
Indite	$FeIn_2S_4$	4.67	4.5	cub., met. white

Chief ores: it is obtained as a by-product of zinc and lead smelting.

World production/tonnes y^{-1}: 75

Main mining areas: see zinc and lead.

Reserves/tonnes: > 1500

Specimen: available as foil, granules, pieces, powder, rod, shot or wire. *Care!*

Abundances

Sun (relative to H = 1×10^{12}): 44.7
Earth's crust/p.p.m.: 0.049
Seawater/p.p.m.: 1×10^{-7}
Residence time/years: n.a.
Classification: n.a.
Oxidation state: III

I

Atomic number: 53
Relative atomic mass ($^{12}C = 12.0000$)**:** 126.90447

CAS: [7553-56-2]

• CHEMICAL DATA

Description: Iodine is a black, shiny, non-metallic solid (I_2) which sublimes easily on heating to give a purple vapour. It is used as a disinfectant, in pharmaceuticals, food supplements, dyes, catalysts, and photography.

Radii/pm: I^- 196; covalent 133; van der Waals 215
Electronegativity: 2.66 (Pauling); 2.21 (Allred); 6.76 eV (absolute)
Effective nuclear charge: 7.60 (Slater); 11.61 (Clementi); 14.59 (Froese-Fischer)

Standard reduction potentials E°/V

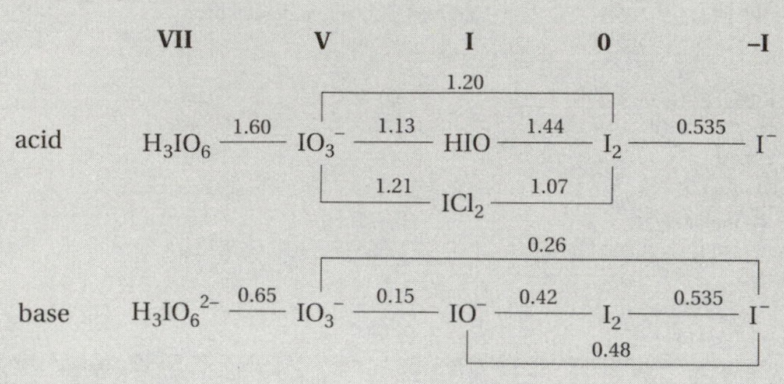

Oxidation states

I^{-I}	[Xe]	I^- (aq), HI, KI, etc.
I^0	s^2p^5	I_2, I_3^-, I_5^-, etc.
I^I	s^2p^4	I_n^+, ICl_2^-, etc.
I^{III}	s^2p^2	I_4O_9 (= $I^{3+}(IO_3^-)_3$), ICl_3
I^V	s^2	I_2O_5, HIO_3, IO_3^- (aq), IF_5, IF_6^-
I^{VII}	d^{10}	H_5IO_6, $H_4IO_6^-$ (aq) etc., HIO_4 IO_4^- (aq), IF_7

Covalent bonds

Bond	r/ pm	E/ kJ mol^{-1}
I—H	161	295
I—C	214	213
I—O	195	201
I—F	191	278
I—Cl	232	208
I—I	267	149

For other bonds to iodine: see other elements

• PHYSICAL DATA

Melting point/K: 386.7
Boiling point/K: 457.50
Critical temperature/K: 819

ΔH_{fusion}/kJ mol^{-1}: 15.27
ΔH_{vap}/kJ mol^{-1}: 41.67

Thermodynamic properties (298.15 K, 0.1 MPa)

State	$\Delta_f H^\circ$/kJ mol^{-1}	$\Delta_f G^\circ$/kJ mol^{-1}	S°/J K^{-1} mol^{-1}	C_p/J K^{-1} mol^{-1}
Solid	0	0	116.135	54.438
Gas (I_2)	62.438	19.327	260.69	36.90
Gas (atoms)	106.838	70.250	180.791	20.786

Density/kg m^{-3}: 4930 [293 K]
Molar volume/cm^3: 25.74
Thermal conductivity/W m^{-1} K^{-1}: 0.449 [300 K]
Coefficient of linear thermal expansion/K^{-1}: n.a.
Electrical resistivity/Ω m: 1.37×10^7 [293 K]
Mass magnetic susceptibility/kg^{-1} m^3: -4.40×10^{-9} (s)

• BIOLOGICAL DATA

Biological role

Most iodine exists in nature as iodide ions, I^-, the form in which it is taken into our bodies. Iodine is essential to many species, including humans.

Toxicity

Toxic intake: 2 mg as I_2. Iodides are similar in toxicity to bromides.

Lethal intake: human, oral = 2 g as I_2.
LD_{50} (NaI, oral, rat) = 14 000 mg kg^{-1}

Hazards

Iodine in its elemental form, I_2, is toxic, and its vapour irritates the eyes and lungs. The maximum allowable concentration when working with iodine is 1 mg m^{-3} in air.

Levels in humans

Blood/mg dm^{-3}: 0.057
Bone/p.p.m.: 0.27
Liver/p.p.m.: 0.7
Muscle/p.p.m.: 0.05 – 0.5
Daily dietary intake: 0.1 – 0.2 mg
Total mass of element in average (70 kg) person: 12 – 20 mg

Discovered in 1811 by Bernard Courtois at Dijon, France.
[Greek, *iodes* = violet]
French, *iode*; German, *Iod*; Italian, *iodio*; Spanish, *yodo*

Iodine
[iyo-deen]

• NUCLEAR DATA

Number of isotopes (including nuclear isomers): 37 **Isotope mass range:** 110 → 140

Key isotopes

Nuclide	Atomic mass	Natural abundance (%)	Nuclear spin I	Nuclear magnetic moment μ	Uses
^{127}I	126.904 473	100	5/2+	+2.813 28	NMR

A table of radioactive isotopes is given in Appendix 1, on p242.

NMR [Reference: NaI (aq)]	^{127}I
Relative sensitivity (^1H = 1.00)	0.0934
Receptivity (^{13}C = 1.00)	530
Magnetogyric ratio/rad $T^{-1}s^{-1}$	5.3525×10^7
Nuclear quadrupole moment/m^2	-0.789×10^{-28}
Frequency (^1H = 100 Hz; 2.3488T)/MHz	20.007

• ELECTRON SHELL DATA

Ground state electron configuration: $[Kr]4d^{10}5s^25p^5$
Term symbol: $^3P_{3/2}$
Electron affinity ($M \rightarrow M^-$)/kJ mol^{-1}**:** 295.2

Ionization energies/kJ mol^{-1}**:**

1.	$M \rightarrow M^+$	1008.4
2.	$M^+ \rightarrow M^{2+}$	1845.9
3.	$M^{2+} \rightarrow M^{3+}$	3200
4.	$M^{3+} \rightarrow M^{4+}$	(4100)
5.	$M^{4+} \rightarrow M^{5+}$	(5000)
6.	$M^{5+} \rightarrow M^{6+}$	(7400)
7.	$M^{6+} \rightarrow M^{7+}$	(8700)
8.	$M^{7+} \rightarrow M^{8+}$	(16 400)
9.	$M^{8+} \rightarrow M^{9+}$	(19 300)
10.	$M^{9+} \rightarrow M^{10+}$	(22 100)

Electron binding energies/eV

K	1s	33 169
L_I	2s	5188
L_{II}	$2p_{1/2}$	4852
L_{III}	$2p_{3/2}$	4557
M_I	3s	1072
M_{II}	$3p_{1/2}$	931
M_{III}	$3p_{3/2}$	875
M_{IV}	$3d_{3/2}$	631
M_V	$3d_{5/2}$	620

continued in Appendix 2, p256

Main lines in atomic spectrum
[Wavelength/nm(species)]
511.929 (I)
533.822 (II)
562.569 (II)
804.374 (I)
905.833 (I)
911.391 (I)

• CRYSTAL DATA

Crystal structure (cell dimensions/pm), space group
orthorhombic (a = 726.47, b = 478.57, c = 979.08), Cmca

X-ray diffraction: mass absorption coefficients (μ/ρ)/$cm^2\ g^{-1}$**:** CuK_α 294 MoK_α 37.1
Neutron scattering length, $b/10^{-12}$ cm**:** 0.528
Thermal neutron capture cross-section, σ_a/barns**:** 6.2

• GEOLOGICAL DATA

Minerals

Minerals are very rare. Iodine cycles through the environment, and rain water contains about 0.7 p.p.b.

Mineral	Formula	Density	Hardness	Crystal appearance
Iodargyrite	β-AgI	5.69	1.5	hex., res./adam. colourless
Lautarite	$Ca(IO_3)_2$	4.519	3.5 – 4	mon., col./yellow, transparent

Chief source: from brines, which may have 50 ppm of iodide, and the Chilean nitrate deposits which contain up to 0.3% calcium iodate. Some iodine is also extracted from seaweed.

World production/tonnes y^{-1}**:** 12 000 (elemental iodine)

Producing areas: Chile, Japan.

Reserves/tonnes: 2.6×10^6

Specimen: available as crystals. *Warning!*

Abundances

Sun (relative to H = 1×10^{12})**:** n.a.
Earth's crust/p.p.m.**:** 0.14
Atmosphere/p.p.m. (volume)**:** trace
Seawater/p.p.m.**:**
Atlantic surface: 0.0489
Atlantic deep: 0.056
Pacific surface: 0.043
Pacific deep: 0.058
Residence time/years**:** 300 000
Classification: scavenged as I(–I), recycled as I(V)
Oxidation state: –I and V, mainly V

Ir

Atomic number: 77
Relative atomic mass ($^{12}C = 12.0000$)**:** 192.217

CAS: [7439-88-5]

• CHEMICAL DATA

Description: Iridium is a hard, lustrous, silvery metal of the so-called platinum group. It is unaffected by air, water, and acids, but dissolves in molten alkali. Iridium is used in special alloys and spark plugs.

Radii/pm: Ir^{4+} 66; Ir^{3+} 75; Ir^{2+} 89; atomic 136; covalent 126
Electronegativity: 2.20 (Pauling); 1.55 (Allred); 5.4 eV (absolute)
Effective nuclear charge: 3.90 (Slater); 10.57 (Clementi); 15.33 (Froese-Fischer)

Standard reduction potentials $E^{\ominus}$/V

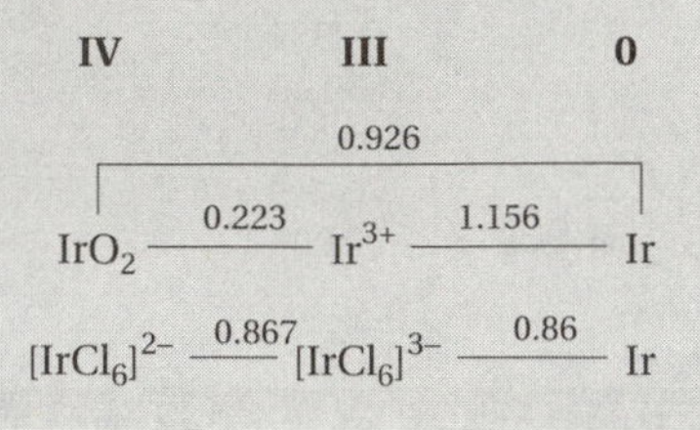

Oxidation states

Ir^{-I}	d^{10}	rare $[Ir(CO)_3(PPh_3)]^-$	**Ir^{III}**	d^6	IrF_3, $IrCl_3$ etc., $[IrCl_6]^{3-}$ (aq)
Ir^0	d^9	rare $[Ir_4(CO)_{12}]$, $[Ir_6(CO)_{16}]$	**Ir^{IV}**	d^5	IrO_2, IrF_4, IrS_2, $[IrCl_6]^{2-}$ (aq)
Ir^I	d^8	$[IrCl(CO)(PPh_3)_2]$	Ir^V	d^4	IrF_5, $[IrF_6]^-$
Ir^{II}	d^7	$IrCl_2$	Ir^{VI}	d^3	IrF_6

• PHYSICAL DATA

Melting point/K: 2683
Boiling point/K: 4403

ΔH_{fusion}/kJ mol^{-1}**:** 26.4
ΔH_{vap}/kJ mol^{-1}**:** 563.6

Thermodynamic properties (298.15 K, 0.1 MPa)

State	$\Delta_f H^{\ominus}$/kJ mol^{-1}	$\Delta_f G^{\ominus}$/kJ mol^{-1}	$S^{\ominus}$/J K^{-1} mol^{-1}	C_p/J K^{-1} mol^{-1}
Solid	0	0	35.48	25.10
Gas	665.3	617.9	193.578	20.786

Density/kg m^{-3}**:** 22 560 [290 K]; 20 000 [liquid at m.p.]
Molar volume/cm^3**:** 8.57
Thermal conductivity/W m^{-1} K^{-1}**:** 147 [300 K]
Coefficient of linear thermal expansion/K^{-1}**:** 6.4×10^{-6}
Electrical resistivity/Ω m**:** 5.3×10^{-8} (s)
Mass magnetic susceptibility/kg^{-1} m^3**:** $+1.67 \times 10^{-9}$ (s)

Young's modulus/GPa**:** 528
Rigidity modulus/GPa**:** 209
Bulk modulus/GPa**:** 371
Poisson's ratio/GPa**:** 0.26

• BIOLOGICAL DATA

Biological role

None.

Toxicity

Toxic intake: the pure metal is clinically inert; data on compounds is sparse.
Lethal intake: LD_{50} (chloride, oral, mouse) = 8.12 mg kg^{-1}

Hazards

Indium chloride is moderately toxic by ingestion, but most compounds are insoluble and not absorbed by the body.

Levels in humans

Blood/mg dm^{-3}**:** n.a., but very low
Bone/p.p.m.**:** n.a.
Liver/p.p.m.**:** n.a.
Muscle/p.p.m.**:** *c.* 2×10^{-5}
Daily dietary intake: n.a. but very low
Total mass of element in average (70 kg) person: n.a

Discovered in 1803 by S. Tennant at London, England.
[Latin, *iris* = rainbow]
French, *iridium*; German, *Iridium*; Italian, *iridio*; Spanish, *iridio*

[irid-iuhm]

• NUCLEAR DATA

Number of isotopes (including nuclear isomers): 40 **Isotope mass range:** 170 → 198

Key isotopes

Nuclide	Atomic mass	Natural abundance (%)	Nuclear spin I	Nuclear magnetic moment μ	Uses
^{191}Ir	190.960 584	37.3	3/2+	+0.146 2	E, NMR
^{193}Ir	192.962 917	62.7	3/2+	+0.159 2	E, NMR

A table of radioactive isotopes is given in Appendix 1, on p242.

NMR [Never used]	^{191}Ir	^{193}Ir
Relative sensitivity (^{1}H = 1.00)	2.53×10^{-5}	3.27×10^{-5}
Receptivity (^{13}C = 1.00)	0.023	0.050
Magnetogyric ratio/rad $T^{-1}s^{-1}$	0.539×10^7	0.391×10^7
Nuclear quadrupole moment/m^2	$+0.816 \times 10^{-28}$	$+0.751 \times 10^{-28}$
Frequency (^{1}H = 100 Hz; 2.3488T)/MHz	1.718	1.871

• ELECTRON SHELL DATA

Ground state electron configuration: [Xe]$4f^{14}5d^76s^2$
Term symbol: $^4F_{9/2}$
Electron affinity ($M \rightarrow M^-$)/kJ mol^{-1}**:** 151

Ionization energies/kJ mol^{-1}**:**

1.	$M \rightarrow M^+$	880
2.	$M^+ \rightarrow M^{2+}$	(1680)
3.	$M^{2+} \rightarrow M^{3+}$	(2600)
4.	$M^{3+} \rightarrow M^{4+}$	(3800)
5.	$M^{4+} \rightarrow M^{5+}$	(5500)
6.	$M^{5+} \rightarrow M^{6+}$	(6900)
7.	$M^{6+} \rightarrow M^{7+}$	(8500)
8.	$M^{7+} \rightarrow M^{8+}$	(10 000)
9.	$M^{8+} \rightarrow M^{9+}$	(11 700)

Electron binding energies/eV

K	1s	76 111
L_I	2s	13 419
L_{II}	$2p_{1/2}$	12 824
L_{III}	$2p_{3/2}$	11 215
M_I	3s	3174
M_{II}	$3p_{1/2}$	2909
M_{III}	$3p_{3/2}$	2551
M_{IV}	$3d_{3/2}$	2116
M_V	$3d_{5/2}$	2040

continued in Appendix 2, p256

Main lines in atomic spectrum
[Wavelength/nm(species)]
203.357 (I)
208.882 (I) (AA)
209.263 (I)
215.805 (I)
254.397 (I)
263.971 (I)
322.078 (I)

• CRYSTAL DATA

Crystal structure (cell dimensions/pm), space group
f.c.c. (a = 383.92), Fm3m

X-ray diffraction: mass absorption coefficients (μ/ρ)/cm^2 g^{-1}**:** CuK_α 193 MoK_α 110
Neutron scattering length, $b/10^{-12}$ cm**:** 1.06
Thermal neutron capture cross-section, σ_a/barns**:** 425

• GEOLOGICAL DATA

Minerals

Found as the native element.

Mineral	Formula	Density	Hardness	Crystal appearance
Iridium	Ir	22.4	6 – 6.5	cub., met. white
Iridosmine	(Os,Ir)	20	6 – 7	hex., met. white/grey
Osmiridium	(Ir,Os)	20	6 – 7	cub., met. white

Chief ores: osmiridium; also found with platinum ores
World production/tonnes y^{-1}**:** 3
Main mining areas: Canada for native element; see also platinum.
Reserves/tonnes**:** n.a.
Specimen: available as foil, powder, sponge or wire. Safe.

Abundances

Sun (relative to H = 1×10^{12})**:** 7.1
Earth's crust/p.p.m.**:** *c.* 3×10^{-6}
Seawater/p.p.m.**:** n.a. but very low
Residence time/years**:** n.a.

Fe

Atomic number: 26
Relative atomic mass ($^{12}C = 12.0000$)**:** 55.845

CAS: [7439-89-6]

• CHEMICAL DATA

Description: Iron, when absolutely pure, is lustrous, silvery and soft (workable). This is the most important of all the metals and it is used chiefly as steel in which there is carbon (up to 1.7%). Stainless steels are alloys with other metals, mainly nickel. Iron rusts in damp air and dissolves readily in dilute acids. Its uses are legion.

Radii/pm: Fe^{3+} 67; Fe^{2+} 82; atomic 124; covalent 116; van der Waals n.a.
Electronegativity: 1.83 (Pauling); 1.64 (Allred); 4.06 eV (absolute)
Effective nuclear charge: 3.75 (Slater); 5.43 (Clementi); 7.40 (Froese-Fischer)

Standard reduction potentials $E^{\ominus}$/V

	VI		III		II		0
acid (pH 0)			Fe^{3+}	0.771	Fe^{2+}	−0.44	Fe
			Fe^{3+}	−0.04			Fe
			$[Fe(CN)_6]^{3-}$	0.361	$[Fe(CN)_6]^{2-}$	−1.16	Fe
base (pH 14)	$[FeO_4]^{2-}$	c. 0.55	FeO_2^-	c. −0.69	$HFeO_2^-$	c. −0.8	Fe

Oxidation states

Fe^{-II}	d^{10}	rare $[Fe(CO)_4]^{2-}$
Fe^{-I}	d^9	rare $[Fe_2(CO)_8]^{2-}$
Fe^0	d^8	$[Fe(CO)_5]$
Fe^I	d^7	rare $[Fe(NO)(OH_2)_5]^{2+}$
Fe^{II}	d^6	FeO, FeS_2 (= $Fe^{II}S_2^{2-}$), $Fe(OH)_2$, $[Fe(OH_2)_6]^{2+}$(aq), FeF_2, $[Fe(\eta\text{-}C_5H_5)_2]$ etc.
Fe^{III}	d^5	Fe_2O_3, Fe_3O_4, (= $Fe^{II}O.Fe^{III}{}_2O_3$), FeF_3, $FeCl_3$, Fe(OH)(O), $[Fe(OH_2)_6]^{3+}$(aq) etc.
Fe^{IV}	d^4	rare, some complexes
Fe^V	d^3	$[FeO_4]^{3-}$?
Fe^{VI}	d^2	$[FeO_4]^{2-}$

• PHYSICAL DATA

Melting point/K: 1808
Boiling point/K: 3023
ΔH_{fusion}/kJ mol^{-1}: 14.9
ΔH_{vap}/kJ mol^{-1}: 351.0

Thermodynamic properties (298.15 K, 0.1 MPa)

State	$\Delta_f H^{\ominus}$/kJ mol^{-1}	$\Delta_f G^{\ominus}$/kJ mol^{-1}	$S^{\ominus}$/J K^{-1} mol^{-1}	C_p/J K^{-1} mol^{-1}
Solid	0	0	27.28	25.10
Gas	416.3	370.7	180.490	25.677

Density/kg m^{-3}: 7874 [293 K]; 7035 [liquid at m.p.]
Molar volume/cm^3: 7.09
Thermal conductivity/W m^{-1} K^{-1}: 80.2 [300 K]
Coefficient of linear thermal expansion/K^{-1}: 12.3×10^{-6}
Electrical resistivity/Ω m: 9.71×10^{-8} [293 K]
Mass magnetic susceptibility/kg^{-1} m^3: ferromagnetic

Young's modulus/GPa: 152.3 (cast iron); 208 (steel)
Rigidity modulus/GPa: 60.0 (cast iron); 81 (steel)
Bulk modulus/GPa: 109.5 (cast iron); 160 (steel)
Poisson's ratio/GPa: 0.27 (cast iron); 0.27 (steel)

• BIOLOGICAL DATA

Biological role

Essential to all species.

Toxicity

Toxic intake: 200 mg. Iron(II) compounds are more toxic than iron(III)
Lethal intake: 7 – 35 g

Hazards

Iron dust poses a moderate fire or explosion hazard; chronic exposure causes iron pneumoconiosis (welders lung). Iron deficiency leads to anaemia, but excess iron in the body causes liver and kidney damage. Some iron compounds are suspected carcinogens.

Levels in humans

Blood/mg dm^{-3}: 447
Bone/p.p.m.: 3 – 380
Liver/p.p.m.: 250 – 1400
Muscle/p.p.m.: 180
Daily dietary intake: 6 – 40 mg
Total mass of element in average (70 kg) person: 4.2 g

Known to ancient civilizations.
[Anglo-Saxon, *iron*; Latin, *ferrum*]
French, *fer*; German, *Eisen*; Italian, *ferro*; Spanish, *hierro*

[iy-on]

• NUCLEAR DATA

Number of isotopes (including nuclear isomers): 16 **Isotope mass range:** 49 → 63

Key isotopes

Nuclide	Atomic mass	Natural abundance (%)	Nuclear spin *I*	Nuclear magnetic moment μ	Uses
^{54}Fe	53.939 612	5.8	0+		E
^{56}Fe	55.934 939	91.72	0+		E
^{57}Fe	56.935 395	2.2	1/2–	+0.090 622 94	E, NMR
^{58}Fe	57.933 277	0.28	0+		E

A table of radioactive isotopes is given in Appendix 1, on p242.

NMR [Reference: $[Fe(CO)_5]$]	^{57}Fe
Relative sensitivity ($^1H = 1.00$)	3.37×10^{-5}
Receptivity ($^{13}C = 1.00$)	4.2×10^{-3}
Magnetogyric ratio/rad $T^{-1} s^{-1}$	0.8661×10^7
Nuclear quadrupole moment/m^2	–
Frequency (^1H = 100 Hz; 2.3488T)/MHz	3.231

• ELECTRON SHELL DATA

Ground state electron configuration: $[Ar]3d^64s^2$
Term symbol: $^5D_4$
Electron affinity ($M \rightarrow M^-$)/kJ mol^{-1}**:** 15.7

Ionization energies/kJ mol^{-1}**:**

1.	$M \rightarrow M^+$	759.3
2.	$M^+ \rightarrow M^{2+}$	1561
3.	$M^{2+} \rightarrow M^{3+}$	2957
4.	$M^{3+} \rightarrow M^{4+}$	5290
5.	$M^{4+} \rightarrow M^{5+}$	7240
6.	$M^{5+} \rightarrow M^{6+}$	9600
7.	$M^{6+} \rightarrow M^{7+}$	12 100
8.	$M^{7+} \rightarrow M^{8+}$	14 575
9.	$M^{8+} \rightarrow M^{9+}$	22 678
10.	$M^{9+} \rightarrow M^{10+}$	25 290

Electron binding energies/eV

K	1s	7112
L_I	2s	844.6
L_{II}	$2p_{1/2}$	719.9
L_{III}	$2p_{3/2}$	706.8
M_I	3s	91.3
M_{II}	$3p_{1/2}$	52.7
M_{III}	$3p_{3/2}$	52.7

Main lines in atomic spectrum
[Wavelength/nm(species)]
248.327 (I) (AA)
248.814 (I)
252.285 (I)
344.061 (I)
371.994 (I)
373.713 (I)
374.556 (I)
385.991 (I)

• CRYSTAL DATA

Crystal structure (cell dimensions/pm), space group
α-Fe b.c.c. (a = 286.645), Im3m
β-Fe not true allotrope
γ-Fe c.c.p. (a = 364.68), Fm3m
δ-Fe b.c.c. (a = 293.22), Im3m
$T(\alpha \rightarrow \gamma)$ = 1183 K; $T(\gamma \rightarrow \delta)$ = 1663 K

X-ray diffraction: mass absorption coefficients (μ/ρ)/$cm^2 g^{-1}$**:** CuK_α 308 MoK_α 38.5
Neutron scattering length, $b/10^{-12}$ cm**:** 0.954
Thermal neutron capture cross-section, σ_a/barns**:** 2.56

• GEOLOGICAL DATA

Minerals

Many iron minerals are known, only a few of which are listed.

Mineral	Formula	Density	Hardness	Crystal appearance
Goethite	α-FeO(OH)	4.28	5 – 5.5	orth., met. earthy brown
Hematite	Fe_2O_3	5.26	5 – 6	rhom., met. earth grey
Lepidocrocite	γ-FeO(OH)	4.09	5	orth., met. reddish-brown
Magnetite	Fe_3O_4	5.175	5.5 – 6.5	cub., met. black
Siderite	$FeCO_3$	3.96	4	rhom., vit. yellow-brown

Chief ores: hematite, magnetite, goethite, lepidocrocite, siderite
World production/tonnes y^{-1}**:** 7.16×10^8
Main mining areas: USA, Canada, Sweden, South Africa, Russia, India, Japan.
Reserves/tonnes: 1.1×10^{11}
Specimen: available as chips, filings, foil, granules, powder and wire. Safe.

Abundances

Sun (relative to $H = 1 \times 10^{12}$)**:** 3.16×10^7
Earth's crust/p.p.m.**:** 41 000
Seawater/p.p.m.**:**
Atlantic surface: 1×10^{-4}
Atlantic deep: 4×10^{-4}
Pacific surface: 0.1×10^{-4}
Pacific deep: 1×10^{-4}
Residence time/years**:** 98
Classification: recycled
Oxidation state: III

Atomic number: 36
Relative atomic mass ($^{12}C = 12.0000$)**:** 83.80

CAS: [7439-90-9]

• CHEMICAL DATA

Discovery: Krypton was discovered in 1898 by Sir William Ramsay and M.W. Travers at London, England.
Description: Krypton is a colourless, odourless gas obtained from liquid air. It is inert towards all other elements and chemicals except fluorine. Isotope ^{86}Kr has an orange-red line in its atomic spectrum which is the fundamental standard of length: 1 metre = 1 650 763.73 wavelengths.

Radii / pm: Kr^+ 169; covalent 189; van der Waals 198
Electronegativity: n.a. (Pauling); 2.94 (Allred); 6.8 eV (absolute) (see Key)
Effective nuclear charge: 8.25 (Slater); 9.77 (Clementi); 11.79 (Froese-Fischer)

Standard reduction potentials $E^{\ominus}$/V

n.a. [KrF_2 decomposes in water]

Oxidation states

Kr^0	[Kr]	clathrates $Kr_8(OH_2)_{46}$, $Kr(quinol)_3$
Kr^{II}	s^2p^4	KrF_2, $[KrF]^+[AsF_6]^-$

Covalent bonds

Bond	r/ pm	E/ kJ mol⁻¹
Kr—F	189	50

• PHYSICAL DATA

Melting point / K: 116.6
Boiling point / K: 120.85
Critical temperature / K: 209.45
Critical pressure / kPa: 5502

ΔH_{fusion} /kJ mol^{-1}: 1.64
ΔH_{vap} /kJ mol^{-1}: 9.05

Thermodynamic properties (298.15 K, 0.1 MPa)

State	$\Delta_f H^{\ominus}$/kJ mol^{-1}	$\Delta_f G^{\ominus}$/kJ mol^{-1}	$S^{\ominus}$/J K^{-1} mol^{-1}	C_p/J K^{-1} mol^{-1}
Gas	0	0	164.082	20.786

Density / kg m^{-3}: 2823 [s., m.p.]; 2413 [liq. b.p.]; 3.7493 [gas, 273 K]
Molar volume / cm^3: 29.68 [116 K]
Thermal conductivity / W m^{-1} K^{-1}: 0.00949 [300 K] (g)
Mass magnetic susceptibility / kg^{-1} m^3: -4.32×10^{-9} (g)

• BIOLOGICAL DATA

Biological role

None.

Toxicity

Non-toxic.

Hazards

Krypton is a harmless gas, although it could asphyxiate if it excluded oxygen from the lungs.

Levels in humans

Blood / mg dm^{-3}: trace
Bone / p.p.m.: nil
Liver / p.p.m.: nil
Muscle / p.p.m.: nil
Daily dietary intake: n.a., but low
Total mass of element in average (70 kg) person: n.a., but small

Discovery: see Chemical Data section.
[Greek, *kryptos* = hidden]
French, *krypton*; German, *Krypton*; Italian, *cripto*; Spanish, *kriptón*

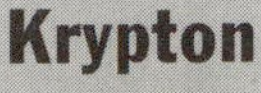

[krip-ton]

• NUCLEAR DATA

Number of isotopes (including nuclear isomers): 27 **Isotope mass range:** $72 \rightarrow 94$

Key isotopes

Nuclide	Atomic mass	Natural abundance (%)	Nuclear spin I	Nuclear magnetic moment μ	Uses
^{78}Kr	77.920 396	0.35	0+		
^{80}Kr	79.916 380	2.25	0+		
^{82}Kr	81.913 482	11.6	0+		
^{83}Kr	82.914 135	11.5	9/2+	–0.970	NMR
^{84}Kr	83.911 507	57.0	0+		
^{86}Kr	85.910 616	17.3	0+		

A table of radioactive isotopes is given in Appendix 1, on p242.

NMR [Rare]

	^{83}Kr
Relative sensitivity (^1H = 1.00)	1.88×10^{-3}
Receptivity (^{13}C = 1.00)	1.23
Magnetogyric ratio/rad $T^{-1}s^{-1}$	-1.029×10^{7}
Nuclear quadrupole moment/m^2	$+0.253 \times 10^{-28}$
Frequency (^1H = 100 Hz; 2.3488T)/MHz	3.847

• ELECTRON SHELL DATA

Ground state electron configuration: $[Ar]3d^{10}4s^24p^6$ = [Kr]
Term symbol: $^1S_0$
Electron affinity ($M \rightarrow M^-$)/kJ mol^{-1}**:** –39 (calc.)

Ionization energies/kJ mol^{-1}**:**

1.	$M \rightarrow M^+$	1350.7
2.	$M^+ \rightarrow M^{2+}$	2350
3.	$M^{2+} \rightarrow M^{3+}$	3565
4.	$M^{3+} \rightarrow M^{4+}$	5070
5.	$M^{4+} \rightarrow M^{5+}$	6240
6.	$M^{5+} \rightarrow M^{6+}$	7570
7.	$M^{6+} \rightarrow M^{7+}$	10 710
8.	$M^{7+} \rightarrow M^{8+}$	12 200
9.	$M^{8+} \rightarrow M^{9+}$	22 229
10.	$M^{9+} \rightarrow M^{10+}$	(28 900)

Electron binding energies/eV

K	1s	14 326
L_I	2s	1921
L_{II}	$2p_{1/2}$	1730.9
L_{III}	$2p_{3/2}$	1678.4
M_I	3s	292.8
M_{II}	$3p_{1/2}$	222.2
M_{III}	$3p_{3/2}$	214.4
M_{IV}	$3d_{3/2}$	95.0
M_V	$3d_{5/2}$	93.8

continued in Appendix 2, p256

Main lines in atomic spectrum
[Wavelength/nm(species)]

473.900 (II)
587.091 (I)
810.436 (I)
811.290 (I)
829.811 (I)
877.675 (I)

• CRYSTAL DATA

Crystal structure (cell dimensions/pm), space group
f.c.c. (80 K) (a = 572.1), Fm3m

X-ray diffraction: mass absorption coefficients (μ/ρ)/cm^2 g^{-1}**:** CuK_α 108 MoK_α 84.9
Neutron scattering length, $b/10^{-12}$ cm**:** 0.785
Thermal neutron capture cross-section, σ_a/barns**:** 25

• GEOLOGICAL DATA

Minerals

None as such.

Chief source: liquid air
World production/tonnes y^{-1}**:** 8
Reserves/tonnes: 1.7×10^{10} (atmosphere)
Specimen: not generally available. Safe.

Abundances

Sun (relative to H = 1×10^{12})**:** n.a.
Earth's crust/p.p.m.**:** *c.* 1×10^{-5}
Atmosphere/p.p.m. (volume)**:** 1.14
Seawater/p.p.m.**:** 8×10^{-5}
Residence time/years**:** n.a.
Oxidation state: 0

La

Atomic number: 57
Relative atomic mass ($^{12}C = 12.0000$)**:** 138.9055

CAS: [7439-91-0]

• CHEMICAL DATA

Description: Lanthanum is a soft, silvery-white metal that rapidly tarnishes in air and burns easily if ignited. It reacts with water to give hydrogen gas. Lanthanum is used in optical glass and for flints. Lanthanum(III) salts are used as biological tracers for calcium.

Radii/pm: La^{3+} 122; atomic 188; covalent 169
Electronegativity: 1.10 (Pauling); 1.08 (Allred); 3.1 eV (absolute)
Effective nuclear charge: 2.85 (Slater); 9.31 (Clementi); 10.43 (Froese-Fischer)

Standard reduction potentials $E^{\ominus}/V$

	III		0
acid	La^{3+}	$\xrightarrow{-2.38}$	La
base	$La(OH)_3$	$\xrightarrow{-2.80}$	La

Oxidation states

La^{III} [Xe] La_2O_3, $La(OH)_3$, $[La(OH_2)_x]^{3+}$ (aq), La^{3+} salts, LaF_3, $LaCl_3$ etc., $LaOCl$, $[La(NCS)_6]^{3-}$, complexes. LaH_2-LaH_3 is probably $La^{3+}H^-$

• PHYSICAL DATA

Melting point/K: 1194
Boiling point/K: 3730
ΔH_{fusion}/kJ mol^{-1}: 10.04
ΔH_{vap}/kJ mol^{-1}: 399.6

Thermodynamic properties (298.15 K, 0.1 MPa)

State	$\Delta_f H^{\ominus}$/kJ mol^{-1}	$\Delta_f G^{\ominus}$/kJ mol^{-1}	$S^{\ominus}$/J K^{-1} mol^{-1}	C_p/J K^{-1} mol^{-1}
Solid	0	0	56.9	27.11
Gas	431.0	393.56	182.377	22.753

Density/kg m^{-3}: 6145 [298 K]
Molar volume/cm^3: 22.60
Thermal conductivity/W m^{-1} K^{-1}: 13.5 [300 K]
Coefficient of linear thermal expansion/K^{-1}: 4.9×10^{-6}
Electrical resistivity/Ω m: 57×10^{-8} [298 K]
Mass magnetic susceptibility/kg^{-1} m^3: $+1.1 \times 10^{-8}$ (s)
Young's modulus/GPa: 37.9
Rigidity modulus/GPa: 14.9
Bulk modulus/GPa: n.a.
Poisson's ratio/GPa: 0.28

• BIOLOGICAL DATA

Biological role

None.

Toxicity

Toxic intake: n.a.
Lethal intake: LD_{50} (chloride, oral, rat) = 4200 mg kg^{-1}

Hazards

Lanthanum is mildly toxic by ingestion, and causes liver injury.

Levels in humans

Blood/mg dm^{-3}: n.a.
Bone/p.p.m.: < 0.08
Liver/p.p.m.: 0.3
Muscle/p.p.m.: 0.0004
Daily dietary intake: n.a. but very low
Total mass of element in average (70 kg) person: *c.* 0.8 mg

Discovered in 1839 by C.G. Mosander at Stockholm, Sweden.
[Greek, *lanthanein* = to lie hidden]
French, *lanthane*; German, *Lanthan*; Italian, *lantanio*; Spanish, *lantano*

Lanthanum

[lan-than-uhm]

• NUCLEAR DATA

Number of isotopes (including nuclear isomers): 26 **Isotope mass range:** 125 → 149

Key isotopes

Nuclide	Atomic mass	Natural abundance (%)	Nuclear spin I	Nuclear magnetic moment μ	Uses
^{138}La*	137.907 105	0.09	5+	+3.713 9	E, NMR
^{139}La	138.906 347	99.91	7/2+	+2.783 2	E, NMR

* ^{138}La is radioactive with a half-life of 1.0×10^{11} y and decay mode β^- (1.04 MeV) 34%, EC (1.75 MeV) 66%; γ.
A table of radioactive isotopes is given in Appendix 1, on p243.

NMR [Reference: 0.01M $LaCl_3$]	^{138}La	^{139}La
Relative sensitivity (^{1}H = 1.00)	0.0919	0.0592
Receptivity (^{13}C = 1.00)	0.43	336
Magnetogyric ratio/rad $T^{-1}s^{-1}$	3.5295×10^7	3.7787×10^7
Nuclear quadrupole moment/m^2	$+0.450 \times 10^{-28}$	$+0.200 \times 10^{-28}$
Frequency (^{1}H = 100 Hz; 2.3488T)/MHz	13.193	14.126

• ELECTRON SHELL DATA

Ground state electron configuration: $[Xe]5d^1 6s^2$
Term symbol: $^2D_{3/2}$
Electron affinity ($M \rightarrow M^-$)/kJ mol^{-1}: *c.* 50

Ionization energies/kJ mol^{-1}:

1.	$M \rightarrow M^+$	538.1
2.	$M^+ \rightarrow M^{2+}$	1067
3.	$M^{2+} \rightarrow M^{3+}$	1850
4.	$M^{3+} \rightarrow M^{4+}$	4819
5.	$M^{4+} \rightarrow M^{5+}$	(6400)
6.	$M^{5+} \rightarrow M^{6+}$	(7600)
7.	$M^{6+} \rightarrow M^{7+}$	(9600)
8.	$M^{7+} \rightarrow M^{8+}$	(11 000)
9.	$M^{8+} \rightarrow M^{9+}$	(12 400)
10.	$M^{9+} \rightarrow M^{10+}$	(15 900)

Electron binding energies/eV

K	1s	38 925
L_I	2s	6266
L_{II}	$2p_{1/2}$	5891
L_{III}	$2p_{3/2}$	5483
M_I	3s	1362
M_{II}	$3p_{1/2}$	1209
M_{III}	$3p_{3/2}$	1128
M_{IV}	$3d_{3/2}$	853
M_V	$3d_{5/2}$	836

continued in Appendix 2, p256

Main lines in atomic spectrum
[Wavelength/nm(species)]
394.910 (II)
408.672 (II)
418.732 (II)
433.374 (II)
550.134 (I) (AA)

• CRYSTAL DATA

Crystal structure (cell dimensions/pm), space group
α-La hexagonal (a = 377.0, c = 121.59), P6₃/mmc
β-La f.c.c. (a = 529.6), Fm3m
γ-La b.c.c. (a = 426), Im3m
$T(\alpha \rightarrow \beta)$ = 583 K; $T(\beta \rightarrow \gamma)$ = 1137 K
X-ray diffraction: mass absorption coefficients (μ/ρ)/$cm^2 g^{-1}$: CuK_α 341 MoK_α 45.8
Neutron scattering length, $b/10^{-12}$ cm: 0.824
Thermal neutron capture cross-section, σ_a/barns: 8.98

• GEOLOGICAL DATA

Minerals

Mineral	Formula	Density	Hardness	Crystal appearance
Allanite	$Ca(Ce,La)(Al,Fe)_3(SiO_4)_3OH$	4.0	5.5 – 6.0	mon., sub-met. black
Bastnäsite-La*	$(La,Ce, etc.)CO_3(F,OH)$	n.a.	n.a.	hex.
Cerite	$(Ce,La,Ca)_9(Mg,Fe)Si_7(O,OH,F)_{28}$	4.75	5	rhom., res. black
Monazite-La*	(La, Ce, Nd, Th, etc.)PO_4	5.20	5 – 5.5	mon., waxy/vit. yellow-brown

*Varieties of these minerals that are particularly rich in lanthanum.

Chief ores: monazite, bastnäsite
World production/tonnes y^{-1}: 12 500
Main mining areas: USA, Brazil, India, Sri Lanka, Australia.
Reserves/tonnes: *c.* 6×10^6
Specimen: available as chips, ingots or powder. *Care!*

Abundances

Sun (relative to H = 1×10^{12}): 13.5
Earth's crust/p.p.m.: 32
Seawater/p.p.m.:
Atlantic surface: 1.8×10^{-6}
Atlantic deep: 3.8×10^{-6}
Pacific surface: 2.6×10^{-6}
Pacific deep: 6.9×10^{-6}
Residence time/years: 200
Classification: recycled
Oxidation state: III

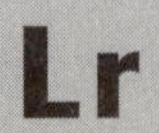

Lr

Atomic number: 103
Relative atomic mass ($^{12}C = 12.0000$)**:** 262.11 (Lr-262)
CAS: [22537-19-5]

• CHEMICAL DATA

Description: Lawrencium is a radioactive metal element which does not occur naturally, and is of research interest only.

Radii/pm: Lr^{4+} 83; Lr^{3+} 88; Lr^{2+} 112; atomic n.a.
Electronegativity: 1.3 (Pauling)
Effective nuclear charge: 1.80 (Slater); n.a. (Clementi); (n.a.) Froese-Fischer

Standard reduction potentials $E^{\ominus}$/V

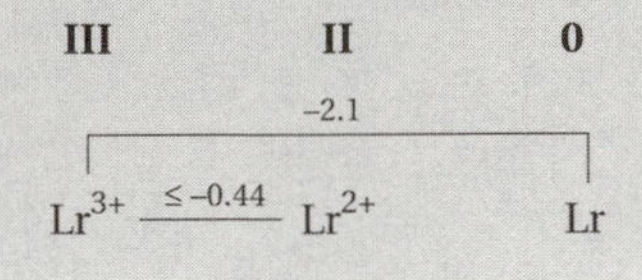

Oxidation states

Lr^{II}	d^1	Lr^{2+} ?
Lr^{III}	$[f^{14}]$	Lr^{3+} (aq)

• PHYSICAL DATA

Melting point/K: n.a.
Boiling point/K: n.a.
ΔH_{fusion}/kJ mol^{-1}: n.a.
ΔH_{vap}/kJ mol^{-1}: n.a.

Thermodynamic properties (298.15 K, 0.1 MPa)

State	$\Delta_f H^{\ominus}$/kJ mol^{-1}	$\Delta_f G^{\ominus}$/kJ mol^{-1}	$S^{\ominus}$/J K^{-1} mol^{-1}	C_p/J K^{-1} mol^{-1}
Solid	0	0	n.a.	n.a.
Gas	n.a.	n.a.	n.a.	n.a.

Density/kg m^{-3}: n.a.
Molar volume/cm^3: n.a.
Thermal conductivity/W m^{-1} K^{-1}: 10 (est.) [300 K]
Coefficient of linear thermal expansion/K^{-1}: n.a.
Electrical resistivity/Ω m: n.a.
Mass magnetic susceptibility/kg^{-1} m^3: n.a.

• BIOLOGICAL DATA

Biological role
None.

Toxicity
Toxic intake: n.a.
Lethal intake: n.a.

Hazards
Lawrencium is never encountered normally, and relatively few atoms have ever been made. It would be dangerous because of its intense radioactivity.

Levels in humans
Nil
Daily dietary intake: nil
Total mass of element in average (70 kg) person: nil

Discovery: see Nuclear Data section.
[Named after Ernest O. Lawrence, inventor of the cyclotron]
French, *lawrencium*; German, *Lawrencium*; Italian, *lawrentio*; Spanish, *lawrencio*

Lawrencium
[law-ren-see-uhm]

• NUCLEAR DATA

Discovery: Lawrencium was prepared in 1961 by A. Ghiorso, T. Sikkeland, A.E. Larsh and R.M. Latimer at Berkeley, California, USA.

Number of isotopes (including nuclear isomers): 10 **Isotope mass range:** 253 → 262

Key isotopes

Nuclide	Atomic mass	Half life ($T_{1/2}$)	Decay mode and energy (MeV)	Nuclear spin *I*	Nucl. mag. moment μ	Uses
^{253}Lr	253.095 190	1.3 s	α			
^{254}Lr	254.096 320	13 s	α			
^{255}Lr	255.096 670	22 s	α (8.80); EC			
^{256}Lr	256.098 490	26 s	α (8.554) 99.7%; SF 0.3%			
^{257}Lr	257.099 480	0.65 s	α (9.30)	7/2+		
^{258}Lr	258.101 710	3.9 s	α (9.00)			
^{259}Lr	259.102 900	6.1 s	α (8.70); SF			
^{260}Lr	260.105 320	3 m	α (8.30); EC			
^{261}Lr		39 m	SF			
^{262}Lr		261 m	EC			

NMR [Not recorded]

• ELECTRON SHELL DATA

Ground state electron configuration: $[Rn]5f^{14}6d^{1}7s^{2}$
Term symbol: $^{2}D_{5/2}$
Electron affinity ($M \rightarrow M^{-}$)/kJ mol^{-1}**:** n.a.

Ionization energies/kJ mol^{-1}**:**
1. $M \rightarrow M^{+}$ n.a.

Electron binding energies/eV
n.a.

Main lines in atomic spectrum
[Wavelength/nm(species)]
n.a.

• CRYSTAL DATA

Crystal structure (cell dimensions/pm), space group
n.a.

X-ray diffraction: mass absorption coefficients (μ/ρ)/cm^2 g^{-1}**:** CuK_α n.a. MoK_α n.a.
Neutron scattering length, $b/10^{-12}$ cm**:** n.a.
Thermal neutron capture cross-section, σ_a/barns**:** n.a.

• GEOLOGICAL DATA

Minerals
Not found on Earth.

Chief source: only a few atoms of lawrencium have been made, by bombarding ^{252}Cf with boron nuclei, or ^{249}Bk with ^{18}O nuclei.
Specimen: not available commercially.

Abundances
Sun (relative to $H = 1 \times 10^{12}$)**:** n.a.
Earth's crust/p.p.m.**:** nil
Seawater/p.p.m.**:** nil

Pb

Atomic number: 82
Relative atomic mass ($^{12}C = 12.0000$)**:** 207.2

CAS: [7439-92-1]

• CHEMICAL DATA

Description: Lead is a soft, weak, ductile, dull grey metal that tarnishes in moist air but is stable to oxygen and water. It dissolves in HNO_3. Lead is used in batteries, cables, glass, solder, radiation shielding etc. A little is still used in paints and petrol but generally this use is being phased out.

Radii/pm: Pb^{4+} 84; Pb^{2+} 132; atomic 175; covalent 154
Electronegativity: 2.33 (Pauling); 1.55 (Allred); 3.90 eV (absolute)
Effective nuclear charge: 5.65 (Slater); 12.39 (Clementi); 15.33 (Froese-Fischer)

Standard reduction potentials $E^{\ominus}/V$

	IV		II		0		–II
acid	Pb^{4+}	1.69	Pb^{2+}	−0.1251	Pb	−1.507	PbH_2

[alkaline solutions contain many different forms]

Oxidation states

Pb^{II}	s^2	PbO, PbF_2, $PbCl_2$ etc., $[Pb(OH)]^+(aq)$, $[Pb(OH_2)_x]^{2+}(aq)$, salts, complexes
Pb^{IV}	d^{10}	PbO_2, Pb_3O_4, (= $2PbO.PbO_2$), PbF_4, $PbCl_4$, $PbBr_4$, $[PbCl_6]^{2-}$, $[Pb(OH)_6]^{2-}(aq)$, [Pb^{4+} does not exist in water], organolead compounds, e.g. $Pb(C_2H_5)_4$, complexes

Covalent bonds

Bond	r/ pm	E/ kJ mol^{-1}
Pb—H	184	180
Pb—C	229	130
Pb—O	192	398
Pb^{IV}—F	213	331
Pb^{IV}—Cl	247	243
Pb—Pb	350	100

• PHYSICAL DATA

Melting point/K: 600.65
Boiling point/K: 2013
ΔH_{fusion}/kJ mol^{-1}: 5.121
ΔH_{vap}/kJ mol^{-1}: 179.4

Thermodynamic properties (298.15 K, 0.1 MPa)

State	$\Delta_f H^{\ominus}$/kJ mol^{-1}	$\Delta_f G^{\ominus}$/kJ mol^{-1}	$S^{\ominus}$/J K^{-1} mol^{-1}	C_p/J K^{-1} mol^{-1}
Solid	0	0	64.81	26.44
Gas	195.0	161.9	175.373	20.786

Density/kg m^{-3}: 11 350 [293 K]; 10 678 [liquid at m.p.]
Molar volume/cm^3: 18.26
Thermal conductivity/W m^{-1} K^{-1}: 35.3 [300 K]
Coefficient of linear thermal expansion/K^{-1}: 29.1×10^{-6}
Electrical resistivity/Ω m: 20.648×10^{-8} [293 K]
Mass magnetic susceptibility/kg^{-1} m^3: -1.39×10^{-9} (s)

Young's modulus/GPa: 16.1
Rigidity modulus/GPa: 5.59
Bulk modulus/GPa: 45.8
Poisson's ratio/GPa: 0.44

• BIOLOGICAL DATA

Biological role

None.

Toxicity

Toxic intake: *c.* 50 mg – 100 g (see health hazards)

Lethal intake: LD_{50} (acetate, intravenous, mouse) = 104 mg kg^{-1}

Hazards

Lead is moderately toxic by ingestion and affects the gut and central nervous system, and causes anaemia. However, most ingested lead passes through the body without being absorbed. It is a cumulative poison and lead poisoning is a common occupational disease. Lead compounds can be carcinogenic and teratogenic.

Levels in humans

Blood/mg dm^{-3}: 0.21
Bone/p.p.m.: 3.6 – 30
Liver/p.p.m.: 3 – 12
Muscle/p.p.m.: 0.23 – 3.3
Daily dietary intake: 0.06 – 0.5 mg
Total mass of element in average (70 kg) person: 120 mg (stored in skeleton)

Known to ancient civilizations.
[Anglo-Saxon, *lead*; Latin, *plumbum*]
French, *plomb*; German, *Blei*; Italian, *piombo*; Spanish, *plomo*

Lead
[led]

• NUCLEAR DATA

Number of isotopes (including nuclear isomers): 41 **Isotope mass range:** 184 → 214

Key isotopes

Nuclide	Atomic mass	Natural abundance (%)	Nuclear spin *I*	Nuclear magnetic moment μ	Uses
^{204}Pb	203.973 020	1.4	0+		E
^{206}Pb	205.974 440	24.1	0+		E
^{207}Pb	206.975 872	22.1	1/2–	+0.582 19	E, NMR
^{208}Pb	207.976 627	52.4	0+		E

A table of radioactive isotopes is given in Appendix 1, on p243.

NMR [Reference: $Pb(CH_3)_4$]	^{207}Pb	^{209}Pb
Relative sensitivity ($^1H = 1.00$)	9.16×10^{-3}	
Receptivity ($^{13}C = 1.00$)	11.8	
Magnetogyric ratio/rad $T^{-1}s^{-1}$	5.5797×10^7	
Nuclear quadrupole moment/m^2	-	-0.269×10^{-28}
Frequency (^1H = 100 Hz; 2.3488T)/MHz	20.921	

• ELECTRON SHELL DATA

Ground state electron configuration: $[Xe]4f^{14}5d^{10}6s^26p^2$
Term symbol: $^3P_0$
Electron affinity ($M \rightarrow M^-$)/kJ mol^{-1}**:** 35.1

Ionization energies/kJ mol^{-1}:

1.	$M \rightarrow M^+$	715.5
2.	$M^+ \rightarrow M^{2+}$	1450.4
3.	$M^{2+} \rightarrow M^{3+}$	3081.5
4.	$M^{3+} \rightarrow M^{4+}$	4083
5.	$M^{4+} \rightarrow M^{5+}$	6640
6.	$M^{5+} \rightarrow M^{6+}$	(8100)
7.	$M^{6+} \rightarrow M^{7+}$	(9900)
8.	$M^{7+} \rightarrow M^{8+}$	(11 800)
9.	$M^{8+} \rightarrow M^{9+}$	(13 700)
10.	$M^{9+} \rightarrow M^{10+}$	(16 700)

Electron binding energies/eV

K	1s	88 005
L_I	2s	15 861
L_{II}	$2p_{1/2}$	15 200
L_{III}	$2p_{3/2}$	13 055
M_I	3s	3851
M_{II}	$3p_{1/2}$	3554
M_{III}	$3p_{3/2}$	3066
M_{IV}	$3d_{3/2}$	2586
M_V	$3d_{5/2}$	2484

continued in Appendix 2, p256

Main lines in atomic spectrum
[Wavelength/nm(species)]

217.000 (I) (AA)
261.418 (I)
283.305 (I)
357.273 (I)
363.957 (I)
368.346 (I)
405.781 (I)

• CRYSTAL DATA

Crystal structure (cell dimensions/pm), space group
f.c.c. ($a = 495.00$), Fm3m

X-ray diffraction: mass absorption coefficients $(\mu/\rho)/cm^2\ g^{-1}$**:** CuK_α 232 MoK_α 120
Neutron scattering length, $b/10^{-12}$ cm: 0.9405
Thermal neutron capture cross-section, σ_a/barns**:** 0.171

• GEOLOGICAL DATA

Minerals

Mineral	Formula	Density	Hardness	Crystal appearance
Anglesite	$PbSO_4$	6.38	2.5 – 3	orth, adam./greasy white
Boulangerite	$Pb_5Sb_4S_{11}$	6.23	2.5 – 3	orth., met. bluish-grey
Bournonite	$PbCuSbS_3$	5.8	2.5 – 3	orth., met. steel-grey
Cerussite	$PbCO_3$	6.55	3 – 3.5	orth., adam./vit. col.
Galena	PbS	7.58	2.6	cub., met. grey
Minium	Pb_3O_4	9.05	2.5	tet., greasy red/brown
Pyromorphite	$Pb_5(PO_4)_3Cl$	7.04	3.5 – 4	hex., barrel-shaped crystals, often hollow

Chief ores: galena is the main ore (with silver as a by-product), pyromorphite, boulangerite and cerussite are minor ores.

World production/tonnes y^{-1}**:** 2.8×10^6

Main mining areas: galena in USA, Australia, Mexico, West Germany; boulangerite in France.

Reserves/tonnes: 85×10^6

Specimen: available as foil, granules, ingots, powder, rod, shot and wire. *Care!*

Abundances

Sun (relative to $H = 1 \times 10^{12}$)**:** 85.1
Earth's crust/p.p.m.**:** 14
Atmosphere/p.p.m. (volume)**:** trace
Seawater/p.p.m.**:**
Atlantic surface: 30×10^{-6}
Atlantic deep: 4.0×10^{-6}
Pacific surface: 10×10^{-6}
Pacific deep: 1×10^{-6}
Residence time/years**:** 50
Classification: scavenged
Oxidation state: II

Li

Atomic number: 3
Relative atomic mass ($^{12}C = 12.0000$)**:** 6.941

CAS: [7439-93-2]

• CHEMICAL DATA

Discovery: Lithium was discovered in 1817 by J.A. Arfvedson at Stockholm, Sweden. Isolated in 1821 by W.T. Brande.
Description: Lithium is a soft, silvery-white, metal that reacts slowly with oxygen and water. It is used in light-weight alloys, especially with aluminium and magnesium, and in greases, batteries, glass, medicine and nuclear bombs.

Radii/pm: Li^+ 78; atomic 152; covalent 123
Electronegativity: 0.98 (Pauling); 0.97 (Allred); 3.01 eV (absolute)
Effective nuclear charge: 1.30 (Slater); 1.28 (Clementi); 1.55 (Froese-Fischer)

Standard reduction potentials $E^{\ominus}$/V

I		0
Li^+	—−3.040—	Li

Oxidation states

Li^{-I}	s^2	Li solutions in liquid ammonia
Li^{I}	[He]	Li_2O, LiOH, LiH, $LiAlH_4$, LiF, LiCl etc., $[Li(OH_2)_4]^+$ (aq), Li_2CO_3, salts of Li^+, some complexes, $[Li(CH_3)]_4$, $[Li(12\text{-crown-}4)]^+$

• PHYSICAL DATA

Melting point/K: 453.69
Boiling point/K: 1620
ΔH_{fusion}/kJ mol^{-1}: 4.60
ΔH_{vap}/kJ mol^{-1}: 134.7

Thermodynamic properties (298.15 K, 0.1 MPa)

State	$\Delta_f H^{\ominus}$/kJ mol^{-1}	$\Delta_f G^{\ominus}$/kJ mol^{-1}	$S^{\ominus}$/J K^{-1} mol^{-1}	C_p/J K^{-1} mol^{-1}
Solid	0	0	29.12	24.77
Gas	159.37	126.66	138.77	20.786

Density/kg m^{-3}: 534 [293 K]; 515 [liquid m.p.]
Molar volume/cm^3: 13.00
Thermal conductivity/W m^{-1} K^{-1}: 84.7 [300 K]
Coefficient of linear thermal expansion/K^{-1}: 56×10^{-6}
Electrical resistivity/Ω m: 8.55×10^{-8} [273 K]
Mass magnetic susceptibility/kg^{-1} m^3: $+2.56 \times 10^{-8}$ (s)

Young's modulus/GPa: 4.91
Rigidity modulus/GPa: 4.24
Bulk modulus/GPa: n.a.
Poisson's ratio/GPa: 0.36

• BIOLOGICAL DATA

Biological role

None; but lithium acts to stimulate metabolism and can control manic-depressive disorders.

Toxicity

Toxic intake: 20 – 200 g (see under hazards)
Lethal intake: LD_{50} (carbonate, oral, rat) = 525 mg kg^{-1}

Hazards

Lithium is moderately toxic by ingestion but there are wide variations of tolerance. Even lithium carbonate, which is used in psychiatry, is prescribed at doses near to the toxic level. Some lithium compounds are carcinogenic and teratogenic.

Levels in humans

Blood/mg dm^{-3}: 0.004
Bone/p.p.m.: 1.3
Liver/p.p.m.: 0.025
Muscle/p.p.m.: 0.023
Daily dietary intake: 0.1 – 2mg
Total mass of element in average (70 kg) person: 7 mg

Discovery: see Chemical Data section.
[Greek, *lithos* = stone]
French, *lithium*; German, *Lithium*; Italian, *litio*; Spanish, *litio*

[lith-iuhm]

• NUCLEAR DATA

Number of isotopes (including nuclear isomers): 5 **Isotope mass range:** 5 → 9

Key isotopes

Nuclide	Atomic mass	Natural abundance (%)	Nuclear spin I	Nuclear magnetic moment μ	Uses
^{6}Li	6.015 121	7.5	1+	+0.822 046 7	E, NMR
^{7}Li	7.016 003	92.5	3/2–	+3.256 424	E, NMR

A table of radioactive isotopes is given in Appendix 1, on p243.

NMR [Reference: LiCl (aq)]	^{6}Li	^{7}Li
Relative sensitivity (^{1}H = 1.00)	8.50×10^{-3}	0.29
Receptivity (^{13}C = 1.00)	3.58	1540
Magnetogyric ratio/rad $T^{-1}s^{-1}$	3.9366×10^{7}	10.3964×10^{7}
Nuclear quadrupole moment/m^2	-0.00082×10^{-28}	-0.041×10^{-28}
Frequency (^{1}H = 100 Hz; 2.3488T)/MHz	14.716	38.863

• ELECTRON SHELL DATA

Ground state electron configuration: [He]$2s^1$
Term symbol: $^{2}S_{1/2}$
Electron affinity ($M \rightarrow M^-$)/kJ mol^{-1}**:** 59.6

Ionization energies/kJ mol^{-1}**:**

1. $M \rightarrow M^+$ 513.3
2. $M^+ \rightarrow M^{2+}$ 7298.0
3. $M^{2+} \rightarrow M^{3+}$ 11 814.8

Electron binding energies/eV

K	1s	54.7

Main lines in atomic spectrum
[Wavelength/nm(species)]

323.266 (I)
548.355 (II)
548.565 (II)
610.362 (I)
670.776 (I)
670.791 (I) (AA)

• CRYSTAL DATA

Crystal structure (cell dimensions/pm), space group
α-Li b.c.c. (a = 351.00), Im3m
β-Li f.c.c. (a = 437.9), Fm3m
α form stable at room temperature; converts to β form at low temperatures
X-ray diffraction: mass absorption coefficients (μ/ρ)/cm^2 g^{-1}**:** CuK$_\alpha$ 0.716 MoK$_\alpha$ 0.217
Neutron scattering length, $b/10^{-12}$ cm**:** –0.190
Thermal neutron capture cross-section, σ_a/barns**:** 70.5

• GEOLOGICAL DATA

Minerals

Lithium is found in small amounts in nearly all rocks, and in many mineral spring waters.

Mineral	Formula	Density	Hardness	Crystal appearance
Amblygonite	$(Li,Na)AlPO_4(F,OH)$	3.1	5.5 – 6	tric., vit./greasy white/grey
Lepidolite	$K(Li,Al)_3(Si,Al)_4O_{10}(F,OH)_2$	2.85	2.5 – 3	mon., pink/lilac, lamellae
Petalite	$LiAlSi_4O_{10}$	2.4	6 – 6.5	mon., vit./pearly colourless/white
Spudomene	$LiAlSi_2O_6$	3.2	6.5 – 7.5	mon., vit. colourless/grey

Chief ores: petalite, lepidolite
World production/tonnes y^{-1}**:** 39 000
Main mining areas: USA; lithium is also recovered from brines of Searles Lake in California.
Reserves/tonnes: 7.3×10^{6}
Specimen: available as chunks, ingot, powder, ribbon, rod, shot or wire. *Care!*

Abundances

Sun (relative to H = 1×10^{12}): 10
Earth's crust/p.p.m.: 20
Seawater/p.p.m.: 0.17
Residence time/years: 2×10^{6}
Classification: accumulating
Oxidation state: I

Atomic number: 71
Relative atomic mass ($^{12}C = 12.0000$): **174.967**

CAS: [7439-94-3]

• CHEMICAL DATA

Discovery: Lutetium was discovered in 1907 by G. Urbain at Paris, France, and independently by C. James at the University of New Hampshire, USA.
Description: Lutetium is the hardest, densest and one of the rarest so-called rare-earth metals. It is little used except in chemical research.

Radii/pm: Lu^{3+} 85; atomic 173; covalent 156
Electronegativity: 1.27 (Pauling); 1.14 (Allred); ≤ 3.0 eV (absolute)
Effective nuclear charge: 3.00 (Slater); 8.80 (Clementi); 12.68 (Froese-Fischer)

Standard reduction potentials $E^{\ominus}/V$

	III		0
acid	Lu^{3+}	−2.30	Lu
base	$Lu(OH)_3$	−2.83	Lu

Oxidation states

Lu^{III} f^{14} — Lu_2O_3, $Lu(OH)_3$, $[Lu(OH_2)_x]^{3+}$(aq), Lu^{3+} salts, LuF_3, $LuCl_3$ etc., $[LuCl_6]^{3-}$, complexes

• PHYSICAL DATA

Melting point/K: 1936
Boiling point/K: 3668
ΔH_{fusion}/kJ mol^{-1}: 19.2
ΔH_{vap}/kJ mol^{-1}: 428

Thermodynamic properties (298.15 K, 0.1 MPa)

State	$\Delta_f H^{\ominus}$/kJ mol^{-1}	$\Delta_f G^{\ominus}$/kJ mol^{-1}	$S^{\ominus}$/J K^{-1} mol^{-1}	C_p/J K^{-1} mol^{-1}
Solid	0	0	50.96	26.86
Gas	427.6	387.8	184.800	20.861

Density/kg m^{-3}: 9840 [298 K]
Molar volume/cm^3: 17.78
Thermal conductivity/W m^{-1} K^{-1}: 16.4 [300 K]
Coefficient of linear thermal expansion/K^{-1}: 8.12×10^{-6}
Electrical resistivity/Ω m: 79.0×10^{-8} [298 K]
Mass magnetic susceptibility/kg^{-1} m^3: $+1.3 \times 10^{-9}$ (s)

Young's modulus/GPa: 68.6
Rigidity modulus/GPa: 27.2
Bulk modulus/GPa: 47.6
Poisson's ratio/GPa: 0.261

• BIOLOGICAL DATA

Biological role
None, but acts to stimulate metabolism.

Toxicity
Toxic intake: n.a.
Lethal intake: LD_{50} (chloride, oral, mouse) = 7100 mg kg^{-1}

Hazards
Lutetium is mildly toxic by ingestion.

Levels in humans
Organs: n.a., but low
Daily dietary intake: n.a.
Total mass of element in average (70 kg) person: n.a., but very low

Discovery: see Chemical Data section.
[Greek, *Lutetia* = Paris]
French, *lutétium*; German, *Lutetium*; Italian, *lutezio*; Spanish, *lutecio*

Lutetium

[loo-tee-shi-uhm]

• NUCLEAR DATA

Number of isotopes (including nuclear isomers): 41 **Isotope mass range:** 154 → 182

Key isotopes

Nuclide	Atomic mass	Natural abundance (%)	Nuclear spin *I*	Nuclear magnetic moment μ	Uses
^{175}Lu	174.940 770	97.41	7/2+	+2.232 7	E, NMR
^{176}Lu*	175.942 679	2.59	7–	+3.19	E

*^{176}Lu is radioactive with a half-life of 2.2×10^{10} y and decay mode β^- (1.02 MeV) with γ.
A table of radioactive isotopes is given in Appendix 1, on p243.

NMR [Reference: Not used]

	^{175}Lu	^{176}Lu
Relative sensitivity (^{1}H = 1.00)	0.0312	
Receptivity (^{13}C = 1.00)	156	
Magnetogyric ratio/rad $T^{-1}s^{-1}$	3.05×10^7	
Nuclear quadrupole moment/m^2	$+3.490 \times 10^{-28}$	$+4.970 \times 10^{-28}$
Frequency (^{1}H = 100 Hz; 2.3488T)/MHz	11.407	

• ELECTRON SHELL DATA

Ground state electron configuration: [Xe]$4f^{14}5d^16s^2$
Term symbol: $^2D_{3/2}$
Electron affinity ($M \rightarrow M^-$)/kJ mol^{-1}**:** ≤ 50

Ionization energies/kJ mol^{-1}**:**

1.	$M \rightarrow M^+$	523.5
2.	$M^+ \rightarrow M^{2+}$	1340
3.	$M^{2+} \rightarrow M^{3+}$	2022
4.	$M^{3+} \rightarrow M^{4+}$	4360
5.	$M^{4+} \rightarrow M^{5+}$	6445

Electron binding energies/eV

K	1s	63 314
L_I	2s	10 870
L_{II}	$2p_{1/2}$	10 349
L_{III}	$2p_{3/2}$	9244
M_I	3s	2491
M_{II}	$3p_{1/2}$	2264
M_{III}	$3p_{3/2}$	2024
M_{IV}	$3d_{3/2}$	1639
M_V	$3d_{5/2}$	1589

continued in Appendix 2, p256

Main lines in atomic spectrum
[Wavelength/nm(species)]

261.542 (II)
291.139 (II)
328.174 (I)
331.211 (I)
335.956 (I) (AA)
350.739 (II)

• CRYSTAL DATA

Crystal structure (cell dimensions/pm), space group

α-Lu h.c.p. (a = 350.31, c = 555.09), $P6_3/mmc$
β-Lu b.c.c. (a = 390), Im3m

X-ray diffraction: mass absorption coefficients (μ/ρ)/cm^2 g^{-1}**:** CuK_α 153 MoK_α 88.2
Neutron scattering length, $b/10^{-12}$ cm**:** 0.721
Thermal neutron capture cross-section, σ_a/barns**:** 84

• GEOLOGICAL DATA

Minerals

Mineral	Formula	Density	Hardness	Crystal appearance
Bastnäsite*	(Ce,La, etc.)CO_3F	4.9	4 – 5.5	hex., vit./greasy yellow
Monazite*	(Ce, La, Nd, Th, etc)PO_4	5.20	5 – 5.5	mon., waxy/vit. yellow-brown

*Although not a major constituent, lutetium is present in extractable amounts.

Chief ores: monazite, bastnäsite
World production/tonnes y^{-1}**:** *c.* 10
Main mining areas: USA, Brazil, India, Sri Lanka, Australia, China.
Reserves/tonnes**:** *c.* 2×10^5
Specimen: available as ingots or powder. Safe.

Abundances

Sun (relative to H = 1×10^{12})**:** 5.8
Earth's crust/p.p.m.**:** 0.51
Seawater/p.p.m.**:**
Atlantic surface: 1.4×10^{-7}
Atlantic deep: 2.0×10^{-7}
Pacific surface: 0.60×10^{-7}
Pacific deep: 4.1×10^{-7}
Residence time/years**:** 4000
Classification: recycled
Oxidation state: III

Mg

Atomic number: 12
Relative atomic mass ($^{12}C = 12.0000$)**:** 24.3050

CAS: [7439-95-4]

• CHEMICAL DATA

Discovery: Magnesium was recognized as an element in 1755 by Joseph Black at Edinburgh, Scotland; isolated by Sir Humphry Davy in 1808.
Description: Magnesium is a silvery white, lustrous and relatively soft metal. It is obtained by the electrolysis of fused $MgCl_2$. Magnesium burns in air when ignited and it reacts with hot water. It is used as the bulk metal and in lightweight alloys. Magnesium as a 'sacrificial' electrode will protect other metals that are exposed to seawater and ground water.

Radii/pm: Mg^{2+} 79; atomic 160; covalent 136
Electronegativity: 1.31 (Pauling); 1.23 (Allred); 3.75 eV (absolute)
Effective nuclear charge: 2.85 (Slater); 3.31 (Clementi); 4.15 (Froese-Fischer)

Standard reduction potentials $E^{\ominus}$/V

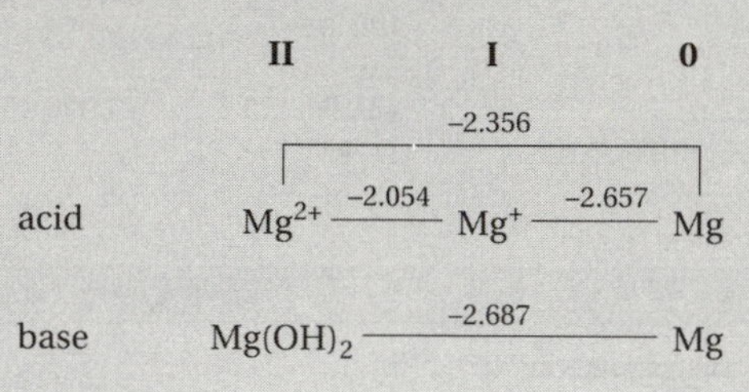

Oxidation states

Mg^{II} [Ne] MgO, MgO_2, $Mg(OH)_2$, $[Mg(OH_2)_6]^{2+}$ (aq), MgH_2, $MgCO_3$, Mg^{2+} salts, MgF_2, $MgCl_2$, etc., CH_3MgI, complexes

• PHYSICAL DATA

Melting point/K: 922.0
Boiling point/K: 1363
ΔH_{fusion}/kJ mol^{-1}: 9.04
ΔH_{vap}/kJ mol^{-1}: 128.7

Thermodynamic properties (298.15 K, 0.1 MPa)

State	$\Delta_f H^{\ominus}$/kJ mol^{-1}	$\Delta_f G^{\ominus}$/kJ mol^{-1}	$S^{\ominus}$/J K^{-1} mol^{-1}	C_p/J K^{-1} mol^{-1}
Solid	0	0	32.68	24.89
Gas	147.70	113.10	148.650	20.786

Density/kg m^{-3}: 1738 [293 K]; 1585 [liquid at m.p.]
Molar volume/cm^3: 13.98
Thermal conductivity/W m^{-1} K^{-1}: 156 [300 K]
Coefficient of linear thermal expansion/K^{-1}: 26.1×10^{-6}
Electrical resistivity/Ω m: 4.38×10^{-8} [293 K]
Mass magnetic susceptibility/kg^{-1} m^3: $+6.8 \times 10^{-9}$ (s)

Young's modulus/GPa: 44.7
Rigidity modulus/GPa: 17.3
Bulk modulus/GPa: 35.6
Poisson's ratio/GPa: 0.291

• BIOLOGICAL DATA

Biological role

Essential to all species.

Toxicity

Toxic intake: low toxicity
Lethal intake: LD_{50} (chloride, oral, rat) = 8100 mg kg^{-1}

Hazards

Magnesium compounds vary in toxicity but there is no evidence that magnesium itself produces systemic poisoning.

Levels in humans

Blood/mg dm^{-3}: 37.8
Bone/p.p.m.: 700 – 1800
Liver/p.p.m.: 590
Muscle/p.p.m.: 900
Daily dietary intake: 250 – 380 mg
Total mass of element in average (70 kg) person: 19 g

Discovery: see Chemical Data section.
[Greek, *Magnesia* = district of Thessaly]
French, *magnésium*; German, *Magnesium*; Italian, *magnesio*; Spanish, *magnesio*

Magnesium

[mag-neez-iuhm]

• NUCLEAR DATA

Number of isotopes (including nuclear isomers): 12 **Isotope mass range:** 20 → 31

Key isotopes

Nuclide	Atomic mass	Natural abundance (%)	Nuclear spin *I*	Nuclear magnetic moment μ	Uses
^{24}Mg	23.985 042	78.99	0+		E
^{25}Mg	24.985 837	10.00	5/2+	–0.855 46	E, NMR
^{26}Mg	25.982 593	11.01	0+		E

A table of radioactive isotopes is given in Appendix 1, on p243.

NMR [Reference: $MgCl_2$ (aq)]	^{25}Mg
Relative sensitivity (^1H = 1.00)	2.67×10^{-3}
Receptivity (^{13}C = 1.00)	1.54
Magnetogyric ratio/rad $T^{-1} s^{-1}$	1.6375×10^7
Nuclear quadrupole moment/m^2	0.1994×10^{-28}
Frequency (^1H = 100 Hz; 2.3488T)/MHz	6.1195

• ELECTRON SHELL DATA

Ground state electron configuration: $[Ne]3s^2$
Term symbol: $^1S_0$
Electron affinity ($M \rightarrow M^-$)/kJ mol^{-1}**:** –21

Ionization energies/kJ mol^{-1}**:**

1.	$M \rightarrow M^+$	737.7
2.	$M^+ \rightarrow M^{2+}$	1450.7
3.	$M^{2+} \rightarrow M^{3+}$	7732.6
4.	$M^{3+} \rightarrow M^{4+}$	10 540
5.	$M^{4+} \rightarrow M^{5+}$	13 630
6.	$M^{5+} \rightarrow M^{6+}$	17 995
7.	$M^{6+} \rightarrow M^{7+}$	21 703
8.	$M^{7+} \rightarrow M^{8+}$	25 656
9.	$M^{8+} \rightarrow M^{9+}$	31 642
10.	$M^{9+} \rightarrow M^{10+}$	35 461

Electron binding energies/eV

K	1s	1303.0
L_I	2s	88.6
L_{II}	$2p_{1/2}$	49.6
L_{III}	$2p_{3/2}$	49.2

Main lines in atomic spectrum
[Wavelength/nm(species)]
279.553 (II)
280.270 (II)
285.213 (I) (AA)
383.829 (I)
518.361 (I)

• CRYSTAL DATA

Crystal structure (cell dimensions/pm), space group
h.c.p. (*a* = 320.94; *c* = 521.03), $P6_3/mmc$

X-ray diffraction: mass absorption coefficients $(\mu/\rho)/cm^2\ g^{-1}$**:** CuK_α 38.6 MoK_α 4.11
Neutron scattering length, $b/10^{-12}$ cm: 0.5375
Thermal neutron capture cross-section, σ_a/barns**:** 0.063

• GEOLOGICAL DATA

Minerals

Many minerals are known which contain magnesium.

Mineral	Formula	Density	Hardness	Crystal appearance
Brucite	$Mg(OH)_2$	2.39	2.5	rhom., waxy/vit. white
Carnallite	$KMgCl_3$	1.602	2.5	orth., greasy, colourless reddish,
Cordierite*	$Mg_2Al_4Si_5O_{18}$	2.57	7	orth., transluscent grey/lilac etc.
Diopside	$CaMgSi_2O_6$	3.3	5.5 – 6.5	mon., vit./resinous white/yellow
Dolomite	$CaMg(CO_3)_2$	2.85	3.4 – 4	rhom., vit. colourless
Enstatite	$Mg_2Si_2O_6$	3.209	6 – 6	orth. pearly/vit., colourless grey

*Also known as 'white sapphire'.
Continued in Appendix 3 on page 260.

Chief sources: seawater; and the ores dolomite, magnesite; carnallite, kieserite and brucite
World production/tonnes y^{-1}**:** 325 000
Main mining areas: Austria, China, Poland, Russia, USA, India, Greece, Canada.
Reserves/tonnes: $> 2 \times 10^{10}$ as ores; and $> 1 \times 10^{24}$ in the sea
Specimen: available as chips, granules, powder, ribbon, rod or turnings. Safe.

Abundances

Sun (relative to H = 1×10^{12})**:** 4.0×10^7
Earth's crust/p.p.m.**:** 23 000
Seawater/p.p.m.**:** 1200
Residence time/years**:** 1×10^7
Classification: accumulating
Oxidation state: II

Atomic number: 25
Relative atomic mass ($^{12}C = 12.0000$)**:** 54.93805

CAS: [7439-96-5]

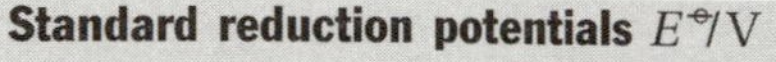

Description: Manganese is a hard, brittle, silvery metal. It is reactive when pure, burns in oxygen, reacts with water and dissolves in dilute acids. It is used in steel production and to make ceramics. Its compounds are used as feed supplements and fertilizer additives.

Radii/pm: Mn^{4+} 52; Mn^{3+} 70; Mn^{2+} 91; atomic 124; covalent 117
Electronegativity: 1.55 (Pauling); 1.60 (Allred); 3.72 eV (absolute)
Effective nuclear charge: 3.60 (Slater); 5.23 (Clementi); 7.17 (Froese-Fischer)

Standard reduction potentials $E^{\ominus}$/V

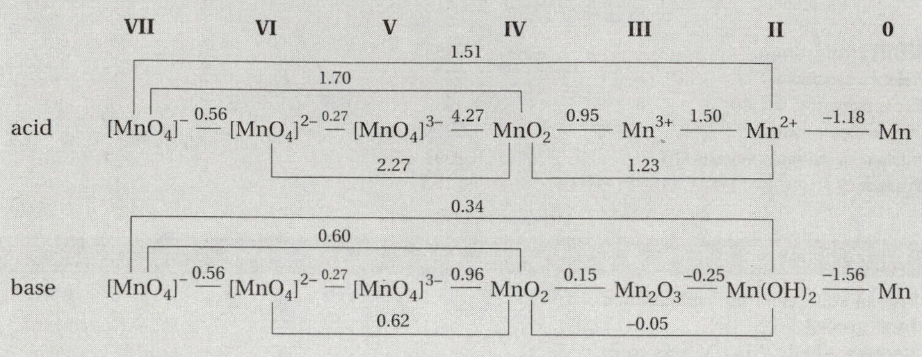

Oxidation states

Mn^{-III}	d^{10}	$[MnCO(NO)_3]$, $[Mn(CO)_4]^{3-}$
Mn^{-II}	d^9	some complexes known
Mn^{-I}	d^8	$[Mn(CO)_5]^-$
Mn^0	d^7	$[Mn_2(CO)_{10}]$
Mn^I	d^6	$[Mn(CN)_6]^-$, $[MnH(CO)_5]$
Mn^{II}	d^5	MnO, Mn_3O_4 ($=Mn^{II}Mn^{III}{}_2O_4$), $[Mn(OH_2)_6]^{2+}$ (aq); MnF_2, $MnCl_2$ etc., salts, complexes
Mn^{III}	d^4	Mn_2O_3, $[Mn(OH_2)_6]^{3+}$ (aq) unstable; MnF_3, $[MnCl_5]^{2-}$
Mn^{IV}	d^3	MnO_2, MnF_4, $[MnF_6]^{2-}$
Mn^V	d^2	$[MnO_4]^{3-}$
Mn^{VI}	d^1	$[MnO_4]^{2-}$
Mn^{VII}	d^0 [Ar]	Mn_2O_7, $[MnO_4]^-$

• PHYSICAL DATA

Melting point/K: 1517
Boiling point/K: 2235

ΔH_{fusion}/kJ mol^{-1}: 14.4
ΔH_{vap}/kJ mol^{-1}: 219.7

Thermodynamic properties (298.15 K, 0.1 MPa)

State	$\Delta_f H^{\ominus}$/kJ mol^{-1}	$\Delta_f G^{\ominus}$/kJ mol^{-1}	$S^{\ominus}$/J K^{-1} mol^{-1}	C_p/J K^{-1} mol^{-1}
Solid	0	0	32.01	26.32
Gas	280.7	238.5	173.70	20.79

Density/kg m^{-3}: 7440 (α) [293 K]; 6430 [liquid at m.p.]
Molar volume/cm^3: 7.38
Thermal conductivity/W m^{-1} K^{-1}: 7.82 [300 K]
Coefficient of linear thermal expansion/K^{-1}: 22×10^{-6}
Electrical resistivity/Ω m: 185.0×10^{-8} [298 K]
Mass magnetic susceptibility/kg^{-1} m^3: $+1.21 \times 10^{-7}$ (s)

Young's modulus/GPa: 191
Rigidity modulus/GPa: 79.5
Bulk modulus/GPa: n.a.
Poisson's ratio/GPa: 0.24

• BIOLOGICAL DATA

Biological role

Essential to all species.

Toxicity

Toxic intake: slightly toxic by ingestion
Lethal intake: LD_{50} (chloride, oral, mouse) = 1715 mg kg^{-1}

Hazards

Few poisonings have been caused by ingesting manganese compounds, but exposure to dust or fumes is a health hazard and working conditions should not exceed 5 mg m^{-3} even for short periods. Its compounds are experimental carcinogens and teratogens.

Levels in humans

Blood/mg dm^{-3}: 0.0016 – 0.075
Bone/p.p.m.: 0.2 – 100
Liver/p.p.m.: 3.6 – 9.6
Muscle/p.p.m.: 0.2 – 2.3
Daily dietary intake: 0.4 – 10 mg
Total mass of element in average (70 kg) person: 12 mg

Isolated in 1774 by J. G. Gahn at Stockholm, Sweden.
[Latin, *magnes* = magnet, or *magnesia nigri* = black magnesia (MnO_2)]
French, *manganèse*; German, *Mangan*; Italian, *manganese*; Spanish, *manganeso*

Manganese

[man-gan-eez]

• NUCLEAR DATA

Number of isotopes (including nuclear isomers): 15 **Isotope mass range:** 49 → 62

Key isotopes

Nuclide	Atomic mass	Natural abundance (%)	Nuclear spin *I*	Nuclear magnetic moment μ	Uses
^{55}Mn	54.938 047	100	5/2–	+3.453 2	NMR

A table of radioactive isotopes is given in Appendix 1, on p244.

NMR [Reference: $KMnO_4$ (aq)]	^{55}Mn
Relative sensitivity (^1H = 1.00)	0.18
Receptivity (^{13}C = 1.00)	994
Magnetogyric ratio/rad $T^{-1}s^{-1}$	6.6195×10^7
Nuclear quadrupole moment/m^2	$+0.330 \times 10^{-28}$
Frequency (^1H = 100 Hz; 2.3488T)/MHz	24.664

• ELECTRON SHELL DATA

Ground state electron configuration: $[Ar]3d^54s^2$
Term symbol: $^6S_{5/2}$
Electron affinity ($M \rightarrow M^-$)/kJ mol^{-1}**:** < 0

Ionization energies/kJ mol^{-1}**:**

1.	$M \rightarrow M^+$	717.4
2.	$M^+ \rightarrow M^{2+}$	1509.0
3.	$M^{2+} \rightarrow M^{3+}$	3248.4
4.	$M^{3+} \rightarrow M^{4+}$	4940
5.	$M^{4+} \rightarrow M^{5+}$	6990
6.	$M^{5+} \rightarrow M^{6+}$	9200
7.	$M^{6+} \rightarrow M^{7+}$	11 508
8.	$M^{7+} \rightarrow M^{8+}$	18 956
9.	$M^{8+} \rightarrow M^{9+}$	21 400
10.	$M^{9+} \rightarrow M^{10+}$	23 960

Electron binding energies/eV

K	1s	6539
L_I	2s	769.1
L_{II}	$2p_{1/2}$	649.9
L_{III}	$2p_{3/2}$	638.7
M_I	3s	82.3
M_{II}	$3p_{1/2}$	47.2
M_{III}	$3p_{3/2}$	47.2

Main lines in atomic spectrum
[Wavelength/nm(species)]

257.610 (I)
279.482 (I) (AA)
279.827 (I)
403.076 (I)
403.307 (I)
403.449 (I)

• CRYSTAL DATA

Crystal structure (cell dimensions/pm), space group

α-Mn b.c.c. (*a* = 891.39), $I\bar{4}3m$
β-Mn b.c.c. (*a* = 631.45), $P4_132$
γ-Mn f.c.c. (*a* = 386.3), Fm3m
δ-Mn b.c.c. (*a* = 308.1), Im3m
$T(\alpha \rightarrow \beta)$ = 973 K; $T(\beta \rightarrow \gamma)$ = 1352 K; $T(\gamma \rightarrow \delta)$ = 1413 K

X-ray diffraction: mass absorption coefficients $(\mu/\rho)/cm^2\ g^{-1}$**:** CuK_α 285 MoK_α 34.7
Neutron scattering length, $b/10^{-12}$ cm**:** –0.373
Thermal neutron capture cross-section, σ_a/barns**:** 13.3

• GEOLOGICAL DATA

Minerals

Many manganese minerals are known.

Mineral	Formula	Density	Hardness	Crystal appearance
Bixbyite	$(Mn,Fe)_2O_3$	4.975	6 – 6.5	cub. met. black
Manganite	γ-MnO(OH)	4.33	4	mon., met. grey-black
Pyrolusite	β-MnO_2	5.06	2 – 6	tet., met. grey-black
Rhodochrosite	$MnCO_3$	3.4 – 3.6	3.5 – 4	rhom., vit. pink
Rhodonite*	$(Mn,Fe,Mg)SiO_3$	3.6	5.5 – 6.5	tric., vit. rose-pink
Romanechite	$BaMn_9O_{16}(OH)_4$	3.7 – 4.7	5 – 6	mon., met. black, fern-like

* used in jewelry

Chief ores: pyrolusite, romanechite (also known as psilomelane), manganite (useful but rare)

World production/tonnes y^{-1}**:** 6.22×10^6

Main mining areas: South Africa, Russia, Gabon, Australia, Brazil

Reserves/tonnes**:** 3.6×10^9 (plus ocean floor nodules which are 24% Mn)

Specimen: available as chips, flake or powder. Safe.

Abundances

Sun (relative to H = 1×10^{12})**:** 2.63×10^5
Earth's crust/p.p.m.**:** 950
Seawater/p.p.m.**:**
Atlantic surface: 1.0×10^{-4}
Atlantic deep: 0.96×10^{-4}
Pacific surface: 1.0×10^{-4}
Pacific deep: 0.4×10^{-4}
Residence time/years**:** 50
Classification: scavenged
Oxidation state: II

Atomic number: 109 **CAS:**
Relative atomic mass ($^{12}C = 12.0000$)**:** (266) [54038-01-6]

• CHEMICAL DATA

Description: Meitnerium is a radioactive metal, of which only a few atoms have ever been made.

Radii/pm: Mt^{3+} 83 (est.)
Electronegativity: n.a.
Effective nuclear charge: n.a.

Standard reduction potentials $E^{\ominus}$/V

III		0
Mt^{3+}	$\xrightarrow{+0.8 \text{ (est.)}}$	Mt

Oxidation states

Mt^{I}	d^7	predicted
Mt^{II}	d^6	predicted, most stable?
Mt^{III}	d^5	predicted
Mt^{IV}	d^4	predicted

• PHYSICAL DATA

Melting point/K: n.a.
Boiling point/K: n.a.
ΔH_{fusion}/kJ mol^{-1}**:** n.a.
ΔH_{vap}/kJ mol^{-1}**:** n.a.
ΔH_{subl}/kJ mol^{-1}**:** 594 (est.)

Thermodynamic properties (298.15 K, 0.1 MPa)

State	$\Delta_f H^{\ominus}$/kJ mol^{-1}	$\Delta_f G^{\ominus}$/kJ mol^{-1}	$S^{\ominus}$/J K^{-1} mol^{-1}	C_p/J K^{-1} mol^{-1}
Solid	0	0	n.a.	n.a.
Gas	n.a.	n.a.	n.a.	n.a.

Density/kg m^{-3}**:** n.a.
Molar volume/cm^3**:** n.a.
Thermal conductivity/W m^{-1} K^{-1}**:** n.a.
Coefficient of linear thermal expansion/K^{-1}**:** n.a.
Electrical resistivity/Ω m**:** n.a.
Mass magnetic susceptibility/kg^{-1} m^3**:** n.a.

• BIOLOGICAL DATA

Biological role

None.

Toxicity

Toxic intake: n.a.
Lethal intake: n.a.

Hazards

Meitnerium is never encountered normally, and only a few atoms have ever been made. It would be dangerous because of its intense radioactivity.

Levels in humans

nil
Daily dietary intake: nil
Total mass of element in average (70 kg) person: nil

Discovery: see Nuclear Data section.
[Named after Lise Meitner, Austrian physicist who first suggested spontaneous nuclear fission]
French, *meitnerium*; German, *Meitnerium*; Italian, *meitnerio*; Spanish, *meitnerio*

Meitnerium

[miyt-neer-iuhm]

• NUCLEAR DATA

Discovery: Meitnerium was first made in 1982 by Peter Armbruster, Gottfried Münzenberg and their co-workers at Gesellschaft für Schwerionenforschung in Darmstadt, Germany.

Number of isotopes (including nuclear isomers): 1

Isotope mass range: 266

Key isotopes

Nuclide	Atomic mass	Half life ($T_{1/2}$)	Decay mode and energy (MeV)	Nuclear spin I	Nucl. mag. moment μ	Uses
^{266}Mt	266.1378	$c. 3.4 \times 10^{-3}$ s	α	n.a.	n.a.	

NMR [Not recorded]

• ELECTRON SHELL DATA

Ground state electron configuration: $[Rn]5f^{14}6d^{7}7s^{2}$
Term symbol: $^{4}F_{9/2}$
Electron affinity ($M \rightarrow M^{-}$)/kJ mol^{-1}**:** n.a.

Ionization energies/kJ mol^{-1}**:**
1. M → M^{+} 840 (est.)

Electron binding energies/eV
n.a.

Main lines in atomic spectrum
[Wavelength/nm(species)]
n.a.

• CRYSTAL DATA

Crystal structure (cell dimensions/pm), space group
n.a.

X-ray diffraction: mass absorption coefficients (μ/ρ)/cm^{2} g^{-1}**:** CuK_{α} n.a. MoK_{α} n.a.
Neutron scattering length, $b/10^{-12}$ cm**:** n.a.
Thermal neutron capture cross-section, σ_{a}/barns**:** n.a.

• GEOLOGICAL DATA

Minerals
Not found on Earth.

Chief source: a single atom of meitnerium was made by the cold fusion method in which a target of bismuth was bombarded with atoms of iron having just the right energy:

$^{209}Bi + {}^{58}Fe \rightarrow {}^{266}Mt + n$

Specimen: not available commercially

Abundances
Sun (relative to $H = 1 \times 10^{12}$)**:** n.a.
Earth's crust/p.p.m.**:** nil
Seawater/p.p.m.**:** nil

Md

Atomic number: 101
Relative atomic mass ($^{12}C = 12.0000$)**:** 258.10 (Md-258)

CAS: [7440-11-1]

• CHEMICAL DATA

Description: Mendelevium is a radioactive metal element which does not occur naturally, and is of research interest only.

Radii/pm: Md^{4+} 84; Md^{3+} 90; Md^{2+} 114
Electronegativity: 1.3 (Pauling); 1.2 (est.) (Allred); n.a. (absolute)
Effective nuclear charge: 1.65 (Slater)

Standard reduction potentials $E^{\ominus}$/V

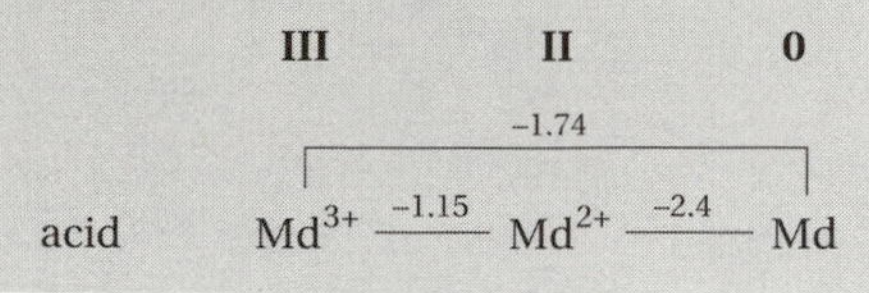

Oxidation states

Md^{I}	$f^{13}s^{1}$	?
Md^{II}	f^{13}	?
Md^{III}	f^{12}	$[Md(OH_2)_x]^{3+}$ (aq)

• PHYSICAL DATA

Melting point/K: n.a.
Boiling point/K: n.a.
ΔH_{fusion}/kJ mol^{-1}: n.a.
ΔH_{vap}/kJ mol^{-1}: n.a.

Thermodynamic properties (298.15 K, 0.1 MPa)

State	$\Delta_f H^{\ominus}$/kJ mol^{-1}	$\Delta_f G^{\ominus}$/kJ mol^{-1}	$S^{\ominus}$/J K^{-1} mol^{-1}	C_p/J K^{-1} mol^{-1}
Solid	0	0	n.a.	n.a.
Gas	n.a	n.a.	n.a.	n.a.

Density/kg m^{-3}: n.a.
Molar volume/cm^{3}: n.a.
Thermal conductivity/W m^{-1} K^{-1}: 10 (est.) [300 K]
Coefficient of linear thermal expansion/K^{-1}: n.a.
Electrical resistivity/Ω m: n.a.
Mass magnetic susceptibility/kg^{-1} m^{3}: n.a.

• BIOLOGICAL DATA

Biological role

None.

Toxicity

Toxic intake: n.a.
Lethal intake: n.a.

Hazards

Mendelevium is never encountered normally, and relatively few atoms have ever been made. It would be dangerous because of its intense radioactivity.

Levels in humans

nil

Daily dietary intake: nil
Total mass of element in average (70 kg) person: nil

Discovery: see Nuclear Data section.
[Named after Dimitri Mendeleyev, Russian chemist]
French, *mendelévium*; German, *Mendelevium*; Italian, *mendelevio*; Spanish, *mendelevio* [men-del-eev-iuhm]

Mendelevium

• NUCLEAR DATA

Discovery: Mendelevium was prepared in 1955 by Albert Ghiorso, B.G. Harvey, G.R. Choppin, S.G. Thompson and G.T. Seaborg at Berkeley, California, USA.

Number of isotopes (including nuclear isomers): 14 **Isotope mass range:** 247 → 260

Key isotopes

Nuclide	Atomic mass	Half life ($T_{1/2}$)	Decay mode and energy (MeV)	Nuclear spin I	Nucl. mag. moment μ	Uses
^{247}Md		3 s	α			
^{248}Md	248.082 750	7 s	EC (5.210) 80%; α (8.60) 20%			
^{249}Md	249.082 590	24 s	EC (3.760) < 80%; α (8.46) > 20%			
^{250}Md	250.084 340	50 s	EC (4.54) 94%; α (8.25) 6%			
^{251}Md	251.084 830	4 m	EC (3.702) > 94%; α (8.05) < 6%			
^{252}Md	252.086 470	2 s	EC (3.73) > 50%; α (7.856) < 50%			
^{253}Md		*c.* 6 m	α			
^{254}Md	254.089 630	30 m	EC (2.600)			
		10 m	EC			
^{255}Md	255.091 081	27 s	EC (1.055) 92%; α (7.911) 8%	7/2–?		
^{256}Md	256.093 960	1.3 h	EC (2.041) 90%; α (7.483) 10%			
^{257}Md	257.095 580	5.5 h	EC (0.450) 90%; α (7.60) 10%	7/2–?		
^{258}Md	258.098 570	57 m	EC			
		52 d	α (7.40)	8–?		
^{259}Md		1.6 h	SF			
^{260}Md		27.8 d	SF			

NMR [Not recorded]

• ELECTRON SHELL DATA

Ground state electron configuration: [Rn]$5f^{13}7s^2$
Term symbol: $^2F_{7/2}$
Electron affinity (M → M^-)/kJ mol^{-1}**:** n.a.

Ionization energies/kJ mol^{-1}**:**

1. M	→ M^+	635

Electron binding energies/eV
n.a.

Main lines in atomic spectrum
[Wavelength/nm(species)]
n.a.

• CRYSTAL DATA

Crystal structure (cell dimensions/pm), space group
n.a.

X-ray diffraction: mass absorption coefficients (μ/ρ)/$cm^2\ g^{-1}$**:** CuK_α n.a. MoK_α n.a.
Neutron scattering length, $b/10^{-12}$ cm**:** n.a.
Thermal neutron capture cross-section, σ_a/barns**:** n.a.

• GEOLOGICAL DATA

Minerals
Not found on Earth.

Chief source: only 17 atoms of mendelevium were originally made in 1955 by bombarding ^{253}Es with α-particles, although several thousand atoms were made in later experiments. Now millions of atoms can be produced.

Specimen: not available commercially

Abundances
Sun (relative to H = 1 × 10^{12})**:** n.a.
Earth's crust/p.p.m.**:** nil
Seawater/p.p.m.**:** nil

Hg

Atomic number: 80
Relative atomic mass ($^{12}C = 12.0000$): **200.59**

CAS: [7439-97-6]

• CHEMICAL DATA

Description: Mercury is a liquid, silvery metal. It is stable in air and with water, and is unreactive towards acids (except concentrated HNO_3) and alkalis. Mercury is used in the electrolysis cell for the manufacture of chlorine and sodium hydroxide by means of the electrolysis of brine, but is being phased out in favour of alternatives. Mercury is used in street lights, fungicides, electrical apparatus, etc.

Radii/pm: Hg^{2+} 112; Hg^{+} 127; atomic 160; covalent 144
Electronegativity: 2.00 (Pauling); 1.44 (Allred); 4.91 eV (absolute)
Effective nuclear charge: 4.35 (Slater); 11.15 (Clementi); 16.22 (Froese-Fischer)

Standard reduction potentials $E^{\ominus}$/V

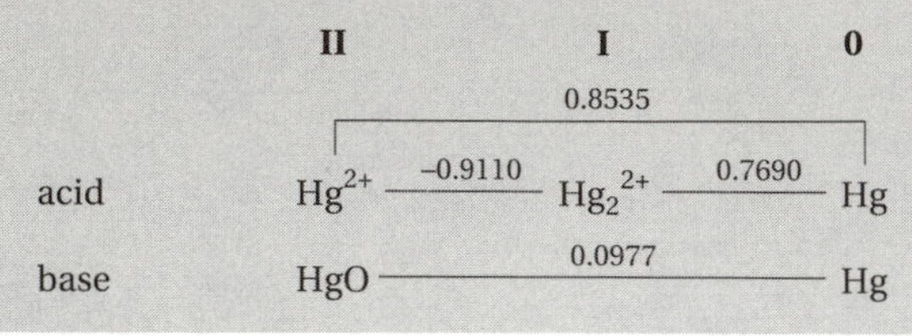

Oxidation states

Hg^{I}	$d^{10}s^{1}$	Hg_2F_2, Hg_2Cl_2 etc., Hg_2^{2+} salts, most are insoluble, except $Hg_2(NO_3)_2.2H_2O$
Hg^{II}	d^{10}	HgO, HgS, HgF_2, $HgCl_2$ etc., Hg^{2+} salts, $[Hg(OH_2)_6]^{2+}$ (aq), $HgN(OH)$, complexes e.g. $[Hg(SCN)_4]^{2-}$, $Hg(CH_3)_2$, etc.

• PHYSICAL DATA

Melting point/K: 234.28
Boiling point/K: 629.73
Critical temperature/K: 1750
Critical pressure/KPa: 172000

ΔH_{fusion}/kJ mol^{-1}: 2.331
ΔH_{vap}/kJ mol^{-1}: 59.15

Thermodynamic properties (298.15 K, 0.1 MPa)

State	$\Delta_f H^{\ominus}$/kJ mol^{-1}	$\Delta_f G^{\ominus}$/kJ mol^{-1}	$S^{\ominus}$/J K^{-1} mol^{-1}	C_p/J K^{-1} mol^{-1}
Solid	0	0	76.02	27.983
Gas	61.317	31.820	174.96	20.786

Density/kg m^{-3}: 13 546 [293 K]
Molar volume/cm^3: 14.81
Thermal conductivity/W m^{-1} K^{-1}: 8.34 [300 K]
Coefficient of cubical thermal expansion/K^{-1}: 18.1×10^{-5}
Electrical resistivity/Ω m: 94.1×10^{-8} [273 K]
Mass magnetic susceptibility/kg^{-1} m^3: -2.095×10^{-9} (l)

• BIOLOGICAL DATA

Biological role

None, although it is present in everything we eat.

Toxicity

Toxic intake: (metal vapour, human exposure) = 44 mg m^{-3} (8 hours)
Lethal intake: LC_{50} (metal vapour, inhalation, rabbit) = 29 mg m^{-3} (30 hours).

Hazards

Mercury vapour is poisonous by inhalation and should not exceed 0.1 mg m^{-3} in air. All mercury compounds are toxic, methyl mercury extremely so. Mercury affects the central nervous system and is teratogenic.

Levels in humans

Blood/mg dm^{-3}: 0.0078
Bone/p.p.m.: 0.45
Liver/p.p.m.: 0.018 – 3.7
Muscle/p.p.m.: 0.02 – 0.7
Daily dietary intake: 0.004 – 0.02 mg
Total mass of element in average (70 kg) person: 6 mg

Known to ancient civilizations.
[Named after the planet Mercury; Latin, *hydrargyrum* = liquid silver]
French, *mercure*; German, *Quecksilber*; Italian, *mercurio*; Spanish, *mercurio*

Mercury

[merk-uhr-ee]

• NUCLEAR DATA

Number of isotopes (including nuclear isomers): 37

Isotope mass range: 178 → 206

Key isotopes

Nuclide	Atomic mass	Natural abundance (%)	Nuclear spin I	Nuclear magnetic moment μ	Uses
^{196}Hg	195.965 807	0.14	0+		E
^{198}Hg	197.966 743	10.02	0+		E
^{199}Hg	198.968 254	16.84	1/2–	+0.505 885 2	E, NMR
^{200}Hg	199.968 300	23.13	0+		E
^{201}Hg	200.970 277	13.22	3/2–	–0.560 225	E, NMR
^{202}Hg	201.970 617	29.80	0+		E
^{204}Hg	203.973 467	6.85	0+		E

A table of radioactive isotopes is given in Appendix 1, on p244.

NMR [Reference: $Hg(CH_3)_2$]	^{199}Hg	^{201}Hg
Relative sensitivity (^1H = 1.00)	5.67×10^{-3}	1.44×10^{-3}
Receptivity (^{13}C = 1.00)	5.42	1.08
Magnetogyric ratio/rad $T^{-1}s^{-1}$	4.7912×10^7	-1.7686×10^7
Nuclear quadrupole moment/m^2	–	$+0.386 \times 10^{-28}$
Frequency (^1H = 100 Hz; 2.3488T)/MHz	17.827	6.599

• ELECTRON SHELL DATA

Ground state electron configuration: $[Xe]4f^{14}5d^{10}6s^2$
Term symbol: $^1S_0$
Electron affinity ($M \rightarrow M^-$)/kJ mol^{-1}**:** –18

Ionization energies/kJ mol^{-1}**:**

1.	$M \rightarrow M^+$	1007.0
2.	$M^+ \rightarrow M^{2+}$	1809.7
3.	$M^{2+} \rightarrow M^{3+}$	3300
4.	$M^{3+} \rightarrow M^{4+}$	(4400)
5.	$M^{4+} \rightarrow M^{5+}$	(5900)
6.	$M^{5+} \rightarrow M^{6+}$	(7400)
7.	$M^{6+} \rightarrow M^{7+}$	(9100)
8.	$M^{7+} \rightarrow M^{8+}$	(11 600)
9.	$M^{8+} \rightarrow M^{9+}$	(13 400)
10.	$M^{9+} \rightarrow M^{10+}$	(15 300)

Electron binding energies/eV

K	1s	83 102
L_I	2s	14 839
L_{II}	$2p_{1/2}$	14 209
L_{III}	$2p_{3/2}$	12 284
M_I	3s	3562
M_{II}	$3p_{1/2}$	3279
M_{III}	$3p_{3/2}$	2847
M_{IV}	$3d_{3/2}$	2385
M_V	$3d_{5/2}$	2295

continued in Appendix 2, p256

Main lines in atomic spectrum
[Wavelength/nm(species)]
253.652 (I) (AA)
365.015 (I)
404.656 (I)
435.833 (I)
1013.975 (I)

• CRYSTAL DATA

Crystal structure (cell dimensions/pm), space group
α-Hg rhombohedral (a = 299.25, α = 70° 44.6'), $R\bar{3}m$
β-Hg tetragonal (a = 399.5, c = 282.5), I4/mmm
$\alpha \rightarrow \beta$ high pressure

X-ray diffraction: mass absorption coefficients (μ/ρ)/$cm^2\ g^{-1}$**:** CuK_α 216 MoK_α 117
Neutron scattering length, $b/10^{-12}$ cm**:** 1.266
Thermal neutron capture cross-section, σ_a/barns**:** 374

• GEOLOGICAL DATA

Minerals

Native mercury occurs naturally as tiny drops of the liquid metal usually associated with cinnabar deposits but also found in some volcanic rocks.

Mineral	Formula	Density	Hardness	Crystal appearance
Cinnabar	HgS	8.09	2 – 2.5	rhom. microcrystalline, scarlet mass

Chief ore: cinnabar
World production/tonnes y^{-1}**:** 8400
Main mining areas: Spain, Italy, Yugoslavia.
Reserves/tonnes: 590 000
Specimen: available as liquid of varying grades of purity of up to 99.9999%. *Warning!*

Abundances

Sun (relative to H = 1×10^{12})**:** < 125
Earth's crust/p.p.m.: 0.05
Seawater/p.p.m.:
Atlantic surface: 4.9×10^{-7}
Atlantic deep: 4.9×10^{-7}
Pacific surface: 3.3×10^{-7}
Pacific deep: 3.3×10^{-7}
Residence time/years**:** n.a.
Classification: scavenged
Oxidation state: II

Mo

Atomic number: 42
Relative atomic mass ($^{12}C = 12.0000$): **95.94**

CAS: [7439-98-7]

• CHEMICAL DATA

Description: Molybdenum is a lustrous, silvery metal which is fairly soft when pure. It is usually obtained as a grey powder. It is attacked slowly by acids. It is used in alloys, electrodes and catalysts.

Radii/pm: Mo^{6+} 62; Mo^{2+} 92; atomic 136; covalent 129
Electronegativity: 2.16 (Pauling); 1.30 (Allred); 3.9 eV (absolute)
Effective nuclear charge: 3.45 (Slater); 6.98 (Clementi); 9.95 (Froese-Fischer)

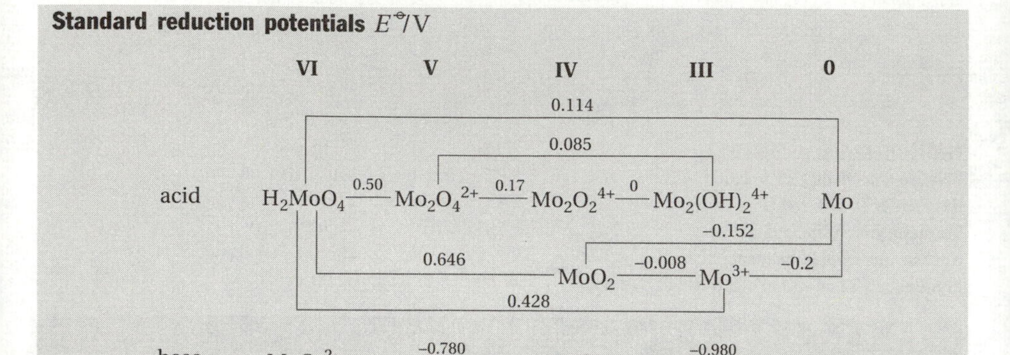

Oxidation states

Mo^{-II}	d^8	rare $[Mo(CO)_5]^{2-}$
Mo^0	d^6	rare $[Mo(CO)_6]$
Mo^I	d^5	rare $[Mo(\eta\text{-}C_6H_6)_2]^+$, $[Mo(CO)_3(\eta\text{-}C_5H_5)]_2$
Mo^{II}	d^4	Mo_6Cl_{12}, $[Mo_2Cl_8]^{4-}$, Mo_2^{4+} (aq)
Mo^{III}	d^3	MoF_3, $MoCl_3$ etc., $[Mo(OH_2)_6]^{3+}$ (aq)
Mo^{IV}	d^2	MoO_2, MoS_2, MoF_4, $MoCl_4$ etc.
Mo^V	d^1	Mo_2O_5, MoF_5, $MoCl_5$
$\mathbf{Mo^{VI}}$	d^0 [Kr]	MoO_3, $[MoO_4]^{2-}$ (aq), MoF_6, $[MoF_8]^{2-}$, MoF_4O

• PHYSICAL DATA

Melting point/K: 2890
Boiling point/K: 4885

ΔH_{fusion}/kJ mol^{-1}: 27.6
ΔH_{vap}/kJ mol^{-1}: 594.1

Thermodynamic properties (298.15 K, 0.1 MPa)

State	$\Delta_f H^{\ominus}$/kJ mol^{-1}	$\Delta_f G^{\ominus}$/kJ mol^{-1}	$S^{\ominus}$/J K^{-1} mol^{-1}	C_p/J K^{-1} mol^{-1}
Solid	0	0	28.66	24.06
Gas	658.1	612.5	181.950	20.786

Density/kg m^{-3}: 10 220 [293 K]; 9330 [liquid at m.p.]
Molar volume/cm^3: 9.39
Thermal conductivity/W m^{-1} K^{-1}: 138 [300 K]
Coefficient of linear thermal expansion/K^{-1}: 5.43×10^{-6}
Electrical resistivity/Ω m: 5.2×10^{-8} [273 K]
Mass magnetic susceptibility/kg^{-1} m^3: $+1.2 \times 10^{-8}$ (s)

Young's modulus/GPa: 324.8
Rigidity modulus/GPa: 125.6
Bulk modulus/GPa: 261.2
Poisson's ratio/GPa: 0.293

• BIOLOGICAL DATA

Biological role
Essential to all species.

Toxicity
Toxic intake: n.a.
Lethal intake: LD_{50} (MoO_2, subcutaneous, mouse) = 318 mg kg^{-1}

Hazards
Animal experiments show molybdenum compounds to be highly toxic and teratogenic, but there is little human data.

Levels in humans
Blood/mg dm^{-3}: *c.* 0.001
Bone/p.p.m.: < 0.7
Liver/p.p.m.: 1.3 – 5.8
Muscle/p.p.m.: 0.018
Daily dietary intake: 0.05 – 0.35 mg
Total mass of element in average (70 kg) person: 5 mg

Isolated in 1781 by P.J. Hjelm at Uppsala, Sweden.
[Greek, *molybdos* = lead]
French, *molybdène*; German, *Molybdän*; Italian, *molibdeno*; Spanish, *molibdeno*

Molybdenum

[mol-ib-den-uhm]

• NUCLEAR DATA

Number of isotopes (including nuclear isomers): 23

Isotope mass range: 88 → 106

Key isotopes

Nuclide	Atomic mass	Natural abundance (%)	Nuclear spin I	Nuclear magnetic moment μ	Uses
^{92}Mo	91.906 809	14.84	0+		E
^{94}Mo	93.905 085	9.25	0+		E
^{95}Mo	94.905 841	15.92	5/2+	–0.914 2	E, NMR
^{96}Mo	95.904 678	16.68	0+		E
^{97}Mo	96.906 020	9.55	5/2+	–0.933 5	E, NMR
^{98}Mo	97.905 407	24.13	0+		E
^{100}Mo	99.907 477	9.63	0+		E

A table of radioactive isotopes is given in Appendix 1, on p244.

NMR [Reference: $[MoO_4]^{2-}$ (aq)]	^{95}Mo	^{97}Mo
Relative sensitivity (^1H = 1.00)	3.23×10^{-3}	3.43×10^{-3}
Receptivity (^{13}C = 1.00)	2.88	1.84
Magnetogyric ratio/rad $T^{-1}s^{-1}$	1.7433×10^7	-1.7799×10^7
Nuclear quadrupole moment/m^2	-0.022×10^{-28}	$+0.255 \times 10^{-28}$
Frequency (^1H = 100 Hz; 2.3488T)/MHz	6.514	6.652

• ELECTRON SHELL DATA

Ground state electron configuration: $[Kr]4d^55s^1$
Term symbol: $^7S_3$
Electron affinity ($M \rightarrow M^-$)/kJ mol^{-1}**:** 72.0

Ionization energies/kJ mol^{-1}**:**

1.	$M \rightarrow M^+$	685.0
2.	$M^+ \rightarrow M^{2+}$	1558
3.	$M^{2+} \rightarrow M^{3+}$	2621
4.	$M^{3+} \rightarrow M^{4+}$	4480
5.	$M^{4+} \rightarrow M^{5+}$	5900
6.	$M^{5+} \rightarrow M^{6+}$	6560
7.	$M^{6+} \rightarrow M^{7+}$	12 230
8.	$M^{7+} \rightarrow M^{8+}$	14 800
9.	$M^{8+} \rightarrow M^{9+}$	(16 800)
10.	$M^{9+} \rightarrow M^{10+}$	(19 700)

Electron binding energies/eV

K	1s	20 000
L_I	2s	2866
L_{II}	$2p_{1/2}$	2625
L_{III}	$2p_{3/2}$	2520
M_I	3s	506.3
M_{II}	$3p_{1/2}$	411.6
M_{III}	$3p_{3/2}$	394.0
M_{IV}	$3d_{3/2}$	231.1
M_V	$3d_{5/2}$	227.9

continued in Appendix 2, p256

Main lines in atomic spectrum
[Wavelength/nm(species)]
201.511 (II)
202.030 (II)
203.844 (II)
313.259 (I) (AA)
379.825 (I)
386.411 (I)
390.296 (I)

• CRYSTAL DATA

Crystal structure (cell dimensions/pm), space group
b.c.c. (a = 314.700), Im3m

X-ray diffraction: mass absorption coefficients (μ/ρ)/$cm^2\ g^{-1}$**:** CuK_α 162 MoK_α 18.4
Neutron scattering length, $b/10^{-12}$ cm**:** 0.6715
Thermal neutron capture cross-section, σ_a/barns**:** 2.60

• GEOLOGICAL DATA

Minerals

Mineral	Formula	Density	Hardness	Crystal appearance
Molybdenite	MoS_2	4.7	1 – 1.5	hex. met. grey
Wulfenite	$PbMoO_4$	6.78	2.7 – 3	tet., res. adam. orange

Chief ores: molybdenite; wulfenite to lesser extent; also obtained as a by-product of copper production.
World production/tonnes y^{-1}**:** 80 000
Main mining areas: USA, Australia, Italy, Norway, Bolivia
Reserves/tonnes**:** 5×10^6
Specimen: available as foil, powder, rod or wire. Safe.

Abundances

Sun (relative to H = 1×10^{12})**:** 145
Earth's crust/p.p.m.**:** 1.5
Seawater/p.p.m.**:** 0.0100
Residence time/years**:** 600 000
Classification: accumulating
Oxidation state: VI

Nd

Atomic number: 60
Relative atomic mass ($^{12}C = 12.0000$): 144.24

CAS: [7440-00-8]

• CHEMICAL DATA

Description: Neodymium is a silvery-white metal of the so-called rare earth group (more correctly termed the lanthanides). It tarnishes in air, reacts slowly with cold water, rapidly with hot. Neodymium is used in alloys for permanent magnets, lasers, flints, glazes and glass.

Radii/pm: Nd^{3+} 104; atomic 182; covalent 164
Electronegativity: 1.14 (Pauling); 1.07 (Allred); ≤ 3.0 eV (absolute)
Effective nuclear charge: 2.85 (Slater); 9.31 (Clementi); 10.83 (Froese-Fischer)

Standard reduction potentials E°/V

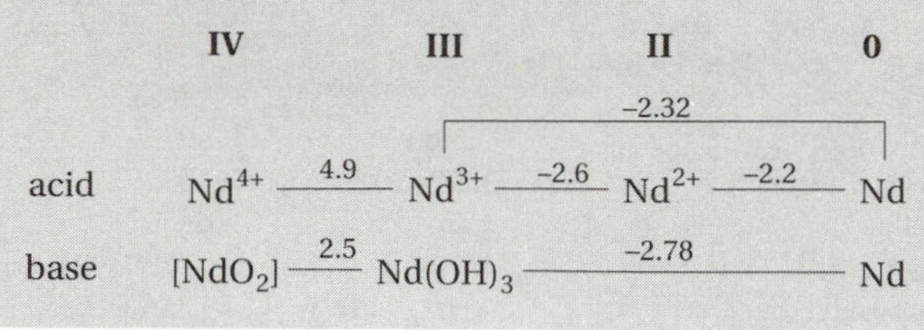

Oxidation states

Nd^{II}	f^4	NdO, $NdCl_2$, NdI_2
Nd^{III}	f^3	Nd_2O_3, $Nd(OH)_3$, $[Nd(OH_2)_x]^{3+}$(aq), Nd^{3+} salts, NdF_3, $NdCl_3$ etc., complexes
Nd^{IV}	f^2	Cs_3NdF_7

• PHYSICAL DATA

Melting point/K: 1294
Boiling point/K: 3341

ΔH_{fusion}/kJ mol^{-1}: 7.113
ΔH_{vap}/kJ mol^{-1}: 283.7

Thermodynamic properties (298.15 K, 0.1 MPa)

State	$\Delta_f H^{\circ}$/kJ mol^{-1}	$\Delta_f G^{\circ}$/kJ mol^{-1}	S°/J K^{-1} mol^{-1}	C_p/J K^{-1} mol^{-1}
Solid	0	0	71.5	27.45
Gas	327.6	292.4	189.406	22.092

Density/kg m^{-3}: 7007 [293 K]
Molar volume/cm^3: 20.59
Thermal conductivity/W m^{-1} K^{-1}: 16.5 [300 K]
Coefficient of linear thermal expansion/K^{-1}: 6.7×10^{-6}
Electrical resistivity/Ω m: 64.0×10^{-8} [293 K]
Mass magnetic susceptibility/kg^{-1} m^3: $+4.902 \times 10^{-7}$ (s)

Young's modulus/GPa: 41.4
Rigidity modulus/GPa: 16.3
Bulk modulus/GPa: 31.8
Poisson's ratio/GPa: 0.281

• BIOLOGICAL DATA

Biological role

None.

Toxicity

Toxic intake: n.a.
Lethal intake: LD_{50} (chloride, oral, mouse) = 5250 mg kg^{-1}

Hazards

Neodymium is mildly toxic by ingestion, but is a skin and eye irritant.

Levels in humans

Organs: n.a., but very low
Daily dietary intake: n.a.
Total mass of element in average (70 kg) person: n.a., but very low

Separated in 1885 by Baron Auer von Welsbach at Vienna, Austria.
[Greek, *neos didymos* = new twin]
French, *néodyme*; German, *Neodym*; Italian, *neodimio*; Spanish, *neodimio*

Neodymium

[nee-o-dim-ium]

• NUCLEAR DATA

Number of isotopes (including nuclear isomers): 24

Isotope mass range: 133 → 154

Key isotopes

Nuclide	Atomic mass	Natural abundance (%)	Nuclear spin *I*	Nuclear magnetic moment μ	Uses
^{142}Nd	141.907 719	27.13	0+		E
^{143}Nd	142.909 810	12.18	7/2–	–1.065	E, NMR
^{144}Nd*	143.910 083	23.80	0+		E
^{145}Nd	144.912 570	8.30	7/2–	–0.656	E, NMR
^{146}Nd	145.913 113	17.19	0+		E
^{148}Nd	147.916 889	5.76	0+		E
^{150}Nd	149.920 887	5.64	0+		E

^{144}Nd is radioactive with a half-life of 2.1×10^{15} y and decay mode α (1.83 MeV).
A table of radioactive isotopes is given in Appendix 1, on p244.

NMR [Reference: not recorded]	^{143}Nd	^{145}Nd
Relative sensitivity (^{1}H = 1.00)	3.38×10^{-3}	7.86×10^{-4}
Receptivity (^{13}C = 1.00)	2.43	0.393
Magnetogyric ratio/rad $T^{-1}s^{-1}$	-1.474×10^{7}	-0.913×10^{7}
Nuclear quadrupole moment/m^2	-0.630×10^{-28}	-0.330×10^{-28}
Frequency (^{1}H = 100 Hz; 2.3488T)/MHz	5.437	3.346

• ELECTRON SHELL DATA

Ground state electron configuration: [Xe]$4f^4 6s^2$
Term symbol: $^5I_4$
Electron affinity ($M \rightarrow M^-$)/kJ mol^{-1}**:** ≤ 50

Ionization energies/kJ mol^{-1}**:**

1.	$M \rightarrow M^{+}$	529.6
2.	$M^{+} \rightarrow M^{2+}$	1035
3.	$M^{2+} \rightarrow M^{3+}$	2130
4.	$M^{3+} \rightarrow M^{4+}$	3899
5.	$M^{4+} \rightarrow M^{5+}$	5790

Electron binding energies/eV

K	1s	43 569
L_I	2s	7126
L_{II}	$2p_{1/2}$	6722
L_{III}	$2p_{3/2}$	6208
M_I	3s	1575
M_{II}	$3p_{1/2}$	1403
M_{III}	$3p_{3/2}$	1297
M_{IV}	$3d_{3/2}$	1003.3
M_V	$3p_{5/2}$	980.4

continued in Appendix 2, p256

Main lines in atomic spectrum
[Wavelength/nm(species)]

386.333 (II)
395.116 (II)
401.225 (II)
406.109 (II)
430.358 (II)
495.453 (I) (AA)

• CRYSTAL DATA

Crystal structure (cell dimensions/pm), space group
α-Nd hexagonal (*a* = 365.79, *c* = 1179.92), $P6_3$/mmc
β-Nd b.c.c. (*a* = 413), Im3m
$T(\alpha \rightarrow \beta)$ = 1135 K
high pressure form: f.c.c. (*a* = 480), Fm3m

X-ray diffraction: mass absorption coefficients (μ/ρ)/cm^2 g^{-1}**:** CuK_α 374 MoK_α 53.2
Neutron scattering length, $b/10^{-12}$ cm**:** 0.769
Thermal neutron capture cross-section, σ_a/barns**:** 49

• GEOLOGICAL DATA

Minerals

Mineral	Formula	Density	Hardness	Crystal appearance
Bastnäsite*	(Ce, La, etc.)CO_3F	4.90	4 – 4.5	hex., vit./greasy yellow
Monazite	(Ce, La, Nd, Th, etc.)PO_4	5.20	5 – 5.5	mon., waxy/vit. yellow-brown

*Although not a major constituent, neodymium is present in extractable amounts.

Chief ores: monazite, bastnäsite
World production/tonnes y^{-1}**:** 7300
Main mining areas: USA, Brazil, India, Sri Lanka, Australia
Reserves/tonnes: *c.* 8×10^6
Specimen: available as chips, or ingots. Safe. Neodymium is also available as a powder but dust in the eye is an irritant. *Care!*

Abundances

Sun (relative to H = 1×10^{12})**:** 17.0
Earth's crust/p.p.m.**:** 38
Seawater/p.p.m.**:**
Atlantic surface: 1.8×10^{-6}
Atlantic deep: 3.2×10^{-6}
Pacific surface: 1.8×10^{-6}
Pacific deep: 4.8×10^{-6}
Residence time/years**:** 500
Classification: recycled
Oxidation state: III

Atomic number: 10
Relative atomic mass (^{12}C = 12.0000): **20.1797**

CAS: [7440-01-9]

• CHEMICAL DATA

Discovery: Neon was discovered in 1898 by Sir William Ramsay and M.W. Travers at London, England.
Description: Neon is a colourless, odourless gas obtained from liquid air. It is inert towards all other elements and chemicals. Neon is used mainly in ornamental illuminated lighting ('neon signs').

Radii/pm: van der Waals 160
Electronegativity: n.a. (Pauling); 4.84 (Allred); [10.6 eV (absolute) - see Key]
Effective nuclear charge: 5.85 (Slater); 5.76 (Clementi); 5.18 (Froese-Fischer)

Oxidation states

Ne^0 [Ne] only Ne as gas

Covalent bonds

forms no bonds

• PHYSICAL DATA

Melting point/K: 24.48
Boiling point/K: 27.10
Critical temperature/K: 44.5
Critical pressure/ kPa: 2721

ΔH_{fusion}/kJ mol^{-1}: 0.324
ΔH_{vap}/kJ mol^{-1}: 1.736

Thermodynamic properties (298.15 K, 0.1 MPa)

State	$\Delta_f H^\circ$/kJ mol^{-1}	$\Delta_f G^\circ$/kJ mol^{-1}	S°/J K^{-1} mol^{-1}	C_p/J K^{-1} mol^{-1}
Gas	0	0	146.328	20.786

Density/kg m^{-3}: 1444 [s., m.p.]; 1207.3 [liq., b.p.]; 0.89994 [g., 273 K]
Molar volume/cm^3: 13.97 [24 K]
Thermal conductivity/W m^{-1} K^{-1}: 0.0493 [300K] (g)
Mass magnetic susceptibility/kg^{-1} m^3: -4.2×10^{-9} (g)

• BIOLOGICAL DATA

Biological role

None.

Toxicity

Non-toxic.

Toxic intake: n.a.

Lethal intake: n.a.

Hazards

Neon is a harmless gas, although it could asphyxiate if it excluded oxygen from the lungs.

Levels in humans

Blood/mg dm^{-3}: trace
Bone/p.p.m.: nil
Liver/p.p.m.: nil
Muscle/p.p.m.: nil
Daily dietary intake: n.a. but low
Total mass of element in average (70 kg) person: n.a. but tiny

Discovery: see Chemical Data section.
[Greek, *neon* = new]
French, *néon*; German, *Neon*; Italian, *neo*; Spanish, *neón*

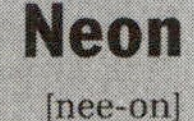

Neon

[nee-on]

• NUCLEAR DATA

Number of isotopes (including nuclear isomers): 9 **Isotope mass range:** 17 → 25

Key isotopes

Nuclide	Atomic mass	Natural abundance (%)	Nuclear spin I	Nuclear magnetic moment μ	Uses
^{20}Ne	19.992 435	90.48	0+	0	
^{21}Ne	20.993 843	0.27	3/2+	–0.661 796	NMR
^{22}Ne	21.991 383	9.25	0+	0	

A table of radioactive isotopes is given in Appendix 1, on p245.

NMR [Reference: no known compounds]	^{21}Ne
Relative sensitivity (^{1}H = 1.00)	2.50×10^{-3}
Receptivity (^{13}C = 1.00)	0.0359
Magnetogyric ratio/rad $T^{-1} s^{-1}$	-2.1118×10^7
Nuclear quadrupole moment/m^2	0.10155×10^{-28}
Frequency (^{1}H = 100 Hz; 2.3488T)/MHz	7.894

• ELECTRON SHELL DATA

Ground state electron configuration: [He]$2s^2 2p^6$ = [Ne]
Term symbol: $^1S_0$
Electron affinity ($M \rightarrow M^-$)/kJ mol^{-1}**:** –29 (calc.)

Ionization energies/kJ mol^{-1}**:**

1.	$M \rightarrow M^+$	2080.6
2.	$M^+ \rightarrow M^{2+}$	3952.2
3.	$M^{2+} \rightarrow M^{3+}$	6122
4.	$M^{3+} \rightarrow M^{4+}$	9370
5.	$M^{4+} \rightarrow M^{5+}$	12 177
6.	$M^{5+} \rightarrow M^{6+}$	15 238
7.	$M^{6+} \rightarrow M^{7+}$	19 998
8.	$M^{7+} \rightarrow M^{8+}$	23 069
9.	$M^{8+} \rightarrow M^{9+}$	115 377
10.	$M^{9+} \rightarrow M^{10+}$	131 429

Electron binding energies/eV

K	1s	870.2
L_I	2s	48.5
L_{II}	$2p_{1/2}$	21.7
L_{III}	$2p_{3/2}$	21.6

Main lines in atomic spectrum
[Wavelength/nm(species)]

837.761 (I)
865.438 (I)
878.062 (I)
878.375 (I)
885.387 (I)

• CRYSTAL DATA

Crystal structure (cell dimensions/pm), space group
f.c.c (a = 445.462), Fm3m
h.c.p. (a = 314.5, c = 514), $P6_3/mmc$ [3 K]

X-ray diffraction: mass absorption coefficients (μ/ρ)/cm^2 g^{-1}**:** CuK_α 22.9 MoK_α 2.47
Neutron scattering length, $b/10^{-12}$ cm**:** 0.455
Thermal neutron capture cross-section, σ_a/barns**:** 0.040

• GEOLOGICAL DATA

Minerals

None.

Chief source: liquid air
World production/tonnes y^{-1}**:** *c.* 1
Reserves/tonnes: 6.5×10^{10} (atmosphere)
Specimen: available in small pressurized canisters. Safe.

Abundances

Sun (relative to H = 1×10^{12})**:** 3.72×10^7
Earth's crust/p.p.m.**:** 7×10^{-5}
Atmosphere/p.p.m. (volume)**:** 18
Seawater/p.p.m.**:** 2×10^{-4}
Residence time/years**:** n.a.
Oxidation state: 0

Np

Atomic number: 93
Relative atomic mass ($^{12}C = 12.0000$): **237.0482**

CAS: [7439-99-8]

• CHEMICAL DATA

Description: Neptunium is a silvery, radioactive metal which occurs naturally in minute amounts in uranium ores. It is attacked by oxygen, steam and acids, but not alkalis. The metal is produced by reacting NpF_3 with either lithium or barium at 1200 °C. Neptunium has been used in neutron detectors.

Radii/pm: Np^{6+} 82; Np^{5+} 88; Np^{4+} 95; Np^{3+} 110; atomic 150
Electronegativity: 1.36 (Pauling); 1.22 (Allred); n.a. (absolute)
Effective nuclear charge: 1.80 (Slater)

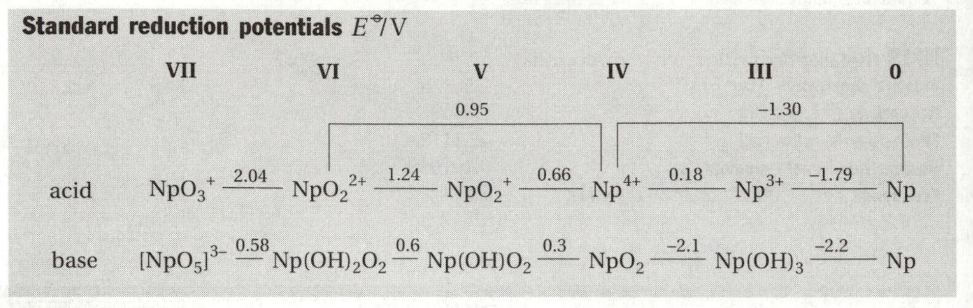

Oxidation states

Np^{II}	f^4d^1	NpO
Np^{III}	f^4	NpF_3, $NpCl_3$ etc., $[NpCl_6]^{3-}$, $[Np(OH_2)_x]^{3+}$ (aq)
Np^{IV}	f^4	NpO_2, $[Np(OH_2)_x]^{4+}$ (aq), NpF_4, $NpCl_4$, $NpBr_4$, $[NpCl_6]^{2-}$, complexes
Np^{V}	f^2	Np_2O_5, NpF_5, $CsNpF_6$, Na_3NpF_8, NpO_2^+ (aq)
Np^{VI}	f^1	$NpO_3.H_2O$, NpO_2^{2+} (aq), NpF_6
Np^{VII}	[Rn]	Li_5NpO_6

• PHYSICAL DATA

Melting point/K: 913
Boiling point/K: 4175
ΔH_{fusion}/kJ mol^{-1}: 9.46
ΔH_{vap}/kJ mol^{-1}: 336.6

Thermodynamic properties (298.15 K, 0.1 MPa)

State	$\Delta_f H^{\ominus}$/kJ mol^{-1}	$\Delta_f G^{\ominus}$/kJ mol^{-1}	$S^{\ominus}$/J K^{-1} mol^{-1}	C_p/J K^{-1} mol^{-1}
Solid	0	0	n.a.	n.a.
Gas	n.a.	n.a.	n.a.	n.a.

Density/kg m^{-3}: 20 250 [293 K]
Molar volume/cm^3: 11.71
Thermal conductivity/W m^{-1} K^{-1}: 6.3 [300 K]
Coefficient of linear thermal expansion/K^{-1}: 27.5×10^{-6}
Electrical resistivity/Ω m: 122×10^{-8} [293 K]
Mass magnetic susceptibility/kg^{-1} m^3: n.a.

• BIOLOGICAL DATA

Biological role
None.

Toxicity
Toxic intake: n.a.
Lethal intake: n.a.

Hazards
Neptunium is never encountered normally. It is dangerous because of its radioactivity. This element is only to be found inside nuclear facilities or research laboratories.

Levels in humans
nil
Daily dietary intake: nil
Total mass of element in average (70 kg) person: nil

Discovery: see Nuclear Data section.
[Named after the planet Neptune]
French, *neptunium*; German, *Neptunium*; Italian, *nettunio*; Spanish, *neptunio*

Neptunium

[nep-tyoon-iuhm]

• NUCLEAR DATA

Discovery: Neptunium was prepared in 1940 by Edwin M. McMillan and Philip H. Abelson at Berkeley, California, USA.

Number of isotopes (including nuclear isomers): 21 **Isotope mass range:** 228 → 242

Key isotopes

Nuclide	Atomic mass	Half life ($T_{1/2}$)	Decay mode and energy (MeV)	Nuclear spin I	Nucl. mag. moment μ	Uses
^{234}Np	234.042 888	4.4 d	β^+, EC (1.81); γ	0+		
^{235}Np	235.044 056	1.058 y	EC (0.124) 99.9%; α (5.191); γ	5/2+		
^{236}Np	236.046 550	1.55×10^5 y	EC (0.99) 91%; β (0.54) 9%; γ	6–		
^{237}Np	237.048 167	2.14×10^6 y	α (4.957); γ	5/2+	+3.14	R
^{238}Np	238.050 941	2.117 d	β^- (1.292); γ	2+		
^{239}Np	239.052 933	2.355 d	β^- (0.722); γ	5/2+		
^{240}Np	240.056 050	1.032 h	β^- (2.20); γ	5+		

Other isotopes of neptunium have half-lives shorter than 1 hour.

NMR [Reference: n.a.]

	^{237}Np
Relative sensitivity (^{1}H = 1.00)	–
Receptivity (^{13}C = 1.00)	–
Magnetogyric ratio/rad $T^{-1}s^{-1}$	3.1×10^7
Nuclear quadrupole moment/m^2	$+3.886 \times 10^{-28}$
Frequency (^{1}H = 100 Hz; 2.3488T)/MHz	11.25

• ELECTRON SHELL DATA

Ground state electron configuration: $[Rn]5f^4 6d^1 7s^2$
Term symbol: $^6L_{11/2}$
Electron affinity ($M \rightarrow M^-$)/kJ mol^{-1}**:** n.a.

Ionization energies/kJ mol^{-1}**:**
1. $M \rightarrow M^+$ 597

Electron binding energies/eV
n.a.

Main lines in atomic spectrum
[Wavelength/nm(species)]
901.618 (I)
1009.199 (I)
1081.745 (I)
1169.515 (I)
1177.664 (I)
1214.818 (I)
1237.742 (I)
1240.799 (I)
1383.433 ()

• CRYSTAL DATA

Crystal structure (cell dimensions/pm), space group
α-Np orthorhombic (a = 472.3, b = 488.7, c = 666.3), Pmcn
β-Np tetragonal (a = 489.7, b = 338.8), P42$_1$2
γ-Np cubic (a = 352.4), Im3m
$T(\alpha \rightarrow \beta)$ = 551 K; $T(\beta \rightarrow \gamma)$ = 850 K

X-ray diffraction: mass absorption coefficients (μ/ρ)/cm^2 g^{-1}**:** CuK$_\alpha$ n.a. MoK$_\alpha$ n.a.
Neutron scattering length, $b/10^{-12}$ cm**:** 1.055
Thermal neutron capture cross-section, σ_a/barns**:** 180

• GEOLOGICAL DATA

Minerals

Although present in uranium minerals, none is extracted from this source.
Neptunium-237 is obtained in kg quantities from uranium fuel elements where it is produced by neutron capture: $^{238}U + n \rightarrow ^{239}U$ (β emission) $\rightarrow ^{237}Np$.

Specimen: commercially available, under licence – see Key.

Abundances

Sun (relative to H = 1×10^{12})**:** n.a.
Earth's crust/p.p.m.**:** Neptunium is present in minute quantities in uranium ores, and is formed when one of the emitted neutrons, from a uranium atom undergoing fission, is captured by another uranium nucleus.
Seawater/p.p.m.**:** nil

Ni

Atomic number: 28 **CAS:** [7440-02-0]

Relative atomic mass ($^{12}C = 12.0000$)**:** 58.6934

• CHEMICAL DATA

Description: Nickel is a silvery-white, lustrous, malleable and ductile metal. It resists corrosion, but is soluble in acids (except concentrated HNO_3), yet unaffected by alkalis. It is used in alloys, especially stainless steel, in coins, metal plating, and catalysts.

Radii/pm: Ni^{3+} 62; Ni^{2+} 78; atomic 125; covalent 115

Electronegativity: 1.91 (Pauling); 1.75 (Allred); 4.40 eV (absolute)

Effective nuclear charge: 4.05 (Slater); 5.71 (Clementi); 7.86 (Froese-Fischer)

Standard reduction potentials $E^{\ominus}$/V

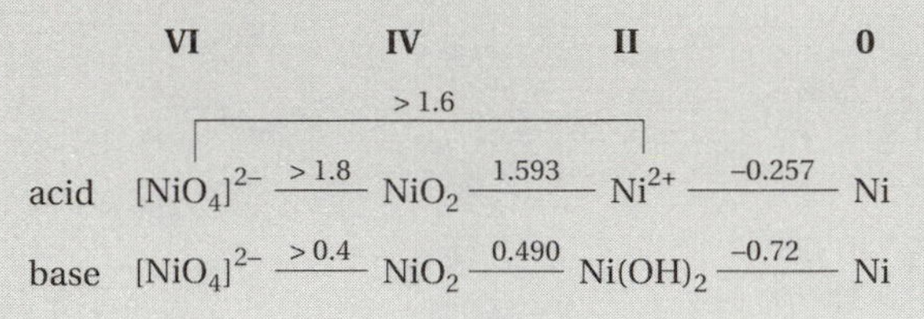

Oxidation states

Ni^{-I}	$d^{10}s^1$	$[Ni_2(CO)_6]^{2-}$
Ni^0	d^{10}	$[Ni(CO)_4]$, $K_4[Ni(CN)_4]$
Ni^I	d^9	$[NiBr(PPh_3)_3]$
Ni^{II}	d^8	NiO, $Ni(OH)_2$, $[Ni(OH_2)_6]^{2+}$ (aq), NiF_2, $NiCl_2$, etc., salts, $K_2[Ni(CN)_4]$, $[NiCl_4]^{2-}$, complexes, $[Ni(\eta\text{-}C_5H_5)_2]$
Ni^{III}	d^7	Ni(OH)O, NiF_3 ?, $[NiF_6]^{3-}$
Ni^{IV}	d^6	NiO_2?, $[NiF_6]^{2-}$
Ni^{VI}	d^4	K_2NiO_4 ?

• PHYSICAL DATA

Melting point/K: 1726

Boiling point/K: 3005

Curie temperature/K: 633

ΔH_{fusion}/kJ mol^{-1}: 17.6

ΔH_{vap}/kJ mol^{-1}: 371.8

Thermodynamic properties (298.15 K, 0.1 MPa)

State	$\Delta_f H^{\ominus}$/kJ mol^{-1}	$\Delta_f G^{\ominus}$/kJ mol^{-1}	$S^{\ominus}$/J K^{-1} mol^{-1}	C_p/J K^{-1} mol^{-1}
Solid	0	0	29.87	26.07
Gas	429.7	384.5	182.193	23.359

Density/kg m^{-3}: 8902 [298 K]; 7780 [liquid at m.p.]

Molar volume/cm^3: 6.59

Thermal conductivity/W m^{-1} K^{-1}: 90.7 [300 K]

Coefficient of linear thermal expansion/K^{-1}: 13.3×10^{-6}

Electrical resistivity/Ω m: 6.84×10^{-8} [293 K]

Mass magnetic susceptibility/kg^{-1} m^3: ferromagnetic

Young's modulus/GPa: 199.5

Rigidity modulus/GPa: 76.0

Bulk modulus/GPa: 177.3

Poisson's ratio/GPa: 0.312

• BIOLOGICAL DATA

Biological role

Essential to some species, and can act to stimulate metabolism.

Toxicity

Toxic intake: 1 – 3 mg kg^{-1}

Lethal intake: LD_{50} (acetate, oral, rat) = 350 mg kg^{-1}

Hazards

Nickel and many of its compounds are poisonous, carcinogenic, and teratogenic. Nickel carbonyl is extremely toxic.

Levels in humans

Blood/mg dm^{-3}: 0.01 – 0.05

Bone/p.p.m.: < 0.7

Liver/p.p.m.: 0.02 – 1.8

Muscle/p.p.m.: 1 – 2

Daily dietary intake: 0.3 – 0.5 mg

Total mass of element in average (70 kg) person: 15 mg

Discovered in 1751 by A.F. Cronstedt at Stockholm, Sweden.
[German, comes from *kupfernickel* meaning either Devil's copper or St Nicholas's copper]
French, *nickel*; German, *Nickel*; Italian, *nichel*; Spanish, *niquel*

Nickel

[nik-el]

• NUCLEAR DATA

Number of isotopes (including nuclear isomers): 14 **Isotope mass range:** 53 → 67

Key isotopes

Nuclide	Atomic mass	Natural abundance (%)	Nuclear spin I	Nuclear magnetic moment μ	Uses
^{58}Ni	57.935 346	68.27	0+		E
^{60}Ni	59.930 788	26.10	0+		E
^{61}Ni	60.931 057	1.13	3/2–	–0.750 02	E, NMR
^{62}Ni	61.928 346	3.59	0+		E
^{64}Ni	63.927 967	0.91	0+		E

A table of radioactive isotopes is given in Appendix 1, on p245.

NMR [Reference: not recorded]	^{61}Ni
Relative sensitivity (^{1}H = 1.00)	3.57×10^{-3}
Receptivity (^{13}C = 1.00)	0.242
Magnetogyric ratio/rad T^{-1}s^{-1}	-2.3948×10^7
Nuclear quadrupole moment/m^2	$+0.162 \times 10^{-28}$
Frequency (^{1}H = 100 Hz; 2.3488T)/MHz	8.936

• ELECTRON SHELL DATA

Ground state electron configuration: [Ar]$3d^84s^2$
Term symbol: $^3F_4$
Electron affinity ($M \rightarrow M^-$)/kJ mol^{-1}**:** 156

Ionization energies/kJ mol^{-1}**:**

1.	$M \rightarrow M^+$	736.7
2.	$M^+ \rightarrow M^{2+}$	1753.0
3.	$M^{2+} \rightarrow M^{3+}$	3393
4.	$M^{3+} \rightarrow M^{4+}$	5300
5.	$M^{4+} \rightarrow M^{5+}$	7280
6.	$M^{5+} \rightarrow M^{6+}$	10 400
7.	$M^{6+} \rightarrow M^{7+}$	12 800
8.	$M^{7+} \rightarrow M^{8+}$	15 600
9.	$M^{8+} \rightarrow M^{9+}$	18 600
10.	$M^{9+} \rightarrow M^{10+}$	21 660

Electron binding energies/eV

K	1s	8333
L_I	2s	1008.6
L_{II}	$2p_{1/2}$	870.0
L_{III}	$2p_{3/2}$	852.7
M_I	3s	110.8
M_{II}	$3p_{1/2}$	68.0
M_{III}	$3p_{3/2}$	66.2

Main lines in atomic spectrum
[Wavelength/nm(species)]

232.003 (I) (AA)
341.476 (I)
349.296 (I)
351.505 (I)
352.454 (I)
361.939 (I)

• CRYSTAL DATA

Crystal structure (cell dimensions/pm), space group
f.c.c. (a = 352.38), Fm3m
'hexagonal' nickel (an impure form of nickel) (a = 266, c = 432), $P6_3/mmc$
X-ray diffraction: mass absorption coefficients (μ/ρ)/cm^2 g^{-1}**:** CuK$_\alpha$ 45.7 MoK$_\alpha$ 46.6
Neutron scattering length, $b/10^{-12}$ cm**:** 1.03
Thermal neutron capture cross-section, σ_a/barns**:** 4.49

• GEOLOGICAL DATA

Minerals

A large nickel-iron alloy deposit occurs naturally in Greenland, and some meteorites are nickel-iron.

Mineral	Formula	Density	Hardness	Crystal appearance
Garnierite*	$(Ni,Mg)_6Si_4O_{10}(OH)_8$	2.3–2.5	2 – 4	mon., aggregates, bright green
Millerite	β-NiS	5.27	3 – 3.5	rhom. met. brassy-yellow
Nickeline	NiAs	7.78	5 – 5.5	hex., met. copper-red
Pentlandite	$(Ni,Fe)_9S_8$	4.8	3.5 – 4	cub., met. bronze-yellow

*General term for hydrous nickel silicates.

Chief ores: garnierite, pentlanite (nickeline is rare but was the first mineral from which nickel was extracted)

World production/tonnes y^{-1}**:** 510 000

Main mining areas: garnierite in Russia, South Africa, USA; pentlandite in Canada, South Africa

Reserves/tonnes: 70×10^6

Specimen: available as foil, powder, rod, slugs, spheres and wire. Safe.

Abundances

Sun (relative to H = 1×10^{12})**:** 1.91×10^6
Earth's crust/p.p.m.**:** *c.* 80
Seawater/p.p.m.**:**
Atlantic surface: 1×10^{-4}
Atlantic deep: 4.0×10^{-4}
Pacific surface: 1×10^{-4}
Pacific deep: 5.7×10^{-4}
Residence time/years**:** 80 000
Classification: recycled
Oxidation state: II

Nb

Atomic number: 41
Relative atomic mass ($^{12}C = 12.0000$): 92.90638

CAS: [7440-03-1]

• CHEMICAL DATA

Description: Niobium is a shiny, silvery metal, which is soft when pure. It resists corrosion due to an oxide film on the surface. Niobium is attacked by hot, concentrated acids, but resists attack by alkalis, even when they are molten. Niobium is used in stainless steels.

Radii/pm: Nb^{5+} 69; Nb^{4+} 74; atomic 143; covalent 134
Electronegativity: 1.6 (Pauling); 1.23 (Allred); 4.0 eV (absolute)
Effective nuclear charge: 3.30 (Slater); 6.70 (Clementi); 9.60 (Froese-Fischer)

Standard reduction potentials $E^{\ominus}$/V

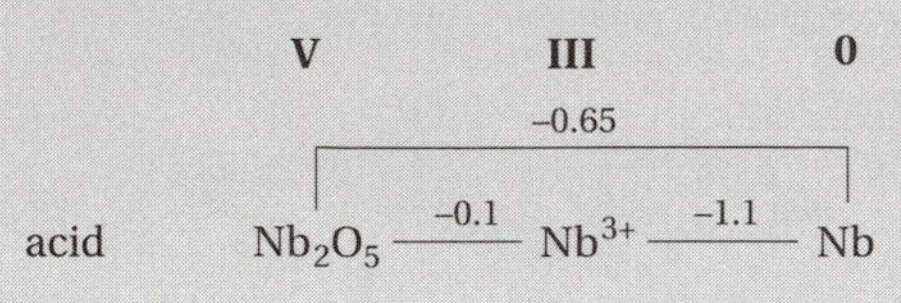

Oxidation states

Nb^{-III}	d^8	$[Nb(CO)_5]^{3-}$
Nb^{-I}	d^6	$[Nb(CO)_6]^-$
Nb^0	d^5	
Nb^I	d^4	$[Nb(CO)_4(\eta\text{-}C_5H_5)]$
Nb^{II}	d^3	NbO
Nb^{III}	d^2	$LiNbO_2$, $NbCl_3$, $NbBr_3$, NbI_3, $[Nb(CN)_8]^{5-}$
Nb^{IV}	d^1	NbO_2, NbF_4, $NbCl_4$ etc., $NbCl_2O$
Nb^V	d^0 [Kr]	Nb_2O_5, $[Nb_6HO_{19}]^-$ (aq), NbF_5, $NbCl_5$ etc., $NbFO_2$, $NbCl_3O$

• PHYSICAL DATA

Melting point/K: 2741
Boiling point/K: 5015

ΔH_{fusion}/kJ mol^{-1}: 27.2
ΔH_{vap}/kJ mol^{-1}: 696.6

Thermodynamic properties (298.15 K, 0.1 MPa)

State	$\Delta_f H^{\ominus}$/kJ mol^{-1}	$\Delta_f G^{\ominus}$/kJ mol^{-1}	$S^{\ominus}$/J K^{-1} mol^{-1}	C_p/J K^{-1} mol^{-1}
Solid	0	0	36.40	24.60
Gas	725.9	681.1	186.256	30.158

Density/kg m^{-3}: 8570 [293 K]; 7830 [liquid at m.p.]
Molar volume/cm^3: 10.84
Thermal conductivity/W m^{-1} K^{-1}: 53.7 [300 K]
Coefficient of linear thermal expansion/K^{-1}: 7.07×10^{-6}
Electrical resistivity/Ω m: 12.5×10^{-8} [273 K]
Mass magnetic susceptibility/kg^{-1} m^3: $+2.76 \times 10^{-8}$ (s)

Young's modulus/GPa: 104.9
Rigidity modulus/GPa: 37.5
Bulk modulus/GPa: 170.3
Poisson's ratio/GPa: 0.397

• BIOLOGICAL DATA

Biological role

None.

Toxicity

Toxic intake: n.a.
Lethal intake: LD_{50} (chloride, oral, rat) = 1500 mg kg^{-1}

Hazards

Niobium and its compounds may be toxic but there are no reports of humans being poisoned. Niobium dust causes eye and skin irritation.

Levels in humans

Blood/mg dm^{-3}: 0.005
Bone/p.p.m.: < 0.07
Liver/p.p.m.: 0.14
Muscle/p.p.m.: 0.14
Daily dietary intake: 0.02 – 0.6 mg
Total mass of element in average (70 kg) person: 1.5 mg

Discovered in 1801 by C. Hatchett at London, England.
[Greek, *Niobe* = daughter of Tantalus]
French, *niobium*; German, *Niob*; Italian, *niobio*; Spanish, *niobio*

Niobium

[niy-o-bi-uhm]

• NUCLEAR DATA

Number of isotopes (including nuclear isomers): 31 **Isotope mass range:** 86 → 103

Key isotopes

Nuclide	Atomic mass	Natural abundance (%)	Nuclear spin *I*	Nuclear magnetic moment μ	Uses
^{93}Nb	92.906 377	100	9/2+	+6.170 5	NMR

A table of radioactive isotopes is given in Appendix 1, on p245.

NMR [Reference: $[NbF_6]^-$ in conc. HF]	^{93}Nb
Relative sensitivity (^1H = 1.00)	0.48
Receptivity (^{13}C = 1.00)	2740
Magnetogyric ratio/rad $T^{-1} s^{-1}$	6.5476×10^7
Nuclear quadrupole moment/m^2	-0.320×10^{-28}
Frequency (^1H = 100 Hz; 2.3488T)/MHz	24.442

• ELECTRON SHELL DATA

Ground state electron configuration: $[Kr]4d^4 5s^1$
Term symbol: $^6D_{1/2}$
Electron affinity ($M \rightarrow M^-$)/kJ mol^{-1}**:** 86.2

Ionization energies/kJ mol^{-1}**:**

1.	$M \rightarrow M^+$	664
2.	$M^+ \rightarrow M^{2+}$	1382
3.	$M^{2+} \rightarrow M^{3+}$	2416
4.	$M^{3+} \rightarrow M^{4+}$	3695
5.	$M^{4+} \rightarrow M^{5+}$	4877
6.	$M^{5+} \rightarrow M^{6+}$	9899
7.	$M^{6+} \rightarrow M^{7+}$	12 100

Electron binding energies/eV

K	1s	18 986
L_I	2s	2698
L_{II}	$2p_{1/2}$	2465
L_{III}	$2p_{3/2}$	2371
M_I	3s	466.6
M_{II}	$3p_{1/2}$	376.1
M_{III}	$3p_{3/2}$	360.6
M_{IV}	$3d_{3/2}$	205.0
M_V	$3d_{5/2}$	202.3

continued in Appendix 2, p256

Main lines in atomic spectrum
[Wavelength/nm(species)]
334.371 (I) (AA)
358.027 (I)
405.894 (I)
407.973 (I)
410.092 (I)
412.381 (I)

• CRYSTAL DATA

Crystal structure (cell dimensions/pm), space group
b.c.c. (*a* = 329.86), Im3m

X-ray diffraction: mass absorption coefficients (μ/ρ)/$cm^2 g^{-1}$**:** CuK_α 153 MoK_α 17.1
Neutron scattering length, $b/10^{-12}$ cm**:** 0.7054
Thermal neutron capture cross-section, σ_a/barns**:** 1.15

• GEOLOGICAL DATA

Minerals

Mineral	Formula	Density	Hardness	Crystal appearance
Betafite	$(Ca,U)_2(Ti,Nb,Ta)_2O_6(OH)$	4.5	4 – 5.5	cub., waxy/vit. black/greenish
Columbite*	$(Fe,Mn)(Nb,Ta)_2O_6$	varies	*c.* 6	orth., opaque sub-met. lustre
Fergusonite	$YNbO_4$	5.7	5.5 – 6.5	tet., vitreous, sub-met. black
Samarskite	$(Y,Ce,U,Fe)_3(Nb,Ta,Ti)_5O_{16}$	5.69	5 – 6	orth., vitreous/resinous black

*This is a group of minerals such as ferrocolumbite, manganocolumbite, etc.

Chief ores: columbite, samarskite, betafite; obtained as a by-product of tin-extraction.

World production/tonnes y^{-1}**:** 15 000

Main mining areas: Australia, Zaire, Brazil, Russia, Norway, Canada, Madagascar

Reserves/tonnes: n.a.

Specimen: available as foil, powder, rod, turnings or wire. Safe.

Abundances

Sun (relative to H = 1×10^{12})**:** 79
Earth's crust/p.p.m.**:** 20
Seawater/p.p.m.**:** 9×10^{-7}
Residence time/years**:** n.a.
Oxidation state: V

N

Atomic number: 7
Relative atomic mass ($^{12}C = 12.0000$)**:** **14.00674**

CAS: [7727-37-9]

• CHEMICAL DATA

Description: Nitrogen is a colourless, odourless gas (N_2) which is generally unreactive at normal temperatures. It has an extensive inorganic and organic chemistry. Nitrogen compounds are used as fertilizers, as acids (HNO_3), explosives, plastics, dyes, etc.

Radii/pm: atomic 71; covalent 70 (single bond); van der Waals 154

Electronegativity: 3.04 (Pauling); 3.07 (Allred); 7.30 eV (absolute)

Effective nuclear charge: 3.90 (Slater); 3.83 (Clementi); 3.46 (Froese-Fischer)

Standard reduction potentials $E^{\ominus}/V$

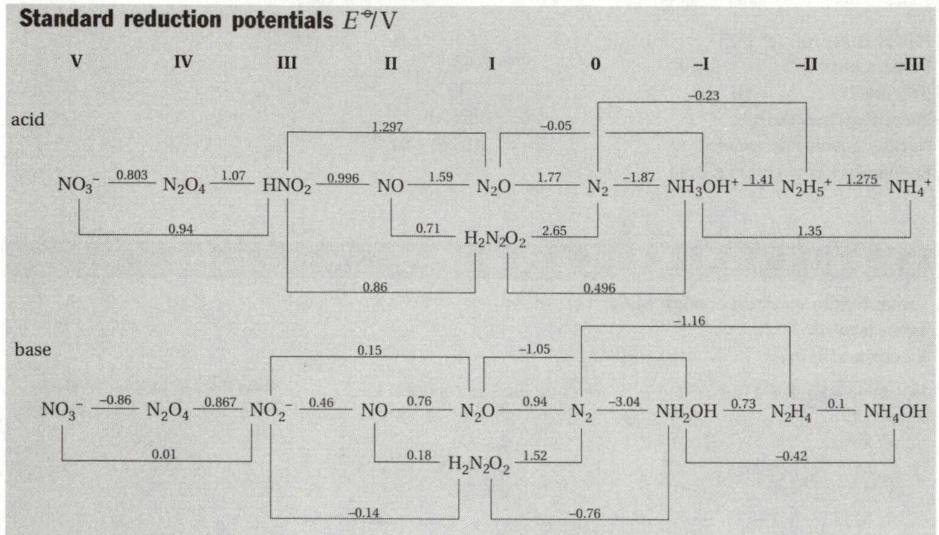

Oxidation states

N^{-III}	[Ne]	NH_3, NH_4^+ (aq)
N^{-II}	s^2p^5	N_2H_4, $N_2H_5^+$ (aq)
N^{-I}	s^2p^4	NH_2OH
N^0	s^2p^3	N_2
N^{II}	s^2p^1	NO
N^{III}	s^2	HNO_2, NO_2^- (aq), NF_3
N^{IV}	s^1	$N_2O_4 \rightleftharpoons 2NO_2$
N^V	[He]	HNO_3, NO_3^- (aq)

Covalent bonds

Bond	r/ pm	E/ kJ mol^{-1}
N–H	101	386
N–N	145	167
N=N	125	418
N≡N	110	942
N–F	136	283
N–Cl	175	313

• PHYSICAL DATA

Melting point/K: 63.29

Boiling point/K: 77.4

Critical temperature/K: 126.05

Critical pressure/ kPa: 3394

ΔH_{fusion}/kJ mol^{-1}: 0.720

ΔH_{vap}/kJ mol^{-1}: 5.577

Thermodynamic properties (298.15 K, 0.1 MPa)

State	$\Delta_f H^{\ominus}$/kJ mol^{-1}	$\Delta_f G^{\ominus}$/kJ mol^{-1}	$S^{\ominus}$/J K^{-1} mol^{-1}	C_p/J K^{-1} mol^{-1}
Gas (N_2)	0	0	191.61	29.125
Gas (atoms)	472.704	455.563	153.298	20.786

Density/kg m^{-3}: 1026 [s., 21 K]; 880 [liq., b.p.]; 1.2506 [gas, 273 K]

Molar volume/cm^3: 13.65 [21 K]

Thermal conductivity/W m^{-1} K^{-1}: 0.02598 [300 K] (g)

Mass magnetic susceptibility/kg^{-1} m^3: -5.4×10^{-9} (g)

• BIOLOGICAL DATA

Biological role

Constituent element of DNA and amino acids; the nitrogen cycle in nature is very important.

Toxicity

Non-toxic as N_2 gas, but some nitrogen compounds (NO_2, HCN, NH_3, etc.) are toxic.

Hazards

N_2 is a harmless gas, but it could asphyxiate if it excluded oxygen from the lungs.

Levels in humans

Blood/mg dm^{-3}: 34 300

Bone/p.p.m.: 43 000

Liver/p.p.m.: 72 000

Muscle/p.p.m.: 72 000

Daily dietary intake: n.a. but high

Total mass of element in average (70 kg) person: 1.8 kg

Discovered by D. Rutherford in 1772 at Edinburgh, Scotland.
[Greek, *nitron genes* = nitre forming. Nitre is a common name for potassium nitrate]
French, *azote*; German, *Stickstoff*; Italian, *azoto*; Spanish, *nitrógeno*

Nitrogen

[niy-troh-jen]

• NUCLEAR DATA

Number of isotopes (including nuclear isomers): 8 **Isotope mass range:** $12 \rightarrow 18$

Key isotopes

Nuclide	Atomic mass	Natural abundance (%)	Nuclear spin *I*	Nuclear magnetic moment μ	Uses
^{14}N	14.003 074 002	99.634	1+	+0.403 760 7	NMR
^{15}N	15.000 108 97	0.366	1/2–	–0.283 189 2	NMR

A table of radioactive isotopes is given in Appendix 1, p245.

NMR [Reference: CH_3NO_2 or NO_3^-]	^{14}N	^{15}N
Relative sensitivity ($^1H = 1.00$)	1.01×10^{-3}	1.04×10^{-3}
Receptivity ($^{13}C = 1.00$)	5.69	0.0219
Magnetogyric ratio/rad $T^{-1}s^{-1}$	1.9331×10^7	-2.7116×10^7
Nuclear quadrupole moment/m^2	0.0202×10^{-28}	–
Frequency (^1H = 100 Hz; 2.3488T)/MHz	7.224	10.133

• ELECTRON SHELL DATA

Ground state electron configuration: $[He]2s^22p^3$
Term symbol: $^4S_{3/2}$
Electron affinity ($M \rightarrow M^-$)/kJ mol^{-1}**:** –7

Ionization energies/kJ mol^{-1}**:**

1.	$M \rightarrow M^+$	1402.3
2.	$M^+ \rightarrow M^{2+}$	2856.1
3.	$M^{2+} \rightarrow M^{3+}$	4578.0
4.	$M^{3+} \rightarrow M^{4+}$	7474.9
5.	$M^{4+} \rightarrow M^{5+}$	9440.0
6.	$M^{5+} \rightarrow M^{6+}$	53 265.6
7.	$M^{6+} \rightarrow M^{7+}$	64 358.7

Electron binding energies/eV

K	1s	409.9
L_I	2s	37.3

Main lines in atomic spectrum
[Wavelength/nm(species)]

399.500 (II)
463.054 (II)
500.515 (II)
567.956 (II)
746.831 (I)
1 246.962 (I)

• CRYSTAL DATA

Crystal structure (cell dimensions/pm), space group
α-N_2 cubic (a = 564.4), $P2_13$
β-N_2 h.c.p. (a = 404.2, c = 660.1), $P6_3/mmc$
$T(\alpha \rightarrow \beta)$ = 35 K

X-ray diffraction: mass absorption coefficients $(\mu/\rho)/cm^2\ g^{-1}$**:** CuK_α 7.52 MoK_α 0.916
Neutron scattering length, $b/10^{-12}$ cm**:** 0.936
Thermal neutron capture cross-section, σ_a/barns**:** 1.91

• GEOLOGICAL DATA

Minerals

Nitrate minerals are uncommon.

Mineral	Formula	Density	Hardness	Crystal appearance
Nitratine*	$NaNO_3$	2.26	3.250	rhom., vit., colourless/white
Nitrammite	NH_4NO_3	1.72	n.a.	orth.
Nitrobarite	$Ba(NO_3)_2$	3.250	n.a.	cub., transparent colourless
Nitrocalcite	$Ca(NO_3)_2.4H_2O$	1.90	n.a.	mon., transparent white/grey
Nitromagnesite	$Mg(NO_3)_2.6H_2O$	1.58	n.a.	mon., transparent col./white

*Also known as Chile saltpetre.

Chief source: liquid air
World production/tonnes y^{-1}**:** 44×10^6
Reserves/tonnes**:** 3.9×10^{15} (atmosphere)
Specimen: available in small pressurized canisters. Safe.

Abundances

Sun (relative to $H = 1 \times 10^{12}$)**:** 8.71×10^7
Earth's crust/p.p.m.**:** 25
Atmosphere/p.p.m. (volume)**:** 780 900
Seawater/p.p.m.**:**
Atlantic surface: 0.00008
Atlantic deep: 0.27
Pacific surface: 0.00008
Pacific deep: 0.54
Residence time/years**:** 6000
Classification: recycled
Oxidation state: V

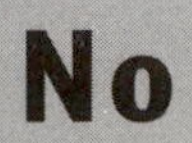

No

Atomic number: 102
Relative atomic mass ($^{12}C = 12.0000$)**: 259.1009 (No-259)**

CAS: [10028-14-5]

• CHEMICAL DATA

Description: Nobelium is a radioactive metal element which does not occur naturally, and is of research interest only.

Radii/pm: No^{4+} 83; No^{3+} 95; No^{2+} 113
Electronegativity: 1.3 (Pauling); 1.2 (est.) (Allred); n.a. (absolute)
Effective nuclear charge: 1.65 (Slater)

Standard reduction potentials $E^{\ominus}$/V

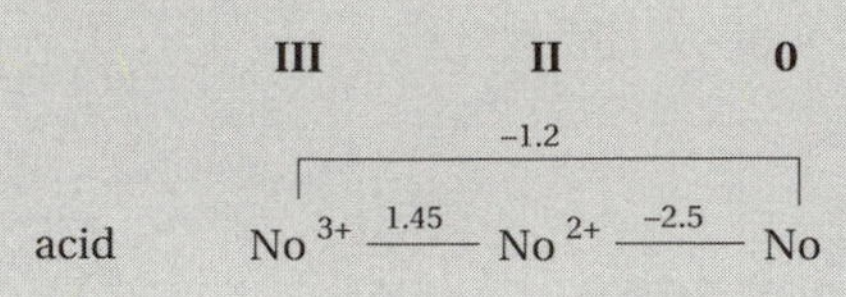

Oxidation states

No^{II}	f^{14}	$[No(OH_2)_x]^{2+}$ (aq)
No^{III}	f^{13}	$[No(OH_2)_x]^{3+}$ (aq)

• PHYSICAL DATA

Melting point/K: n.a.
Boiling point/K: n.a.
ΔH_{fusion}/kJ mol^{-1}: n.a.
ΔH_{vap}/kJ mol^{-1}: n.a.

Thermodynamic properties (298.15 K, 0.1 MPa)

State	$\Delta_f H^{\ominus}$/kJ mol^{-1}	$\Delta_f G^{\ominus}$/kJ mol^{-1}	$S^{\ominus}$/J K^{-1} mol^{-1}	C_p/J K^{-1} mol^{-1}
Solid	0	0	n.a.	n.a.
Gas	n.a.	n.a.	n.a.	n.a.

Density/kg m^{-3}: n.a.
Molar volume/cm^3: n.a.
Thermal conductivity/W m^{-1} K^{-1}: 10 (est.) [300 K]
Coefficient of linear thermal expansion/K^{-1}: n.a.
Electrical resistivity/Ω m: n.a.
Mass magnetic susceptibility/kg^{-1} m^3: n.a.

• BIOLOGICAL DATA

Biological role

None.

Toxicity

Toxic intake: n.a.
Lethal intake: n.a.

Hazards

Nobelium is never encountered normally, and relatively few atoms have ever been made. It would be dangerous because of its intense radioactivity.

Levels in humans

nil
Daily dietary intake: nil
Total mass of element in average (70 kg) person: nil

Discovery: see Nuclear Data section.
[Named after Alfred Nobel, Swedish chemist and founder of the Nobel Prizes]
French, *nobélium*; German, *Nobelium*; Italian, *nobelio*; Spanish, *nobelio*

Nobelium

[no-beel-iuhm]

• NUCLEAR DATA

Discovery: Nobelium was conclusively identified in 1958 by A. Ghiorso, T. Sikkeland, J.R. Walton and G.T. Seaborg at Berkeley, California, USA.

Number of isotopes (including nuclear isomers): 14

Isotope mass range: 250 → 262

Key isotopes

Nuclide	Atomic mass	Half life ($T_{1/2}$)	Decay mode and energy (MeV)	Nuclear spin I	Nucl. mag. moment μ	Uses
^{250}No		3.25×10^{-4} s	SF	0+		
^{251}No	251.088 870	0.8 s	α (8.600)			
^{252}No	252.088 949	2.3 s	α (8.551) 73%; SF 27%	0+		
^{253}No	253.090 530	1.7 m	α; EC (3.1)	9/2–?		
^{254}No	254.090 953	55 s	α; EC	0+		
^{254m}No		0.28 s	IT			
^{255}No	255.093 260	3.1 m	α 62%; EC (2.01) 38%	1/2+		
^{256}No	256.094 252	2.9 s	α	0+		
^{257}No	257.096 850	25 s	α	7/2+		
^{258}No	258.098 150	1.2×10^{-3} s	SF	0+		
^{259}No	259.100 931	58 m	α (7.794) 78%; EC 22%	9/2+?		
^{262}No		8×10^{-3} s	SF			

NMR [Not recorded]

• ELECTRON SHELL DATA

Ground state electron configuration: [Rn]$5f^{14}7s^2$

Term symbol: $^1S_0$

Electron affinity ($M \rightarrow M^-$)/kJ mol^{-1}: n.a.

Ionization energies/kJ mol^{-1}:

1. $M \rightarrow M^+$ 642

Electron binding energies/eV

n.a.

Main lines in atomic spectrum
[Wavelength/nm(species)]

n.a.

• CRYSTAL DATA

Crystal structure (cell dimensions/pm), space group

n.a.

X-ray diffraction: mass absorption coefficients (μ/ρ)/cm^2 g^{-1}: CuK_α n.a. MoK_α n.a.

Neutron scattering length, $b/10^{-12}$ cm: n.a.

Thermal neutron capture cross-section, σ_a/barns: n.a.

• GEOLOGICAL DATA

Minerals

Not found on Earth.

Chief source: atoms of nobelium were originally made one-at-a-time by bombarding ^{246}Cm with ^{12}C nuclei, although it is now possible to make several thousand atoms within 10 minutes by bombarding ^{249}Cf with ^{12}C.

Specimen: not available commercially

Abundances

Sun (relative to H = 1×10^{12}): n.a.

Earth's crust/p.p.m.: nil

Seawater/p.p.m.: nil

Os

Atomic number: 76
Relative atomic mass ($^{12}C = 12.0000$)**:** 190.23

CAS: [7440-04-2]

• CHEMICAL DATA

Description: Osmium is a lustrous, silvery metal of the so-called platinum group. It is unaffected by air, water, and acids, but dissolves in molten alkalis. Osmium metal gives off a recognizable smell due to the formation of volatile osmium tetroxide, OsO_4. The metal is used in alloys and catalysts.

Radii/pm: Os^{4+} 67; Os^{3+} 81; Os^{2+} 89; atomic 135; covalent 126
Electronegativity: 2.2 (Pauling); 1.52 (Allred); 4.9 eV (absolute)
Effective nuclear charge: 3.75 (Slater); 10.32 (Clementi); 14.90 (Froese-Fischer)

Standard reduction potentials $E^{\ominus}$/V

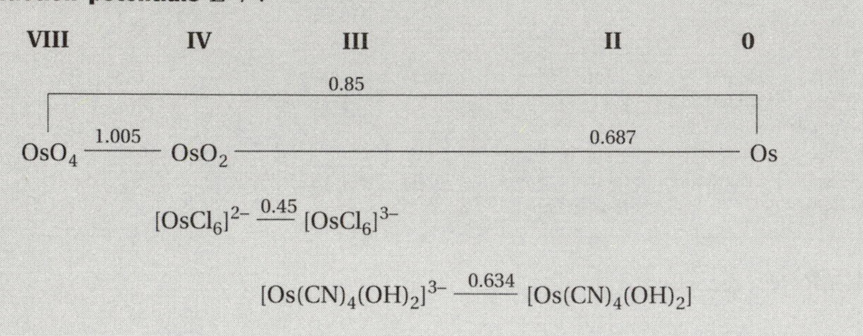

Oxidation states

Os^{-II}	d^{10}	$[Os(CO)_4]^{2-}$	**Os^{IV}**	d^4	OsO_2, OsO_2 (aq), OsF_4, $OsCl_4$, $OsBr_4$, $[OsCl_6]^{2-}$, complexes
Os^0	d^8	$[Os(CO)_5]$, $[Os_2(CO)_9]$, $[Os_3(CO)_{12}]$	Os^V	d^3	OsF_5, $OsCl_5$
Os^I	d^7	OsI	Os^{VI}	d^2	OsO_3?, OsF_6
Os^{II}	d^6	$OsCl_2$, OsI_2	Os^{VII}	d^1	OsF_7
Os^{III}	d^5	$OsCl_3$, $OsBr_3$, OsI_3, complexes	Os^{VIII}	$d^0[f^{14}]$	OsO_4, $[OsO_4(OH)_2]^{2-}$ (aq)

• PHYSICAL DATA

Melting point/K: 3327
Boiling point/K: 5300
ΔH_{fusion}/kJ mol^{-1}: 29.3
ΔH_{vap}/kJ mol^{-1}: 627.6

Thermodynamic properties (298.15 K, 0.1 MPa)

State	$\Delta_f H^{\ominus}$/kJ mol^{-1}	$\Delta_f G^{\ominus}$/kJ mol^{-1}	$S^{\ominus}$/J K^{-1} mol^{-1}	C_p/J K^{-1} mol^{-1}
Solid	0	0	32.6	24.7
Gas	791	745	192.573	20.786

Density/kg m^{-3}: 22 590 [293 K]; 20 100 [liquid at m.p.]
Molar volume/cm^3: 8.43
Thermal conductivity/W m^{-1} K^{-1}: 87.6 [300 K]
Coefficient of linear thermal expansion/K^{-1}: 4.3×10^{-6} (*a* axis); 6.1×10^{-6} (*b* axis); 6.8×10^{-6} (*c* axis)
Electrical resistivity/Ω m: 8.12×10^{-8} [273 K]
Mass magnetic susceptibility/kg^{-1} m^3: $+6.5 \times 10^{-10}$ (s)

Young's modulus/GPa: 559
Rigidity modulus/GPa: 223
Bulk modulus/GPa: 373
Poisson's ratio/GPa: 0.25

• BIOLOGICAL DATA

Biological role

None.

Toxicity

The metal itself is not toxic but its volatile oxide, OsO_4, is.

Toxic intake: oxide, inhalation, human = 133 μg m^{-3}
Lethal intake: LD_{50} (oxide, oral, rat) = 162 mg kg^{-1}

Hazards

Osmium dust is an irritant to eyes, skin and mucous membranes.

Levels in humans

Organs: n.a., but low
Daily dietary intake: n.a.
Total mass of element in average (70 kg) person: n.a., but very low

Discovered in 1803 by S. Tennant at London, England.
[Greek, *osme* = smell]
French, *osmium*; German, *Osmium*; Italian, *osmio*; Spanish, *osmio*

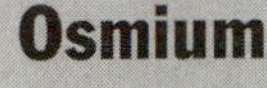

[oz-mi-uhm]

• NUCLEAR DATA

Number of isotopes (including nuclear isomers): 37 **Isotope mass range:** 166 → 196

Key isotopes

Nuclide	Atomic mass	Natural abundance (%)	Nuclear spin I	Nuclear magnetic moment μ	Uses
^{184}Os	183.952 488	0.02	0+		E
^{186}Os*	185.953 830	1.58	0+		E
^{187}Os	186.955 741	1.6	1/2–	+0.064 651	E, NMR
^{188}Os	187.955 830	13.3	0+		E
^{189}Os	188.958 137	16.1	3/2+	+0.659 933	E, NMR
^{190}Os	189.958 436	26.4	0+		E
^{192}Os	191.961 467	41.0	0+		E

*^{186}Os is radioactive with a half-life of 2×10^{15} y and decay mode α.
A table of radioactive isotopes is given in Appendix 1, on p246.

NMR [Reference: OsO_4]	^{187}Os	^{189}Os
Relative sensitivity (^1H = 1.00)	1.22×10^{-5}	2.34×10^{-3}
Receptivity (^{13}C = 1.00)	0.00114	2.13
Magnetogyric ratio/rad $T^{-1}s^{-1}$	0.6105×10^7	2.0773×10^7
Nuclear quadrupole moment/m^2	–	$+0.856 \times 10^{-28}$
Frequency (^1H = 100 Hz; 2.3488T)/MHz	2.282	7.758

• ELECTRON SHELL DATA

Ground state electron configuration: $[Xe]4f^{14}5d^66s^2$
Term symbol: $^5D_4$
Electron affinity ($M \rightarrow M^-$)/kJ mol^{-1}**:** 106

Ionization energies/kJ mol^{-1}**:**

1.	$M \rightarrow M^+$	840
2.	$M^+ \rightarrow M^{2+}$	(1600)
3.	$M^{2+} \rightarrow M^{3+}$	(2400)
4.	$M^{3+} \rightarrow M^{4+}$	(3900)
5.	$M^{4+} \rightarrow M^{5+}$	(5200)
6.	$M^{5+} \rightarrow M^{6+}$	(6600)
7.	$M^{6+} \rightarrow M^{7+}$	(8100)
8.	$M^{7+} \rightarrow M^{8+}$	(9500)

Electron binding energies/eV

K	1s	73 871
L_I	2s	12 968
L_{II}	$2p_{1/2}$	12 385
L_{III}	$2p_{3/2}$	10 871
M_I	3s	3049
M_{II}	$3p_{1/2}$	2792
M_{III}	$3p_{3/2}$	2457
M_{IV}	$3d_{3/2}$	2031
M_V	$3d_{5/2}$	1960

continued in Appendix 2, p256

Main lines in atomic spectrum
[Wavelength/nm(species)]

201.015 (I)
201.814 (I)
202.026 (I)
203.444 (I)
204.536 (I)
290.906 (I) (AA)

• CRYSTAL DATA

Crystal structure (cell dimensions/pm), space group
h.c.p. (a = 273.43, c = 432.00), $P6_3/mmc$

X-ray diffraction: mass absorption coefficients (μ/ρ)/$cm^2 g^{-1}$**:** CuK_α 186 MoK_α 106
Neutron scattering length, $b/10^{-12}$ cm: 1.07
Thermal neutron capture cross-section, σ_a/barns: 15

• GEOLOGICAL DATA

Minerals

Found in free state and as alloy with iridium.

Mineral	Formula	Density	Hardness	Crystal appearance
Osmiridium	(Ir,Os)	*c.* 20	6 – 7	cub. met. white
Osmium	Os	22.48	7	hex. met. white

Chief ores: obtained as a by-product of nickel refining
World production/tonnes y^{-1}: 0.06
Main mining areas: see nickel
Reserves/tonnes: n.a.
Specimen: available as powder or sponge. *Danger!*

Abundances

Sun (relative to H = 1×10^{12})**:** 5
Earth's crust/p.p.m.: *c.* 1×10^{-4}
Seawater/p.p.m.: n.a. but minute

O

Atomic number: 8
Relative atomic mass ($^{12}C = 12.0000$)**: 15.9994**

CAS: [7782-44-7]

• CHEMICAL DATA

Discovery: Oxygen was discovered in 1774 by J. Priestley at Leeds, England, and independently by C.W. Scheele at Uppsala, Sweden.
Description: Oxygen is a colourless, odourless gas (O_2) which is very reactive and will form oxides with all other elements except He, Ne, Ar and Kr. Oxygen is moderately soluble in water (30 cm^3 per dm^3) at 293 K. Oxygen gas is used in steel-making, metal cutting, the chemicals industry and in medical treatment.

Radii/pm: O^+ 22; O^{2-} 132; covalent 66 (single bond); van der Waals 140
Electronegativity: 3.44 (Pauling); 3.50 (Allred); 7.54 eV (absolute)
Effective nuclear charge: 4.55 (Slater); 4.45 (Clementi); 4.04 (Froese-Fischer)

Standard reduction potentials $E^{\ominus}$/V

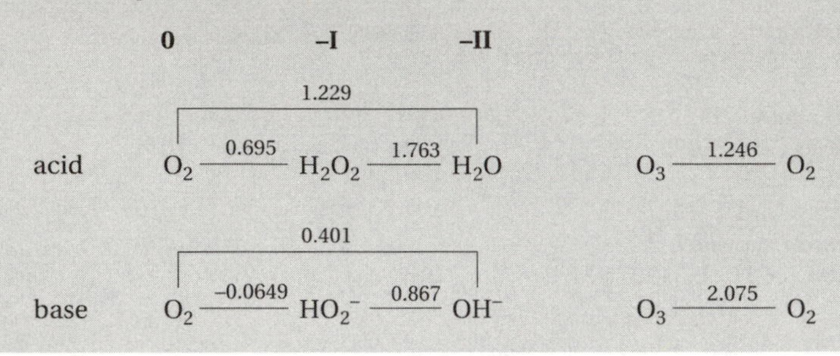

Oxidation states

O^{-II}	[Ne]	H_2O, H_3O^+, OH^-, oxides
O^{-I}	s^2p^5	H_2O_2, peroxides
O^0	s^2p^4	O_2, O_3
O^I	s^2p^3	O_2F_2
O^{II}	s^2p^2	OF_2

Covalent bonds

Bond	r/ pm	E/ kJ mol^{-1}
O–H	96	459
O–O	148	142
O=O (O_2)	121	494
O–N	146	200
O=N	115	678
O≡N (NO)	106	1063

For other bonds to oxygen see other elements

• PHYSICAL DATA

Melting point/K: 54.8
Boiling point/K: 90.188
Critical temperature/K: 154.58
Critical pressure/ kPa: 5043
ΔH_{fusion}/kJ mol^{-1}: 0.444
ΔH_{vap}/kJ mol^{-1}: 6.82

Thermodynamic properties (298.15 K, 0.1 MPa)

State	$\Delta_f H^{\ominus}$/kJ mol^{-1}	$\Delta_f G^{\ominus}$/kJ mol^{-1}	$S^{\ominus}$/J K^{-1} mol^{-1}	C_p/J K^{-1} mol^{-1}
Gas (O_2)	0	0	205.138	29.355
Gas (atoms)	249.170	231.731	161.055	21.912

Density/kg m^{-3}: 2000 [s., m.p.]; 1140 [liq., b.p.]; 1.429 [gas, 273 K]
Molar volume/cm^3: 8.00 [54 K]
Thermal conductivity/W m^{-1} K^{-1}: 0.2674 [300 K]
Mass magnetic susceptibility/kg^{-1} m^3: $+1.355 \times 10^{-6}$ (g)

• BIOLOGICAL DATA

Biological role
Constituent element of DNA and of most other biologically important compounds.

Toxicity
Non-toxic as O_2, but toxic as O_3.

Lethal intake: LC_{50} (ozone, inhalation, rat) = 4800 ppm for 4 hours

Hazards
Oxygen within a few percent of its natural concentration in air is harmless, but too little and it cannot sustain life, too much and it can cause pulmonary changes and teratogenic effects. Oxygen-enriched air poses a fire hazard because it dramatically increases the burning rate of combustible materials.

Levels in humans
Blood/mg dm^{-3}: constituent of water
Bone/p.p.m.: 285 000
Liver/p.p.m.: 160 000
Muscle/p.p.m.: 160 000
Daily dietary intake: mainly as water
Total mass of element in average (70 kg) person: 43 kg, mainly as water

Discovery: see Chemical Data section.
[Greek, *oxy genes* = acid forming]
French, *oxygène*; German, *Sauerstoff*; Italian, *ossigeno*; Spanish, *oxígeno*

Oxygen

[oksi-jen]

• NUCLEAR DATA

Number of isotopes (including nuclear isomers): 8 **Isotope mass range:** 13 → 20

Key isotopes

Nuclide	Atomic mass	Natural abundance (%)	Nuclear spin *I*	Nuclear magnetic moment *μ*	Uses
^{16}O	15.994 914 63	99.762	0+	0	
^{17}O	16.999 131 2	0.038	5/2+	–1.893 80	NMR
^{18}O	17.999 160 3	0.200	0+	0	

A table of radioactive isotopes is given in Appendix 1, on p246.

NMR [Reference: H_2O]

	^{17}O
Relative sensitivity (^1H = 1.00)	0.0291
Receptivity (^{13}C = 1.00)	0.061
Magnetogyric ratio/rad $T^{-1}s^{-1}$	-3.6264×10^7
Nuclear quadrupole moment/m^2	$-0.025\,58 \times 10^{-28}$
Frequency (^1H = 100 Hz; 2.3488T)/MHz	13.557

• ELECTRON SHELL DATA

Ground state electron configuration: $[He]2s^22p^4$
Term symbol: $^3P_2$
Electron affinity ($M \rightarrow M^-$)/kJ mol^{-1}**:** 141

Ionization energies/kJ mol^{-1}**:**

1.	$M \rightarrow M^+$	1313.9
2.	$M^+ \rightarrow M^{2+}$	3388.2
3.	$M^{2+} \rightarrow M^{3+}$	5300.3
4.	$M^{3+} \rightarrow M^{4+}$	7469.1
5.	$M^{4+} \rightarrow M^{5+}$	10 989.3
6.	$M^{5+} \rightarrow M^{6+}$	13 326.2
7.	$M^{6+} \rightarrow M^{7+}$	71 333.3
8.	$M^{7+} \rightarrow M^{8+}$	84 076.3

Electron binding energies/eV

K	1s	543.1
L_I	2s	41.6

Main lines in atomic spectrum
[Wavelength/nm(species)]

777.194 (I)
777.417 (I)
844.625 (I)
844.636 (I)
844.676 (I)

• CRYSTAL DATA

Crystal structure (cell dimensions/pm), space group
α-O_2 orthorhombic (a = 540.3, b = 342.9, c = 508.6, β = 132.53°), C2/m
β-O_2 rhombohedral (a = 330.7, c = 1125.6), $R\bar{3}m$
γ-O_2 cubic (a = 683), Pm3n
$T(\alpha \rightarrow \beta)$ = 23.8 K
$T(\beta \rightarrow \gamma)$ = 43.8 K

X-ray diffraction: mass absorption coefficients (μ/ρ)/$cm^2\ g^{-1}$**:** CuK_α 11.5 MoK_α 1.31
Neutron scattering length, $b/10^{-12}$ cm**:** 0.5803
Thermal neutron capture cross-section, σ_a/barns**:** 0.00019

• GEOLOGICAL DATA

Minerals

Oxygen is the most abundant element on the surface of the Earth, occurring as oxygen gas in the atmosphere, water in the oceans, and in minerals as oxides, or in combination with other elements as silicates, carbonates, phosphates, sulfates, etc.

Chief source: liquid air
World production/tonnes y^{-1}**:** 1×10^8
Reserves/tonnes: 1.2×10^{15} (in atmosphere)
Specimen: available in small pressurized canisters. Safe, but be aware of possible dangers.

Abundances

Sun (relative to H = 1×10^{12})**:** 6.92×10^8
Earth's crust/p.p.m.**:** 474 000
Atmosphere/p.p.m. (volume)**:** 209 500
Seawater/p.p.m.**:** constituent element of water

Pd

Atomic number: 46
Relative atomic mass ($^{12}C = 12.0000$)**:** 106.42

CAS: [7440-05-3]

• CHEMICAL DATA

Description: Palladium is a lustrous, silvery-white, malleable and ductile metal of the so-called platinum group. It resists corrosion, but dissolves in oxidising acids and in molten alkalis. Palladium metal has the unusual ability of allowing hydrogen gas to filter through it. It is mainly used as a catalyst.

Radii/pm: Pd^{4+} 64; Pd^{2+} 86; atomic 138; covalent 128
Electronegativity: 2.20 (Pauling); 1.35 (Allred); 4.45 eV (absolute)
Effective nuclear charge: 4.05 (Slater); 7.84 (Clementi); 11.11 (Froese-Fischer)

Standard reduction potentials $E^{\ominus}$/V

	VI		IV		II		0
acid			PdO_2	1.263	Pd^{2+}	0.915	Pd
base	'PdO_3'	2.03	PdO_2	1.283	$Pd(OH)_2$	−0.19	Pd

Oxidation states

Pd^0	d^{10}	$[Pd(PPh_3)_3]$, $[Pd(PF_3)_4]$
Pd^{II}	d^8	PdO, $[Pd(OH_2)_4]^{2+}$ (aq), PdF_2, $PdCl_2$ etc., $[PdCl_4]^{2-}$, salts, complexes
Pd^{IV}	d^6	PdO_2, PdF_4, $[PdCl_6]^{2-}$

• PHYSICAL DATA

Melting point/K: 1825
Boiling point/K: 3413
ΔH_{fusion}/kJ mol^{-1}: 17.2
ΔH_{vap}/kJ mol^{-1}: 393.3

Thermodynamic properties (298.15 K, 0.1 MPa)

State	$\Delta_f H^{\ominus}$/kJ mol^{-1}	$\Delta_f G^{\ominus}$/kJ mol^{-1}	$S^{\ominus}$/J K^{-1} mol^{-1}	C_p/J K^{-1} mol^{-1}
Solid	0	0	37.57	25.98
Gas	378.2	339.7	167.05	20.786

Density/kg m^{-3}: 12 020 [293 K]; 10 379 [liquid at m.p.]
Molar volume/cm^3: 8.85
Thermal conductivity/W m^{-1} K^{-1}: 71.8 [300 K]
Coefficient of linear thermal expansion/K^{-1}: 11.2×10^{-6}
Electrical resistivity/Ω m: 10.8×10^{-8} [293 K]
Mass magnetic susceptibility/kg^{-1} m^3: $+6.702 \times 10^{-8}$ (s)

Young's modulus/GPa: 121
Rigidity modulus/GPa: 43.6
Bulk modulus/GPa: 187
Poisson's ratio/GPa: 0.39

• BIOLOGICAL DATA

Biological role

None.

Toxicity

Toxic intake: n.a.
Lethal intake: LD_{50} (chloride, oral, rat) = 25 mg kg^{-1}

Hazards

Palladium is poorly absorbed by the body when ingested and $PdCl_2$ was formerly prescribed as a treatment for tuberculosis at the rate of 65 mg per day (approximately 1 mg kg^{-1}) without apparent ill effects. Palladium at higher intakes is poisonous and is an experimental carcinogen.

Levels in humans

Organs: n.a. but very low
Daily dietary intake: n.a.
Total mass of element in average (70 kg) person: n.a.

Discovered in 1803 by W.H. Wollaston at London, England.
[Named after the asteroid Pallas]
French, *palladium*; German, *Palladium*; Italian, *palladio*; Spanish, *paladio*

Palladium

[pal-ayd-iuhm]

• NUCLEAR DATA

Number of isotopes (including nuclear isomers): 25

Isotope mass range: 96 → 116

Key isotopes

Nuclide	Atomic mass	Natural abundance (%)	Nuclear spin *I*	Nuclear magnetic moment μ	Uses
^{102}Pd	101.905 634	1.02	0+		E
^{104}Pd	103.904 029	11.14	0+		E
^{105}Pd	104.905 079	22.33	5/2+	–0.642	E, NMR
^{106}Pd	105.903 478	27.33	0+		E
^{108}Pd	107.903 895	26.46	0+		E
^{110}Pd	109.905 167	11.72	0+		E

A table of radioactive isotopes is given in Appendix 1, on p246.

NMR [Only K_2PdCl_6 recorded]

	^{105}Pd
Relative sensitivity (^1H = 1.00)	1.12×10^{-3}
Receptivity (^{13}C = 1.00)	1.41
Magnetogyric ratio/rad $T^{-1} s^{-1}$	-0.756×10^7
Nuclear quadrupole moment/m^2	$+0.660 \times 10^{-28}$
Frequency (^1H = 100 Hz; 2.3488T)/MHz	4.576

• ELECTRON SHELL DATA

Ground state electron configuration: $[Kr]4d^{10}$
Term symbol: $^1S_0$
Electron affinity ($M \rightarrow M^-$)/kJ mol^{-1}**:** 53.7

Ionization energies/kJ mol^{-1}**:**

1.	$M \rightarrow M^+$	805
2.	$M^+ \rightarrow M^{2+}$	1875
3.	$M^{2+} \rightarrow M^{3+}$	3177
4.	$M^{3+} \rightarrow M^{4+}$	(4700)
5.	$M^{4+} \rightarrow M^{5+}$	(6300)
6.	$M^{5+} \rightarrow M^{6+}$	(8700)
7.	$M^{6+} \rightarrow M^{7+}$	(10 700)
8.	$M^{7+} \rightarrow M^{8+}$	(12 700)
9.	$M^{8+} \rightarrow M^{9+}$	(15 000)
10.	$M^{9+} \rightarrow M^{10+}$	(17 200)

Electron binding energies/eV

K	1s	24 350
L_I	2s	3604
L_{II}	$2p_{1/2}$	3330
L_{III}	$2p_{3/2}$	3173
M_I	3s	671.6
M_{II}	$3p_{1/2}$	559.9
M_{III}	$3p_{3/2}$	532.3
M_{IV}	$3d_{3/2}$	340.5
M_V	$3d_{5/2}$	335.2

continued in Appendix 2, p256

Main lines in atomic spectrum
[Wavelength/nm(species)]

247.642 (I) (AA)
340.458 (I)
342.124 (I)
351.694 (I)
355.308 (I)
360.955 (I)
363.470 (I)

• CRYSTAL DATA

Crystal structure (cell dimensions/pm), space group
f.c.c. (a = 389.08), Fm3m

X-ray diffraction: mass absorption coefficients (μ/ρ)/$cm^2 g^{-1}$**:** CuK_α 206 MoK_α 24.1
Neutron scattering length, $b/10^{-12}$ cm**:** 0.591
Thermal neutron capture cross-section, σ_a/barns**:** 6.9

• GEOLOGICAL DATA

Minerals

Found as the native element.

Mineral	Formula	Density	Hardness	Crystal appearance
Palladium	Pd	11.9	4.5 – 5	cub. met. white/grey

Chief ores: specimens of the native metal are found in Brazil. Most is extracted as a by-product from copper and zinc refining.

World production/tonnes y^{-1}**:** 24

Main mining areas: see copper and zinc

Reserves/tonnes**:** 24 000

Specimen: available as foil, granules, powder, rod, shot, sponge or wire. Safe.

Abundances

Sun (relative to H = 1×10^{12})**:** 32
Earth's crust/p.p.m.**:** *c.* 6×10^{-4}
Seawater/p.p.m.**:**
Atlantic surface: n.a.
Atlantic deep: n.a.
Pacific surface: 1.9×10^{-8}
Pacific deep: 6.8×10^{-8}
Residence time/years**:** 50 000
Classification: recycled
Oxidation state: II

P

Atomic number: 15
Relative atomic mass ($^{12}C = 12.0000$): **30.973762**

CAS: [7723-14-0]

• CHEMICAL DATA

Description: White phosphorus (P_4) is soft and flammable, while red phosphorus is powdery and usually non-flammable. Neither form reacts with water or dilute acids, but alkalis react to form phosphine gas. Phosphorus compounds are used in fertilizers, insecticides, metal treatments, detergents, foods, etc.

Radii/pm: P^{3-} 212; atomic 115 (red) 93 (white); covalent single bond 110; van der waals 190
Electronegativity: 2.19 (Pauling); 2.06 (Allred); 5.62 eV (absolute)
Effective nuclear charge: 4.80 (Slater); 4.89 (Clementi); 5.28 (Froese-Fischer)

Standard reduction potentials $E^\ominus$/V

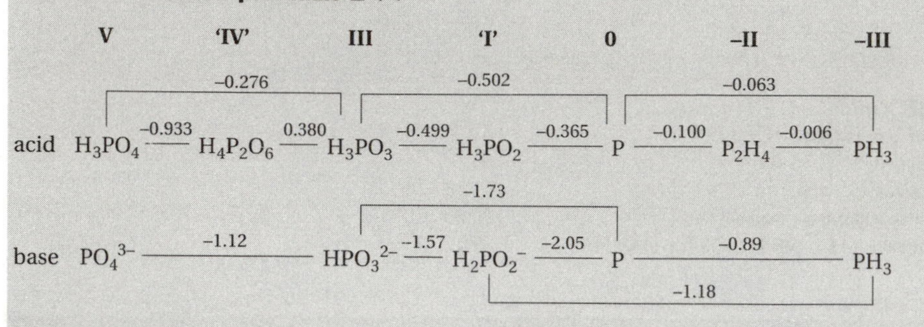

Oxidation states

P^{-III}	[Ar]	PH_3, Ca_3P_2
P^{-II}	s^2p^5	P_2H_4
P^0	s^2p^3	P_4
P^I	s^2p^2	H_3PO_2 (aq), $H_2PO_2^-$ (aq)
P^{II}	s^2p^1	P_2I_4
P^{III}	s^2	P_4O_6, H_3PO_3 (aq), PF_3, PCl_3 etc.
P^V	[Ne]	P_4O_{10}, H_3PO_4 (aq), $H_2PO_4^-$ (aq) etc. PF_5, PCl_5, $POCl_3$, phosphates

Covalent bonds

Bond	r/ pm	E/ kJ mol^{-1}
P–H	144	322
P–C	184	264
P–O	163	407
P=O	145	560
P–F	154	490
P–Cl	203	326
P–Br	218	264
P–I	252	184
P–P	221	201

• PHYSICAL DATA

Melting point/K: 317.3 (P_4); 683 (red) under pressure
Boiling point/K: 553 (P_4)
Critical temperature/K: 994

ΔH_{fusion}/kJ mol^{-1}: 2.51 (P_4)
ΔH_{vap}/kJ mol^{-1}: 51.9 (P_4)

Thermodynamic properties (298.15 K, 0.1 MPa)

State	$\Delta_f H^\ominus$/kJ mol^{-1}	$\Delta_f G^\ominus$/kJ mol^{-1}	$S^\ominus$/J K^{-1} mol^{-1}	C_p/J K^{-1} mol^{-1}
Solid (P_4)	0	0	41.09	23.840
Solid (red)	–17.6	–12.1	22.80	21.21
Gas	314.64	278.25	163.193	20.786

Density/kg m^{-3}: 1820 (P_4); 2200 (red); 2690 (black) [293 K]
Molar volume/cm^3: 17.02 (P_4)
Thermal conductivity/W m^{-1} K^{-1}: 0.235 (P_4); 12.1 (black) [300 K]
Coefficient of linear thermal expansion/K^{-1}: 124.5×10^{-6} (P_4)
Electrical resistivity/Ω m: 1×10^9 (P_4) [293 K]
Mass magnetic susceptibility/kg^{-1} m^3: -1.1×10^{-8} (P_4); -8.4×10^{-9} (red)

• BIOLOGICAL DATA

Biological role

Phosphorus is a constituent of DNA, ATP and many other biochemical molecules. The phosphate cycle in nature is very important and this nutrient is often the limiting factor, e.g. in the oceans.

Toxicity

Toxic intake: (white P, oral, rat) = 11 μg kg^{-1}

Lethal intake: (white P, oral, human) = 100 mg

Hazards

White phosphorus is much more toxic than red phosphorus. Chronic poisoning of those working unprotected with the former leads to necrosis of the jaw (phossy-jaw).

Levels in humans

Blood/mg dm^{-3}: 345
Bone/p.p.m.: 67 000 – 71 000
Liver/p.p.m.: 3 – 8.5
Muscle/p.p.m.: 3000 – 8500
Daily dietary intake: 900 – 1900 mg
Total mass of element in average (70 kg) person: 780 g

Discovered in 1669 by Hennig Brandt at Hamburg, Germany.
[Greek, *phosphoros* = bringer of light]
French, *phosphore*; German, *Phosphor*; Italian, *fosforo*; Spanish, *fósforo*

Phosphorus

[fos-for-us]

• NUCLEAR DATA

Number of isotopes (including nuclear isomers): 10 **Isotope mass range:** 26 → 36

Key isotopes

Nuclide	Atomic mass	Natural abundance (%)	Nuclear spin *I*	Nuclear magnetic moment μ	Uses
^{31}P	30.973 762 0	100	1/2+	+1.131 60	NMR

A table of radioactive isotopes is given in Appendix 1, on p246.

NMR [Reference: 85% H_3PO_4]	^{31}P
Relative sensitivity (^1H = 1.00)	0.0663
Receptivity (^{13}C = 1.00)	377
Magnetogyric ratio/rad $T^{-1}s^{-1}$	10.8289×10^7
Nuclear quadrupole moment/m^2	–
Frequency (^1H = 100 Hz; 2.3488T)/MHz	40.481

• ELECTRON SHELL DATA

Ground state electron configuration: $[Ne]3s^23p^3$
Term symbol: $^4S_{3/2}$
Electron affinity ($M \rightarrow M^-$)/kJ mol^{-1}**:** 44

Ionization energies/kJ mol^{-1}**:**

1.	$M \rightarrow M^+$	1011.7
2.	$M^+ \rightarrow M^{2+}$	1903.2
3.	$M^{2+} \rightarrow M^{3+}$	2912
4.	$M^{3+} \rightarrow M^{4+}$	4956
5.	$M^{4+} \rightarrow M^{5+}$	6273
6.	$M^{5+} \rightarrow M^{6+}$	21 268
7.	$M^{6+} \rightarrow M^{7+}$	25 397
8.	$M^{7+} \rightarrow M^{8+}$	28 854
9.	$M^{8+} \rightarrow M^{9+}$	35 867
10.	$M^{9+} \rightarrow M^{10+}$	40 958

Electron binding energies/eV

K	1s	2145.5
L_I	2s	189
L_{II}	$2p_{1/2}$	136
L_{III}	$2p_{3/2}$	135

Main lines in atomic spectrum
[Wavelength/nm(species)]

213.618 (I) (AA)
952.573 (I)
956.344 (I)
979.685 (I)
1648.292 (I)

• CRYSTAL DATA

Crystal structure (cell dimensions/pm), space group
α-P_4 white, cubic (a = 1851), I$\bar{4}$3m
β-P_4 white, rhombohedral (a = 337.7; c = 880.6), R3m [high pressure form]
γ-P_4 white, cubic (a = 237.7), Pm3m [high pressure form]
Red, cubic (a = 1131), Pm3m or P$\bar{4}$3
Hittorf's phosphorus (purple), monoclinic (a = 921, b = 915, c = 2260, β = 106.1°), P2/c
Black, orthorhombic (a = 331.36, b = 1047.8, c = 437.63), Cmca

X-ray diffraction: mass absorption coefficients (μ/ρ)/$cm^2 g^{-1}$**:** CuK_α 74.1 MoK_α 7.89
Neutron scattering length, $b/10^{-12}$ cm**:** 0.513
Thermal neutron capture cross-section, σ_a/barns**:** 0.172

• GEOLOGICAL DATA

Minerals

Mineral	Formula	Density	Hardness	Crystal appearance
Apatite*	$Ca_5(PO_4)_3(F,OH)$	3.2	5	hex., vit. var. colours (esp. yellow)
Phosphophyllite	$Zn_2(Fe,Mn)(PO_4)_2.4H_2O$	3.109	3 – 3.5	mon., vit. colourless/blue green
Turquoise	$CuAl_6(PO_4)_4(OH)_8.5H_2O$	2.84	5 – 6	tric., vit. blue/blue-green
Vivianite	$Fe_3(PO_4)_2.8H_2O$	2.68	1.5 – 2	mon., vit./earthy deep blue

*The fluoride form is also known as fluoroapatite.

Chief ores: apatite/fluoroapatite, of which there are vast deposits; turquoise (ornamental stone)

World production/tonnes y^{-1}**:** 153×10^6

Main mining areas: Russia, USA, Morocco, Tunisia, Togo, Nauru

Reserves/tonnes**:** 5.7×10^9

Specimen: available as white phosphorus sticks (*Danger!*) and red phosphorus lumps or powder (*Care!*)

Abundances

Sun (relative to H = 1×10^{12})**:** 3.16×10^5
Earth's crust/p.p.m.**:** 1000
Seawater/p.p.m.**:**
Atlantic surface: 0.0015
Atlantic deep: 0.042
Pacific surface: 0.0015
Pacific deep: 0.084
Residence time/years**:** 100 000
Classification: recycled
Oxidation state: V

Pt

Atomic number: 78
Relative atomic mass ($^{12}C = 12.0000$)**:** 195.08

CAS: [7440-06-4]

• CHEMICAL DATA

Description: Platinum is a lustrous, silvery-white, malleable and ductile metal. It is unaffected by air and water, and will only dissolve in aqua regia (HCl/HNO_3) and molten alkali. Platinum is used in jewellery, anti-cancer drugs, catalysts and catalytic convertors.

Radii/pm: Pt^{4+} 70; Pt^{2+} 85; atomic 138; covalent 129
Electronegativity: 2.28 (Pauling); 1.44 (Allred); 5.6 eV (absolute)
Effective nuclear charge: 4.05 (Slater); 10.75 (Clementi); 15.65 (Froese-Fischer)

Standard reduction potentials $E^{\ominus}$/V

VI		IV		II		0
PtO_3	2.0	PtO_2	1.045	PtO	0.980	Pt
		PtO_2	0.837	Pt^{2+}	1.188	Pt
		$[PtCl_6]^{2-}$	0.726	$[PtCl_4]^{2-}$	0.758	Pt

Oxidation states

Pt^0	d^{10}	$[Pt(PPh_3)_3]$, $[Pt(PF_3)_4]$
Pt^{II}	d^8	PtO, $PtCl_2$, $PtBr_2$, PtI_2, $[PtCl_4]^{2-}$, $[Pt(CN)_4]^{2-}$, complexes
Pt^{IV}	d^6	PtO_2, $[Pt(OH)_6]^{2-}$ (aq), PtF_4, $PtCl_4$ etc., $[PtCl_6]^{2-}$, complexes
Pt^V	d^5	$(PtF_5)_4$, $[PtF_6]^-$
Pt^{VI}	d^4	PtO_3, PtF_6

• PHYSICAL DATA

Melting point/K: 2045
Boiling point/K: 4100 ± 100
ΔH_{fusion}/kJ mol^{-1}: 19.7
ΔH_{vap}/kJ mol^{-1}: 510.5

Thermodynamic properties (298.15 K, 0.1 MPa)

State	$\Delta_f H^{\ominus}$/kJ mol^{-1}	$\Delta_f G^{\ominus}$/kJ mol^{-1}	$S^{\ominus}$/J K^{-1} mol^{-1}	C_p/J K^{-1} mol^{-1}
Solid	0	0	41.63	25.86
Gas	565.3	520.5	192.406	25.531

Density/kg m^{-3}: 21 450 [293 K]
Molar volume/cm^3: 9.10
Thermal conductivity/W m^{-1} K^{-1}: 71.6 [300 K]
Coefficient of linear thermal expansion/K^{-1}: 9.0×10^{-6}
Electrical resistivity/Ω m: 10.6×10^{-8} [293 K]
Mass magnetic susceptibility/kg^{-1} m^3: $+1.301 \times 10^{-8}$ (s)

Young's modulus/GPa: 170
Rigidity modulus/GPa: 60.9
Bulk modulus/GPa: 276
Poisson's ratio/GPa: 0.39

• BIOLOGICAL DATA

Biological role

None.

Toxicity

Toxic intake: regarded as non-toxic
Lethal intake: LD_{50} ($PtCl_2$, oral, rat) = 17.5 mg kg^{-1}

Hazards

Platinum implants are generally tolerated by the body. Platinum salts are poisonous.

Levels in humans

Organs: n.a., but low
Daily dietary intake: n.a.
Total mass of element in average (70 kg) person: n.a.

Known to pre-Columbian South Americans and taken to Europe about 1750. [Spanish, *platina* = silver]
French, *platine*; German, *Platin*; Italian, *platino*; Spanish, *platino*

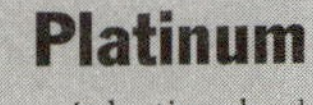

[pla-tin-uhm]

• NUCLEAR DATA

Number of isotopes (including nuclear isomers): 36 **Isotope mass range:** 172 → 201

Key isotopes

Nuclide	Atomic mass	Natural abundance (%)	Nuclear spin I	Nuclear magnetic moment μ	Uses
^{190}Pt*	189.959 917	0.01	0+		E
^{192}Pt**	191.961 019	0.79	0+		E
^{194}Pt	193.962 655	32.9	0+		E
^{195}Pt	194.964 766	33.8	1/2–	+0.609 50	E, NMR
^{196}Pt	195.964 926	25.3	0+		E
^{198}Pt	197.967 869	7.2	0+		E

*^{190}Pt is radioactive with a half-life of 6.9×10^{11} y and decay mode α (3.18 MeV).
**^{192}Pt is radioactive with a half-life of *c.* 1×10^{10} y and decay mode α.
A table of radioactive isotopes is given in Appendix 1, on p246.

NMR [Reference: $[Pt(CN)_6]^{2-}$]	^{195}Pt
Relative sensitivity (^1H = 1.00)	9.94×10^{-3}
Receptivity (^{13}C = 1.00)	19.1
Magnetogyric ratio/rad $T^{-1} s^{-1}$	5.7412×10^7
Nuclear quadrupole moment/m^2	–
Frequency (^1H = 100 Hz; 2.3488T)/MHz	21.449

• ELECTRON SHELL DATA

Ground state electron configuration: $[Xe]4f^{14}5d^96s^1$
Term symbol: $^3D_3$
Electron affinity ($M \rightarrow M^-$)/kJ mol^{-1}**:** 205.3

Ionization energies/kJ mol^{-1}**:**

1.	$M \rightarrow M^+$	870
2.	$M^+ \rightarrow M^{2+}$	1791
3.	$M^{2+} \rightarrow M^{3+}$	(2800)
4.	$M^{3+} \rightarrow M^{4+}$	(3900)
5.	$M^{4+} \rightarrow M^{5+}$	(5300)
6.	$M^{5+} \rightarrow M^{6+}$	(7200)
7.	$M^{6+} \rightarrow M^{7+}$	(8900)
8.	$M^{7+} \rightarrow M^{8+}$	(10 500)
9.	$M^{8+} \rightarrow M^{9+}$	(12 300)
10.	$M^{9+} \rightarrow M^{10+}$	(14 100)

Electron binding energies/eV

K	1s	78 395
L_I	2s	13 880
L_{II}	$2p_{1/2}$	13 273
L_{III}	$2p_{3/2}$	11 564
M_I	3s	3296
M_{II}	$3p_{1/2}$	3027
M_{III}	$3p_{3/2}$	2645
M_{IV}	$3d_{3/2}$	2202
M_V	$3d_{5/2}$	2122

continued in Appendix 2, p256

Main lines in atomic spectrum
[Wavelength/nm(species)]
204.937 (I)
208.459 (I)
214.432 (I)
265.945 (I) (AA)
270.240 (I)
299.767 (I)
306.471 (I)

• CRYSTAL DATA

Crystal structure (cell dimensions/pm), space group
f.c.c. (a = 392.40), Fm3m

X-ray diffraction: mass absorption coefficients (μ/ρ)/$cm^2 g^{-1}$**:** CuK_α 200 MoK_α 113
Neutron scattering length, $b/10^{-12}$ cm: 0.960
Thermal neutron capture cross-section, σ_a/barns**:** 10.3

• GEOLOGICAL DATA

Minerals

Native platinum occurs naturally.

Mineral	Formula	Density	Hardness	Crystal appearance
Platiniridium	(Ir,Pt)	22.7	6 – 7	cub., met. white
Platinum	Pt	*c.* 21	4 – 4.5	cub., metallic white/grey

Chief ores: platinum ore; some platinum is extracted as a by-product of copper and nickel refining.

World production/tonnes y^{-1}**:** 30

Main mining areas: native platinum occurs naturally, mostly as nuggets, in the rivers of the Urals in Russia, and in deposits in Canada, South Africa, Columbia and Peru.

Reserves/tonnes**:** n.a.

Specimen: available as foil, gauze, sponge, powder or wire. Safe.

Abundances

Sun (relative to H = 1×10^{12})**:** 56.2
Earth's crust/p.p.m.**:** *c.* 0.001
Seawater/p.p.m.**:**
Atlantic surface: n.a.
Atlantic deep: n.a.
Pacific surface: 1.1×10^{-7}
Pacific deep: 2.7×10^{-7}
Residence time/years**:** n.a.
Oxidation state: II

Pu

Atomic number: 94
Relative atomic mass ($^{12}C = 12.0000$)**:** 244.0642 (Pu-244)
CAS: [7440-07-5]

• CHEMICAL DATA

Description: Plutonium is a silvery, radioactive metal which occurs naturally in minute amounts in uranium ores. It is attacked by oxygen, steam and acids, but not alkalis. The metal is produced by reacting PuF_3 with either lithium or barium at 1200 °C. A piece of plutonium is warm to the touch because of the energy given off by the α-decay. Plutonium can be used as a nuclear explosive or as a nuclear fuel; 1 kg of plutonium produces an explosion equivalent to 20 000 tonnes of TNT. Plutonium was used on the Apollo flights to power seismic devices and other equipment left on the Moon.

Radii/pm: Pu^{6+} 81; Pu^{5+} 87; Pu^{4+} 93; Pu^{3+} 108
Electronegativity: 1.28 (Pauling); 1.22 (Allred); n.a. (absolute)
Effective nuclear charge: 1.65 (Slater)

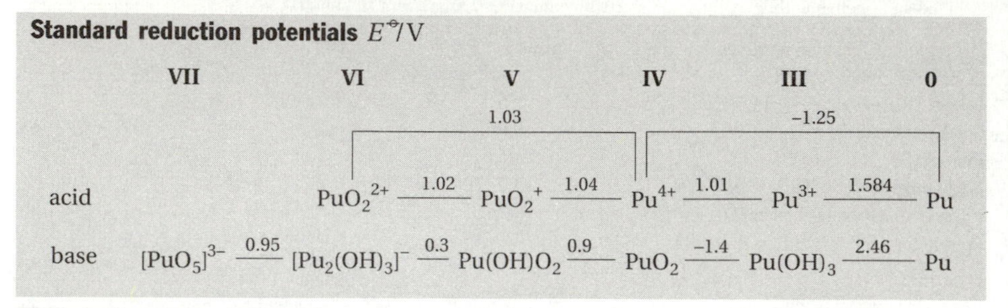

Oxidation states

Pu^{II}	f^6	PuO, PuH_2
Pu^{III}	f^5	Pu_2O_3, PuF_3, $PuCl_3$ etc., $[Pu(OH_2)_x]^{3+}$ (aq), Pu^{3+} salts, complexes, $[Pu(C_5H_5)_3]$
Pu^{IV}	f^4	PuO_2, PuF_4, $[PuCl_6]^{2-}$, $[Pu(OH_2)_x]^{4+}$ (aq) unstable, complexes
Pu^{V}	f^3	PuO_2^{+} (aq) unstable , $CsPuF_6$
Pu^{VI}	f^2	PuO_2^{2+} (aq), PuF_6
Pu^{VII}	f^1	Li_5PuO_6, $[PuO_5]^{3-}$ (aq)

Mixed valence oxide: Pu_3O_5

• PHYSICAL DATA

Melting point/K: 914
Boiling point/K: 3505
ΔH_{fusion}/kJ mol^{-1}: 2.8
ΔH_{vap}/kJ mol^{-1}: 343.5

Thermodynamic properties (298.15 K, 0.1 MPa)

State	$\Delta_f H^{\ominus}$/kJ mol^{-1}	$\Delta_f G^{\ominus}$/kJ mol^{-1}	$S^{\ominus}$/J K^{-1} mol^{-1}	C_p/J K^{-1} mol^{-1}
Solid	0	0	n.a.	n.a.
Gas	n.a.	n.a.	n.a.	n.a.

Density/kg m^{-3}: 19 840 (α) [298 K]; 16 623 [liquid at m.p.]
Molar volume/cm^3: 12.3
Thermal conductivity/W m^{-1} K^{-1}: 6.74 [300 K]
Coefficient of linear thermal expansion/K^{-1}: 55×10^{-6}
Electrical resistivity/Ω m: 146×10^{-8} [273 K]
Mass magnetic susceptibility/kg^{-1} m^3: $+3.17 \times 10^{-8}$ (s)

Young's modulus/GPa: 87.5
Rigidity modulus/GPa: 34.5
Bulk modulus/GPa: n.a.
Poisson's ratio/GPa: 0.18

• BIOLOGICAL DATA

Biological role

None.

Toxicity

The very high radiotoxicity of plutonium overrides any other toxicity considerations.

Hazards

Never normally encountered outside the laboratory or nuclear industry, but is highly dangerous because of its intense radioactivity. Inside the human body the element tends to accumulate in bone marrow and there it is highly dangerous because of the α-particles it emits. The permitted levels of plutonium are the lowest of any of the radioactive elements.

Levels in humans

Organs: n.a., but extremely low
Daily dietary intake: nil
Total mass of element in average (70 kg) person: n.a., but extremely low

Discovery: see Nuclear Data section.
[Named after the planet Pluto]
French, *plutonium*; German, *Plutonium*; Italian, *plutonio*; Spanish, *plutonio*

Plutonium

[ploo-toh-nee-uhm]

• NUCLEAR DATA

Discovery: Plutonium was discovered in 1940 by G.T. Seaborg, A.C. Wahl and J.W. Kennedy at Berkeley, California, USA.

Number of isotopes (including nuclear isomers): 15 **Isotope mass range:** 232 → 246

Key isotopes

Nuclide	Atomic mass	Half life ($T_{1/2}$)	Decay mode and energy (MeV)	Nuclear spin I	Nucl. mag. moment μ	Uses
^{234}Pu	234.043 299	8.8 h	EC (0.38) 94%; α (6.310) 6%; no γ	0+		
^{236}Pu	236.046 032	2.87 y	α (5.867); no γ	0+		
^{237}Pu	237.048 401	45.2 d	EC (0.22) 99.9%; α (5.747) 0.03%; γ	7/2–		
^{238}Pu	238.049 554	87.74 y	α (5.593); γ	0+		
^{239}Pu	239.052 157	24 110 y	α (5.244); γ	1/2+	+0.203	NMR
^{240}Pu	240.053 808	6537 y	α (5.255); γ	0+		
^{241}Pu	241.056 845	14.4 y	β^- (0.021) >99%; α (5.139) 0.02%; γ	5/2+	–0.683	
^{242}Pu	242.058 737	3.76×10^5 y	α (4.983); γ	0+		
^{243}Pu	243.061 998	4.956 h	β^- (0.582); γ	7/2+		
^{244}Pu	244.064 199	8.2×10^7 y	α (4.665) 99.9%; SF 0.1%; γ	0+		
^{245}Pu	245.067 820	10.5 h	β^- (1.28); γ	9/2–?		
^{246}Pu	246.070 171	10.85 d	β^- (0.40); γ	0+		

Other isotopes of plutonium have half-lives shorter than 1 hour.

NMR [Reference: n.a.]	^{239}Pu	^{241}Pu
Relative sensitivity (^1H = 1.00)	–	
Receptivity (^{13}C = 1.00)	–	
Magnetogyric ratio/rad $T^{-1}s^{-1}$	0.972×10^7	
Nuclear quadrupole moment/m^2	–	5.600×10^{-28}
Frequency (^1H = 100 Hz; 2.3488T)/MHz	3.63	

• ELECTRON SHELL DATA

Ground state electron configuration: [Rn]$5f^6 7s^2$
Term symbol: $^7F_0$
Electron affinity ($M \rightarrow M^-$)/kJ mol^{-1}**:** n.a.

Ionization energies/kJ mol^{-1}**:**
1. M → M^+ 585

Electron binding energies/eV
n.a.

Main lines in atomic spectrum
[Wavelength/nm(species)]
321.508 (I)
324.416 (I)
325.208 (I)
327.524 (I)
329.256 (I)
329.361 (I)
329.691 (I)

• CRYSTAL DATA

Crystal structure (cell dimensions/pm), space group
α-Pu monoclinic (a = 618.3, b = 482.2, c = 1096.3; β = 101.79°), $P2_1m$
β-Pu monoclinic (a = 928.4, b = 1046.3, c = 758.9; β = 92.13°), I2/m
γ-Pu orthorhombic (a = 315.87, b = 576.82, c = 1016.2; β = 92.13°), Fddd
δ-Pu f.c.c. (a = 463.71), Fm3m
δ'-Pu tetragonal (a = 333.9, c = 444.6), I4/mmm
ε-Pu b.c.c. (a = 363.48), Im3m
$T(\alpha \rightarrow \beta)$ = 395 K; $T(\beta \rightarrow \gamma)$ = 473 K; $T(\gamma \rightarrow \delta)$ = 583 K; $T(\delta \rightarrow \delta')$ = 725 K; $T(\delta' \rightarrow \varepsilon)$ = 753 K

X-ray diffraction: mass absorption coefficients $(\mu/\rho)/cm^2\,g^{-1}$**:** CuK_α n.a. MoK_α n.a.
Neutron scattering length, $b/10^{-12}$ cm**:** 0.77 (^{239}Pu)
Thermal neutron capture cross-section, σ_a/barns**:** 1017.3 (^{239}Pu)

• GEOLOGICAL DATA

Minerals
none

Chief source: plutonium-239 is obtained in tonne quantities from uranium fuel elements where it is produced by neutron capture: $^{238}U + n \rightarrow {}^{239}U$ (β emission) → ^{239}Np (β emission) → ^{239}Pu.

World production/tonnes y^{-1}**:** 20

Reserves/tonnes**:** > 500

Specimen: commercially available, under licence – see Key.

Abundances
Sun (relative to H = 1×10^{12})**:** n.a.
Earth's crust/p.p.m.**:** plutonium is present in minute quantities in uranium ores, and is formed when one of the emitted neutrons, from a uranium atom undergoing fission, is captured by another uranium nucleus which then undergoes β-emission to give first neptunium, then plutonium.
Seawater/p.p.m.**:** nil

Po

Atomic number: 84
Relative atomic mass ($^{12}C = 12.0000$)**:** 208.9824 (Po-209)

CAS: [7440-08-6]

• CHEMICAL DATA

Description: Polonium is a reactive, silvery-grey metal that dissolves in dilute acids. It is fairly volatile and about half will evaporate within two days if kept at 55 °C. A gram capsule of polonium will reach 500 °C because of the intense α-radiation, and for this reason polonium is used as a lightweight heat supply for space satellites. It is also used as a source of α-radiation for research.

Radii/pm: Po^{4+} 65; Po^{2-} 230; atomic 167; covalent 153
Electronegativity: 2.0 (Pauling); 1.76 (Allred); 5.16 eV (absolute)
Effective nuclear charge: 6.95 (Slater); 14.22 (Clementi); 18.31 (Froese-Fischer)

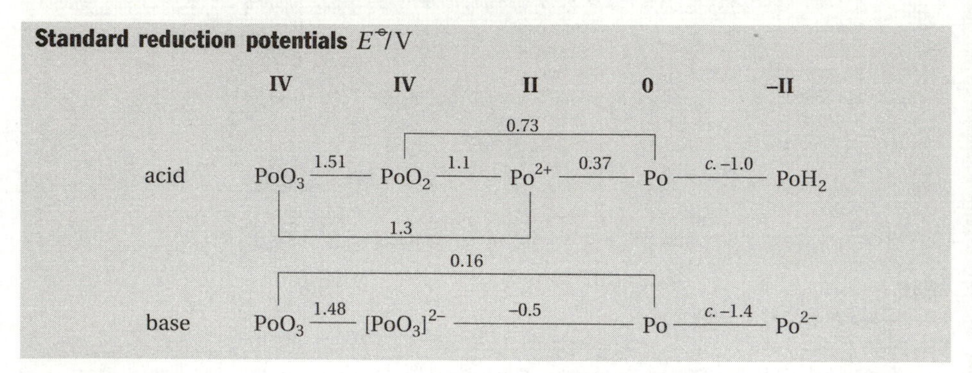

Oxidation states

Po^{-II}	[Rn]	H_2Po, Na_2Po
Po^{II}	s^2p^2	PoO, $PoCl_2$, $PoBr_2$
Po^{IV}	s^2	PoO_2, $[PoO_3]^{2-}$ (aq), $PoCl_4$, $PoBr_4$, PoI_4, $[PoI_6]^{2-}$
Po^{VI}	d^{10}	PoO_3 ?, PoF_6

Covalent bonds

Bond	r/ pm	E/ kJ mol^{-1}
Po—Cl	238	n.a.
Po—Po	335	n.a.

• PHYSICAL DATA

Melting point/K: 527
Boiling point/K: 1235

ΔH_{fusion}/kJ mol^{-1}: 10
ΔH_{vap}/kJ mol^{-1}: 100.8

Thermodynamic properties (298.15 K, 0.1 MPa)

State	$\Delta_f H^{\ominus}$/kJ mol^{-1}	$\Delta_f G^{\ominus}$/kJ mol^{-1}	$S^{\ominus}$/J K^{-1} mol^{-1}	C_p/J K^{-1} mol^{-1}
Solid	0	0	n.a.	26.1
Gas	146 ?	n.a.	n.a.	n.a.

Density/kg m^{-3}: 9320 (α) [293 K]
Molar volume/cm^3: 22.4
Thermal conductivity/W m^{-1} K^{-1}: 20 [300 K]
Coefficient of linear thermal expansion/K^{-1}: 23.0×10^{-6}
Electrical resistivity/Ω m: n.a.
Mass magnetic susceptibility/kg^{-1} m^3: n.a.

• BIOLOGICAL DATA

Biological role
None.

Toxicity
The high radiotoxicity of polonium overrides other toxicity considerations.

Hazards
Polonium is very dangerous because of its α-emission.

Levels in humans
nil

Daily dietary intake: nil
Total mass of element in average (70 kg) person: nil

Discovered in 1898 by Marie Curie at Paris, France.
[Named after Poland]
French, *polonium*; German, *Polonium*; Italian, *polonio*; Spanish, *polonio*

Polonium

[pol-oh-nee-uhm]

• NUCLEAR DATA

Number of isotopes (including nuclear isomers): 33 **Isotope mass range:** 194 → 218

Key isotopes

Nuclide	Atomic mass	Half life ($T_{1/2}$)	Decay mode and energy (MeV)	Nuclear spin I	Nucl. mag. moment μ	Uses
^{204}Po	203.980 280	3.53 h	EC (2.4); γ	0+		
^{206}Po	205.980 456	8.8 d	EC (1.85) 95%; α 5%; γ	0+		
^{207}Po	206.981 570	5.80 h	EC; β^+ (2.91); γ	5/2–	+0.79	
^{208}Po	207.981 222	2.898 y	α (5.213); γ	0+		
^{209}Po	208.982 404	102 y	α (4.976); γ	1/2–	+0.77	R
^{210}Po*	209.982 848	138.38 d	α (5.407); γ	0+		tracer, fuel
^{211}Po*	210.986 627	0.516 s	α (7.594); γ	9/2+		
^{216}Po*	216.001 889	0.145 s	α (6.906); no γ	0+		
^{218}Po*		3.04 m	α (6.114); no γ	0+		

*Traces of these isotopes occur naturally.
Other isotopes of polonium have half-lives shorter than 2 hours.

NMR [Reference: not reported]	^{209}Po
Relative sensitivity (^{1}H = 1.00)	–
Receptivity (^{13}C = 1.00)	–
Magnetogyric ratio/rad $T^{-1}s^{-1}$	7.4×10^7
Nuclear quadrupole moment/m^2	–
Frequency (^{1}H = 100 Hz; 2.3488T)/MHz	28

• ELECTRON SHELL DATA

Ground state electron configuration: [Xe]$4f^{14}5d^{10}6s^26p^4$
Term symbol: $^3P_2$
Electron affinity ($M \rightarrow M^-$)/kJ mol^{-1}**:** 183

Ionization energies/kJ mol^{-1}**:**

1.	$M \rightarrow M^+$	812
2.	$M^+ \rightarrow M^{2+}$	(1800)
3.	$M^{2+} \rightarrow M^{3+}$	(2700)
4.	$M^{3+} \rightarrow M^{4+}$	(3700)
5.	$M^{4+} \rightarrow M^{5+}$	(5900)
6.	$M^{5+} \rightarrow M^{6+}$	(7000)
7.	$M^{6+} \rightarrow M^{7+}$	(10 800)
8.	$M^{7+} \rightarrow M^{8+}$	(12 700)
9.	$M^{8+} \rightarrow M^{9+}$	(14 900)
10.	$M^{9+} \rightarrow M^{10+}$	(17 000)

Electron binding energies/eV

K	1s	93 105
L_I	2s	16 939
L_{II}	$2p_{1/2}$	16 244
L_{III}	$2p_{3/2}$	13 814
M_I	3s	4149
M_{II}	$3p_{1/2}$	3854
M_{III}	$3p_{3/2}$	3302
M_{IV}	$3d_{3/2}$	2798
M_V	$3d_{5/2}$	2683

continued in Appendix 2, p256

Main lines in atomic spectrum
[Wavelength/nm(species)]

245.008 (I)
255.801 (I)
300.321 (I)
417.052 (I)

• CRYSTAL DATA

Crystal structure (cell dimensions/pm), space group

α-Po cubic (a = 335.2), Pm3m
β-Po rhombohedral (a = 336.6, α = 98° 13'), R$\bar{3}$m
$T(\alpha \rightarrow \beta)$ = 309 K

X-ray diffraction: mass absorption coefficients (μ/ρ)/cm^2 g^{-1}**:** CuK_α n.a. MoK_α n.a.
Neutron scattering length, $b/10^{-12}$ cm**:** n.a.
Thermal neutron capture cross-section, σ_a/barns**:** < 0.03 (^{210}Po)

• GEOLOGICAL DATA

Minerals

Uranium ores contain about 100×10^{-6} g of polonium per tonne, and Mme. Curie obtained the first sample of the element from this source. Polonium-210 is made in g quantities by bombarding bismuth with neutrons.

World production: n.a. but probably *c.* 100 g per year.

Specimen: commercially available, under licence – see Key.

Abundances

Sun (relative to H = 1×10^{12})**:** n.a.
Earth's crust/p.p.m.**:** traces in uranium ores
Seawater/p.p.m.**:** nil

K

Atomic number: 19
Relative atomic mass ($^{12}C = 12.0000$)**:** 39.0983

CAS: [7440-09-7]

• CHEMICAL DATA

Description: Potassium is a soft white metal which is silvery when first cut but oxidizes rapidly in air. It reacts violently with water. Potassium is obtained from the reaction of sodium metal with potassium chloride. The metal itself is little used, but potassium compounds are important in fertilizers, chemicals and glass.

Radii/pm: K^+ 133; atomic 227; covalent 203; van der Waals 231
Electronegativity: 0.82 (Pauling); 0.91 (Allred); 2.42 eV (absolute)
Effective nuclear charge: 2.20 (Slater); 3.50 (Clementi); 4.58 (Froese-Fischer)

Standard reduction potentials $E^\ominus$/V

I		0
K^+	−2.924	K

Oxidation states

K^{-I}	s^2	solution in liquid ammonia
K^{I}	[Ar]	K_2O, K_2O_2 (peroxide), KO_2 (superoxide), KO_3 (ozonide), KOH, $[K(OH_2)_4]^+$ (aq), KH, KF,KCl etc., K^+ salts, K_2CO_3,complexes, [K(18-crown-6)]$^+$

• PHYSICAL DATA

Melting point/K: 336.80
Boiling point/K: 1047
ΔH_{fusion}/kJ mol^{-1}: 2.40
ΔH_{vap}/kJ mol^{-1}: 77.53

Thermodynamic properties (298.15 K, 0.1 MPa)

State	$\Delta_f H^\ominus$/kJ mol^{-1}	$\Delta_f G^\ominus$/kJ mol^{-1}	$S^\ominus$/J K^{-1} mol^{-1}	C_p/J K^{-1} mol^{-1}
Solid	0	0	64.18	29.58
Gas	89.24	60.59	160.336	20.786

Density/kg m^{-3}: 862 [293 K]; 828 [liquid at m.p.]
Molar volume/cm^3: 45.36
Thermal conductivity/W m^{-1} K^{-1}: 102.4 [300 K]
Coefficient of linear thermal expansion/K^{-1}: 83×10^{-6}
Electrical resistivity/Ω m: 6.15×10^{-8} [273 K]
Mass magnetic susceptibility/kg^{-1} m^3: $+6.7 \times 10^{-9}$ (s)

Young's modulus/GPa: 3.53 [83 K]
Rigidity modulus/GPa: 1.30
Bulk modulus/GPa: n.a.
Poisson's ratio/GPa: 0.35 [83 K]

• BIOLOGICAL DATA

Biological role

Essential to all living things.

Toxicity

Toxic intake: KCl = *c.* 4 g
Lethal intake: LD_{50} (chloride, oral rat) = 2600 mg kg^{-1}

Hazards

The toxicity of potassium compounds is almost always that of the anion, not of the K^+. However, although KCl is often used as a nutrient or dietary supplement, there are rare cases of excess ingestion by humans proving fatal.

Levels in humans

Blood/mg dm^{-3}: 1620
Bone/p.p.m.: 2100
Liver/p.p.m.: 16 000
Muscle/p.p.m.: 16 000
Daily dietary intake: 1400 – 7400 mg
Total mass of element in average (70 kg) person: 140 g

Isolated in 1807 by Sir Humphry Davy at London, England.
[English, *potash*; Latin, *kalium*]
French, *potassium*; German, *Kalium*; Italian, *potassio*; Spanish, *potasio*

Potassium

[poh-tass-ium]

• NUCLEAR DATA

Number of isotopes (including nuclear isomers): 18 **Isotope mass range:** 35 → 51

Key isotopes

Nuclide	Atomic mass	Natural abundance (%)	Nuclear spin I	Nuclear magnetic moment μ	Uses
^{39}K	38.963 707	93.258 1	3/2+	+0.391 465	E, NMR
^{40}K *	39.963 999	0.011 7	4–	–1.298 009	E
^{41}K	40.961 825	6.730 2	3/2+	+0.214 869	E, NMR

^{40}K is radioactive with a half-life of 1.25×10^9 y and decay mode β^- (1.32 MeV); EC; γ.
A table of radioactive isotopes is given in Appendix 1, on p247.

NMR [Reference: K^+ (aq)]	^{39}K	^{40}K	^{41}K
Relative sensitivity ($^1H = 1.00$)	5.08×10^{-4}		8.40×10^{-4}
Receptivity ($^{13}C = 1.00$)	2.69		0.0328
Magnetogyric ratio/rad $T^{-1}s^{-1}$	1.2483×10^7		0.6851×10^7
Nuclear quadrupole moment/m^2	0.0601×10^{-28}	-0.0749×10^{-28}	0.0733×10^{-28}
Frequency (^1H = 100 Hz; 2.3488T)/MHz	4.667		2.561

• ELECTRON SHELL DATA

Ground state electron configuration: [Ar]$4s^1$
Term symbol: $^2S_{1/2}$
Electron affinity ($M \rightarrow M^-$)/kJ mol^{-1}**:** 48.4

Ionization energies/kJ mol^{-1}**:**

		kJ mol⁻¹
1.	$M \rightarrow M^+$	418.8
2.	$M^+ \rightarrow M^{2+}$	3051.4
3.	$M^{2+} \rightarrow M^{3+}$	4411
4.	$M^{3+} \rightarrow M^{4+}$	5877
5.	$M^{4+} \rightarrow M^{5+}$	7975
6.	$M^{5+} \rightarrow M^{6+}$	9649
7.	$M^{6+} \rightarrow M^{7+}$	11 343
8.	$M^{7+} \rightarrow M^{8+}$	14 942
9.	$M^{8+} \rightarrow M^{9+}$	16 964
10.	$M^{9+} \rightarrow M^{10+}$	48 575

Electron binding energies/eV

		eV
K	1s	3608.4
L_I	2s	378.6
L_{II}	$2p_{1/2}$	297.3
L_{III}	$2p_{3/2}$	294.6
M_I	3s	34.8
M_{II}	$3p_{1/2}$	18.3
M_{III}	$3p_{3/2}$	18.3

Main lines in atomic spectrum
[Wavelength/nm(species)]

404.414 (I)
691.108 (I)
693.877 (I)
766.491 (I)
769.896 (I)

• CRYSTAL DATA

Crystal structure (cell dimensions/pm), space group
b.c.c. (a = 533.4), Im3m
X-ray diffraction: mass absorption coefficients (μ/ρ)/cm^2 g^{-1}**:** CuK_α 143 MoK_α 15.8
Neutron scattering length, $b/10^{-12}$ cm**:** 0.367
Thermal neutron capture cross-section, σ_a/barns**:** 2.1

• GEOLOGICAL DATA

Minerals

Potassium occurs in many minerals.

Mineral	Formula	Density	Hardness	Crystal appearance
Alunite	$KAl_3(SO_4)_2(OH)_6$	2.69	3.5 – 4	rhom., vit. white/grey
Carnallite	$KCl.MgCl_2.6H_2O$	1.602	2.5	orth., greasy colourless-red
Orthoclase*	$KAlSi_3O_8$	2.563	6 – 6.5	mon., vit. colourless/white
Sylvite	KCl	1.993	2	cub., vit. colourless/white

*Mined on a large scale for porcelain, ceramics and glass.

Chief ores: sylvite, carnallite, alunite
World production/tonnes y^{-1}**:** 200 (potassium metal); 51×10^6 (potassium salts)
Main mining areas: Germany, Spain, Canada, USA, Italy
Reserves/tonnes: vast, $> 1 \times 10^{10}$
Specimen: available as chunks (in mineral oil) or ingots (in ampoules). *Warning!*

Abundances

Sun (relative to H = 1×10^{12})**:** 1.45×10^5
Earth's crust/p.p.m.**:** 21 000
Residence time/years**:** 5×10^6
Classification: accumulating
Oxidation state: I

Atomic number: 59 **CAS:** [7440-10-0]

Relative atomic mass ($^{12}C = 12.0000$): 140.90765

• CHEMICAL DATA

Description: Praseodymium is a soft, malleable, silvery metal of the so-called rare earth group (more correctly termed the lanthanides). It reacts slowly with oxygen, and rapidly with water. Praseodymium is used in alloys for permanent magnets and flints. It is used to make the yellow glass for eye protection for welders, etc.

Radii/pm: Pr^{4+} 92; Pr^{3+} 106; atomic 183; covalent 165

Electronegativity: 1.13 (Pauling); 1.07 (Allred); ≤ 3.0 eV (absolute)

Effective nuclear charge: 2.85 (Slater); 7.75 (Clementi); 10.70 (Froese-Fischer)

Standard reduction potentials $E^{\ominus}/V$

	IV		III		0
acid	Pr^{4+}	3.2	Pr^{3+}	−2.35	Pr
base	PrO_2	0.8	$Pr(OH)_3$	−2.79	Pr

Oxidation states

Pr^{III}	f^2	Pr_2O_3, $Pr(OH)_3$, $[Pr(OH_2)_x]^{3+}$ (aq), Pr^{3+} salts, PrF_3, $PrCl_3$ etc., complexes
Pr^{IV}	f^1	PrO_2, PrF_4, Na_2PrF_6

• PHYSICAL DATA

Melting point/K: 1204

Boiling point/K: 3785

ΔH_{fusion}/kJ mol^{-1}: 11.3

ΔH_{vap}/kJ mol^{-1}: 332.6

Thermodynamic properties (298.15 K, 0.1 MPa)

State	$\Delta_f H^{\ominus}$/kJ mol^{-1}	$\Delta_f G^{\ominus}$/kJ mol^{-1}	$S^{\ominus}$/J K^{-1} mol^{-1}	C_p/J K^{-1} mol^{-1}
Solid	0	0	73.2	27.20
Gas	355.6	320.9	189.808	21.359

Density/kg m^{-3}: 6773 [293 K]

Molar volume/cm^3: 20.80

Thermal conductivity/W m^{-1} K^{-1}: 12.5 [300 K]

Coefficient of linear thermal expansion/K^{-1}: 6.79×10^{-6}

Electrical resistivity/Ω m: 68×10^{-8} [298 K]

Mass magnetic susceptibility/kg^{-1} m^3: $+4.47 \times 10^{-7}$ (s)

Young's modulus/GPa: 37.3

Rigidity modulus/GPa: 14.8

Bulk modulus/GPa: 28.8

Poisson's ratio/GPa: 0.281

• BIOLOGICAL DATA

Biological role

None, but acts to stimulate metabolism.

Toxicity

Toxic intake: n.a.

Lethal intake: LD_{50} (chloride, oral, rat) = 4200 mg kg^{-1}

Hazards

Praseodymium is mildly toxic by ingestion, and is a skin and eye irritant.

Levels in humans

Organs: n.a., but very low

Daily dietary intake: n.a.

Total mass of element in average (70 kg) person: n.a., but very low

Separated in 1885 by Baron Auer von Welsbach at Vienna, Austria.
[Greek, *prasios didymos* = green twin]
French, *praséodyme*; German, *Praseodym*; Italian, *praseodimio*; Spanish, *praseodimio* [prah-zee-o-dim-iuhm]

Praseodymium

• NUCLEAR DATA

Number of isotopes (including nuclear isomers): 26 **Isotope mass range:** 132 → 152

Key isotopes

Nuclide	Atomic mass	Natural abundance (%)	Nuclear spin I	Nuclear magnetic moment μ	Uses
^{141}Pr	140.907 647	100	5/2+	+4.136	NMR

A table of radioactive isotopes is given in Appendix 1, on p247.

NMR [Reference: not recorded]	^{141}Pr
Relative sensitivity (^1H = 1.00)	0.29
Receptivity (^{13}C = 1.00)	1620
Magnetogyric ratio/rad $T^{-1}s^{-1}$	7.765×10^7
Nuclear quadrupole moment/m^2	-0.589×10^{-28}
Frequency (^1H = 100 Hz; 2.3488T)/MHz	29.291

• ELECTRON SHELL DATA

Ground state electron configuration: $[Xe]4f^36s^2$
Term symbol: $^4I_{9/2}$
Electron affinity ($M \rightarrow M^-$)/kJ mol^{-1}**:** ≤ 50

Ionization energies/kJ mol^{-1}**:**

1.	$M \rightarrow M^+$	523.1
2.	$M^+ \rightarrow M^{2+}$	1018
3.	$M^{2+} \rightarrow M^{3+}$	2086
4.	$M^{3+} \rightarrow M^{4+}$	3761
5.	$M^{4+} \rightarrow M^{5+}$	5543

Electron binding energies/eV

K	1s	41 991
L_I	2s	6835
L_{II}	$2p_{1/2}$	6440
L_{III}	$2p_{3/2}$	5964
M_I	3s	1511
M_{II}	$3p_{1/2}$	1337
M_{III}	$3p_{3/2}$	1242
M_{IV}	$3d_{3/2}$	848.3
M_V	$3d_{5/2}$	928.8

continued in Appendix 2, p257

Main lines in atomic spectrum
[Wavelength/nm(species)]

406.282 (II)
410.072 (II)
417.939 (II)
422.293 (II)
422.535 (II)
495.137 (I) (AA)

• CRYSTAL DATA

Crystal structure (cell dimensions/pm), space group
α-Pr h.c.p. (a = 367.25, c = 1183.5), $P6_3/mmc$
β-Pr b.c.c. (a = 413), Im3m
$T(\alpha \rightarrow \beta)$ = 1065 K

X-ray diffraction: mass absorption coefficients (μ/ρ)/$cm^2\ g^{-1}$**:** CuK_α 363 MoK_α 50.7
Neutron scattering length, $b/10^{-12}$ cm**:** 0.445
Thermal neutron capture cross-section, σ_a/barns**:** 11.5

• GEOLOGICAL DATA

Minerals

Mineral	Formula	Density	Hardness	Crystal appearance
Bastnäsite*	(Ce,La, etc.)CO_3F	4.9	4 – 4.5	hex., vit. greasy yellow
Monazite*	(Ce, La, Nd, Th, etc.)PO_4	5.20	5 – 5.5	mon., waxy/vit. yellow-brown

*Although not a major constituent, praseodymium is present in extractable amounts.

Chief ores: monazite, bastnäsite
World production/tonnes y^{-1}**:** 2400
Main mining areas: USA, Brazil, India, Sri Lanka, Australia
Reserves/tonnes: $c.2 \times 10^6$
Specimen: available as chips, foil and ingots. Safe. Pr powder is a skin and eye irritant. *Care!*

Abundances

Sun (relative to H = 1×10^{12})**:** 4.6
Earth's crust/p.p.m.**:** 9.5
Seawater/p.p.m.**:**
Atlantic surface: 4×10^{-7}
Atlantic deep: 7×10^{-7}
Pacific surface: 4.4×10^{-7}
Pacific deep: 10×10^{-7}
Residence time/years**:** n.a.
Oxidation state: III

Pm

Atomic number: 61 **CAS:** [7440-12-2]
Relative atomic mass ($^{12}C = 12.0000$)**:** 144.9127 (Pm-145)

• CHEMICAL DATA

Description: Promethium is a radioactive metal of the so-called rare earth group (more correctly termed the lanthanides). It is used in specialized miniature batteries.

Radii/pm: Pm^{3+} 106; atomic 181
Electronegativity: n.a. (Pauling); 1.07 (Allred); ≤ 3.0 eV (absolute)
Effective nuclear charge: 2.85 (Slater); 9.40 (Clementi); 10.94 (Froese-Fischer)

Standard reduction potentials $E^{\ominus}$/V

	III		0
acid	Pm^{3+}	—— −2.29 ——	Pm
base	$Pm(OH)_3$	—— −2.76 ——	Pm

Oxidation states

Pm^{III}	f^4	Pm_2O_3, $Pm(OH)_3$, $[Pm(OH_2)_x]^{3+}$ (aq), PmF_3, some complexes

• PHYSICAL DATA

Melting point/K: 1441
Boiling point/K: *c.* 3000
ΔH_{fusion}/kJ mol^{-1}: 12.6
ΔH_{vap}/kJ mol^{-1}: n.a.

Thermodynamic properties (298.15 K, 0.1 MPa)

State	$\Delta_f H^{\ominus}$/kJ mol^{-1}	$\Delta_f G^{\ominus}$/kJ mol^{-1}	$S^{\ominus}$/J K^{-1} mol^{-1}	C_p/J K^{-1} mol^{-1}
Solid	0	0	n.a.	26.8
Gas	n.a.	n.a.	187.101	24.255

Density/kg m^{-3}: 7220 [298 K]
Molar volume/cm^3: 20.1
Thermal conductivity/W m^{-1} K^{-1}: 17.9 (est.) [300 K]
Coefficient of linear thermal expansion/K^{-1}: 16×10^{-6}
Electrical resistivity/Ω m: 50×10^{-8} (est.) [273 K]
Mass magnetic susceptibility/kg^{-1} m^3: n.a.

Young's modulus/GPa: 46 (est.)
Rigidity modulus/GPa: 18 (est.)
Bulk modulus/GPa: 33 (est.)
Poisson's ratio/GPa: 0.28 (est.)

• BIOLOGICAL DATA

Biological role
None.

Toxicity
The radiotoxicity of promethium overrides other toxicity considerations.

Hazards
Never normally encountered outside the laboratory or nuclear industry, but it is hazardous because of its radioactivity.

Levels in humans
Organs: nil
Daily dietary intake: nil
Total mass of element in average (70 kg) person: nil

Discovery: see Nuclear Data section.
[Named after Prometheus of Greek mythology, who stole fire from the gods]
French, *prométhium*; German, *Promethium*; Italian, *prometo*; Spanish, *prometio*

Promethium

[proh-mee-thi-uhm]

• NUCLEAR DATA

Discovery: Promethium was produced in 1945 by J.A. Marinsky, L.E. Glendenin and C.D. Coryell at Oak Ridge, Tennessee, USA.

Number of isotopes (including nuclear isomers): 27 **Isotope mass range:** 134 → 155

Key isotopes

Nuclide	Atomic mass	Half life ($T_{1/2}$)	Decay mode and energy (MeV)	Nuclear spin I	Nucl. mag. moment μ	Uses
^{141}Pm	140.913 600	20.9 m	β^+ (3.73) 52%; EC 48%; γ	5/2+		
^{143}Pm	142.910 930	265 d	EC (1.042); γ	5/2+	3.8	
^{144}Pm	143.912 588	360 d	EC (2.333); γ	5–	1.7	
^{145}Pm	144.912 743	17.7 y	EC (0.164); γ	5/2+		
^{146}Pm	145.914 708	5.53 y	EC (1.48) 63%; β^- (1.54) 37%; γ	3–		
^{147}Pm	146.915 135	2.6234 y	β^- (0.224); weak γ	7/2+	+2.6	R
^{148}Pm	147.917 473	5.37 d	β^- (2.47); γ	1–	+2.0	
^{148m}Pm		41.3 d	β^- (2.6); 95%; IT (0.137) 5%; γ	6–	1.8	
^{149}Pm	148.918 332	2.212 d	β^- (1.073); γ	7/2+	3.3	
^{150}Pm	149.920 981	2.68 h	β^- (3.45); γ	1–?		
^{151}Pm	150.921 203	1.183d	β^- (1.197); γ	5/2+	+1.8	

Other isotopes of promethium have half-lives shorter than 10 minutes.

NMR [Reference: not recorded]	^{147}Pm
Relative sensitivity ($^1H = 1.00$)	–
Receptivity ($^{13}C = 1.00$)	–
Magnetogyric ratio/rad $T^{-1}s^{-1}$	3.613×10^7
Nuclear quadrupole moment/m^2	–
Frequency (^1H = 100 Hz; 2.3488T)/MHz	13.51

• ELECTRON SHELL DATA

Ground state electron configuration: $[Xe]4f^5 6s^2$
Term symbol: $^6H_{5/2}$
Electron affinity ($M \rightarrow M^-$)/kJ mol^{-1}**:** ≤ 50

Ionization energies/kJ mol^{-1}**:**

1.	$M \rightarrow M^+$	535.9
2.	$M^+ \rightarrow M^{2+}$	1052
3.	$M^{2+} \rightarrow M^{3+}$	2150
4.	$M^{3+} \rightarrow M^{4+}$	3970
5.	$M^{4+} \rightarrow M^{5+}$	5953

Electron binding energies/eV

K	1s	45 184
L_I	2s	7428
L_{II}	$2p_{1/2}$	7013
L_{III}	$2p_{3/2}$	6459
M_I	3s	–
M_{II}	$3p_{1/2}$	1471.4
M_{III}	$3p_{3/2}$	1357
M_{IV}	$3d_{3/2}$	1052
M_V	$3d_{5/2}$	1027

continued in Appendix 2, p257

Main lines in atomic spectrum
[Wavelength/nm(species)]
389.215 (II)
391.026 (II)
391.910 (II)
395.774 (II)
399.896 (II)
441.796 (II)

• CRYSTAL DATA

Crystal structure (cell dimensions/pm), space group
hexagonal

X-ray diffraction: mass absorption coefficients (μ/ρ)/$cm^2 g^{-1}$**:** CuK_α 386 MoK_α 55.9
Neutron scattering length, $b/10^{-12}$ cm**:** 0.126
Thermal neutron capture cross-section, σ_a/barns**:** 168.4

• GEOLOGICAL DATA

Minerals

All the original promethium present when the Earth formed has long since undergone radioactive decay. There is a tiny amount present in uranium ores as a fission product.

Chief ore: none
World production: obtained in mg quantities from the fission products of nuclear reactors.
Reserves/tonnes**:** n.a.
Specimen: commercially available, under licence – see Key

Abundances

Sun (relative to $H = 1 \times 10^{12}$)**:** n.a.
Earth's crust/p.p.m.**:** traces in uranium ores
Seawater/p.p.m.**:** nil

Pa

Atomic number: 91
Relative atomic mass ($^{12}C = 12.0000$): 231.03588

CAS: [7440-13-3]

• CHEMICAL DATA

Description: Protactinium is a silvery, radioactive metal found naturally in uranium ores. It is attacked by oxygen, steam and acids, but not alkalis. The element becomes superconducting below 1.4 K. It has no commercial use.

Radii/pm: Pa^{5+} 89; Pa^{4+} 98; Pa^{3+} 113; atomic 161
Electronegativity: 1.5 (Pauling); 1.014 (Allred); n.a. (absolute)
Effective nuclear charge: 1.80 (Slater)

Standard reduction potentials $E^{\ominus}$/V

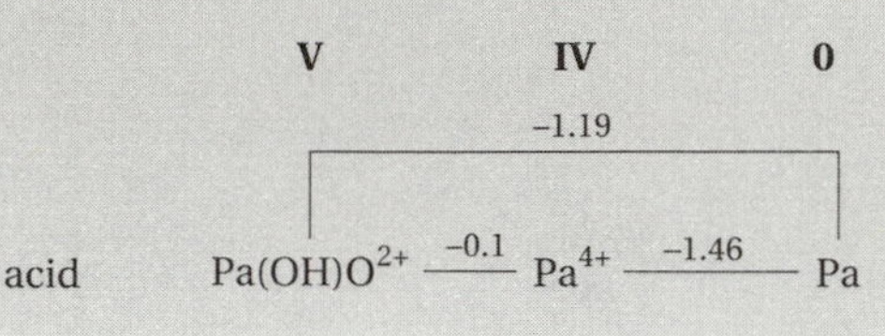

Oxidation states

Pa^{III}	f^2	PaI_3
Pa^{IV}	f^1	PaO_2, $[Pa(OH_2)_x]^{4+}$ (aq), PaF_4, $PaCl_4$ etc. $[Pa(C_5H_5)_4]$
Pa^{V}	[Rn]	Pa_2O_5, PaO_2^+ (compounds) Pa^V (aq) unstable, PaF_5, $PaCl_5$, etc., $[PaF_6]^-$, $[PaF_7]^{2-}$, $[PaF_8]^{3-}$

• PHYSICAL DATA

Melting point/K: 2113
Boiling point/K: *c.* 4300

ΔH_{fusion}/kJ mol^{-1}: 16.7
ΔH_{vap}/kJ mol^{-1}: 481

Thermodynamic properties (298.15 K, 0.1 MPa)

State	$\Delta_f H^{\ominus}$/kJ mol^{-1}	$\Delta_f G^{\ominus}$/kJ mol^{-1}	$S^{\ominus}$/J K^{-1} mol^{-1}	C_p/J K^{-1} mol^{-1}
Solid	0	0	51.9	28
Gas	607	563	198.05	22.93

Density/kg m^{-3}: 15 370 (est.)
Molar volume/cm^3: 15.0
Thermal conductivity/W m^{-1} K^{-1}: 47 (est.) [300 K]
Coefficient of linear thermal expansion/K^{-1}: 7.3×10^{-6}
Electrical resistivity/Ω m: 17.7×10^{-8} [273 K]
Mass magnetic susceptibility/kg^{-1} m^3: n.a.

• BIOLOGICAL DATA

Biological role

None, acts to stimulate metabolism.

Toxicity

The radiotoxicity of protactinium generally overrides other toxicity considerations, but conventional toxicity is thought to be low.

Hazards

Never encountered naturally outside nuclear facilities or research laboratories. Special precautions have to be taken when dealing with it because it is an α-emitter.

Levels in humans

Organs: nil
Daily dietary intake: nil
Total mass of element in average (70 kg) person: nil

Discovery: see Nuclear Data section.
[Greek, *protos* = first]
French, *protactinium*; German, *Protactinium*; Italian, *protoattinio*; Spanish, *protactinio* [pro-tak-tin-iuhm]

Protactinium

• NUCLEAR DATA

Discovery: Protactinium was discovered in 1913 by Kasimir Fajans and O.H. Göhring as the short lived isotope ^{234}Pa. Isotope ^{231}Pa was discovered in 1917 by Lise Meitner and Otto Hahn at Berlin (Germany), K. Fajans at Karlsruhe (Germany) and by F. Soddy, J.A. Cranston and A. Fleck at Glasgow, Scotland.

Number of isotopes (including nuclear isomers): 21 **Isotope mass range:** 216 → 238

Key isotopes

Nuclide	Atomic mass	Half life ($T_{1/2}$)	Decay mode and energy (MeV)	Nuclear spin *I*	Nucl. mag. moment μ	Uses
^{228}Pa	228.030 773	22 h	EC (2.11) 98%; α 2%; γ	3+?	+3.5	
^{229}Pa	229.032 073	1.5 d	EC (0.31) 99.8%; α (5.836) 0.2%; γ	5/2?		
^{230}Pa	230.034 527	17.4 d	EC (1.309) 90%; β^- (0.57) 10%; γ	2–?	2.0	
^{231}Pa*	231.035 880	32 500 y	α (5.148); γ	3/2–	2.01	R
^{232}Pa	232.038 565	1.31 d	β^- (1.34); γ	2–?		
^{233}Pa	233.040 242	27.0 d	β^- (0.571); γ	3/2–	+4.0	
^{234}Pa*	234.043 303	6.69 h	β^- (2.19); γ	4+?		

*Traces of these isotopes occur naturally.

NMR [Reference: n.a.]

	^{231}Pa
Relative sensitivity (^{1}H = 1.00)	–
Receptivity (^{13}C = 1.00)	–
Magnetogyric ratio/rad $T^{-1}s^{-1}$	3.21×10^7
Nuclear quadrupole moment/m^2	-1.720×10^{-28}
Frequency (^{1}H = 100 Hz; 2.3488T)/MHz	12.0

• ELECTRON SHELL DATA

Ground state electron configuration: [Rn]$5f^2 6d^1 7s^2$
Term symbol: $^4K_{11/2}$
Electron affinity ($M \rightarrow M^-$)/kJ mol^{-1}**:** n.a.

Ionization energies/kJ mol^{-1}**:**

1. M → M^+	568

Electron binding energies/eV

K	1s	112 601
L_I	2s	21 105
L_{II}	$2p_{1/2}$	20 314
L_{III}	$2p_{3/2}$	16 733
M_I	3s	5367
M_{II}	$3p_{1/2}$	5001
M_{III}	$3p_{3/2}$	4174
M_{IV}	$3d_{3/2}$	3611
M_V	$3d_{5/2}$	3442

continued in Appendix 2, p257

Main lines in atomic spectrum
[Wavelength/nm(species)]
363.652 (I)
398.223 (I)
694.572 (I)
711.489 (I)
736.825 (I)
749.315 (I)
760.820 (I)

• CRYSTAL DATA

Crystal structure (cell dimensions/pm), space group
α-Pa tetragonal (a = 393.2, c = 323.8), I4/mmm
β-Pa b.c.c.
$T(\alpha \rightarrow \beta)$ = 1440 K

X-ray diffraction: mass absorption coefficients (μ/ρ)/cm^2 g^{-1}**:** CuK_α n.a. MoK_α n.a.
Neutron scattering length, $b/10^{-12}$ cm**:** 0.91 (^{231}Pa)
Thermal neutron capture cross-section, σ_a/barns**:** 200.6 (^{231}Pa)

• GEOLOGICAL DATA

Minerals

^{231}Pa is a short-lived member of the ^{238}U decay series and so occurs naturally in uranium ores such as pitchblende, to the extent of 3 p.p.m. in some ores.

Chief source: in 1961 the UK Atomic Energy Authority extracted 125 g of pure protactinium from uranium fuel elements, and this is the major world stock of this element.

World production/tonnes y^{-1}**:** n.a.

Specimen: commercially available, under licence – see Key.

Abundances

Sun (relative to H = 1×10^{12})**:** n.a.
Earth's crust/p.p.m.**:** traces
Seawater/p.p.m.**:** 2×10^{-11}
Oxidation state: V

Ra

Atomic number: 88
Relative atomic mass ($^{12}C = 12.0000$)**:** 226.0254 (Ra-226)

CAS: [7440-14-4]

• CHEMICAL DATA

Description: Radium is a silvery, lustrous, soft, radioactive metal of the alkaline-earth group. It is bright when freshly prepared but darkens on exposure to air. Radium reacts with oxygen and water. Radium salts luminesce. The curie (Ci) is a unit of radioactivity and is defined as that amount of radioactivity which has the same disintegration rate as 1 g of ^{226}Ra, which is 3.7×10^{10} disintegrations per second. Radium was used to treat cancer and for luminous paints, but these uses are now largely superseded.

Radii/pm: Ra^{2+} 152; atomic 223
Electronegativity: 0.89 (Pauling); 0.97 (Allred); n.a. (absolute)
Effective nuclear charge: 1.65 (Slater)

Standard reduction potentials $E^{\ominus}$/V

II → 0

$$Ra^{2+} \xrightarrow{-2.916} Ra$$

$$RaO \xrightarrow{-1.319} Ra$$

Oxidation states

Ra^{II}	[Rn]	RaO, $Ra(OH)_2$, $[Ra(OH_2)_x]^{2+}$ (aq), Ra^{2+} salts

• PHYSICAL DATA

Melting point/K: 973
Boiling point/K: 1413

ΔH_{fusion}/kJ mol^{-1}**:** 7.15
ΔH_{vap}/kJ mol^{-1}**:** 136.8

Thermodynamic properties (298.15 K, 0.1 MPa)

State	$\Delta_f H^{\ominus}$/kJ mol^{-1}	$\Delta_f G^{\ominus}$/kJ mol^{-1}	$S^{\ominus}$/J K^{-1} mol^{-1}	C_p/J K^{-1} mol^{-1}
Solid	0	0	71	27.1
Gas	159	130	176.47	20.79

Density/kg m^{-3}**:** *c.* 5000 [293 K]
Molar volume/cm^3**:** 45.2
Thermal conductivity/W m^{-1} K^{-1}**:** 18.6 (est.) [300 K]
Coefficient of linear thermal expansion/K^{-1}**:** 20.2×10^{-6}
Electrical resistivity/Ω m**:** 100×10^{-8} (s)
Mass magnetic susceptibility/kg^{-1} m^3**:** n.a.

• BIOLOGICAL DATA

Biological role

None.

Toxicity

The high radiotoxicity of radium overrides other toxicity considerations, but chemically it would be toxic in the same way as barium.

Hazards

Radium is very dangerous if absorbed by the body because of its radioactivity; the maximum permissible body burden for ^{226}Ra is 7400 becquerel.

Levels in humans

Blood/mg dm^{-3}**:** 6.6×10^{-9}
Bone/p.p.m.**:** 4×10^{-9}
Liver/p.p.m.**:** 0.023×10^{-9}
Muscle/p.p.m.**:** 0.23×10^{-9}
Daily dietary intake: 2×10^{-9} mg
Total mass of element in average (70 kg) person: 31×10^{-9} mg

Discovery: see Nuclear Data section.
[Latin, *radius* = ray]
French, *radium*; German, *Radium*; Italian, *radio*; Spanish, *radio*

Radium

[ray-dee-uhm]

• NUCLEAR DATA

Discovery: Radium was discovered in 1898 by Pierre and Marie Curie at Paris, France. Isolated as the metal by Mme. Curie in 1911 by the electrolysis of $RaCl_2$.

Number of isotopes (including nuclear isomers): 25 **Isotope mass range:** 213 → 230

Key isotopes

Nuclide	Atomic mass	Half life ($T_{1/2}$)	Decay mode and energy (MeV)	Nuclear spin *I*	Nucl. mag. moment *μ*	Uses
^{223}Ra*	223.018 501	11.435 d	α (5.979); γ	3/2+?	+0.271	
^{224}Ra*	224.020 186	3.66 d	α (5.789); γ	0+		
^{226}Ra*	226.025 402	1599 y	α (4.780); γ	0+		R, T
^{227}Ra	227.029 170	42 m	$β^-$ (1.324); γ	3/2+?	−0.404	
^{228}Ra*	228.031 064	5.76 y	$β^-$ (0.046); no γ	0+		
^{230}Ra	230.036 990	1.5 h	$β^-$ (0.9); γ	0+		

* Traces of these isotopes occur naturally.
Other isotopes of radium have half lives shorter than 5 minutes.

NMR [Not recorded]

• ELECTRON SHELL DATA

Ground state electron configuration: $[Rn]7s^2$
Term symbol: $^1S_0$
Electron affinity ($M \rightarrow M^-$)/kJ mol^{-1}**:** n.a.

Ionization energies/kJ mol^{-1}**:**

1.	$M \rightarrow M^+$	509.3
2.	$M^+ \rightarrow M^{2+}$	979.0
3.	$M^{2+} \rightarrow M^{3+}$	(3300)
4.	$M^{3+} \rightarrow M^{4+}$	(4400)
5.	$M^{4+} \rightarrow M^{5+}$	(5700)
6.	$M^{5+} \rightarrow M^{6+}$	(7300)
7.	$M^{6+} \rightarrow M^{7+}$	(8600)
8.	$M^{7+} \rightarrow M^{8+}$	(9900)
9.	$M^{8+} \rightarrow M^{9+}$	(13 500)
10.	$M^{9+} \rightarrow M^{10+}$	(15 100)

Electron binding energies/eV

K	1s	103 922
L_I	2s	19 237
L_{II}	$2p_{1/2}$	18 484
L_{III}	$2p_{3/2}$	15 444
M_I	3s	4822
M_{II}	$3p_{1/2}$	4490
M_{III}	$3p_{3/2}$	3792
M_{IV}	$3d_{3/2}$	3248
M_V	$3d_{5/2}$	3105

continued in Appendix 2, p257

Main lines in atomic spectrum
[Wavelength/nm(species)]
364.955 (II)
381.442 (II)
434.064 (II)
468.228 (II)
482.591 (I)

• CRYSTAL DATA

Crystal structure (cell dimensions/pm), space group
b.c.c. (*a* = 515)

X-ray diffraction: mass absorption coefficients (μ/ρ)/cm^2 g^{-1}**:** CuK_α 304 MoK_α 172
Neutron scattering length, $b/10^{-12}$ cm**:** 1.00 (^{226}Ra)
Thermal neutron capture cross-section, σ_a/barns**:** 12.8 (^{226}Ra)

• GEOLOGICAL DATA

Minerals

All uranium minerals contain radium, and there is about 1.5 g of radium in 10 tonnes of the uranium ore, pitchblende. At one time it was separated from this source but this is no longer undertaken. It has been estimated that each square kilometre of soil (to a depth of a 40 cm) contains 1 g of radium.

World production: n.a. but probably very little
Main mining areas: not really applicable, but see uranium.
Specimen: commercially available, under licence – see Key.

Abundances

Sun (relative to H = 1×10^{12})**:** n.a.
Earth's crust/p.p.m.**:** 6×10^{-7}
Seawater/p.p.m.**:** 2×10^{-11}
Oxidation state: II

Atomic number: 86 **CAS:** [10043-92-2]

Relative atomic mass ($^{12}C = 12.0000$)**:** 222.0176 (Rn-222)

• CHEMICAL DATA

Description: Radon is a colourless, odourless gas produced by radium-226. When radon is cooled below its freezing point, it phosphoresces brightly. It is little studied, partly because it is a noble gas and is therefore reluctant to form molecules, and partly because its intense radiation would destroy any compound that might form. Radon is sometimes used in hospitals to treat cancer.

Radii/pm: n.a.

Electronegativity: n.a. (Pauling); 2.06 (Allred); [5.1 eV (absolute) - see Key]

Effective nuclear charge: 8.25 (Slater); 16.08 (Clementi); 20.84 (Froese-Fischer)

Oxidation states

Rn^{0}	[Rn]	Rn gas
Rn^{II}	s^2p^4	RnF_2

Covalent bonds

Bond	r/ pm	E/ kJ mol^{-1}
Rn—F	n.a.	n.a.

• PHYSICAL DATA

Melting point/K: 202

Boiling point/K: 211.4

Critical temperature/K: 377

Critical pressure/ kPa: 6300

ΔH_{fusion}/kJ mol^{-1}: 2.7 (est.)

ΔH_{vap}/kJ mol^{-1}: 18.1

Thermodynamic properties (298.15 K, 0.1 MPa)

State	$\Delta_f H^{\circ}$/kJ mol^{-1}	$\Delta_f G^{\circ}$/kJ mol^{-1}	S°/J K^{-1} mol^{-1}	C_p/J K^{-1} mol^{-1}
Solid	0	0	n.a.	n.a.
Gas	n.a.	n.a.	176.21	20.786

Density/kg m^{-3}: n.a. [solid]; 4400 [liq. at b.p.]; 9.73 [gas, 273 K]

Molar volume/cm^3: 50.5 [211 K]

Thermal conductivity/W m^{-1} K^{-1}: 0.00364 (est.) [300 K] (g)

Coefficient of linear thermal expansion/K^{-1}: n.a.

Electrical resistivity/Ω m: n.a.

Mass magnetic susceptibility/kg^{-1} m^3: n.a.

• BIOLOGICAL DATA

Biological role

None.

Toxicity

The high radiotoxicity of radon overrides other toxicity considerations, but chemically it would be inert like the other noble gases.

Hazards

Radon is very dangerous because it is an α-emitter, and the maximum permissible concentration in air is 3×10^{-4} Bq cm^{-3} for an 8 hour day, 40 hour week. Radon is a hazard in uranium mines, and worrying concentrations have been detected inside homes in certain regions.

Levels in humans

Organs: virtually nil

Daily dietary intake: nil

Total mass of element in average (70 kg) person: virtually nil

Discovered in 1900 by F.E. Dorn at Halle, Germany.
[Named after the element radium]
French, *radon*; German, *Radon*; Italian, *radon* (*emanio*); Spanish, *radón*

Radon

[ray-don]

• NUCLEAR DATA

Number of isotopes (including nuclear isomers): 28 **Isotope mass range:** 200 → 226

Key isotopes

Nuclide	Atomic mass	Half life ($T_{1/2}$)	Decay mode and energy (MeV)	Nuclear spin I	Nucl. mag. moment μ	Uses
^{208}Rn	207.989 610	24.3 m	α (6.260) 60%; EC (2.9) 40%	0+		
^{209}Rn	208.990 370	29 m	β^+ (3.93) 83%; α 17%; γ	5/2–	+0.8388	
^{210}Rn	209.989 669	2.4 h	α (6.157) 96%; EC (2.368) 4%; γ	0+		
^{211}Rn	210.990 576	14.6 h	β^+ EC (2.89) 74%; α (5.964) 26%; γ	1/2–	+0.60	
^{212}Rn	211.990 697	24 m	α (6.385)	0+		
^{220}Rn*	220.011 368	55.6 s	α (6.404); γ	0+		
^{221}Rn	221.015 470	25 m	α (6.148) 22%; β^- (1.150) 78%; γ	7/2+	–0.020	
^{222}Rn*	222.017 570	3.8235 d	α (5.590); γ	0+		T
^{223}Rn		23 m	β^-; γ	0+		
^{224}Rn		1.8 h	β^-; γ	0+		

*Traces of these isotopes occur naturally.
Other isotopes of radon have half lives shorter than 10 minutes.

NMR [Not recorded]

• ELECTRON SHELL DATA

Ground state electron configuration: $[Xe]4f^{14}5d^{10}6s^26p^6$ = [Rn]
Term symbol: $^1S_0$
Electron affinity ($M \rightarrow M^-$)/kJ mol^{-1}**:** –41 (est.)

Ionization energies/kJ mol^{-1}**:**

1.	$M \rightarrow M^+$	1040
2.	$M^+ \rightarrow M^{2+}$	1930
3.	$M^{2+} \rightarrow M^{3+}$	2890
4.	$M^{3+} \rightarrow M^{4+}$	4250
5.	$M^{4+} \rightarrow M^{5+}$	5310

Electron binding energies/eV

K	1s	98 404
L_I	2s	18 049
L_{II}	$2p_{1/2}$	17 337
L_{III}	$2p_{3/2}$	14 619
M_I	3s	4482
M_{II}	$3p_{1/2}$	4159
M_{III}	$3p_{3/2}$	3538
M_{IV}	$3d_{3/2}$	3022
M_V	$3d_{5/2}$	2892

continued in Appendix 2, p257

Main lines in atomic spectrum
[Wavelength/nm(species)]

434.960 (I)
705.542 (I)
726.811 (I)
745.000 (I)
780.982 (I)
809.951 (I)
827.096 (I)
860.007 (I)

• CRYSTAL DATA

Crystal structure (cell dimensions/pm), space group
f.c.c.

X-ray diffraction: mass absorption coefficients (μ/ρ)/cm^2 g^{-1}**:** CuK_α n.a. MoK_α n.a.
Neutron scattering length, $b/10^{-12}$ cm**:** n.a.
Thermal neutron capture cross-section, σ_a/barns**:** 0.72 (^{222}Rn)

• GEOLOGICAL DATA

Minerals

Radon emanates from thorium and uranium minerals. It collects over samples of radium, ^{226}Ra, in sealed tubes, and 1 g of radium produces 0.0001 cm^3 of radon per day. Some spring waters, such as those at Hot Springs, Arkansas, contain dissolved radon gas.

Chief source: obtained from ^{226}Ra ampoules

Specimen: commercially available, under licence – see Key.

Abundances

Sun (relative to H = 1×10^{12})**:** n.a.
Earth's crust/p.p.m.**:** traces
Atmosphere/p.p.m. (volume)**:** 1×10^{-15}
Seawater/p.p.m.**:** *c.* 1×10^{-14}
Oxidation state: 0

Re

Atomic number: 75
Relative atomic mass ($^{12}C = 12.0000$): **186.207**

CAS: [7440-15-5]

• CHEMICAL DATA

Description: Rhenium is a silvery metal, but is usually obtained as a grey powder. It resists corrosion but slowly tarnishes in moist air. Rhenium dissolves in HNO_3 and H_2SO_4. It is used in filaments, thermistors and catalysts.

Radii/pm: Re^{7+} 60; Re^{6+} 61; Re^{4+} 72; atomic 137; covalent 128
Electronegativity: 1.9 (Pauling); 1.46 (Allred); 4.02 eV (absolute)
Effective nuclear charge: 3.60 (Slater); 10.12 (Clementi); 14.62 (Froese-Fischer)

Standard reduction potentials E°/V

VII — VI — IV — III — 0 — –I

acid:
$[ReO_4]^-$ —0.768— ReO_3 —0.63— ReO_2 —0.22— Re —0.10— Re^-
$[ReO_4]^-$ → Re: 0.34
$[ReO_4]^-$ → ReO_2: 0.51
$[ReO_4]^-$ —0.12— $[ReCl_6]^{2-}$ —0.51— Re

base:
$[ReO_4]^-$ —−0.890— ReO_3 —−0.446— $ReO_2.2H_2O$ —−1.25— Re_2O_3 —0.333— Re
$[ReO_4]^-$ → Re_2O_3: −0.808
$[ReO_4]^-$ → $ReO_2.2H_2O$: −0.594
$ReO_2.2H_2O$ → Re: −0.564
$[ReO_4]^-$ → Re: −0.604

Oxidation states

Re^{-III}	d^{10}	$[Re(CO)_4]^{3-}$
Re^{-I}	d^8	$[Re(CO)_5]^-$
Re^0	d^7	$[Re_2(CO)_{10}]$
Re^I	d^6	$[ReCl(CO)_5]$, $K_5[Re(CN)_6]$
Re^{II}	d^5	ReF_2, $ReCl_2$ etc.
Re^{III}	d^4	$Re_2O_3.xH_2O$, Re_3Cl_9, Re_3Br_9, Re_3I_9, $[Re_2Cl_8]^{2-}$, $[Re(CN)_7]^{4-}$, complexes
Re^{IV}	d^3	ReO_2, ReF_4, $ReCl_4$ etc., complexes
Re^V	d^2	Re_2O_5, ReF_5, $ReCl_5$, $ReBr_5$, ReF_3O, complexes
Re^{VI}	d^1	ReO_3, ReF_6, $ReCl_6$, $[ReF_8]^{2-}$, $ReCl_4O$, complexes
Re^{VII}	$d^0[f^{14}]$	Re_2O_7, ReF_7, $ReFO_3$, ReF_5O, $[ReO_4]^-$ (aq), $[ReH_9]^{2-}$, complexes

• PHYSICAL DATA

Melting point/K: 3453
Boiling point/K: 5900
ΔH_{fusion}/kJ mol^{-1}: 33.1
ΔH_{vap}/kJ mol^{-1}: 707.1

Thermodynamic properties (298.15 K, 0.1 MPa)

State	$\Delta_f H^{\circ}$/kJ mol^{-1}	$\Delta_f G^{\circ}$/kJ mol^{-1}	S°/J K^{-1} mol^{-1}	C_p/J K^{-1} mol^{-1}
Solid	0	0	36.86	25.48
Gas	769.9	724.6	188.938	20.786

Density/kg m^{-3}: 21 020 [293 K]; 18 900 [liquid at m.p.]
Molar volume/cm^3: 8.86
Thermal conductivity/W m^{-1} K^{-1}: 47.9 [300 K]
Coefficient of linear thermal expansion/K^{-1}: 6.63×10^{-6}
Electrical resistivity/Ω m: 19.3×10^{-8} [293 K]
Mass magnetic susceptibility/kg^{-1} m^3: $+4.56 \times 10^{-9}$ (s)
Young's modulus/GPa: 466
Rigidity modulus/GPa: 181
Bulk modulus/GPa: 334
Poisson's ratio/GPa: 0.26

• BIOLOGICAL DATA

Biological role

None.

Toxicity

Not established, but believed to be low.

Toxic intake: n.a.
Lethal intake: LD_{50} ($ReCl_3$, intraperitoneal, mouse) = 280 mg kg^{-1}

Hazards

There are no reported cases of humans being affected by rhenium. Like other powdered metals, rhenium dust could pose a moderate fire or explosion hazard.

Levels in humans

Organs: n.a., but very low
Daily dietary intake: n.a.
Total mass of element in average (70 kg) person: n.a., but very low

Discovered in 1925 by W. Noddack, Ida Tacke and O. Berg at Berlin, Germany.
[Greek, *Rhenus* = river Rhine]
French, *rhénium*; German, *Rhenium*; Italian, *renio*; Spanish, *renio*

Rhenium

[ree-ni-uhm]

• NUCLEAR DATA

Number of isotopes (including nuclear isomers): 34   **Isotope mass range:** 162 → 192

Key isotopes

Nuclide	Atomic mass	Natural abundance (%)	Nuclear spin I	Nuclear magnetic moment μ	Uses
^{185}Re	184.952 951	37.40	5/2+	+3.187 1	E, NMR
^{187}Re*	186.955 744	62.60	5/2+	+3.219 7	E, NMR

*^{187}Re is radioactive with a half-life of 4.5×10^{10} y and decay mode β^- (0.0025 MeV).
A table of radioactive isotopes is given in Appendix 1, on p247.

NMR [Reference: $Na[ReO_4]$ (aq)]	^{185}Re	^{187}Re
Relative sensitivity (^1H = 1.00)	0.13	0.13
Receptivity (^{13}C = 1.00)	280	490
Magnetogyric ratio/rad $T^{-1}s^{-1}$	6.0255×10^7	6.0862×10^7
Nuclear quadrupole moment/m^2	2.180×10^{-28}	2.070×10^{-28}
Frequency (^1H = 100 Hz; 2.3488T)/MHz	22.513	22.744

• ELECTRON SHELL DATA

Ground state electron configuration: $[Xe]4f^{14}5d^56s^2$
Term symbol: $^6S_{5/2}$
Electron affinity ($M \rightarrow M^-$)/kJ mol^{-1}: 14

Ionization energies/kJ mol^{-1}:

1.	$M \rightarrow M^+$	760
2.	$M^+ \rightarrow M^{2+}$	1260
3.	$M^{2+} \rightarrow M^{3+}$	2510
4.	$M^{3+} \rightarrow M^{4+}$	3640
5.	$M^{4+} \rightarrow M^{5+}$	(4900)
6.	$M^{5+} \rightarrow M^{6+}$	(6300)
7.	$M^{6+} \rightarrow M^{7+}$	(7600)

Electron binding energies/eV

K	1s	71 676
L_I	2s	12 527
L_{II}	$2p_{1/2}$	11 959
L_{III}	$2p_{3/2}$	10 535
M_I	3s	2932
M_{II}	$3p_{1/2}$	2682
M_{III}	$3p_{3/2}$	2367
M_{IV}	$3d_{3/2}$	1949
M_V	$3d_{5/2}$	1883

continued in Appendix 2, p257

Main lines in atomic spectrum
[Wavelength/nm(species)]
200.353 (I)
204.908 (I)
345.188 (I)
346.046 (I) (AA)
346.473 (I)

• CRYSTAL DATA

Crystal structure (cell dimensions/pm), space group
h.c.p. (a = 276.09, c = 445.76), $P6_3/mmc$

X-ray diffraction: mass absorption coefficients (μ/ρ)/cm^2 g^{-1}: CuK_α 179 MoK_α 103
Neutron scattering length, $b/10^{-12}$ cm: 0.92
Thermal neutron capture cross-section, σ_a/barns: 89.7

• GEOLOGICAL DATA

Minerals

Rhenium does not occur in nature as the free metal, nor has a distinct rhenium mineral species been found. Gadolinite (see beryllium) and molybdenite (see molybdenum) may contain up to 0.2% rhenium.

Chief ores: rhenium is extracted from the flue dusts of molybdenum smelters.
World production/tonnes y^{-1}: 4.5
Main mining areas: see molybdenum
Reserves/tonnes: 3500
Specimen: available as foil, powder, ribbon or wire. *Care!*

Abundances

Sun (relative to H = 1×10^{12}): < 2
Earth's crust/p.p.m.: 4×10^{-4}
Seawater/p.p.m.: 4×10^{-6}
Oxidation state: III

Atomic number: 45 **CAS:** [7440-16-6]

Relative atomic mass ($^{12}C = 12.0000$): **102.90550**

• CHEMICAL DATA

Description: Rhodium is a rare, lustrous, silvery, hard metal of the so-called platinum group. It is unaffected by air and water up to 875 K, and unaffected by acids, but is attacked by molten alkalis. Rhodium is used as a catalyst.

Radii/pm: Rh^{4+} 67; Rh^{3+} 75; Rh^{2+} 86; atomic 134; covalent 125
Electronegativity: 2.28 (Pauling); 1.45 (Allred); 4.30 eV (absolute)
Effective nuclear charge: 3.90 (Slater); 7.64 (Clementi); 10.85 (Froese-Fischer)

Standard reduction potentials $E^\ominus$/V

III 0

$$Rh^{3+} \xrightarrow{0.76} Rh$$

Oxidation states

Rh^{-I}	d^{10}	$[Rh(CO)_4]^-$
Rh^0	d^9	$[Rh_4(CO)_{12}]$, $[Rh_6(CO)_{16}]$
Rh^I	d^8	$[RhCl(PPh_3)_3]$
Rh^{II}	d^7	RhO
Rh^{III}	d^6	Rh_2O_3, RhF_3, $RhCl_3$, etc., $[RhCl_6]^{3-}$, $[Rh(OH_2)_x]^{3+}$ (aq)
Rh^{IV}	d^5	RhO_2, RhF_4, $[RhCl_6]^{2-}$
Rh^V	d^4	$[RhF_5]_4$, $[RhF_6]^-$
Rh^{VI}	d^3	RhF_6

• PHYSICAL DATA

Melting point/K: 2239
Boiling point/K: 4000
ΔH_{fusion}/kJ mol^{-1}: 21.55
ΔH_{vap}/kJ mol^{-1}: 495.4

Thermodynamic properties (298.15 K, 0.1 MPa)

State	$\Delta_f H^\ominus$/kJ mol^{-1}	$\Delta_f G^\ominus$/kJ mol^{-1}	$S^\ominus$/J K^{-1} mol^{-1}	C_p/J K^{-1} mol^{-1}
Solid	0	0	31.51	24.98
Gas	556.9	510.8	185.808	21.012

Density/kg m^{-3}: 12 410 [293 K]; 10 650 [liquid at m.p.]
Molar volume/cm^3: 8.29
Thermal conductivity/W m^{-1} K^{-1}: 150 [300 K]
Coefficient of linear thermal expansion/K^{-1}: 8.40×10^{-6}
Electrical resistivity/Ω m: 4.51×10^{-8} [293 K]
Mass magnetic susceptibility/kg^{-1} m^3: $+1.36 \times 10^{-8}$ (s)
Young's modulus/GPa: 379
Rigidity modulus/GPa: 147
Bulk modulus/GPa: 276
Poisson's ratio/GPa: 0.26

• BIOLOGICAL DATA

Biological role

None.

Toxicity

Toxic intake: most rhodium compounds are slightly toxic by ingestion
Lethal intake: LD_{50} ($RhCl_3$, oral, rat) = 12.6 mg kg^{-1}

Hazards

There are few reported cases of humans being affected by rhodium, but it is an experimental carcinogen.

Levels in humans

Organs: n.a. but very low
Daily dietary intake: n.a.
Total mass of element in average (70 kg) person: n.a.

Discovered in 1803 by W.H. Wollaston at London, England.
[Greek, *rhodon* = rose]
French, *rhodium*; German, *Rhodium*; Italian, *rodio*; Spanish, *rodio*

Rhodium

[roh-diuhm]

• NUCLEAR DATA

Number of isotopes (including nuclear isomers): 34 **Isotope mass range:** 94m → 112

Key isotopes

Nuclide	Atomic mass	Natural abundance (%)	Nuclear spin *I*	Nuclear magnetic moment μ	Uses
^{103}Rh	102.905 500	100	1/2–	–0.088 40	NMR

A table of radioactive isotopes is given in Appendix 1, on p248.

NMR [Reference: $RhCl_3\{S(CH_3)_2\}_3$]	^{103}Rh
Relative sensitivity (^1H = 1.00)	3.11×10^{-5}
Receptivity (^{13}C = 1.00)	0.177
Magnetogyric ratio/rad $T^{-1}s^{-1}$	-0.8520×10^7
Nuclear quadrupole moment/m^2	–
Frequency (^1H = 100 Hz; 2.3488T)/MHz	3.172

• ELECTRON SHELL DATA

Ground state electron configuration: $[Kr]4d^85s^1$
Term symbol: $^4F_{9/2}$
Electron affinity ($M \rightarrow M^-$)/kJ mol^{-1}**:** 109.7

Ionization energies/kJ mol^{-1}**:**

1.	$M \rightarrow M^+$	720
2.	$M^+ \rightarrow M^{2+}$	1744
3.	$M^{2+} \rightarrow M^{3+}$	2997
4.	$M^{3+} \rightarrow M^{4+}$	(4400)
5.	$M^{4+} \rightarrow M^{5+}$	(6500)
6.	$M^{5+} \rightarrow M^{6+}$	(8200)
7.	$M^{6+} \rightarrow M^{7+}$	(10 100)
8.	$M^{7+} \rightarrow M^{8+}$	(12 200)
9.	$M^{8+} \rightarrow M^{9+}$	(14 200)
10.	$M^{9+} \rightarrow M^{10+}$	(22 000)

Electron binding energies/eV

K	1s	23 220
L_I	2s	3412
L_{II}	$2p_{1/2}$	3146
L_{III}	$2p_{3/2}$	3004
M_I	3s	628.1
M_{II}	$3p_{1/2}$	521.3
M_{III}	$3p_{3/2}$	496.5
M_{IV}	$3d_{3/2}$	311.9
M_V	$3d_{5/2}$	307.2

continued in Appendix 2, p257

Main lines in atomic spectrum
[Wavelength/nm(species)]

343.489 (I) (AA)
350.252 (I)
352.802 (I)
365.799 (I)
369.236 (I)
370.091 (I)

• CRYSTAL DATA

Crystal structure (cell dimensions/pm), space group
f.c.c. (*a* = 380.36), Fm3m
X-ray diffraction: mass absorption coefficients (μ/ρ)/$cm^2 g^{-1}$**:** CuK_α 194 MoK_α 22.6
Neutron scattering length, $b/10^{-12}$ cm**:** 0.588
Thermal neutron capture cross-section, σ_a/barns**:** 144.8

• GEOLOGICAL DATA

Minerals

Rhodium occurs as rare deposits of the native metal, and even rarer minerals.

Mineral	Formula	Density	Hardness	Crystal appearance
Rhodium	Rh	*c.* 12	3.5	cub., met. white
Rhodplumsite	$Rh_3Pb_2S_2$	9.74	n.a.	rhom., met. cream-pink/grey-blue

Chief source: native rhodium is found in Montana, USA; certain copper and nickel ores contain up to 0.1% rhodium and these are the main source of the element.

World production/tonnes y^{-1}**:** 3

Main mining areas: see copper and nickel

Reserves/tonnes**:** n.a.

Specimen: available as foil, sponge or wire. Safe.

Abundances

Sun (relative to H = 1×10^{12})**:** 25.1
Earth's crust/p.p.m.**:** *c.* 2×10^{-4}
Seawater/p.p.m.**:** n.a. but minute

Atomic number: 37
Relative atomic mass ($^{12}C = 12.0000$)**:** 85.4678

CAS: [7440-17-7]

• CHEMICAL DATA

Discovery: Rubidium was discovered in 1861 by R.W. Bunsen and G. Kirchhoff at the University of Heidelberg, Germany.
Description: Rubidium is a soft, white, metal which is silvery when first cut but it oxidises rapidly in air and ignites. It reacts violently with water. Rubidium is obtained by the reaction of calcium or potassium metal with rubidium chloride. The metal and its compounds are rarely used commercially, and only a little is used for research purposes.

Radii/pm: Rb^+ 149; atomic 247.5; van der Waals 244
Electronegativity: 0.82 (Pauling); 0.89 (Allred); 2.34 eV (absolute)
Effective nuclear charge: 2.20 (Slater); 4.98 (Clementi); 6.66 (Froese-Fischer)

Standard reduction potentials $E^\ominus$/V

I		0
Rb^+	−2.924	Rb

Oxidation states

Rb^{-1}	s^2	solution of metal in liquid ammonia
Rb^{I}	[Kr]	Rb_2O, Rb_2O_2 (peroxide), RbO_2 (superoxide), RbOH, RbH, RbF, RbCl etc., $[Rb(OH_2)_x]^+$ (aq), Rb_2CO_3, many salts, some complexes

• PHYSICAL DATA

Melting point/K: 312.2
Boiling point/K: 961
ΔH_{fusion}/kJ mol^{-1}: 2.20
ΔH_{vap}/kJ mol^{-1}: 69.2

Thermodynamic properties (298.15 K, 0.1 MPa)

State	$\Delta_f H^\ominus$/kJ mol^{-1}	$\Delta_f G^\ominus$/kJ mol^{-1}	$S^\ominus$/J K^{-1} mol^{-1}	C_p/J K^{-1} mol^{-1}
Solid	0	0	76.78	31.062
Gas	80.88	53.06	170.089	20.786

Density/kg m^{-3}: 1532 [293 K]; 1475 [liquid at m.p.]
Molar volume/cm^3: 55.79
Thermal conductivity/W m^{-1} K^{-1}: 58.2 [300 K]
Coefficient of linear thermal expansion/K^{-1}: 90×10^{-6}
Electrical resistivity/Ω m: 12.5×10^{-8} [293 K]
Mass magnetic susceptibility/kg^{-1} m^3: $+2.49 \times 10^{-9}$ (s)
Young's modulus/GPa: 2.35
Rigidity modulus/GPa: 0.91
Bulk modulus/GPa: n.a.
Poisson's ratio/GPa: 0.30

• BIOLOGICAL DATA

Biological role

Rubidium has no known role; its salts have a stimulatory effect.

Toxicity

Toxic intake: can be toxic by ingestion
Lethal intake: LD_{50} (chloride, oral, mouse) = 3800 mg kg^{-1}

Hazards

Rubidium salts are generally inert, and their toxicity is almost always that of the anion, not of the Rb^+. However, in the body, rubidium substitutes for potassium and too much can be dangerous.

Levels in humans

Blood/mg dm^{-3}: 2.49
Bone/p.p.m.: 0.1 – 5
Liver/p.p.m.: 20 – 70
Muscle/p.p.m.: 20 – 70
Daily dietary intake: 1.5 – 6 mg
Total mass of element in average (70 kg) person: 680 mg

Discovery: see Chemical Data section.
[Latin, *rubidius* = deepest red]
French, *rubidium*; German, *Rubidium*; Italian, *rubidio*; Spanish, *rubidio*

Rubidium

[roo-bid-iuhm]

• NUCLEAR DATA

Number of isotopes (including nuclear isomers): 30 **Isotope mass range:** 75 → 98

Key isotopes

Nuclide	Atomic mass	Natural abundance (%)	Nuclear spin *I*	Nuclear magnetic moment μ	Uses
^{85}Rb	84.911 794	72.165	5/2–	+1.353 03	E, NMR
^{87}Rb*	86.909 187	27.835	3/2–	+2.751 24	E, NMR

* ^{87}Rb is radioactive with a half-life of 4.9×10^{10} y and decay mode β^- (0.273 MeV); no γ.
A table of radioactive isotopes is given in Appendix 1, on p248.

NMR [Reference: RbCl (aq)]	^{85}Rb	^{87}Rb
Relative sensitivity (^1H = 1.00)	0.0105	0.17
Receptivity (^{13}C = 1.00)	43	277
Magnetogyric ratio/rad $T^{-1}s^{-1}$	2.5828×10^7	8.7532×10^7
Nuclear quadrupole moment/m^2	$+0.274 \times 10^{-28}$	$+0.132 \times 10^{-28}$
Frequency (^1H = 100 Hz; 2.3488T)/MHz	9.655	32.721

• ELECTRON SHELL DATA

Ground state electron configuration: $[Kr]5s^1$
Term symbol: $^2S_{1/2}$
Electron affinity ($M \rightarrow M^-$)/kJ mol^{-1}**:** 46.9

Ionization energies/kJ mol^{-1}:

1.	$M \rightarrow M^+$	403.0
2.	$M^+ \rightarrow M^{2+}$	2632
3.	$M^{2+} \rightarrow M^{3+}$	3900
4.	$M^{3+} \rightarrow M^{4+}$	5080
5.	$M^{4+} \rightarrow M^{5+}$	6850
6.	$M^{5+} \rightarrow M^{6+}$	8140
7.	$M^{6+} \rightarrow M^{7+}$	9570
8.	$M^{7+} \rightarrow M^{8+}$	13 100
9.	$M^{8+} \rightarrow M^{9+}$	14 800
10.	$M^{9+} \rightarrow M^{10+}$	26 740

Electron binding energies/eV

K	1s	15 200
L_I	2s	2065
L_{II}	$2p_{1/2}$	1864
L_{III}	$2p_{3/2}$	1804
M_I	3s	326.7
M_{II}	$3p_{1/2}$	248.7
M_{III}	$3p_{3/2}$	239.1
M_{IV}	$3d_{3/2}$	113.0
M_V	$3d_{5/2}$	112

continued in Appendix 2, p257

Main lines in atomic spectrum
[Wavelength/nm(species)]

214.383 (II)
247.220 (II)
424.440 (II)
477.595 (II)
780.027 (I) (AA)
794.760 (I)

• CRYSTAL DATA

Crystal structure (cell dimensions/pm), space group
b.c.c. (*a* = 562), Im3m

X-ray diffraction: mass absorption coefficients $(\mu/\rho)/cm^2\ g^{-1}$**:** CuK_α 117 MoK_α 90.0
Neutron scattering length, $b/10^{-12}$ cm**:** 0.709
Thermal neutron capture cross-section, σ_a/barns**:** 0.38

• GEOLOGICAL DATA

Minerals

No minerals as such are known, but rubidium is present in significant amounts in lepidolite (see lithium), pollucite (see caesium) and carnallite (see potassium).

World production/tonnes y^{-1}**:** n.a.
Reserves/tonnes**:** n.a.
Specimen: available as ingots in sealed ampoules. *Danger!*

Abundances

Sun (relative to H = 1×10^{12})**:** 400
Earth's crust/p.p.m.**:** 90
Seawater/p.p.m.**:** 0.12
Residence time/years**:** 800 000
Classification: accumulating
Oxidation state: I

Ru

Atomic number: 44
Relative atomic mass ($^{12}C = 12.0000$)**:** 101.07

CAS: [7440-18-8]

• CHEMICAL DATA

Discovery: Ruthenium was discovered in 1808 by J.A. Sniadecki at the University of Vilno, Poland. Rediscovered in 1828 by G.W. Osann at the University of Tartu, Russia.
Description: Ruthenium is a lustrous, silvery metal of the so-called platinum group. It is unaffected by air, water and acids, but dissolves in molten alkalis. Ruthenium is used to harden platinum and palladium metals, and as a catalyst.

Radii/pm: Ru^{5+} 54; Ru^{4+} 65; Ru^{3+} 77; atomic 134; covalent 124
Electronegativity: 2.2 (Pauling); 1.42 (Allred); 4.5 eV (absolute)
Effective nuclear charge: 3.75 (Slater); 7.45 (Clementi); 10.57 (Froese-Fischer)

Standard reduction potentials $E^{\ominus}$/V

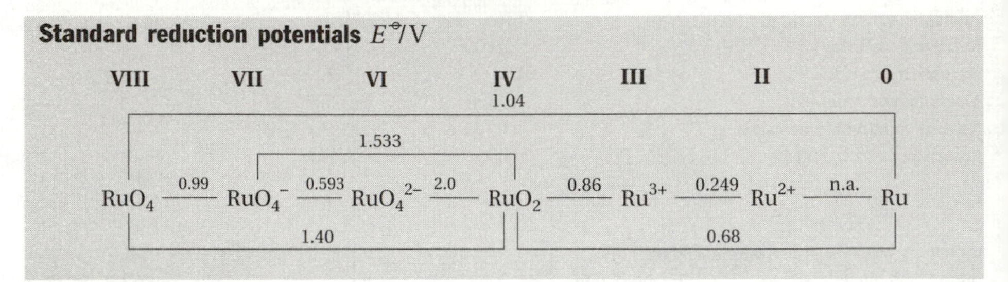

Oxidation states

Ru^{-II}	d^{10}	rare $[Ru(CO)_4]^{2-}$
Ru^0	d^8	rare $[Ru(CO)_5]$
Ru^I	d^7	some complexes, e.g. $[Ru(CO)_2(\eta\text{-}C_5H_5)]_2$
Ru^{II}	d^6	$[Ru(OH_2)_6]^{2+}$ (aq), $RuCl_2$, $RuBr_2$, RuI_2, $[Ru(CN)_6]^{2-}$, complexes
Ru^{III}	d^5	Ru_2O_3, $[Ru(OH_2)_6]^{3+}$ (aq), RuF_3, $RuCl_3$ etc., $[RuCl_6]^{3-}$
Ru^{IV}	d^4	RuO_2, RuF_4, $[RuCl_6]^{2-}$
Ru^V	d^3	RuF_5, $[RuF_6]^-$
Ru^{VI}	d^2	RuO_3, $[RuO_4]^{2-}$ (aq), RuF_6
Ru^{VII}	d^1	$[RuO_4]^-$ (aq)
Ru^{VIII}	d^0 [Kr]	RuO_4

• PHYSICAL DATA

Melting point/K: 2583
Boiling point/K: 4173

ΔH_{fusion}/kJ mol^{-1}: 23.7
ΔH_{vap}/kJ mol^{-1}: 567.8

Thermodynamic properties (298.15 K, 0.1 MPa)

State	$\Delta_f H^{\ominus}$/kJ mol^{-1}	$\Delta_f G^{\ominus}$/kJ mol^{-1}	$S^{\ominus}$/J K^{-1} mol^{-1}	C_p/J K^{-1} mol^{-1}
Solid	0	0	28.53	24.06
Gas	642.7	595.8	186.507	21.522

Density/kg m^{-3}: 12 370 [293 K]; 10 900 [liquid at m.p.]
Molar volume/cm^3: 8.14
Thermal conductivity/W m^{-1} K^{-1}: 117 [300 K]
Coefficient of linear thermal expansion/K^{-1}: 9.1×10^{-6}
Electrical resistivity/Ω m: 7.6×10^{-8} [273 K]
Mass magnetic susceptibility/kg^{-1} m^3: $+5.37 \times 10^{-9}$ (s)

Young's modulus/GPa: 432
Rigidity modulus/GPa: 173
Bulk modulus/GPa: 286
Poisson's ratio/GPa: 0.25

• BIOLOGICAL DATA

Biological role
None.

Toxicity
Toxic intake: most ruthenium compounds are poisonous.
Lethal intake: LD_{50} (RuO_2, oral, rat) = 4580 mg kg^{-1}

Hazards
Ingested ruthenium is retained in the bones for a long time. The volatile oxide, RuO_4, is highly toxic by inhalation.

Levels in humans
Organs: n.a. but very low
Daily dietary intake: n.a.
Total mass of element in average (70 kg) person: n.a.

Discovery: see Chemical Data section.
[Latin, *Ruthenia* = Russia]
French, *ruthénium*; German, *Ruthenium*; Italian, *rutenio*; Spanish, *rutenio*

Ruthenium

[roo-thee-ni-uhm]

• NUCLEAR DATA

Number of isotopes (including nuclear isomers): 20 **Isotope mass range:** 92 → 110

Key isotopes

Nuclide	Atomic mass	Natural abundance (%)	Nuclear spin I	Nuclear magnetic moment μ	Uses
^{96}Ru	95.907 599	5.52	0+		
^{98}Ru	97.905 287	1.88	0+		
^{99}Ru	98.905 939	12.7	5/2+	−6.413	NMR
^{100}Ru	99.904 219	12.6	0+		
^{101}Ru	100.905 582	17.0	5/2+	−0.718 9	NMR
^{102}Ru	101.904 348	31.6	0+		
^{104}Ru	103.905 424	18.7	0+		

A table of radioactive isotopes is given in Appendix 1, on p248.

NMR [Reference: RuO_4]	^{99}Ru	^{101}Ru
Relative sensitivity (^1H = 1.00)	1.95×10^{-4}	1.41×10^{-3}
Receptivity (^{13}C = 1.00)	0.83	1.56
Magnetogyric ratio/rad $T^{-1}s^{-1}$	-1.2343×10^7	-1.3834×10^7
Nuclear quadrupole moment/m^2	$+0.079 \times 10^{-28}$	$+0.457 \times 10^{-28}$
Frequency (^1H = 100 Hz; 2.3488T)/MHz	3.389	4.941

• ELECTRON SHELL DATA

Ground state electron configuration: [Kr]$4d^7 5s^1$
Term symbol: $^5F_5$
Electron affinity ($M \to M^-$)/kJ mol^{-1}**:** 101

Ionization energies/kJ mol^{-1}**:**

1.	$M \to M^+$	711
2.	$M^+ \to M^{2+}$	1617
3.	$M^{2+} \to M^{3+}$	2747
4.	$M^{3+} \to M^{4+}$	(4500)
5.	$M^{4+} \to M^{5+}$	(6100)
6.	$M^{5+} \to M^{6+}$	(7800)
7.	$M^{6+} \to M^{7+}$	(9600)
8.	$M^{7+} \to M^{8+}$	(11 500)
9.	$M^{8+} \to M^{9+}$	(18 700)
10.	$M^{9+} \to M^{10+}$	(20 900)

Electron binding energies/eV

K	1s	22 117
L_I	2s	3224
L_{II}	$2p_{1/2}$	2967
L_{III}	$2p_{3/2}$	2838
M_I	3s	586.1
M_{II}	$3p_{1/2}$	483.7
M_{III}	$3p_{3/2}$	461.5
M_{IV}	$3d_{3/2}$	284.2
M_V	$3d_{5/2}$	280.0

continued in Appendix 2, p257

Main lines in atomic spectrum
[Wavelength/nm(species)]

349.894 (I)
372.693 (I)
372.803 (I) (AA)
379.890(I)
379.935 (I)
419.990 (I)

• CRYSTAL DATA

Crystal structure (cell dimensions/pm), space group
h.c.p. (a = 270.58, c = 428.11), $P6_3/mmc$

X-ray diffraction: mass absorption coefficients (μ/ρ)/$cm^2\ g^{-1}$**:** CuK_α 183 MoK_α 21.1
Neutron scattering length, $b/10^{-12}$ cm: 0.721
Thermal neutron capture cross-section, σ_a/barns**:** 2.56

• GEOLOGICAL DATA

Minerals

Few minerals as such; ruthenium metal is found in the free state, sometimes alloyed with osmium and iridium.

Mineral	Formula	Density	Hardness	Crystal appearance
Ruthenarsenite	(Ru,Ni)As	10	6	orth., met. orange-brown
Ruthenium	Ru	12.2	6.5	hex., met. white

Chief source: from the wastes of nickel refining.
World production/tonnes y^{-1}**:** 0.12
Main mining areas: South Africa, Russia, Canada, USA and Zimbabwe.
Reserves/tonnes**:** n.a.
Specimen: available as powder or sponge.
Warning!

Abundances

Sun (relative to H = 1×10^{12})**:** 67.6
Earth's crust/p.p.m.**:** *c.* 0.001
Seawater/p.p.m.**:** n.a. but minute

Atomic number: 104 **CAS:**
Relative atomic mass ($^{12}C = 12.0000$)**:** 261.11 (Rf-261) [53850-36-5]

• CHEMICAL DATA

Description: Rutherfordium is a radioactive metal which does not occur naturally and is of research interest only.

Radii/pm: Rf^{4+} 67 (est.); atomic 150 (est.)
Electronegativity: n.a.
Effective nuclear charge: n.a.

Standard reduction potentials $E^{\ominus}$/V

IV		0
Rf^{4+}	$\xrightarrow{-1.8\ (\text{est.})}$	Rf

Oxidation states

Rf^{III}	d^1	?
Rf^{IV}	$[f^{14}]$	most stable

• PHYSICAL DATA

Melting point/K: 2400 (est.)
Boiling point/K: 5800 (est.)

ΔH_{fusion}/kJ mol^{-1}**:** n.a.
ΔH_{vap}/kJ mol^{-1}**:** n.a.
ΔH_{subl}/kJ mol^{-1}**:** 694 (est.)

Thermodynamic properties (298.15 K, 0.1 MPa)

State	$\Delta_f H^{\ominus}$/kJ mol^{-1}	$\Delta_f G^{\ominus}$/kJ mol^{-1}	$S^{\ominus}$/J K^{-1} mol^{-1}	C_p/J K^{-1} mol^{-1}
Solid	0	0	n.a.	n.a.
Gas	n.a.	n.a.	n.a.	n.a.

Density/kg m^{-3}**:** 23 000 (est.)
Molar volume/cm^3**:** n.a.
Thermal conductivity/W m^{-1} K^{-1}**:** n.a.
Coefficient of linear thermal expansion/K^{-1}**:** n.a.
Electrical resistivity/Ω m**:** n.a.
Mass magnetic susceptibility/kg^{-1} m^3**:** n.a.

• BIOLOGICAL DATA

Biological role

None.

Toxicity

Toxic intake: n.a.
Lethal intake: n.a.

Hazards

Rutherfordium is never encountered normally, and relatively few atoms have ever been made. It would be dangerous because of its intense radioactivity.

Levels in humans

nil
Daily dietary intake: nil
Total mass of element in average (70 kg) person: nil

Discovery: see Nuclear Data section.
[Named after Lord Rutherford, New Zealand physicist and chemist]
French, *rutherfordium*; German, *Rutherfordium*; Italian, *rutherfordio*; Spanish, *rutherfordio* [ruther-ford-iuhm]

Rutherfordium

• NUCLEAR DATA

Discovery: Isotope 260 was reported in 1964 by a group of of scientists at Dubna, near Moscow, Russia but the claim was disputed in 1969 by a group of scientists led by A. Ghiorso at Berkeley, California, USA who reported isotope 257. IUPAC concluded in 1992 that credit for the discovery should be shared between both groups.

Number of isotopes (including nuclear isomers): 10

Isotope mass range: 253 → 262

Key isotopes

Nuclide	Atomic mass	Half life ($T_{1/2}$)	Decay mode and energy (MeV)	Nuclear spin I	Nucl. mag. moment μ	Uses
^{253}Rf		1.5 s	SF; α?			
^{254}Rf?		5×10^{-4} s	SF			
^{255}Rf		1.4 s	SF; α?			
^{256}Rf		7×10^{-3} s	α; SF			
^{257}Rf	257.102 950	4.8 s	α (9.20); SF			
^{258}Rf	258.103 430	0.013 s	SF; α?			
^{259}Rf	259.105 530	3.0 s	α (9.20); SF			
^{260}Rf	260.106 300	0.020 s	SF			
^{261}Rf	261.108 690	65 s	α (8.60); SF?			
^{262}Rf		0.047 s	SF			

NMR [Not recorded]

• ELECTRON SHELL DATA

Ground state electron configuration: $[Rn]5f^{14}6d^{2}7s^{2}$
Term symbol: $^{3}F_{2}$
Electron affinity $(M \rightarrow M^{-})/kJ\ mol^{-1}$**:** n.a.

Ionization energies/kJ mol^{-1}**:**
1. M → M^{+} 490 (est.)

Electron binding energies/eV
n.a.

Main lines in atomic spectrum
[Wavelength/nm(species)]
n.a.

• CRYSTAL DATA

Crystal structure (cell dimensions/pm), space group
n.a.

X-ray diffraction: mass absorption coefficients $(\mu/\rho)/cm^{2}\ g^{-1}$**:** CuK_{α} n.a. MoK_{α} n.a.
Neutron scattering length, $b/10^{-12}$ **cm:** n.a.
Thermal neutron capture cross-section, σ_{a}/**barns:** n.a.

• GEOLOGICAL DATA

Minerals
Not found on Earth.

Chief source: several thousand atoms of rutherfordium ^{261}Rf have been made by bombarding ^{249}Cf with ^{12}C nuclei, or ^{248}Cm with ^{18}O nuclei.

Specimen: not available commercially.

Abundances
Sun (relative to $H = 1 \times 10^{12}$)**:** n.a.
Earth's crust/p.p.m.**:** nil
Seawater/p.p.m.**:** nil

Sm

Atomic number: 62
Relative atomic mass ($^{12}C = 12.0000$)**:** 150.36

CAS: [7440-19-19]

• CHEMICAL DATA

Description: Samarium is a silvery-white metal of the so-called rare earth group (more correctly termed the lanthanides). It is relatively stable in dry air, but in moist air an oxide coating forms. Samarium is used in permanent magnets, organic reagents, special glass, catalysts, ceramics and electronics.

Radii/pm: Sm^{3+} 100; Sm^{2+} 111; atomic 180; covalent 166
Electronegativity: 1.17 (Pauling); 1.07 (Allred); ≤ 3.1 eV (absolute)
Effective nuclear charge: 2.85 (Slater); 8.01 (Clementi); 11.06 (Froese-Fischer)

Standard reduction potentials $E^{\ominus}$/V

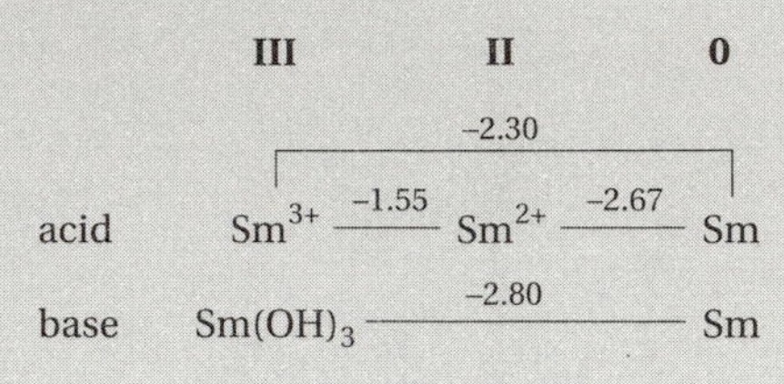

Oxidation states

Sm^{II}	f^6	SmO, SmS, SmF_2, $SmCl_2$ etc.
Sm^{III}	f^5	Sm_2O_3, $Sm(OH)_3$, $[Sm(OH_2)_x]^{3+}$ (aq), Sm^{3+} salts, SmF_3, $SmCl_3$ etc., complexes

• PHYSICAL DATA

Melting point/K: 1350
Boiling point/K: 2064

ΔH_{fusion}/kJ mol^{-1}: 10.9
ΔH_{vap}/kJ mol^{-1}: 191.6

Thermodynamic properties (298.15 K, 0.1 MPa)

State	$\Delta_f H^{\ominus}$/kJ mol^{-1}	$\Delta_f G^{\ominus}$/kJ mol^{-1}	$S^{\ominus}$/J K^{-1} mol^{-1}	C_p/J K^{-1} mol^{-1}
Solid	0	0	69.58	29.54
Gas	206.7	172.8	183.042	30.355

Density/kg m^{-3}: 7520 [293K]
Molar volume/cm^3: 20.00
Thermal conductivity/W m^{-1} K^{-1}: 13.3 [300 K]
Coefficient of linear thermal expansion/K^{-1}: 10.4×10^{-6}
Electrical resistivity/Ω m: 94.0×10^{-8} [298 K]
Mass magnetic susceptibility/kg^{-1} m^3: $+1.52 \times 10^{-7}$ (s)

Young's modulus/GPa: 49.7
Rigidity modulus/GPa: 19.5
Bulk modulus/GPa: 37.8
Poisson's ratio/GPa: 0.274

• BIOLOGICAL DATA

Biological role

None, but acts to stimulate metabolism.

Toxicity

Toxic intake: n.a.
Lethal intake: LD_{50} (nitrate, oral, rat) = 2900 mg kg^{-1}

Hazards

Samarium is mildly toxic by ingestion, and is a skin and eye irritant.

Levels in humans

Blood/mg dm^{-3}: 0.008
Organs: n.a., but very low
Daily dietary intake: n.a.
Total mass of element in average (70 kg) person: *c.* 0.05 mg

Discovered in 1879 by P.E. Lecoq de Boisbaudran at Paris, France.
[Named after the mineral samarskite]
French, *samarium*; German, *Samarium*; Italian, *samario*; Spanish, *samario*

Samarium

[sam-ayr-iuhm]

• NUCLEAR DATA

Number of isotopes (including nuclear isomers): 24 **Isotope mass range:** 138 → 158

Key isotopes

Nuclide	Atomic mass	Natural abundance (%)	Nuclear spin I	Nuclear magnetic moment μ	Uses
^{144}Sm	143.911 998	3.1	0+		E
^{147}Sm*	146.914 895	15.0	7/2–	–0.814 9	E, NMR
^{148}Sm**	147.914 820	11.3	0+		E
^{149}Sm***	148.917 181	13.8	7/2–	–0.671 8	E, NMR
^{150}Sm	149.917 273	7.4	0+		E
^{152}Sm	151.919 729	26.7	0+		E
^{154}Sm	153.922 206	22.7	0+		E

*^{147}Sm is radioactive with a half-life of 1.08×10^{11} y and decay mode α (2.23 MeV).
**^{148}Sm is radioactive with a half-life of 7×10^{15} y and decay mode α (1.96 MeV).
***^{149}Sm is radioactive with a half-life of 1×10^{16} y and decay mode α.
A table of other radioactive isotopes is given in Appendix 1, on p248.

NMR [Reference: not given]

	^{147}Sm	^{149}Sm
Relative sensitivity (^{1}H = 1.00)	1.48×10^{-3}	7.47×10^{-4}
Receptivity (^{13}C = 1.00)	1.28	0.665
Magnetogyric ratio/rad T^{-1} s^{-1}	-1.1124×10^{7}	-0.9175×10^{7}
Nuclear quadrupole moment/m^2	-0.259×10^{-28}	$+0.094 \times 10^{-28}$
Frequency (^{1}H = 100 Hz; 2.3488T)/MHz	4.128	3.289

• ELECTRON SHELL DATA

Ground state electron configuration: [Xe]$4f^6 6s^2$
Term symbol: $^7F_0$
Electron affinity (M → M$^-$)/kJ mol^{-1}**:** ≤ 50

Ionization energies/kJ mol^{-1}**:**

1.	M → M$^+$	543.3
2.	M$^+$ → M^{2+}	1068
3.	M^{2+} → M^{3+}	2260
4.	M^{3+} → M^{4+}	3990
5.	M^{4+} → M^{5+}	6046

Electron binding energies/eV

K	1s	46 834
L_I	2s	7737
L_{II}	$2p_{1/2}$	7312
L_{III}	$2p_{3/2}$	6716
M_I	3s	1723
M_{II}	$3p_{1/2}$	1541
M_{III}	$3p_{3/2}$	1419.8
M_{IV}	$3d_{3/2}$	1110.9
M_V	$3d_{5/2}$	1083.4

continued in Appendix 2, p257

Main lines in atomic spectrum
[Wavelength/nm(species)]

356.827 (II)
359.260 (II)
363.429 (II)
373.912 (II)
388.529 (II)
429.674 (I) (AA)

• CRYSTAL DATA

Crystal structure (cell dimensions/pm), space group
α-Sm rhombohedral (a = 899.6, α = 23° 13'), R$\bar{3}$m
β-Sm cubic (a = 407), Im3m
$T(\alpha \rightarrow \beta)$ = 1190 K
high pressure form: h.c.p. (a = 361.8, c = 1 166), P6_3/mmc

X-ray diffraction: mass absorption coefficients (μ/ρ)/cm^2 g^{-1}**:** CuK$_\alpha$ 397 MoK$_\alpha$ 58.6
Neutron scattering length, $b/10^{-12}$ cm**:** 0.080
Thermal neutron capture cross-section, σ_a/barns**:** 5922

• GEOLOGICAL DATA

Minerals

Mineral	Formula	Density	Hardness	Crystal appearance
Monazite*	(Ce, La, Nd, Th, etc)PO_4	5.20	5 – 5.5	mon., waxy/vit. yellow-brown

*Although not a major constituent, samarium is present in extractable amounts.

Chief ore: monazite
World production/tonnes y^{-1}**:** *c.* 700
Main mining areas: USA, Brazil, India, Sri Lanka, Australia
Reserves/tonnes: *c.* 2×10^6
Specimen: available as chips or ingots. Safe.

Abundances

Sun (relative to H = 1×10^{12}): 5.2
Earth's crust/p.p.m.: 7.9
Seawater/p.p.m.:
Atlantic surface: 4.0×10^{-7}
Atlantic deep: 6.4×10^{-7}
Pacific surface: 4.0×10^{-7}
Pacific deep: 10×10^{-7}
Residence time/years: 200
Classification: recycled
Oxidation state: III

Sc

Atomic number: 21
Relative atomic mass ($^{12}C = 12.0000$)**:** 44.955910

CAS: [7440-20-2]

• CHEMICAL DATA

Description: Scandium is a soft, silvery-white metal which tarnishes in air and burns easily, once ignited. It reacts with water to form hydrogen gas. There are only a few, rather specialised, uses for scandium such as in mercury vapour lights for high intensity lighting when a sunlight effect is required.

Radii/pm: Sc^{3+} 83; atomic 161; covalent 144
Electronegativity: 1.36 (Pauling); 1.20 (Allred); 3.34 eV (absolute)
Effective nuclear charge: 3.00 (Slater); 4.63 (Clementi); 6.06 (Froese-Fischer)

Standard reduction potentials $E^{\ominus}$/V

III		0
Sc^{3+}	−2.03	Sc

Oxidation states

Sc^{II}	d^1	$CsScCl_3$
Sc^{III}	[Ar]	Sc_2O_3, ScO.OH, '$Sc(OH)_3$', $[Sc(H_2O)_7]^{3+}$ $[Sc(OH)_6]^{3-}$, ScF_3, $ScCl_3$, etc., $[ScF_6]^{3-}$, complexes ScH_2 is probably $Sc^{III}H^-$ with complex bonding

• PHYSICAL DATA

Melting point/K: 1814
Boiling point/K: 3104

ΔH_{fusion}/kJ mol^{-1}: 15.9
ΔH_{vap}/kJ mol^{-1}: 304.8

Thermodynamic properties (298.15 K, 0.1 MPa)

State	$\Delta_f H^{\ominus}$/kJ mol^{-1}	$\Delta_f G^{\ominus}$/kJ mol^{-1}	$S^{\ominus}$/J K^{-1} mol^{-1}	C_p/J K^{-1} mol^{-1}
Solid	0	0	34.64	25.52
Gas	377.8	336.03	174.79	22.09

Density/kg m^{-3}: 2989 [273 K]
Molar volume/cm^3: 15.04
Thermal conductivity/W m^{-1} K^{-1}: 15.8 [300 K]
Coefficient of linear thermal expansion/K^{-1}: 10.0×10^{-6}
Electrical resistivity/Ω m: 61.0×10^{-8} [295 K]
Mass magnetic susceptibility/kg^{-1} m^3: $+8.8 \times 10^{-8}$ (s)

Young's modulus/GPa: 74.4
Rigidity modulus/GPa: 29.1
Bulk modulus/GPa: 56.6
Poisson's ratio/GPa: 0.279

• BIOLOGICAL DATA

Biological role

None.

Toxicity

Toxic intake: n.a.
Lethal intake: LD_{50} (chloride, oral, mouse) = 4000 mg kg^{-1}

Hazards

Scandium is mildly toxic by ingestion, and scandium salts are suspected of being carcinogenic.

Levels in humans

Blood/mg dm^{-3}: *c.* 0.008
Bone/p.p.m.: *c.* 0.001
Liver/p.p.m.: 0.0004 – 0.0014
Muscle/p.p.m.: n.a.
Daily dietary intake: *c.* 0.00005 mg
Total mass of element in average (70 kg) person: *c.* 0.2 mg

Discovered in 1879 by L.F. Nilson at Uppsala, Sweden.
[Latin, *Scandia* = Scandinavia]
French, *scandium*; German, *Scandium*; Italian, *scandio*; Spanish, *escandio*

Scandium

[skan-dium]

• NUCLEAR DATA

Number of isotopes (including nuclear isomers): 15 **Isotope mass range:** 40 → 51

Key isotopes

Nuclide	Atomic mass	Natural abundance (%)	Nuclear spin I	Nuclear magnetic moment μ	Uses
^{45}Sc	44.955 910	100	7/2–	+4.756 483	NMR

A table of radioactive isotopes is given in Appendix 1, on p248.

NMR [Reference: $Sc(ClO_4)_3$ (aq)]	^{45}Sc
Relative sensitivity (^1H = 1.00)	0.30
Receptivity (^{13}C = 1.00)	1710
Magnetogyric ratio/rad $T^{-1}s^{-1}$	6.4982×10^7
Nuclear quadrupole moment/m^2	-0.220×10^{-28}
Frequency (^1H = 100 Hz; 2.3488T)/MHz	24.290

• ELECTRON SHELL DATA

Ground state electron configuration: $[Ar]3d^14s^2$
Term symbol: $^2D_{3/2}$
Electron affinity ($M \to M^-$)/kJ mol^{-1}: 18.1

Ionization energies/kJ mol^{-1}:

1.	$M \to M^+$	631
2.	$M^+ \to M^{2+}$	1235
3.	$M^{2+} \to M^{3+}$	2389
4.	$M^{3+} \to M^{4+}$	7089
5.	$M^{4+} \to M^{5+}$	8844
6.	$M^{5+} \to M^{6+}$	10 720
7.	$M^{6+} \to M^{7+}$	13 320
8.	$M^{7+} \to M^{8+}$	15 310
9.	$M^{8+} \to M^{9+}$	17 369
10.	$M^{9+} \to M^{10+}$	21 740

Electron binding energies/eV

K	1s	4492
L_I	2s	498.0
L_{II}	$2p_{1/2}$	403.6
L_{III}	$2p_{3/2}$	398.7
M_I	3s	51.1
M_{II}	$3p_{1/2}$	28.3
M_{III}	$3p_{3/2}$	28.3

Main lines in atomic spectrum
[Wavelength/nm(species)]

361.384 (I)
363.075 (II)
390.749 (I)
391.181 (I) (AA)
402.040 (I)
402.369 (I)

• CRYSTAL DATA

Crystal structure (cell dimensions/pm), space group
α-Sc h.c.p. (a = 330.90, c = 527.3), $P6_3/mmc$
β-Sc cubic, Im3m

$T(\alpha \to \beta)$ = 1223 K

X-ray diffraction: mass absorption coefficients (μ/ρ)/cm^2 g^{-1}**:** CuK_α 184 MoK_α 21.1
Neutron scattering length, $b/10^{-12}$ cm**:** 1.23
Thermal neutron capture cross-section, σ_a/barns**:** 27.2

• GEOLOGICAL DATA

Minerals

Mineral	Formula	Density	Hardness	Crystal appearance
Thortveitite	$(Sc,Y)_2Si_2O_7$	3.57	6 – 7	mon., greyish-green/black

Chief ores: thortveitite (rare); also present in euxenite and gadolinite; extracted from uranium mill tailings.
World production/tonnes y^{-1}**:** 0.05
Main mining areas: Iceland, Norway, Malagasy (Madagascar)
Reserves/tonnes: n.a.
Specimen: available as foil, powder and pieces. *Care!*

Abundances

Sun (relative to H = 1×10^{12})**:** 1100
Earth's crust/p.p.m.**:** 16
Seawater/p.p.m.**:**
Atlantic surface: 6.1×10^{-7}
Atlantic deep: 8.8×10^{-7}
Pacific surface: 3.5×10^{-7}
Pacific deep: 7.9×10^{-7}
Residence time/years**:** 5000
Classification: recycled
Oxidation state: III

Sg

Atomic number: 106
Relative atomic mass (^{12}C = 12.0000)**:** 263.118 (Sg-263)

CAS: [54038-81-2]

• CHEMICAL DATA

Description: Seaborgium is a radioactive metal which does not occur naturally, and is of research interest only. Its chemistry resembles that of tungsten.

Radii/pm: Sg^{5+} 86 (est.); atomic 132 (est.)
Electronegativity: n.a.
Effective nuclear charge: n.a.

Standard reduction potentials $E^{\ominus}$/V

IV		0
Sg^{4+}	—— −0.6 (est.) ——	Sg

Oxidation states

Sg^{IV}	d^2	predicted,
Sg^{V}	d^1	predicted
$\mathbf{Sg^{VI}}$	$[f^{14}]$	SgO_4^{2-}, SgO_2Cl_2

• PHYSICAL DATA

Melting point/K: n.a.
Boiling point/K: n.a.

ΔH_{fusion}/kJ mol^{-1}: n.a.
ΔH_{vap}/kJ mol^{-1}: n.a.
ΔH_{subl}/kJ mol^{-1}: 858 (est.)

Thermodynamic properties (298.15 K, 0.1 MPa)

State	$\Delta_f H^{\ominus}$/kJ mol^{-1}	$\Delta_f G^{\ominus}$/kJ mol^{-1}	$S^{\ominus}$/J K^{-1} mol^{-1}	C_p/J K^{-1} mol^{-1}
Solid	0	0	n.a.	n.a.
Gas	n.a.	n.a.	n.a.	n.a.

Density/kg m^{-3}: 35 000 (est.)
Molar volume/cm^3: n.a.
Thermal conductivity/W m^{-1} K^{-1}: n.a.
Coefficient of linear thermal expansion/K^{-1}: n.a.
Electrical resistivity/Ω m: n.a.
Mass magnetic susceptibility/kg^{-1} m^3: n.a.

• BIOLOGICAL DATA

Biological role

None.

Toxicity

Toxic intake: n.a.
Lethal intake: n.a.

Hazards

Seaborgium is never encountered normally, and only a few atoms have ever been made. It would be dangerous because of its intense radioactivity.

Levels in humans

nil
Daily dietary intake: nil
Total mass of element in average (70 kg) person: nil

Discovery: see Nuclear Data section.
[Named after Glenn T. Seaborg, American nuclear chemist and Nobel prizewinner]
French, *seaborgium*; German, *Seaborgium*; Italian, *seaborgio*; Spanish, *seaborgio*

Seaborgium

[see-borg-iuhm]

• NUCLEAR DATA

Discovery: Isotope 263 was made and identified in 1974 by American scientists led by Albert Ghiorso at the Lawrence Berkeley Laboratory in California and the Livermore National Laboratory, USA.

Number of isotopes (including nuclear isomers): 6

Isotope mass range: 259 → 266

Key isotopes

Nuclide	Atomic mass	Half life ($T_{1/2}$)	Decay mode and energy (MeV)	Nuclear spin *I*	Nucl. mag. moment μ	Uses
^{259}Sg	259.1144	0.5 s	α; SF			
^{260}Sg		4×10^{-3} s	α; SF			
^{261}Sg	261.1161	0.3 s	α; SF?			
^{263}Sg	263.1182	0.9 s	SF; α			
^{265}Sg		2.8 s	α (8.81)			
^{266}Sg		27.3 s	α (8.82)			

NMR [Not recorded]

• ELECTRON SHELL DATA

Ground state electron configuration: $[Rn]5f^{14}6d^{4}7s^{2}$
Term symbol: $^{5}D_{0}$
Electron affinity ($M \rightarrow M^{-}$)/kJ mol^{-1}**:** n.a.

Ionization energies/kJ mol^{-1}**:**
1. M → M^{+} 730 (est.)

Electron binding energies/eV
n.a.

Main lines in atomic spectrum
[Wavelength/nm(species)]
n.a.

• CRYSTAL DATA

Crystal structure (cell dimensions/pm), space group
n.a.

X-ray diffraction: mass absorption coefficients (μ/ρ)/cm^{2} g^{-1}**:** CuK_{α} n.a. MoK_{α} n.a.
Neutron scattering length, $b/10^{-12}$ cm**:** n.a.
Thermal neutron capture cross-section, σ_a/barns**:** n.a.

• GEOLOGICAL DATA

Minerals
Not found on Earth.

Chief source: several atoms of seaborgium have been made from ^{249}Cf by bombarding it with ^{18}O nuclei: $^{249}Cf + ^{18}O \rightarrow ^{263}Sg + 4n$. This was done with an 88-inch-diameter cyclotron which produces about a billion atoms per hour of which only one is element 106. Bombarding ^{248}Cm with ^{22}Ne produced the heavier isotopes, ^{265}Sg and ^{266}Sg, which have longer-than-expected half-lives of 2.8 sec and 27.3 sec respectively.

Specimen: not available commercially.

Abundances
Sun (relative to H = 1×10^{12})**:** n.a.
Earth's crust/p.p.m.**:** nil
Seawater/p.p.m.**:** nil

Se

Atomic number: 34
Relative atomic mass ($^{12}C = 12.0000$)**:** 78.96

CAS: [7782-49-2]

• CHEMICAL DATA

Description: Selenium is obtained in either a silvery metallic form (grey Se) or a red amorphous powder, which is less stable. Selenium burns in air, is unaffected by water, but dissolves in concentrated HNO_3 and alkalis. It is used in photoelectric cells, photocopiers, solar cells and semiconductors.

Radii/pm: Se^{4+} 69; Se^{2-} 191; atomic 215; covalent 117; van der Waals 200
Electronegativity: 2.55 (Pauling); 2.48 (Allred); 5.89 eV (absolute)
Effective nuclear charge: 6.95 (Slater); 8.29 (Clementi); 9.96 (Froese-Fischer)

Standard reduction potentials $E^{\ominus}$/V

	VI		IV		0		–II
acid	SeO_4^{2-}	1.1	H_2SeO_3	0.74	Se	–0.11	H_2Se
base	SeO_4^{2-}	0.03	SeO_3^{2-}	–0.36	Se	–0.67	Se^{2-}

Oxidation states

Se^{-II}	[Kr]	H_2Se
Se^{0-I}	s^2p^4,	Se cluster cations, e.g. Se_4^{2+}, Se_8^{2+}
Se^{I}	s^2p^3	Se_2Cl_2, Se_2Br_2
Se^{II}	s^2p^2	?
Se^{IV}	s^2	SeO_2, H_2SeO_3, $[SeO_3]^{2-}$ (aq), SeF_2O, $SeCl_2O$, $SeBr_2O$, SeF_4, $SeCl_4$, $SeBr_4$, $[SeBr_6]^{2-}$
Se^{VI}	d^{10}	SeO_3?, H_2SeO_4, $[SeO_4]^{2-}$ (aq), SeO_2F_2, SeO_2Cl_2, SeF_6

Covalent bonds

Bond	r/ pm	E/ kJ mol^{-1}
Se—H	146	305
Se—C	198	245
Se—O	161	343
Se—F	170	285
Se—Cl	220	245
Se—Se (Se_8)	232	330

• PHYSICAL DATA

Melting point/K: 490
Boiling point/K: 958.1
Critical temperature/K: 1766
Critical pressure/kPa: 27200

ΔH_{fusion}/kJ mol^{-1}: 5.1
ΔH_{vap}/kJ mol^{-1}: 26.32

Thermodynamic properties (298.15 K, 0.1 MPa)

State	$\Delta_f H^{\ominus}$/kJ mol^{-1}	$\Delta_f G^{\ominus}$/kJ mol^{-1}	$S^{\ominus}$/J K^{-1} mol^{-1}	C_p/J K^{-1} mol^{-1}
Solid (α)	0	0	42.442	25.363
Gas	227.07	187.03	176.72	20.820

Density/kg m^{-3}: 4790 (grey) [293 K]; 3987 [liquid at m.p.]
Molar volume/cm^3: 16.48
Thermal conductivity/W m^{-1} K^{-1}: 2.04 [300 K]
Coefficient of linear thermal expansion/K^{-1}: 36.9×10^{-6}
Electrical resistivity/Ω m: 0.01 [293 K]
Mass magnetic susceptibility/kg^{-1} m^3: -4.0×10^{-9} (s)

Young's modulus/GPa: 58
Rigidity modulus/GPa: n.a.
Bulk modulus/GPa: n.a.
Poisson's ratio/GPa: 0.447

• BIOLOGICAL DATA

Biological role

Essential to some species, including humans, although only in tiny amounts. Selenium acts to stimulate the metabolism.

Toxicity

Toxic intake: human, Se metal = *c.* 10 – 35 mg
Lethal intake: LD_{50} (Se metal, oral, rat) = 6700 mg kg^{-1}. A dose of 5 mg per day can be lethal for many humans. LD_{50} (H_2SeO_3, intravenous, mouse) = 11 mg kg^{-1}.

Hazards

Selenium compounds are toxic by inhalation and intravenous routes. They are also considered to be experimental carcinogens, and teratogens.

Levels in humans

Blood/mg dm^{-3}: 0.171
Bone/p.p.m.: 1 – 9
Liver/p.p.m.: 0.35 – 2.4
Muscle/p.p.m.: 0.42 – 1.9
Daily dietary intake: 0.006 – 0.2 mg
Total mass of element in average (70 kg) person: *c.* 15 mg (wide range possible, 10–65 mg)

Discovered in 1817 by J.J. Berzelius at Stockholm, Sweden.
[Greek, *selene* = moon]
French, *sélénium*; German, *Selen*; Italian, *selenio*; Spanish, *selenio*

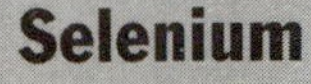

[sel-een-iuhm]

• NUCLEAR DATA

Number of isotopes (including nuclear isomers): 26 **Isotope mass range:** 69 → 89

Key isotopes

Nuclide	Atomic mass	Natural abundance (%)	Nuclear spin I	Nuclear magnetic moment μ	Uses
^{74}Se	73.922 474 6	0.9	0+		E
^{76}Se	75.919 212 0	9.0	0+		E
^{77}Se	76.919 912 5	7.6	1/2–	+0.535 06	E, NMR
^{78}Se	77.917 307 6	23.6	0+		E
^{80}Se	79.916 519 6	49.7	0+		E
^{82}Se	81.916 697 8	9.2	0+		E

A table of radioactive isotopes is given in Appendix 1, on p249 .

NMR [Reference: $Se(CH_3)_2$]	^{77}Se
Relative sensitivity (^1H = 1.00)	6.93×10^{-3}
Receptivity (^{13}C = 1.00)	2.98
Magnetogyric ratio/rad $T^{-1}s^{-1}$	5.1018×10^7
Nuclear quadrupole moment/m^2	–
Frequency (^1H = 100 Hz; 2.3488T)/MHz	19.092

• ELECTRON SHELL DATA

Ground state electron configuration: $[Ar]3d^{10}4s^24p^4$
Term symbol: $^4P_2$
Electron affinity ($M \rightarrow M^-$)/kJ mol^{-1}**:** 195.0

Ionization energies/kJ mol^{-1}**:**

1.	$M \rightarrow M^+$	940.9
2.	$M^+ \rightarrow M^{2+}$	2044
3.	$M^{2+} \rightarrow M^{3+}$	2974
4.	$M^{3+} \rightarrow M^{4+}$	4144
5.	$M^{4+} \rightarrow M^{5+}$	6590
6.	$M^{5+} \rightarrow M^{6+}$	7883
7.	$M^{6+} \rightarrow M^{7+}$	14 990
8.	$M^{7+} \rightarrow M^{8+}$	(19 500)
9.	$M^{8+} \rightarrow M^{9+}$	(23 300)
10.	$M^{9+} \rightarrow M^{10+}$	(27 200)

Electron binding energies/eV

K	1s	12 658
L_I	2s	1652.0
L_{II}	$2p_{1/2}$	1474.3
L_{III}	$2p_{3/2}$	1433.9
M_I	3s	229.6
M_{II}	$3p_{1/2}$	166.5
M_{III}	$3p_{3/2}$	160.7
M_{IV}	$3d_{3/2}$	55.5
M_V	$3d_{5/2}$	54.6

Main lines in atomic spectrum
[Wavelength/nm(species)]

196.026 (I) (AA)
241.350 (I)
1032.726 (I)
1038.636 (I)
2144.256 (I)

• CRYSTAL DATA

Crystal structure (cell dimensions/pm), space group
grey hexagonal (a = 436.56, c = 495.90), $P3_121$, metallic form
α-Se monoclinic (a = 906.4, b = 907.2, c = 115.6, β = 90° 52'), $P2_1/a$, Se_8
β-Se monoclinic (a = 1285, b = 807, c = 931, β = 93° 8'), $P2_1/a$, Se_8
α'-Se cubic (a = 297.0), Pm3m
β'-Se cubic (a = 604), Fd3m

X-ray diffraction: mass absorption coefficients (μ/ρ)/$cm^2\ g^{-1}$**:** CuK_α 91.4 MoK_α 74.7
Neutron scattering length, $b/10^{-12}$ cm**:** 0.797
Thermal neutron capture cross-section, σ_a/barns**:** 11.7

• GEOLOGICAL DATA

Minerals

Native selenium is occasionally found; several minerals are known, but all are very rare, and generally they occur together with sulfides of metals such as copper, zinc and lead.

Mineral	Formula	Density	Hardness	Crystal appearance
Clausthalite	PbSe	7.8	2.5 – 3	cub., met. grey
Crooksite	Cu_7TlSe_4	6.90	2.5 – 3	tetragonal, met. grey

Note: the mineral called selenite is in fact calcium sulfate.

Chief ores: by-product of the electro-refining of copper.

World production/tonnes y^{-1}**:** 1 600

Main mining areas: main mining areas: Canada, USA, Bolivia and Russia

Reserves/tonnes**:** n.a.

Specimen: available as powder or pellets. *Care!*

Abundances

Sun (relative to H = 1×10^{12})**:** n.a.
Earth's crust/p.p.m.**:** 0.05
Seawater/p.p.m.**:**
Atlantic surface: 0.46×10^{-7}
Atlantic deep: 1.8×10^{-7}
Pacific surface: 0.15×10^{-7}
Pacific deep: 1.65×10^{-7}
Residence time/years**:** 3000
Classification: recycled
Oxidation state: VI (mainly) and IV

Si

Atomic number: 14
Relative atomic mass ($^{12}C = 12.0000$)**:** 28.0855

CAS: [7440-21-3]

• CHEMICAL DATA

Description: Black amorphous silicon is obtained by the reduction of sand (SiO_2) with carbon. Ultrapure crystals of silicon have a blue-grey metallic sheen. Bulk silicon is unreactive towards oxygen, water and acids (except HF), but dissolves in hot alkali. Silicon is used in semiconductors, alloys and polymers.

Radii/pm: Si^{4+} 26; Si^{4-} 271; atomic 117; covalent 117; van der Waals 200
Electronegativity: 1.90 (Pauling); 1.74 (Allred); 4.77 eV (absolute)
Effective nuclear charge: 4.15 (Slater); 4.29 (Clementi); 4.48 (Froese-Fischer)

Standard reduction potentials $E^\ominus$/V

	IV		II		0		–IV
acid	SiO_2	–0.967	'SiO'	–0.808	Si	–0.143	SiH_4

[Alkaline solutions contain many different forms]

Oxidation states

Si^{II}	s^2	SiF_2 (gas)
Si^{IV}	[Ne]	SiO_2, 'H_4SiO_4', silicates, zeolites etc., SiH_4 and other silanes, SiF_4, $SiCl_4$etc., $[SiF_6]^{2-}$, metal silicides, e.g. Ca_2Si, CaSi, organosilicon compounds

Covalent bonds

Bond	r/ pm	E/ kJ mol^{-1}
Si—H	148	318
Si—C	187	301
Si—O	166	453
Si—F	157	565
Si—Cl	202	381
Si—Br	215	310
Si—I	243	234
Si—Si	235	222

• PHYSICAL DATA

Melting point/K: 1683
Boiling point/K: 2628

ΔH_{fusion}/kJ mol^{-1}: 39.6
ΔH_{vap}/kJ mol^{-1}: 383.3

Thermodynamic properties (298.15 K, 0.1 MPa)

State	$\Delta_f H^\ominus$/kJ mol^{-1}	$\Delta_f G^\ominus$/kJ mol^{-1}	$S^\ominus$/J K^{-1} mol^{-1}	C_p/J K^{-1} mol^{-1}
Solid	0	0	18.83	20.00
Gas	455.6	411.3	167.97	22.251

Density/kg m^{-3}: 2329 [293 K]; 2525 [liquid at m.p.]
Molar volume/cm^3: 12.06
Thermal conductivity/W m^{-1} K^{-1}: 148 [300 K]
Coefficient of linear thermal expansion/K^{-1}: 4.2×10^{-6}
Electrical resistivity/Ω m: 0.001 [273 K]
Mass magnetic susceptibility/kg^{-1} m^3: -1.8×10^{-9} (s)

Young's modulus/GPa: 113
Rigidity modulus/GPa: 39.7
Bulk modulus/GPa: n.a.
Poisson's ratio/GPa: 0.42

• BIOLOGICAL DATA

Biological role

Silicon is essential to some species and possibly to humans.

Toxicity

Non-toxic as silicon, silicon dioxide or silicate.

Hazards

The fibres of some silicates, such as asbestos-type minerals, are carcinogenic.

Levels in humans

Blood/mg dm^{-3}: 3.9
Bone/p.p.m.: 17
Liver/p.p.m.: 12 – 120
Muscle/p.p.m.: 100 – 200
Daily dietary intake: 18 – 1200 mg
Total mass of element in average (70 kg) person: *c.* 1 g

Discovered in 1824 by J.J. Berzelius at Stockholm, Sweden.
[Latin, *silicis* = flint]
French, *silicium*; German, *Silicium*; Italian, *silicio*; Spanish, *silicio*

Silicon

[sil-i-kon]

• NUCLEAR DATA

Number of isotopes (including nuclear isomers): 11 **Isotope mass range:** 24 → 34

Key isotopes

Nuclide	Atomic mass	Natural abundance (%)	Nuclear spin I	Nuclear magnetic moment μ	Uses
^{28}Si	27.976 927 1	92.23	0+	0	E
^{29}Si	28.976 494 9	4.67	1/2+	–0.555 29	E, NMR
^{30}Si	29.973 770 7	3.10	0+	0	E

A table of radioactive isotopes is given in Appendix 1, on p249.

NMR [Reference: $Si(CH_3)_4$]	^{29}Si
Relative sensitivity (^{1}H = 1.00)	7.84×10^{-3}
Receptivity (^{13}C = 1.00)	2.09
Magnetogyric ratio/rad $T^{-1} s^{-1}$	-5.3146×10^7
Nuclear quadrupole moment/m^2	–
Frequency (^{1}H = 100 Hz; 2.3488T)/MHz	19.865

• ELECTRON SHELL DATA

Ground state electron configuration: [Ne]$3s^2 3p^2$
Term symbol: $^3P_0$
Electron affinity ($M \rightarrow M^-$)/kJ mol^{-1}**:** 133.6

Ionization energies/kJ mol^{-1}**:**

1.	$M \rightarrow M^+$	786.5
2.	$M^+ \rightarrow M^{2+}$	1577.2
3.	$M^{2+} \rightarrow M^{3+}$	3231.4
4.	$M^{3+} \rightarrow M^{4+}$	4355.5
5.	$M^{4+} \rightarrow M^{5+}$	16 091
6.	$M^{5+} \rightarrow M^{6+}$	19 784
7.	$M^{6+} \rightarrow M^{7+}$	23 786
8.	$M^{7+} \rightarrow M^{8+}$	29 252
9.	$M^{8+} \rightarrow M^{9+}$	33 876
10.	$M^{9+} \rightarrow M^{10+}$	38 732

Electron binding energies/eV

K	1s	1839
L_I	2s	149.7
L_{II}	$2p_{1/2}$	99.8
L_{III}	$2p_{3/2}$	99.2

Main lines in atomic spectrum
[Wavelength/nm(species)]

251.611 (I) (AA)
288.156 (I)
504.103 (II)
505.598 (I)
566.956 (I)
634.710 (II)
637.136 (II)

• CRYSTAL DATA

Crystal structure (cell dimensions/pm), space group
cubic (a = 543.07), Fd3m, diamond structure
High pressure forms:
(a = 468.6; c = 285.5), $I4_1$/amd
(a = 664), Ia3
(a = 380; c = 628), $P6_3$/mmc

X-ray diffraction: mass absorption coefficients (μ/ρ)/cm^2 g^{-1}**:** CuK$_\alpha$ 60.6 MoK$_\alpha$ 6.44
Neutron scattering length, $b/10^{-12}$ cm**:** 0.41543
Thermal neutron capture cross-section, σ_a/barns**:** 0.171

• GEOLOGICAL DATA

Minerals

The earth's crust is composed primarily of silicate minerals. In this table only varieties of silicon dioxide (silica) are listed. Examples of metal silicates are given on page 260.

Mineral	Formula	Density	Hardness	Crystal appearance
Cristobalite	SiO_2	2.33	n.a.	tet., vitreous/white
Opal*	$SiO_2.nH_2O$	2	5.5 – 6.5	amor., transp./col./white
Quartz	α-SiO_2	2.655	7	rhom., vit. colourless
Tridymite	SiO_2	2.26	7	mon., vit., colourless/white

*Can be iridescent and used as a gemstone

Chief ores: quartz (most common mineral on Earth); also talc, mica.

World production/tonnes y^{-1}**:** 5000 (electronic grade, i.e. pure, silicon); 480 000 (metallurgical grade silicon); 3.4×10^6 (ferrosilicon)

Main mining areas: talc in Austria, Italy, India, South Africa, Australia; mica in Canada, USA, India, Brazil

Reserves/tonnes**:** unlimited

Specimen: available as powder, pieces or lumps. Safe.

Abundances

Sun (relative to H = 1×10^{12})**:** 4.47×10^7
Earth's crust/p.p.m.**:** 277 100
Seawater/p.p.m.**:**
Atlantic surface: 0.03
Atlantic deep: 0.82
Pacific surface: 0.03
Pacific deep: 4.09
Residence time/years**:** 30 000
Classification: recycled
Oxidation state: IV

Ag

Atomic number: 47
Relative atomic mass ($^{12}C = 12.0000$): **107.8682**

CAS: [7440-22-4]

• CHEMICAL DATA

Description: Silver is a soft, malleable metal with a characteristic silver sheen when polished. It is stable to water and oxygen but is slowly attacked by sulfur compounds in the air to form a black sulfide layer. Silver dissolves in H_2SO_4 and HNO_3. The metal is used for silverware, jewellery, mirrors, and in the electrical industry because of its excellent conductivity; silver salts are used in photography.

Radii/pm: Ag^{2+} 89; Ag^+ 113; atomic 144; covalent 134
Electronegativity: 1.93 (Pauling); 1.42 (Allred); 4.44 eV (absolute)
Effective nuclear charge: 4.20 (Slater); 8.03 (Clementi); 11.35 (Froese-Fischer)

Standard reduction potentials $E^\ominus$/V

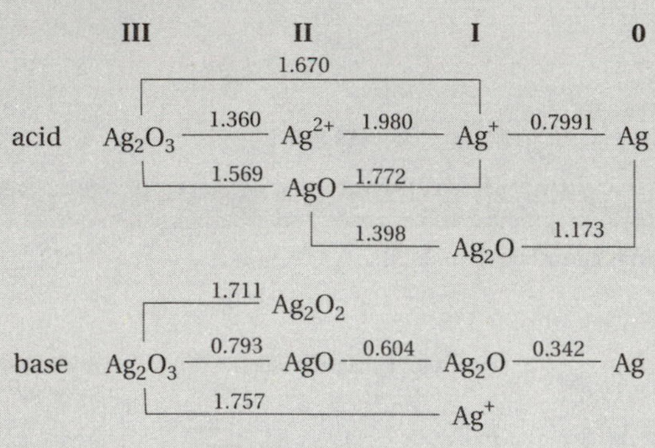

Oxidation states

Ag^0	$d^{10}s^1$	rare $[Ag(CO)_3]$ at 10 K
Ag^I	d^{10}	Ag_2O, $[Ag(OH)_2]^-$ (aq), $[Ag(OH_2)_4]^+$ (aq), AgF, AgCl etc., Ag^+ salts e.g. $AgNO_3$, Ag_2S, $[Ag(CN)_2]^-$, and other complexes
Ag^{II}	d^9	AgF_2, $[Ag(NC_5H_5)_2]^+$, AgO is not Ag^{II} but $Ag^IAg^{III}O_2$
Ag^{III}	d^8	rare $[AgF_4]^-$, $[AgF_6]^{3-}$

• PHYSICAL DATA

Melting point/K: 1235.08
Boiling point/K: 2485

ΔH_{fusion}/kJ mol^{-1}: 11.3
ΔH_{vap}/kJ mol^{-1}: 255.1

Thermodynamic properties (298.15 K, 0.1 MPa)

State	$\Delta_f H^\ominus$/kJ mol^{-1}	$\Delta_f G^\ominus$/kJ mol^{-1}	$S^\ominus$/J K^{-1} mol^{-1}	C_p/J K^{-1} mol^{-1}
Solid	0	0	42.55	25.351
Gas	284.55	245.65	172.997	20.786

Density/kg m^{-3}: 10 500 [293 K]; 9345 [liquid at m.p.]
Molar volume/cm^3: 10.27
Thermal conductivity/W m^{-1} K^{-1}: 429 [300 K]
Coefficient of linear thermal expansion/K^{-1}: 19.2×10^{-6}
Electrical resistivity/Ω m: 1.59×10^{-8} [293 K]
Mass magnetic susceptibility/kg^{-1} m^3: -2.27×10^{-9} (s)

Young's modulus/GPa: 82.7
Rigidity modulus/GPa: 30.3
Bulk modulus/GPa: 103.6
Poisson's ratio/GPa: 0.367

• BIOLOGICAL DATA

Biological role

None; especially toxic to lower organisms.

Toxicity

Toxic intake: soluble salts, ingestion, human = 1 g
Lethal intake: LD_{50} (nitrate, oral, mouse) = 50 mg kg^{-1}

Hazards

Soluble silver salts irritate the skin and mucous membranes and can cause death if ingested even in small doses. Silver is a suspected carcinogen.

Levels in humans

Blood/mg dm^{-3}: < 0.003
Bone/p.p.m.: 0.01 – 0.44
Liver/p.p.m.: 0.005 – 0.25
Muscle/p.p.m.: 0.009 – 0.28
Daily dietary intake: 0.0014 – 0.08 mg
Total mass of element in average (70 kg) person: 2 mg

Known to ancient civilizations.
[Anglo-Saxon, *siolfur* = silver; Latin, *argentum*]
French, *argent*; German, *Silber*; Italian, *argento*; Spanish, *plata*

Silver

[sil-ver]

• NUCLEAR DATA

Number of isotopes (including nuclear isomers): 46 **Isotope mass range:** 96 → 122

Key isotopes

Nuclide	Atomic mass	Natural abundance (%)	Nuclear spin *I*	Nuclear magnetic moment *μ*	Uses
^{107}Ag	106.905 092	51.839	1/2–	–0.113 570	E, NMR
^{109}Ag	108.904 756	48.161	1/2–	–0.130 690 5	E, NMR

A table of radioactive isotopes is given in Appendix 1, on p249.

NMR [Reference: Ag^+ (aq)]	^{107}Ag	^{109}Ag
Relative sensitivity (^1H = 1.00)	6.62×10^{-5}	1.01×10^{-4}
Receptivity (^{13}C = 1.00)	0.195	0.276
Magnetogyric ratio/rad $T^{-1}s^{-1}$	-1.0828×10^7	-1.2448×10^7
Nuclear quadrupole moment/m^2	–	–
Frequency (^1H = 100 Hz; 2.3488T)/MHz	4.046	4.652

• ELECTRON SHELL DATA

Ground state electron configuration: [Kr]$4d^{10}$ $5s^1$
Term symbol: $^2S_{1/2}$
Electron affinity ($M \rightarrow M^-$)/kJ mol^{-1}**:** 125.7

Ionization energies/kJ mol^{-1}**:**

1.	$M \rightarrow M^+$	731.0
2.	$M^+ \rightarrow M^{2+}$	2073
3.	$M^{2+} \rightarrow M^{3+}$	3361
4.	$M^{3+} \rightarrow M^{4+}$	(5000)
5.	$M^{4+} \rightarrow M^{5+}$	(6700)
6.	$M^{5+} \rightarrow M^{6+}$	(8600)
7.	$M^{6+} \rightarrow M^{7+}$	(11 200)
8.	$M^{7+} \rightarrow M^{8+}$	(13 400)
9.	$M^{8+} \rightarrow M^{9+}$	(15 600)
10.	$M^{9+} \rightarrow M^{10+}$	(18 000)

Electron binding energies/eV

K	1s	25 514
L_I	2s	3806
L_{II}	$2p_{1/2}$	3524
L_{III}	$2p_{3/2}$	3351
M_I	3s	719.0
M_{II}	$3p_{1/2}$	603.8
M_{III}	$3p_{3/2}$	573.0
M_{IV}	$3d_{3/2}$	374.0
M_V	$3d_{5/2}$	368.0

continued in Appendix 2, p257

Main lines in atomic spectrum
[Wavelength/nm(species)]
328.068 (I) (AA)
338.289 (I)
520.908 (I)
546.550 (I)
827.352 (I)

• CRYSTAL DATA

Crystal structure (cell dimensions/pm), space group
f.c.c. (*a* = 408.626), Fm3m

X-ray diffraction: mass absorption coefficients (μ/ρ)/cm^2 g^{-1}**:** CuK_α 218 MoK_α 25.8
Neutron scattering length, $b/10^{-12}$ cm**:** 0.597
Thermal neutron capture cross-section, σ_a/barns**:** 63.6

• GEOLOGICAL DATA

Minerals

Native silver occurs naturally as crystals, but more generally as a compact mass, and there are small deposits in Norway, Germany, Mexico, Chile, Canada, Australia, Sardinia and the USA.

Mineral	Formula	Density	Hardness	Crystal appearance
Acanthite	Ag_2S	7.2	2 – 2.5	mon., met. black - lead grey
Chlorargyrite	AgCl	5.556	2.5	cub., res./adam. col./grey
Polybasite	$(Cu,Ag)_{16}Sb_2S_{11}$	6.1	2 – 3	mon., met. black
Stephanite	Ag_5SbS_4	6.26	2 – 2.5	orth., met. black

Chief ores: acanthite, stephanite
World production/tonnes y^{-1}**:** 9950
Main mining areas: acanthite in Mexico, Bolivia, Honduras, stephanite in Canada. Silver is obtained as a by-product in the refining of other metals such as copper.
Reserves/tonnes: *c.* 1×10^6
Specimen: available as crystals, flake, foil, granules, powder, rod, wire or wool. Safe.

Abundances

Sun (relative to H = 1×10^{12})**:** 7.1
Earth's crust/p.p.m.**:** 0.07
Seawater/p.p.m.**:**
Atlantic surface: n.a.
Atlantic deep: n.a.
Pacific surface: 1×10^{-7}
Pacific deep: 24×10^{-7}
Residence time/years**:** 5000
Classification: recycled
Oxidation state: I

Na

Atomic number: 11
Relative atomic mass ($^{12}C = 12.0000$)**:** 22.989768

CAS: [7440-23-5]

• CHEMICAL DATA

Description: Sodium is a soft, silvery-white, metal which oxidises rapidly when cut, and reacts vigorously with water. It is produced in large quantities by the electrolysis of molten sodium chloride. Sodium metal is used in industry in the manufacture of other chemicals and metals. It is also used in heat exchangers for nuclear reactors.

Radii/pm: Na^+ 98; atomic 154; van der Waals 231
Electronegativity: 0.93 (Pauling); 1.01 (Allred); 2.85 eV (absolute)
Effective nuclear charge: 2.20 (Slater); 2.51 (Clementi); 3.21 (Froese-Fischer)

Standard reduction potentials $E^\ominus$/V

I		0
Na^+	$\xrightarrow{-2.713}$	Na

Oxidation states

Na^{-I}	s^2	solution in liquid ammonia
Na^{I}	[Ne]	Na_2O, Na_2O_2 (peroxide), NaOH, NaH, NaF, NaCl etc., $[Na(OH_2)_4]^+$ (aq), $NaHCO_3$,Na_2CO_3, Na^+ salts, some complexes, e.g. $[Na(15\text{-crown-}5)]^+$

• PHYSICAL DATA

Melting point/K: 370.96
Boiling point/K: 1156.1

ΔH_{fusion}/kJ mol^{-1}: 2.64
ΔH_{vap}/kJ mol^{-1}: 89.04

Thermodynamic properties (298.15 K, 0.1 MPa)

State	$\Delta_f H^\ominus$/kJ mol^{-1}	$\Delta_f G^\ominus$/kJ mol^{-1}	$S^\ominus$/J K^{-1} mol^{-1}	C_p/J K^{-1} mol^{-1}
Solid	0	0	51.21	28.24
Gas	107.32	76.761	153.712	20.786

Density/kg m^{-3}: 971 [293 K]; 928 [liquid at m.p.]
Molar volume/cm^3: 23.68
Thermal conductivity/W m^{-1} K^{-1}: 141 [300 K]
Coefficient of linear thermal expansion/K^{-1}: 70.6×10^{-6}
Electrical resistivity/Ω m: 4.2×10^{-8} [273 K]
Mass magnetic susceptibility/kg^{-1} m^3: $+8.8 \times 10^{-9}$ (s)

Young's modulus/GPa: 6.80
Rigidity modulus/GPa: 2.53
Bulk modulus/GPa: n.a.
Poisson's ratio/GPa: 0.34

• BIOLOGICAL DATA

Biological role

Essential to most species including humans.

Toxicity

Toxic intake: chloride, oral, human = 12 g kg^{-1}
Lethal intake: LD_{50} (chloride, oral, rat) = 3000 mg kg^{-1}

Hazards

Sodium compounds are not hazardous insofar as their sodium content is concerned, but excess sodium chloride can be toxic by ingestion. A daily intake in excess of the necessary 2–3 g is not advisable for those people suffering heart disease.

Levels in humans

Blood/mg dm^{-3}: 1970
Bone/p.p.m.: 10 000
Liver/p.p.m.: 2000 – 4000
Muscle/p.p.m.: 2600 – 7800
Daily dietary intake: 2 – 15 g
Total mass of element in average (70 kg) person: 100 g

Isolated by Sir Humphry Davy in 1807 at the Royal Institution, London, England.
[English, *soda*; Latin, *natrium*]
French, *sodium*; German, *Natrium*; Italian, *sodio*; Spanish, *sodio*

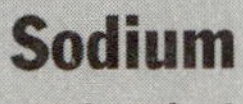

Sodium

[so-dee-uhm]

• NUCLEAR DATA

Number of isotopes (including nuclear isomers): 14 **Isotope mass range:** $19 \rightarrow 31$

Key isotopes

Nuclide	Atomic mass	Natural abundance (%)	Nuclear spin I	Nuclear magnetic moment μ	Uses
^{23}Na	22.989 767	100	3/2+	+2.217 520	NMR

A table of radioactive isotopes is given in Appendix 1, on p249.

NMR [Reference: NaCl (aq)]	^{23}Na
Relative sensitivity (^{1}H = 1.00)	0.0925
Receptivity (^{13}C = 1.00)	525
Magnetogyric ratio/rad $T^{-1} s^{-1}$	7.0761×10^7
Nuclear quadrupole moment/m^2	0.1089×10^{-28}
Frequency (^{1}H = 100 Hz; 2.3488T)/MHz	26.451

• ELECTRON SHELL DATA

Ground state electron configuration: [Ne]$3s^1$
Term symbol: $^2S_{1/2}$
Electron affinity ($M \rightarrow M^-$)/kJ mol^{-1}**:** 52.9

Ionization energies/kJ mol^{-1}**:**

1.	$M \rightarrow M^+$	495.8
2.	$M^+ \rightarrow M^{2+}$	4562.4
3.	$M^{2+} \rightarrow M^{3+}$	6912
4.	$M^{3+} \rightarrow M^{4+}$	9543
5.	$M^{4+} \rightarrow M^{5+}$	13 353
6.	$M^{5+} \rightarrow M^{6+}$	16 610
7.	$M^{6+} \rightarrow M^{7+}$	20 114
8.	$M^{7+} \rightarrow M^{8+}$	25 490
9.	$M^{8+} \rightarrow M^{9+}$	28 933
10.	$M^{9+} \rightarrow M^{10+}$	141 360

Electron binding energies/eV

K	1s	1070.8
L_I	2s	63.5
L_{II}	$2p_{1/2}$	30.4
L_{III}	$2p_{3/2}$	30.5

Main lines in atomic spectrum
[Wavelength/nm(species)]

313.548 (II)
588.995 (I) (AA)
589.592 (I)
818.326 (I)
819.482 (I)

• CRYSTAL DATA

Crystal structure (cell dimensions/pm), space group

α-Na hexagonal (a = 376.7, c = 615.4), $P6_3/mmc$
β-Na b.c.c. (a = 429.06), Im3m
T(b.c.c. $\rightarrow$ hexagonal) = 5 K

X-ray diffraction: mass absorption coefficients $(\mu/\rho)/cm^2\ g^{-1}$**:** CuK_α 30.1 MoK_α 3.21
Neutron scattering length, $b/10^{-12}$ cm**:** 0.358
Thermal neutron capture cross-section, σ_a/barns**:** 0.530

• GEOLOGICAL DATA

Minerals

Sodium occurs in many minerals but these are not mined as a source of sodium compounds.

Mineral	Formula	Density	Hardness	Crystal appearance
Halite (rock salt)	NaCl	2.168	2	cub., vit. usually colourless
Trona	$Na_3(CO_3)(HCO_3).2H_2O$	2.14	2.5 – 3	mon., vit. colourless

Chief ores: halite, trona

World production/tonnes y^{-1}**:** *c.* 200 000 (sodium metal); 168×10^6 (salt); 29×10^6 (sodium carbonate)

Main mining areas: halite in Germany, Poland, USA, UK; trona in Kenya, USA

Reserves/tonnes**:** almost unlimited

Specimen: available as ingots or lumps, in sealed ampoules under nitrogen, or spheres and sticks stored under mineral oil. *Warning!*

Abundances

Sun (relative to H = 1×10^{12})**:** 1.91×10^6
Earth's crust/p.p.m.**:** 23 000
Seawater/p.p.m.**:** 10 500
Residence time/years**:** 1×10^8
Classification: accumulating
Oxidation state: I

Sr

Atomic number: 38
Relative atomic mass ($^{12}C = 12.0000$)**:** 87.62

CAS: [7440-24-6]

• CHEMICAL DATA

Discovery: Strontium was recognized as an element in 1790 by A. Crawford at Edinburgh, Scotland. Isolated in 1808 by Sir Humphry Davy at London, England.
Description: Strontium is a silvery-white, relatively soft metal that is obtained by heating strontium oxide (SrO) with aluminium metal. The bulk metal is protected by an oxide film, but it will burn in air if ignited, and is attacked by water. Strontium is used in special glass for televisions and VDUs, and the red colour of fireworks and flares is produced by strontium salts.

Radii/pm: Sr^{2+} 127; atomic 215 (α-form); covalent 192
Electronegativity: 0.95 (Pauling); 0.99 (Allred); 2.0 eV (absolute)
Effective nuclear charge: 2.85 (Slater); 6.07 (Clementi); 8.09 (Froese-Fischer)

Standard reduction potentials $E^{\ominus}$/V

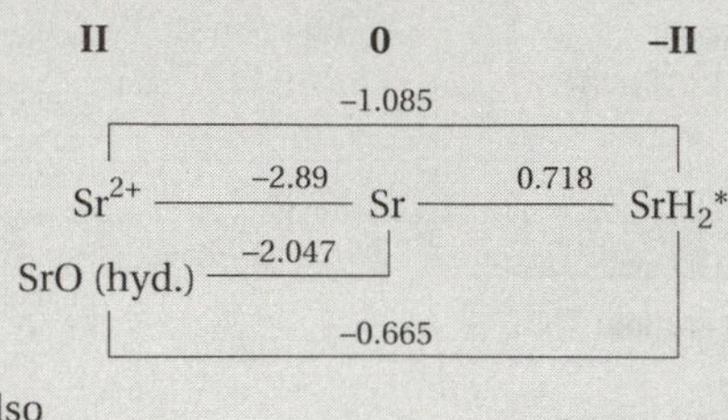

Also

SrO_2* —2.333— Sr^{2+}

SrO_2 —1.492— SrO (hyd.)

* see oxidation states

Oxidation states

Sr^{II} [Kr] SrO, SrO_2 (peroxide), $Sr(OH)_2$, $[Sr(OH_2)_x]^{2+}$ (aq), Sr^{2+} salts, SrF_2, $SrCl_2$ etc., $SrCO_3$, some complexes. SrH_2 is $Sr^{2+}2H^-$

• PHYSICAL DATA

Melting point/K: 1042
Boiling point/K: 1657

ΔH_{fusion}/kJ mol^{-1}: 9.16
ΔH_{vap}/kJ mol^{-1}: 138.91

Thermodynamic properties (298.15 K, 0.1 MPa)

State	$\Delta_f H^{\ominus}$/kJ mol^{-1}	$\Delta_f G^{\ominus}$/kJ mol^{-1}	$S^{\ominus}$/J K^{-1} mol^{-1}	C_p/J K^{-1} mol^{-1}
Solid	0	0	52.3	26.4
Gas	164.4	130.9	164.62	20.786

Density/kg m^{-3}**:** 2540 [293 K]; 2375 [liquid at m.p.]
Molar volume/cm^3**:** 34.50
Thermal conductivity/W m^{-1} K^{-1}**:** 35.3 [300 K]
Coefficient of linear thermal expansion/K^{-1}**:** 23×10^{-6}
Electrical resistivity/Ω m**:** 23.0×10^{-8} [293 K]
Mass magnetic susceptibility/kg^{-1} m^3**:** $+1.32 \times 10^{-8}$ (s)

Young's modulus/GPa**:** 15.7
Rigidity modulus/GPa**:** 6.03
Bulk modulus/GPa**:** 12.0
Poisson's ratio/GPa**:** 0.28

• BIOLOGICAL DATA

Biological role

None.

Toxicity

Toxic intake: not regarded as toxic
Lethal intake: LD_{50} (chloride, oral, rat) = 2250 mg kg^{-1}

Hazards

Strontium resembles calcium in metabolism and behaviour and is absorbed by the body and stored in the skeleton. This also happens with radioactive ^{90}Sr which was produced by above-ground nuclear explosions in the 1950s and is widely disseminated in the environment.

Levels in humans

Blood/mg dm^{-3}**:** 0.031
Bone/p.p.m.**:** 36 – 140
Liver/p.p.m.**:** 0.05 – 0.36
Muscle/p.p.m.**:** 0.12 – 0.35
Daily dietary intake: 0.8 – 5 mg
Total mass of element in average (70 kg) person: 320 mg

Discovery: see Chemical Data section.
[Named after Strontian, Scotland]
French, *strontium*; German, *Strontium*; Italian, *stronzio*; Spanish, *estroncio*

Strontium

[stron-tee-uhm]

• NUCLEAR DATA

Number of isotopes (including nuclear isomers): 23 **Isotope mass range:** 79 → 98

Key isotopes

Nuclide	Atomic mass	Natural abundance (%)	Nuclear spin I	Nuclear magnetic moment μ	Uses
^{84}Sr	83.913 430	0.56	0+		E
^{86}Sr	85.909 267	9.86	0+		
^{87}Sr	86.908 884	7.00	9/2+	–1.092 83	E, NMR
^{88}Sr	87.905 618	82.58	0+		

A table of radioactive isotopes is given in Appendix 1, on p249.

NMR [Reference: Sr^{2+} (aq)]	^{87}Sr
Relative sensitivity (^1H = 1.00)	2.69×10^{-3}
Receptivity (^{13}C = 1.00)	1.07
Magnetogyric ratio/rad $T^{-1}s^{-1}$	-1.1593×10^{7}
Nuclear quadrupole moment/m^2	0.335×10^{-28}
Frequency (^1H = 100 Hz; 2.3488T)/MHz	4.333

• ELECTRON SHELL DATA

Ground state electron configuration: $[Kr]5s^2$
Term symbol: $^1S_0$
Electron affinity ($M \rightarrow M^-$)/kJ mol^{-1}**:** –146

Ionization energies/kJ mol^{-1}**:**

1.	$M \rightarrow M^+$	549.5
2.	$M^+ \rightarrow M^{2+}$	1064.2
3.	$M^{2+} \rightarrow M^{3+}$	4210
4.	$M^{3+} \rightarrow M^{4+}$	5500
5.	$M^{4+} \rightarrow M^{5+}$	6910
6.	$M^{5+} \rightarrow M^{6+}$	8760
7.	$M^{6+} \rightarrow M^{7+}$	10 200
8.	$M^{7+} \rightarrow M^{8+}$	11 800
9.	$M^{8+} \rightarrow M^{9+}$	15 600
10.	$M^{9+} \rightarrow M^{10+}$	17 100

Electron binding energies/eV

K	1s	16 105
L_I	2s	2216
L_{II}	$2p_{1/2}$	2007
L_{III}	$2p_{3/2}$	1940
M_I	3s	358.7
M_{II}	$3p_{1/2}$	280.3
M_{III}	$3p_{3/2}$	270.0
M_{IV}	$3d_{3/2}$	136.0
M_V	$3d_{5/2}$	134.2

continued in Appendix 2, p257

Main lines in atomic spectrum
[Wavelength/nm(species)]

407.771 (II)
421.552 (II)
460.733 (I) (AA)
496.226 (I)
548.084 (I)
640.847 (I)

• CRYSTAL DATA

Crystal structure (cell dimensions/pm), space group

α-Sr f.c.c. (a = 608.49), Fm3m
β-Sr h.c.p. (a = 432, c = 706), $P6_3/mmc$
γ-Sr b.c.c. (a = 485), Im3m
$T(\alpha \rightarrow \beta)$ = 506 K; $T(\beta \rightarrow \gamma)$ = 813 K

X-ray diffraction: mass absorption coefficients $(\mu/\rho)/cm^2\ g^{-1}$**:** CuK_α 125 MoK_α 95.0
Neutron scattering length, $b/10^{-12}$ cm: 0.702
Thermal neutron capture cross-section, σ_a/barns**:** 1.28

• GEOLOGICAL DATA

Minerals

Mineral	Formula	Density	Hardness	Crystal appearance
Celestite*	$SrSO_4$	3.97	3 – 3.5	orth., vit./colourless-pale blue
Strontianite	$SrCO_3$	3.76	3.5	orth., vit./resinous colourless

*Also known as celestine.

Chief ores: celestite, strontianite

World production/tonnes y^{-1}**:** 137 000 (strontium ores)

Main mining areas: UK, Tunisia, Russia, Germany, Mexico, USA

Reserves/tonnes: n.a.

Specimen: available as granules and pieces. *Warning!*

Abundances

Sun (relative to H = 1×10^{12})**:** 790
Earth's crust/p.p.m.**:** 370
Seawater/p.p.m.**:**
Atlantic surface: 7.6
Atlantic deep: 7.7
Pacific surface: 7.6
Pacific deep: 7.7
Residence time/years**:** 4×10^6
Classification: recycled
Oxidation state: II

S

Atomic number: 16
Relative atomic mass ($^{12}C = 12.0000$)**:** 32.066

CAS: [7704-34-9]

• CHEMICAL DATA

Description: There are several forms of sulfur, of which the yellow orthorhombic (S_8) is the most common. Sulfur is stable to air and water, but burns if heated. It is attacked by oxidising acids. It is a key industrial chemical and is the starting point for sulfuric acid.

Radii/pm: S^{6+} 29; S^{4+} 37; S^{2-} 184; atomic 104; covalent 104; van der Waals 185

Electronegativity: 2.58 (Pauling); 2.44 (Allred); 6.22 eV (absolute)

Effective nuclear charge: 5.45 (Slater); 5.48 (Clementi); 6.04 (Froese-Fischer)

Standard reduction potentials $E^{\ominus}$/V

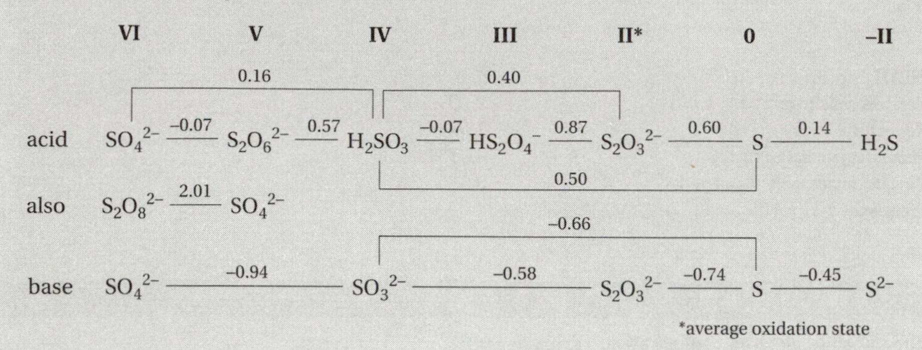

Oxidation states

S^{-II}	[Ar]	H_2S, S^{2-}, polysulfides S_n^{2-}
S^{-I}	s^2p^5	H_2S_2, etc., polysulfides S_n^{2-}
S^0	s^2p^4	S_6, S_8, etc., polysulfides S_n^{2-}
S^I	s^2p^3	S_2O?, S_2F_2, S_2Cl_2
S^{II}	s^2p^2	SF_2, SCl_2
S^{III}	s^2p^1	$Na_2S_2O_4$
S^{IV}	s^2	SO_2, SO_3^{2-} (aq), SF_4, SCl_4, $SOCl_2$
S^V	s^1	$Na_2S_2O_6$, S_2F_{10}
S^{VI}	[Ne]	SO_3, H_2SO_4, SO_4^{2-} (aq), etc., SF_6, HSO_3F, SO_2Cl_2

Covalent bonds

Bond	r/ pm	E/ kJ mol^{-1}
S–H	134	363
S–C	182	272
S=C	160	573
S–O	150	265
S=O	143	532
S–F	156	284
S–Cl	207	255
S–S	205	226

• PHYSICAL DATA

Melting point/K: 386.0 (α); 392.2 (β); 380.0 (γ)

Boiling point/K: 717.824

Critical temperature/K: 1314

Critical pressure/kPa: 20700

ΔH_{fusion}/kJ mol^{-1}: 1.23

ΔH_{vap}/kJ mol^{-1}: 9.62

Thermodynamic properties (298.15 K, 0.1 MPa)

State	$\Delta_f H^{\ominus}$/kJ mol^{-1}	$\Delta_f G^{\ominus}$/kJ mol^{-1}	$S^{\ominus}$/J K^{-1} mol^{-1}	C_p/J K^{-1} mol^{-1}
Solid (α)	0	0	31.80	22.64
Solid (β)	0.33	n.a.	n.a.	n.a.
Gas	278.805	238.250	167.821	23.673

Density/kg m^{-3}: 2070 (α), 1957 (β) [293 K]; 1819 [liquid at 393 K]

Molar volume/cm^3: 15.49

Thermal conductivity/W m^{-1} K^{-1}: 0.269 (α) [300 K]

Coefficient of linear thermal expansion/K^{-1}: 74.33×10^{-6}

Electrical resistivity/Ω m: 2×10^{15} [293 K]

Mass magnetic susceptibility/kg^{-1} m^3: -6.09×10^{-9} (α); -5.83×10^{-9} (β)

• BIOLOGICAL DATA

Biological role

Essential to all living things; part of the amino acids methionine and cysteine.

Toxicity

Elemental sulfur is not very toxic, but simple derivatives (SO_2, H_2S, etc.) are.

Toxic intake: n.a.

Lethal intake: for rabbits, as little as 175 mg kg^{-1} has proved fatal.

Hazards

Elemental sulfur appears to be relatively harmless unless ingested; ignited it emits highly toxic fumes of SO_2. Sulfur dust is a human eye irritant.

Levels in humans

Blood/mg dm^{-3}: 1800

Bone/p.p.m.: 500– 2400

Liver/p.p.m.: 7000– 12 000

Muscle/p.p.m.: 5000 – 11 000

Daily dietary intake: 850 – 930 mg

Total mass of element in average (70 kg) person: 140 g

Known to ancient civilizations.
[Sanskrit, *sulvere* = sulfur; Latin, *sulphurium*]
French, *soufre*; German, *Schwefel*; Italian, *solfo*; Spanish, *azufre*

Sulfur

[sul-fer]

• NUCLEAR DATA

Number of isotopes (including nuclear isomers): 11 **Isotope mass range:** 29 → 39

Key isotopes

Nuclide	Atomic mass	Natural abundance (%)	Nuclear spin *I*	Nuclear magnetic moment *μ*	Uses
^{32}S	31.972 070 70	95.02	0+	0	
^{33}S	32.971 458 43	0.75	3/2+	+0.643821	NMR
^{34}S	33.967 866 65	4.21	0+	0	
^{36}S	35.967 080 62	0.02	0+	0	

A table of radioactive isotopes is given in Appendix 1, on p250.

NMR [Reference: CS_2]

	^{32}S	^{35}S
Relative sensitivity (^1H = 1.00)	2.26×10^{-3}	
Receptivity (^{13}C = 1.00)	0.0973	
Magnetogyric ratio/rad $T^{-1} s^{-1}$	2.0534×10^7	
Nuclear quadrupole moment/m^2	-0.678×10^{-28}	0.0471×10^{-28}
Frequency (^1H = 100 Hz; 2.3488T)/MHz	24.664	

• ELECTRON SHELL DATA

Ground state electron configuration: $[Ne]3s^23p^4$
Term symbol: $^3P_2$
Electron affinity ($M \rightarrow M^-$)/kJ mol^{-1}**:** 200.4

Ionization energies/kJ mol^{-1}**:**

1.	$M \rightarrow M^+$	999.6
2.	$M^+ \rightarrow M^{2+}$	2251
3.	$M^{2+} \rightarrow M^{3+}$	3361
4.	$M^{3+} \rightarrow M^{4+}$	4564
5.	$M^{4+} \rightarrow M^{5+}$	7013
6.	$M^{5+} \rightarrow M^{6+}$	8495
7.	$M^{6+} \rightarrow M^{7+}$	27 106
8.	$M^{7+} \rightarrow M^{8+}$	31 669
9.	$M^{8+} \rightarrow M^{9+}$	36 578
10.	$M^{9+} \rightarrow M^{10+}$	43 138

Electron binding energies/eV

K	1s	2472
L_I	2s	230.9
L_{II}	$2p_{1/2}$	163.6
L_{III}	$2p_{3/2}$	162.5

Main lines in atomic spectrum
[Wavelength/nm(species)]

545.38 (II)
547.36 (II)
550.97 (II)
560.61 (II)
565.99 (II)
792.40 (I)
964.99 (I)

• CRYSTAL DATA

Crystal structure (cell dimensions/pm), space group

α-S_8 orthorhombic (a = 1046.46, b = 1286.60, c = 2448.60), Fddd
β-S_8 monoclinic (a = 1102, b = 1096, c = 1090, β = 96.7°), $P2_1/c$
γ-S_8 monoclinic (a = 857, b = 1305, c = 823, β = 112° 54'), P2/c
ε-S_6 rhombohedral (a = 646, α = 115° 18'), $R\bar{3}$
In addition to the above ring forms there are also S_7, S_{9-12}, S_{18} and S_{20} rings.
Plastic sulfur is long chains of S_n also known in several forms χ, ψ, φ, μ and ω.
$T(\alpha \rightarrow \beta)$ = 366.7 K

X-ray diffraction: mass absorption coefficients (μ/ρ)/$cm^2 g^{-1}$**:** CuK_α 89.1 MoK_α 9.55
Neutron scattering length, $b/10^{-12}$ cm**:** 0.2847
Thermal neutron capture cross-section, σ_a/barns**:** 0.53

• GEOLOGICAL DATA

Minerals

Native sulfur occurs naturally as deposits associated with oil-bearing strata, as in Texas and Louisiana in the USA, and with gypsum ($CaSO_4.2H_2O$) deposits in Sicily and Italy. Many sulfide and sulfate minerals are known. For sulfides consult antimony (stibnite), lead (galena), mercury (cinnabar), zinc (spharelite), etc. For sulfates see barium (barite), calcium (anhydrite, gypsum), magnesium (epsomite, kieserite), strontium (celestite), etc. The table below shows only those which are used as a source of sulfur.

Mineral	Formula	Density	Hardness	Crystal appearance
Marcasite	FeS_2	4.887	6 – 6.5	orth., met. pale yellow
Native sulfur	S_8	2.07	1.5 – 2.5	orth., yellow pyramidal
Pyrite	FeS_2	5.018	6 – 6.5	cub., met. dark yellow

Chief ores: native sulfur, pyrite; a lot of sulfur is recovered from the H_2S of natural gas.
World production/tonnes y^{-1}**:** 54×10^6
Main mining areas: USA (native sulfur), Spain.
Reserves/tonnes: 2.5×10^9
Specimen: available as powder and flake. Safe.

Abundances

Sun (relative to H = 1×10^{12})**:** 1.6×10^7
Earth's crust/p.p.m.**:** 260
Seawater/p.p.m.**:** 870
Residence time/years**:** 8×10^6
Classification: accumulating
Oxidation state: VI

Ta

Atomic number: 73
Relative atomic mass ($^{12}C = 12.0000$): **180.9479**

CAS: [7440-25-7]

• CHEMICAL DATA

Description: Tantalum is a shiny, silvery metal which is soft when pure. It is very resistant to corrosion due to an oxide film on its surface, but it is attacked by HF and molten alkalis. Tantalum is used in electronics, cutting tools, chemical plants and surgery.

Radii/pm: Ta^{5+} 64; Ta^{4+} 68; Ta^{3+} 72; atomic 143; covalent 134
Electronegativity: 1.5 (Pauling); 1.33 (Allred); 4.11 eV (absolute)
Effective nuclear charge: 3.30 (Slater); 9.53 (Clementi); 13.78 (Froese-Fischer)

Standard reduction potentials $E^{\ominus}$/V

V		0
Ta_2O_5	$\xrightarrow{-0.81}$	Ta

Oxidation states

Ta^{-III}	d^8	$[Ta(CO)_5]^{3-}$
Ta^{-I}	d^6	$[Ta(CO)_6]^-$
Ta^{I}	d^4	$[Ta(CO)_4(\eta\text{-}C_5H_5)]$
Ta^{II}	d^3	TaO ?
Ta^{III}	d^2	TaF_3, $TaCl_3$, $TaBr_3$
Ta^{IV}	d^1	TaO_2, $TaCl_4$, $TaBr_4$, TaI_4
Ta^{V}	$d^0[f^{14}]$	Ta_2O_5, $[Ta_6O_{19}]^{8-}$ (aq), TaF_5, $TaCl_5$, $[TaF_6]^-$, $[TaF_7]^{2-}$, TaF_3O, $TaFO_2$

• PHYSICAL DATA

Melting point/K: 3269
Boiling point/K: 5698 ± 100
ΔH_{fusion}/kJ mol^{-1}: 31.4
ΔH_{vap}/kJ mol^{-1}: 753.1

Thermodynamic properties (298.15 K, 0.1 MPa)

State	$\Delta_f H^{\ominus}$/kJ mol^{-1}	$\Delta_f G^{\ominus}$/kJ mol^{-1}	$S^{\ominus}$/J K^{-1} mol^{-1}	C_p/J K^{-1} mol^{-1}
Solid	0	0	41.51	25.36
Gas	782.0	739.3	185.214	20.857

Density/kg m^{-3}: 16 654 [293 K]; 15 000 [liquid at m.p.]
Molar volume/cm^3: 10.87
Thermal conductivity/W m^{-1} K^{-1}: 57.5 [300 K]
Coefficient of linear thermal expansion/K^{-1}: 6.6×10^{-6}
Electrical resistivity/Ω m: 12.45×10^{-8} [298 K]
Mass magnetic susceptibility/kg^{-1} m^3: $+1.07 \times 10^{-8}$ (s)

Young's modulus/GPa: 185.7
Rigidity modulus/GPa: 69.2
Bulk modulus/GPa: 196.3
Poisson's ratio/GPa: 0.342

• BIOLOGICAL DATA

Biological role

None.

Toxicity

Toxic intake: moderately poisonous by ingestion
Lethal intake: LD_{50} (chloride, oral, rat) = 1900 mg kg^{-1}

Hazards

There are no cases of industrial poisoning caused by tantalum or its compounds. However, it is an experimental tumorigen.

Levels in humans

Blood/mg dm^{-3}: n.a., but low
Bone/p.p.m.: *c.* 0.03
Liver/p.p.m.: n.a.
Muscle/p.p.m.: n.a., but low
Daily dietary intake: 0.001 mg
Total mass of element in average (70 kg) person: *c.* 0.2 mg

Discovered in 1802 by A.G. Ekeberg at Uppsala, Sweden.
[Greek, *Tantalos* = father of Niobe of Greek mythology]
French, *tantale*; German, *Tantal*; Italian, *tantalio*; Spanish, *tántalo*

Tantalum

[tan-ta-lum]

• NUCLEAR DATA

Number of isotopes (including nuclear isomers): 28 **Isotope mass range:** 159 → 186

Key isotopes

Nuclide	Atomic mass	Natural abundance (%)	Nuclear spin I	Nuclear magnetic moment μ	Uses
^{180}Ta*	179.947 462	0.012	9–		E
^{181}Ta	180.947 992	99.988	7/2+	+2.371	E, NMR

*^{180}Ta is radioactive with a half-life of >1×10^{13} y.
A table of radioactive isotopes is given in Appendix 1, on p250.

NMR [Reference: $[TaF_6]^-$]	^{181}Ta
Relative sensitivity (^{1}H = 1.00)	0.0260
Receptivity (^{13}C = 1.00)	204
Magnetogyric ratio/rad $T^{-1}s^{-1}$	3.2073×10^7
Nuclear quadrupole moment/m^2	3.170×10^{-28}
Frequency (^{1}H = 100 Hz; 2.3488T)/MHz	11.970

• ELECTRON SHELL DATA

Ground state electron configuration: [Xe]$4f^{14}5d^36s^2$
Term symbol: $^4F_{3/2}$
Electron affinity ($M \rightarrow M^-$)/kJ mol^{-1}**:** 14

Ionization energies/kJ mol^{-1}**:**

1.	$M \rightarrow M^+$	761
2.	$M^+ \rightarrow M^{2+}$	(1500)
3.	$M^{2+} \rightarrow M^{3+}$	(2100)
4.	$M^{3+} \rightarrow M^{4+}$	(3200)
5.	$M^{4+} \rightarrow M^{5+}$	(4300)

Electron binding energies/eV

K	1s	67 416
L_I	2s	11 682
L_{II}	$2p_{1/2}$	11 136
L_{III}	$2p_{3/2}$	9881
M_I	3s	2708
M_{II}	$3p_{1/2}$	2469
M_{III}	$3p_{3/2}$	2194
M_{IV}	$3d_{3/2}$	1793
M_V	$3d_{5/2}$	1735

continued in Appendix 2, p257

Main lines in atomic spectrum
[Wavelength/nm(species)]

240.063 (II)
264.747 (I)
265.327 (I)
271.467 (I) (AA)
285.098 (I)
301.254 (II)

• CRYSTAL DATA

Crystal structure (cell dimensions/pm), space group
b.c.c. (a = 330.29), Im3m

X-ray diffraction: mass absorption coefficients (μ/ρ)/cm^2 g^{-1}**:** CuK$_\alpha$ 166 MoK$_\alpha$ 95.4
Neutron scattering length, b/10^{-12} cm**:** 0.691
Thermal neutron capture cross-section, σ_a/barns**:** 20.6

• GEOLOGICAL DATA

Minerals

Mineral	Formula	Density	Hardness	Crystal appearance
Columbite	$(Fe,Mn)(Ta,Nb)_2O_6$	a group of ores of mixed composition		
Ferrotantalite	$FeTa_2O_6$	7.95	6 – 6.5	orth., black-brownish/black
Manganotantalite	$(Fe,Mn)(Ta,Nb)_2O_6$	6.76	6 – 6.5	orth., black-brownish/black
Microlite	$(Na,Ca)_2Ta_2O_6(O,OH,F)$	6.42	5 – 5.5	cub., vitreous/resinous yellow
Samarskite	$(Y,Ce,U,Fe)_3(Nb,Ta,Ti)_5O_{16}$	5.69	5 – 6	orth., vitreous/resinous black

Chief ores: columbite, samarskite
World production/tonnes y^{-1}**:** 840
Main mining areas: Australia, Zaire, Brazil, Russia, Norway, Canada, Madagascar; mostly obtained as a by-product of tin extraction.
Reserves/tonnes**:** n.a.
Specimen: available as foil, powder, rod or wire. Safe.

Abundances

Sun (relative to H = 1×10^{12})**:** n.a.
Earth's crust/p.p.m.**:** 2
Seawater/p.p.m.**:** 2×10^{-6}
Residence time/years: n.a.
Oxidation state: V

Tc

Atomic number: 43
Relative atomic mass ($^{12}C = 12.0000$): **98.9063 (Tc-99)**

CAS: [7440-26-8]

• CHEMICAL DATA

Description: Technetium is a radioactive metal which does not occur naturally on Earth. The bulk metal is silvery, but it is more commonly obtained as a grey powder. Technetium resists oxidation but slowly tarnishes in moist air, and burns in oxygen. It dissolves in HNO_3 and H_2SO_4.

Radii/pm: Tc^{7+} 56; Tc^{4+} 72; Tc^{2+} 95; atomic 136
Electronegativity: 1.9 (Pauling); 1.36 (Allred); 3.91 eV (absolute)
Effective nuclear charge: 3.60 (Slater); 7.23 (Clementi); 10.28 (Froese-Fischer)

Standard reduction potentials $E^{\ominus}$/V

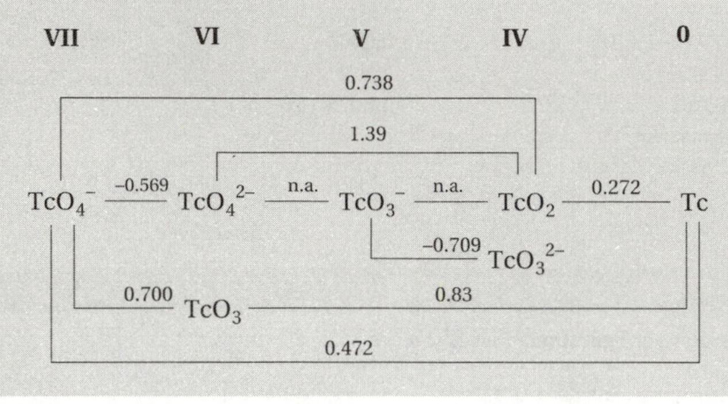

Oxidation states

Tc^{-I}	d^8	$[Tc(CO)_5]^-$
Tc^0	d^7	$[Tc_2(CO)_{10}]$
Tc^I	d^6	$[Tc(CO)_3(\eta\text{-}C_5H_5)]$
Tc^{IV}	d^3	TcO_2, $[TcO_3]^{2-}$ (aq), $TcCl_4$, complexes
Tc^V	d^2	$[TcO_3]^-$ (aq), $TcCl_5$, complexes
Tc^{VI}	d^1	TcO_3 ?, TcF_6, $TcCl_4O$, complexes
Tc^{VII}	d^0 [Kr]	Tc_2O_7, $[TcO_4]^-$ (aq), $TcClO_3$, complexes

• PHYSICAL DATA

Melting point/K: 2445
Boiling point/K: 5150
ΔH_{fusion}/kJ mol^{-1}: 23.81
ΔH_{vap}/kJ mol^{-1}: 585.22

Thermodynamic properties (298.15 K, 0.1 MPa)

State	$\Delta_f H^{\ominus}$/kJ mol^{-1}	$\Delta_f G^{\ominus}$/kJ mol^{-1}	$S^{\ominus}$/J K^{-1} mol^{-1}	C_p/J K^{-1} mol^{-1}
Solid	0	0	n.a.	25
Gas	678	n.a.	181.07	20.79

Density/kg m^{-3}: 11 500 (est.) [293 K]
Molar volume/cm^3: 8.6 (est.)
Thermal conductivity/W m^{-1} K^{-1}: 50.6 [300 K]
Coefficient of linear thermal expansion/K^{-1}: 8.06×10^{-6}
Electrical resistivity/Ω m: 22.6×10^{-8} [393 K]
Mass magnetic susceptibility/kg^{-1} m^3: $+3.1 \times 10^{-8}$ (s)

Young's modulus/GPa: n.a.
Rigidity modulus/GPa: n.a.
Bulk modulus/GPa: n.a.
Poisson's ratio/GPa: n.a.

• BIOLOGICAL DATA

Biological role

None.

Toxicity

The radioactivity of technetium overrides other toxicity considerations, but chemically it would be toxic in the same way as rhenium.

Hazards

Technetium is never encountered normally, but would be dangerous because of its radioactivity.

Levels in humans

nil

Daily dietary intake: nil
Total mass of element in average (70 kg) person: nil

Discovered in 1937 by C.Perrier and E.G. Segré at Palermo, Sicily.
[Greek, *technikos* = artificial]
French, *technétium*; German, *Technetium*; Italian, *tecneto*; Spanish, *tecnecio*

Technetium

[tek-nee-siuhm]

• NUCLEAR DATA

Number of isotopes (including nuclear isomers): 25 **Isotope mass range:** 90 → 108

Key isotopes

Nuclide	Atomic mass	Half life ($T_{1/2}$)	Decay mode and energy (MeV)	Nuclear spin I	Nucl. mag. moment μ	Uses
^{95}Tc	94.907 657	20.0 h	EC (1.69); γ	9/2+	5.89	
^{95m}Tc		61 d	EC 96%; IT 4%; β^+ (0.3); γ	1/2–		
^{96}Tc	95.907 870	4.3 d	EC (2.97); γ	7+	+5.04	
^{97}Tc	96.906 364	2.6×10^6 y	EC (0.320)	9/2+		
^{97m}Tc		90 d	IT (0.0965); γ	1/2–		
^{98}Tc	97.907 215	4.2×10^6 y	β^- (1.80); γ	6+		
^{99}Tc	98.906 254	2.13×10^5 y	β^- (0.293); γ	9/2+	+5.6847	E, NMR
^{99m}Tc		6.01 h	IT (0.142); γ	1/2–		D

Other isotopes of technetium have half-lives shorter than 5 hours.

NMR [Reference: $[TcO_4]^-$]

	^{99}Tc
Relative sensitivity (^1H = 1.00)	–
Receptivity (^{13}C = 1.00)	–
Magnetogyric ratio/rad $T^{-1} s^{-1}$	6.0503×10^7
Nuclear quadrupole moment/m^2	-0.129×10^{-28}
Frequency (^1H = 100 Hz; 2.3488T)/MHz	22.508

• ELECTRON SHELL DATA

Ground state electron configuration: [Kr]$4d^5 5s^2$
Term symbol: $^6S_{5/2}$
Electron affinity ($M \rightarrow M^-$)/kJ mol^{-1}: 96

Ionization energies/kJ mol^{-1}:

1.	$M \rightarrow M^+$	702
2.	$M^+ \rightarrow M^{2+}$	1472
3.	$M^{2+} \rightarrow M^{3+}$	2850
4.	$M^{3+} \rightarrow M^{4+}$	(4100)
5.	$M^{4+} \rightarrow M^{5+}$	(5700)
6.	$M^{5+} \rightarrow M^{6+}$	(7300)
7.	$M^{6+} \rightarrow M^{7+}$	(9100)
8.	$M^{7+} \rightarrow M^{8+}$	(15 600)
9.	$M^{8+} \rightarrow M^{9+}$	(17 800)
10.	$M^{9+} \rightarrow M^{10+}$	(19 900)

Electron binding energies/eV

K	1s	21 044
L_I	2s	3043
L_{II}	$2p_{1/2}$	2793
L_{III}	$2p_{3/2}$	2677
M_I	3s	544
M_{II}	$3p_{1/2}$	447.6
M_{III}	$3p_{3/2}$	417.7
M_{IV}	$3d_{3/2}$	257.6
M_V	$3d_{5/2}$	253.9

continued in Appendix 2, p257

Main lines in atomic spectrum
[Wavelength/nm(species)]
403.163 (I)
409.567 (I)
426.227 (I)
429.706 (I)
485.359 (I)

• CRYSTAL DATA

Crystal structure (cell dimensions/pm), space group
h.c.p. (a = 274.3, c = 440.0), $P6_3/mmc$

X-ray diffraction: mass absorption coefficients (μ/ρ)/$cm^2 g^{-1}$: CuK_α 172 MoK_α 19.7
Neutron scattering length, $b/10^{-12}$ cm: 0.68 (^{99}Tc)
Thermal neutron capture cross-section, σ_a/barns: 20 (^{99}Tc)

• GEOLOGICAL DATA

Minerals

Technetium has never been discovered in terrestrial minerals despite searches. It is produced by, and extracted from, the fission products of nuclear fuels.

World production/tonnes y^{-1}: produced in kg quantities

Specimen: commercially available, under licence. See Key.

Abundances

Sun (relative to H = 1×10^{12}): n.a. (but detected in the spectra of some stars)
Earth's crust/p.p.m.: nil
Seawater/p.p.m.: nil

Te

Atomic number: 52
Relative atomic mass ($^{12}C = 12.0000$): 127.60

CAS: [13494-80-9]

• CHEMICAL DATA

Discovery: Tellurium was discovered in 1783 by Baron Franz Joseph Müller von Reichenstein at Sibiu, Romania.
Description: Tellurium is silvery-white, metallic-looking in bulk, but is usually obtained as a dark grey powder. It is a semi-metal. Tellurium burns in air or oxygen, is unaffected by water or HCl, but dissolves in HNO_3. It is used in alloys to improve machinability; in electronics; and in catalysts.

Radii/pm: Te^{6+} 56; Te^{4+} 97; Te^{2-} 211; atomic 143; covalent 137; van der Waals 220
Electronegativity: 2.1 (Pauling); 2.01 (Allred); 5.49 eV (absolute)
Effective nuclear charge: 6.95 (Slater); 10.81 (Clementi); 13.51 (Froese-Fischer)

Standard reduction potentials $E^{\ominus}$/V

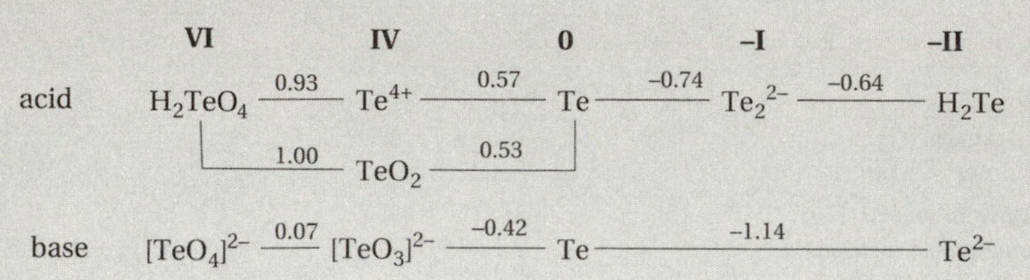

Oxidation states

Te^{-II}	[Xe]	H_2Te, Te^{2-}
Te^{-I}	s^2p^5	Te_2^{2-}
Te^{0-1}	s^2p^4, s^2p^3	cluster cations Te_4^{2+}, Te_6^{4+}
Te^{II}	s^2p^2	TeO, $TeCl_2$, $TeBr_2$
Te^{IV}	s^2	TeO_2, H_2TeO_3, $[TeO_3]^{2-}$ (aq), TeF_4, $TeCl_4$, TeF_5^-
Te^{V}	s^1	Te_2F_{10}
Te^{VI}	d^{10}	TeO_3, H_2TeO_4, $[TeO_4]^{2-}$ (aq), H_6TeO_6, TeF_6, $[TeF_8]^{2-}$

Covalent bonds

Bond	r/ pm	E/ kJ mol^{-1}
Te—H	*c.* 170	*c.* 240
Te—C	205	n.a.
Te—O	200	268
Te—F	185	335
Te—Cl	231	251
Te—Te	286	235

• PHYSICAL DATA

Melting point/K: 722.7
Boiling point/K: 1263.0
ΔH_{fusion}/kJ mol^{-1}: 13.5
ΔH_{vap}/kJ mol^{-1}: 50.63

Thermodynamic properties (298.15 K, 0.1 MPa)

State	$\Delta_f H^{\ominus}$/kJ mol^{-1}	$\Delta_f G^{\ominus}$/kJ mol^{-1}	$S^{\ominus}$/J K^{-1} mol^{-1}	C_p/J K^{-1} mol^{-1}
Solid	0	0	49.71	25.73
Gas	196.73	157.08	182.74	20.786

Density/kg m^{-3}: 6240 [293 K]; 5797 [liquid at m.p.]
Molar volume/cm^3: 20.45
Thermal conductivity/W m^{-1} K^{-1}: 2.35 [300 K]
Coefficient of linear thermal expansion/K^{-1}: 16.75×10^{-6}
Electrical resistivity/Ω m: 4.36×10^{-3} [298 K]
Mass magnetic susceptibility/kg^{-1} m^3: -3.9×10^{-9} (s)
Young's modulus/GPa: 47.1
Rigidity modulus/GPa: 16.7
Bulk modulus/GPa: n.a.
Poisson's ratio/GPa: 0.16 - 0.3

• BIOLOGICAL DATA

Biological role

None.

Toxicity

Toxic intake: elemental tellurium has low toxicity but unpleasant side effects, producing extremely unpleasant breath and body odour.
Lethal intake: 2 g of sodium tellurite has proved fatal to a human.
LD_{50}(Te metal, oral, rat) = 83 mg kg^{-1}

Hazards

Tellurium compounds are toxic by ingestion and intravenous routes. They are also considered to be experimental teratogens.

Levels in humans

Blood/mg dm^{-3}: 0.0055
Bone/p.p.m.: n.a.
Liver/p.p.m.: 0.014
Muscle/p.p.m.: 0.017
Daily dietary intake: *c.* 0.6 mg
Total mass of element in average (70 kg) person: *c.* 0.7 mg

Discovery: see Chemical Data section.
[Latin, *tellus* = earth]
French, *tellure*; German, *Tellur*; Italian, *tellurio*; Spanish, *teluro*

Tellurium

[tel-oor-iuhm]

• NUCLEAR DATA

Number of isotopes (including nuclear isomers): 39 **Isotope mass range:** 108 → 137

Key isotopes

Nuclide	Atomic mass	Natural abundance (%)	Nuclear spin I	Nuclear magnetic moment μ	Uses
^{120}Te	119.904 048	0.09	0+		E
^{122}Te	121.903 050	2.57	0+		E
^{123}Te*	122.904 271	0.89	1/2+	−0.736 79	E, NMR
^{124}Te	123.902 818	4.76	0+		E
^{125}Te	124.904 428	7.10	1/2+	−0.888 28	E, NMR
^{126}Te	125.903 309	18.89	0+		E
^{128}Te	127.904 463	31.73	0+		E
^{130}Te**	129.906 229	33.97	0+		E

*^{123}Te is radioactive with a half-life of 1.3×10^{13} y and decay mode EC (0.052 MeV); no γ.
**^{130}Te is also radioactive with a half-life of 2.4×10^{21} y.
A table of other radioactive isotopes is given in Appendix 1, on p250.

NMR [Reference: $Te(CH_3)_2$]	^{123}Te	^{125}Te
Relative sensitivity (^1H = 1.00)	0.0180	0.0315
Receptivity (^{13}C = 1.00)	0.89	12.5
Magnetogyric ratio/rad $T^{-1}s^{-1}$	-7.0006×10^7	-8.4398×10^7
Nuclear quadrupole moment/m^2	–	–
Frequency (^1H = 100 Hz; 2.3488T)/MHz	26.207	31.596

• ELECTRON SHELL DATA

Ground state electron configuration: $[Kr]4d^{10}5s^25p^4$
Term symbol: $^3P_2$
Electron affinity ($M \rightarrow M^-$)/kJ mol^{-1}**:** 190.2

Ionization energies/kJ mol^{-1}**:**

1.	$M \rightarrow M^+$	869.2
2.	$M^+ \rightarrow M^{2+}$	1795
3.	$M^{2+} \rightarrow M^{3+}$	2698
4.	$M^{3+} \rightarrow M^{4+}$	3610
5.	$M^{4+} \rightarrow M^{5+}$	5668
6.	$M^{5+} \rightarrow M^{6+}$	6822
7.	$M^{6+} \rightarrow M^{7+}$	13200
8.	$M^{7+} \rightarrow M^{8+}$	(15 800)
9.	$M^{8+} \rightarrow M^{9+}$	(18 500)
10.	$M^{9+} \rightarrow M^{10+}$	(21 200)

Electron binding energies/eV

K	1s	31 814
L_I	2s	4939
L_{II}	$2p_{1/2}$	4612
L_{III}	$2p_{3/2}$	4341
M_I	3s	1006
M_{II}	$3p_{1/2}$	870.8
M_{III}	$3p_{3/2}$	820.0
M_{IV}	$3d_{3/2}$	583.4
M_V	$3d_{5/2}$	573.0

continued in Appendix 2, p257

Main lines in atomic spectrum
[Wavelength/nm(species)]
200.202 (I)
214.281 (I) (AA)
972.274 (I)
1005.141 (I)
1108.956 (I)
1148.723 (I)

• CRYSTAL DATA

Crystal structure (cell dimensions/pm), space group
hexagonal (a = 445.65, c = 592.68), $P3_121$ or $P3_221$
High pressure forms: (a = 420.8, c = 1203.6), $R\bar{3}m$; (a = 460.3, c = 382.2), $R\bar{3}m$
X-ray diffraction: mass absorption coefficients (μ/ρ)/$cm^2\ g^{-1}$**:** CuK_α 282 MoK_α 35.0
Neutron scattering length, $b/10^{-12}$ cm: 0.580
Thermal neutron capture cross-section, σ_a/barns**:** 4.7

• GEOLOGICAL DATA

Minerals

Mineral	Formula	Density	Hardness	Crystal appearance
Sylvanite	$AgAuTe_4$	8.16	1.5 – 2	mon., met. grey
Tellurite	TeO_2	5.90	2	orth., sub-adamantine white

Chief ores: none mined as such. Tellurium is obtained from the anode slime of copper refining.

World production/tonnes y^{-1}**:** 215

Areas where minerals found: sylvanite in Australia, USA and Romania

Reserves/tonnes: n.a.

Specimen: available as granules, ingots, pieces or powder. *Danger!*

Abundances

Sun (relative to H = 1×10^{12})**:** n.a.
Earth's crust/p.p.m.**:** *c.* 0.005
Seawater/p.p.m.**:**
Atlantic surface: 1.6×10^{-7}
Atlantic deep: 0.7×10^{-7}
Pacific surface: 1.9×10^{-7}
Pacific deep: 1.7×10^{-7}
Residence time/years**:** n.a.
Classification: scavenged
Oxidation state: IV and VI; mainly VI

Tb

Atomic number: 65
Relative atomic mass ($^{12}C = 12.0000$)**:** 158.92534
CAS: [7440-27-9]

• CHEMICAL DATA

Description: Terbium is a silvery metal, and a particularly rare member of the so-called rare earth group (more correctly termed the lanthanides). It is slowly oxidised by air and reacts with cold water. Terbium is used in solid state devices and lasers.

Radii/pm: Tb^{4+} 81; Tb^{3+} 97; atomic 178; covalent 159
Electronegativity: n.a. (Pauling); 1.10 (Allred); ≤ 3.2 eV (absolute)
Effective nuclear charge: 2.85 (Slater); 8.30 (Clementi); 11.39 (Froese-Fischer)

Standard reduction potentials $E^\ominus$/V

	IV		III		0
acid	Tb^{4+}	3.1	Tb^{3+}	−2.31	Tb
base	TbO_2	0.9	$Tb(OH)_3$	−2.82	Tb

Oxidation states

Tb^{III}	f^8	Tb_2O_3, $Tb(OH)_3$, $[Tb(OH_2)_x]^{3+}$ (aq), Tb^{3+} salts, TbF_3, $TbCl_3$, complexes
Tb^{IV}	f^7	TbO_2, TbF_4

• PHYSICAL DATA

Melting point/K: 1629
Boiling point/K: 3396
ΔH_{fusion}/kJ mol^{-1}: 16.3
ΔH_{vap}/kJ mol^{-1}: 391

Thermodynamic properties (298.15 K, 0.1 MPa)

State	$\Delta_f H^\ominus$/kJ mol^{-1}	$\Delta_f G^\ominus$/kJ mol^{-1}	$S^\ominus$/J K^{-1} mol^{-1}	C_p/J K^{-1} mol^{-1}
Solid	0	0	73.22	28.91
Gas	388.7	349.7	203.58	24.56

Density/kg m^{-3}: 8229 [293 K]
Molar volume/cm^3: 19.31
Thermal conductivity/W m^{-1} K^{-1}: 11.1 [300 K]
Coefficient of linear thermal expansion/K^{-1}: 7.0×10^{-6}
Electrical resistivity/Ω m: 114×10^{-8} [298 K]
Mass magnetic susceptibility/kg^{-1} m^3: $+1.15 \times 10^{-5}$ (s)
Young's modulus/GPa: 55.7
Rigidity modulus/GPa: 22.1
Bulk modulus/GPa: 38.7
Poisson's ratio/GPa: 0.261

• BIOLOGICAL DATA

Biological role

None.

Toxicity

Toxic intake: n.a.
Lethal intake: LD_{50} (chloride, oral, mouse) = > 5100 mg kg^{-1}

Hazards

Terbium is mildly toxic by ingestion, and is a skin and eye irritant.

Levels in humans

Organs: n.a., but low
Daily dietary intake: n.a.
Total mass of element in average (70 kg) person: n.a., but very low

Discovered in 1843 by C.G. Mosander at Stockholm, Sweden.
[Named after Ytterby, Sweden]
French, *terbium*; German, *Terbium*; Italian, *terbio*; Spanish, *terbio*

Terbium

[ter-bi-uhm]

• NUCLEAR DATA

Number of isotopes (including nuclear isomers): 31 **Isotope mass range:** 145 → 165

Key isotopes

Nuclide	Atomic mass	Natural abundance (%)	Nuclear spin I	Nuclear magnetic moment μ	Uses
^{159}Tb	158.925 342	100	3/2+	+2.014	NMR

A table of radioactive isotopes is given in Appendix 1, on p250.

NMR [Reference: not recorded]	^{159}Tb
Relative sensitivity (^1H = 1.00)	0.0583
Receptivity (^{13}C = 1.00)	394
Magnetogyric ratio/rad $T^{-1}s^{-1}$	6.4306×10^7
Nuclear quadrupole moment/m^2	$+1.432 \times 10^{-28}$
Frequency (^1H = 100 Hz; 2.3488T)/MHz	22.678

• ELECTRON SHELL DATA

Ground state electron configuration: $[Xe]4f^96s^2$
Term symbol: $^6H_{15/2}$
Electron affinity ($M \rightarrow M^-$)/kJ mol^{-1}**:** ≤ 50

Ionization energies/kJ mol^{-1}**:**

1.	$M \rightarrow M^+$	564.6
2.	$M^+ \rightarrow M^{2+}$	1112
3.	$M^{2+} \rightarrow M^{3+}$	2114
4.	$M^{3+} \rightarrow M^{4+}$	3839
5.	$M^{4+} \rightarrow M^{5+}$	6413

Electron binding energies/eV

K	1s	51 996
L_I	2s	8708
L_{II}	$2p_{1/2}$	8252
L_{III}	$2p_{3/2}$	7514
M_I	3s	1968
M_{II}	$3p_{1/2}$	1768
M_{III}	$3p_{3/2}$	1611
M_{IV}	$3d_{3/2}$	1276.9
M_V	$3d_{5/2}$	1241.1

continued in Appendix 2, p257

Main lines in atomic spectrum
[Wavelength/nm(species)]

332.440 (II)
350.917 (II)
356.852 (II)
367.635(II)
370.286 (II)
384.873 (II)
387.417 (II)
432.643 (I) (AA)

• CRYSTAL DATA

Crystal structure (cell dimensions/pm), space group
Tb orthorhombic (a = 359.0, b = 626.0, c = 571.5), Cmcm
α-Tb h.c.p. (a = 360.10, c = 569.36), $P6_3/mmc$
β-Tb b.c.c. (a = 402), Im3m
$T(\alpha \rightarrow \text{orthorhombic})$ = 220 K; $T(\alpha \rightarrow \beta)$ = 1590 K

X-ray diffraction: mass absorption coefficients (μ/ρ)/$cm^2\ g^{-1}$**:** CuK_α 273 MoK_α 67.5
Neutron scattering length, $b/10^{-12}$ cm**:** 0.738
Thermal neutron capture cross-section, σ_a/barns**:** 23.4

• GEOLOGICAL DATA

Minerals

Mineral	Formula	Density	Hardness	Crystal appearance
Monazite*	(Ce, La, Nd, Th, etc.)PO_4	5.20	5 – 5.5	mon., waxy/vit. yellow-brown

*Although not a major constituent, terbium is present in extractable amounts.

Chief ore: monazite*
World production/tonnes y^{-1}**:** *c.* 10
Main mining areas: USA, Brazil, India, Sri Lanka, Australia
Reserves/tonnes: *c.* 3×10^5
Specimen: available as chips or ingots. Safe. Terbium powders is very irritating to skin and eyes. *Care!*

Abundances

Sun (relative to H = 1×10^{12})**:** n.a.
Earth's crust/p.p.m.**:** 1.1
Seawater/p.p.m.**:**
Atlantic surface: 1×10^{-7}
Atlantic deep: 1.5×10^{-7}
Pacific surface: 0.8×10^{-7}
Pacific deep: 2.5×10^{-7}
Residence time/years**:** n.a.
Classification: recycled
Oxidation state: III

Tl

Atomic number: 81
Relative atomic mass ($^{12}C = 12.0000$)**:** 204.3833

CAS: [7440-28-0]

• CHEMICAL DATA

Discovery: Thallium was discovered in 1861 by William Crookes at London, England. Isolated in 1862 by C.A. Lamy at Paris, France.
Description: Thallium is a soft, silvery-white metal, which tarnishes readily in moist air and reacts with steam to form TlOH. It is attacked by acids, rapidly so by HNO_3. Thallium is little used because of its toxicity, but is still employed in special types of glass.

Radii/pm: Tl^{3+} 105; Tl^{+} 149; atomic 170; covalent 155
Electronegativity: 1.62 (Tl^{I}) 2.04 (Tl^{III}) (Pauling); 1.44 (Allred); 3.2 eV (absolute)
Effective nuclear charge: 5.00 (Slater); 12.25 (Clementi); 13.50 (Froese-Fischer)

Standard reduction potentials $E^\ominus$/V

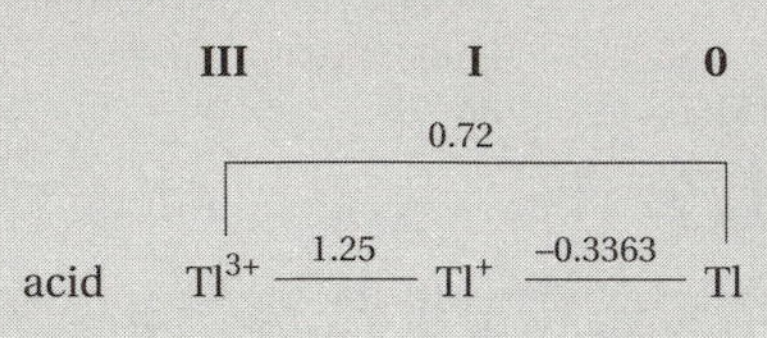

Oxidation states

Tl^{I}	s^2	Tl_2O, TlOH, Tl_2CO_3, $[Tl(OH_2)_6]^+$ (aq), Ti^+ salts, TlF, TlCl, etc.
Tl^{III}	d^{10}	Tl_2O_3, $[Tl(OH_2)_6]^{3+}$ (aq), TlF_3, $TlCl_3$, $TiBr_3$, $[Tl(CH_3)_2l^+$ (aq), $[TlCl_6]^{3-}$

Covalent bonds

Bond	r/ pm	E/ kJ mol^{-1}
Tl^{I}—H	187	185
Tl^{III}—C	230	125
Tl^{III}—O	226	375
Tl^{III}—F	195	460
Tl^{III}—Cl	248	368
Tl^{I}—Br	254	406
Tl—Tl	341	*c.* 63

• PHYSICAL DATA

Melting point/K: 576.7
Boiling point/K: 1730
ΔH_{fusion}/kJ mol^{-1}: 4.31
ΔH_{vap}/kJ mol^{-1}: 162.1

Thermodynamic properties (298.15 K, 0.1 MPa)

State	$\Delta_f H^\ominus$/kJ mol^{-1}	$\Delta_f G^\ominus$/kJ mol^{-1}	$S^\ominus$/J K^{-1} mol^{-1}	C_p/J K^{-1} mol^{-1}
Solid	0	0	64.18	26.32
Gas	182.21	147.41	180.963	20.786

Density/kg m^{-3}: 11 850 [293 K]; 11 290 [liquid at m.p.]
Molar volume/cm^3: 17.24
Thermal conductivity/W m^{-1} K^{-1}: 46.1 [300 K]
Coefficient of linear thermal expansion/K^{-1}: 28×10^{-6}
Electrical resistivity/Ω m: 18.0×10^{-8} [273 K]
Mass magnetic susceptibility/kg^{-1} m^3: -3.13×10^{-9} (s)

Young's modulus/GPa: 7.90
Rigidity modulus/GPa: 2.71
Bulk modulus/GPa: 28.5
Poisson's ratio/GPa: 0.45

• BIOLOGICAL DATA

Biological role

None.

Toxicity

Toxic intake: metal, oral, human = 5.7 mg kg^{-1}
Lethal intake: ingestion of 0.2 – 1 g Tl_2SO_4 for humans. LD_{50} (Tl_2SO_4, oral, mouse) = 29 mg kg^{-1}

Hazards

Thallium compounds are extremely toxic and cumulative; they are also absorbed though the skin. Thallium takes several days to act and affects the nervous system. It is also teratogenic.

Levels in humans

Blood/mg dm^{-3}: 0.00048
Bone/p.p.m.: 0.002
Liver/p.p.m.: 0.004 – 0.033
Muscle/p.p.m.: 0.07
Daily dietary intake: 0.0015 mg
Total mass of element in average (70 kg) person: 0.5 mg

Discovery: see Chemical Data section.
[Greek, *thallos* = green twig]
French, *thallium*; German, *Thallium*; Italian, *tallio*; Spanish, *talio*

Thallium

[thal-iuhm]

• NUCLEAR DATA

Number of isotopes (including nuclear isomers): 41 **Isotope mass range:** 184 → 210

Key isotopes

Nuclide	Atomic mass	Natural abundance (%)	Nuclear spin *I*	Nuclear magnetic moment μ	Uses
^{203}Tl	202.972 320	29.524	1/2+	+1.622 257	NMR
^{205}Tl	204.974 401	70.476	1/2+	+1.638 213	NMR

A table of radioactive isotopes is given in Appendix 1, on p251.

NMR [Reference: $TlNO_3$ (aq)]	^{203}Tl	^{205}Tl
Relative sensitivity (^1H = 1.00)	0.18	0.19
Receptivity (^{13}C = 1.00)	289	769
Magnetogyric ratio/rad $T^{-1}s^{-1}$	15.3078×10^7	15.4584×10^7
Nuclear quadrupole moment/m^2	–	–
Frequency (^1H = 100 Hz; 2.3488T)/MHz	57.149	57.708

• ELECTRON SHELL DATA

Ground state electron configuration: $[Xe]4f^{14}5d^{10}6s^26p^1$
Term symbol: $^2P_{1/2}$
Electron affinity ($M \rightarrow M^-$)/kJ mol^{-1}**:** *c.* 20

Ionization energies/kJ mol^{-1}**:**

1.	$M \rightarrow M^+$	589.3
2.	$M^+ \rightarrow M^{2+}$	1971.0
3.	$M^{2+} \rightarrow M^{3+}$	2878
4.	$M^{3+} \rightarrow M^{4+}$	(4900)
5.	$M^{4+} \rightarrow M^{5+}$	(6100)
6.	$M^{5+} \rightarrow M^{6+}$	(8300)
7.	$M^{6+} \rightarrow M^{7+}$	(9500)
8.	$M^{7+} \rightarrow M^{8+}$	(11 300)
9.	$M^{8+} \rightarrow M^{9+}$	(14 000)
10.	$M^{9+} \rightarrow M^{10+}$	(16 000)

Electron binding energies/eV

K	1s	85 530
L_I	2s	15 347
L_{II}	$2p_{1/2}$	14 698
L_{III}	$2p_{3/2}$	12 658
M_I	3s	3704
M_{II}	$3p_{1/2}$	3416
M_{III}	$3p_{3/2}$	2957
M_{IV}	$3d_{3/2}$	2485
M_V	$3d_{5/2}$	2389

continued in Appendix 2, p257

Main lines in atomic spectrum
[Wavelength/nm(species)]

276.787 (I) (AA)
291.832 (I)
351.924 (I)
352.943 (I)
377.572 (I)
535.046 (I)

• CRYSTAL DATA

Crystal structure (cell dimensions/pm), space group
α-Tl hexagonal (a = 345.6, c = 552.6), $P6_3/mmc$
β-Tl cubic (a = 388.2), Im3m
γ-Tl f.c.c. (a = 485.1), Fm3m
$T(\alpha \rightarrow \beta)$ = 503 K

X-ray diffraction: mass absorption coefficients (μ/ρ)/$cm^2\ g^{-1}$**:** CuK_α 224 MoK_α 119
Neutron scattering length, $b/10^{-12}$ cm**:** 0.8776
Thermal neutron capture cross-section, σ_a/barns**:** 3.43

• GEOLOGICAL DATA

Minerals

Thallium minerals are rare, but the element is dispersed in potassium minerals such as sylvite and the caesium mineral pollucite.

Mineral	Formula	Density	Hardness	Crystal appearance
Crookesite	Cu_7TlSe_4	6.90	2.5 – 3	tet., metallic grey
Hutchinsonite	$(Tl,Pb)_2As_5S_9$	4.6	1.5 – 2	orth., adam. red
Lorandite	$TlAsS_2$	5.53	2 – 2.5	mon., met. adamantine red/grey
Thalcusite	$Cu_3FeTl_2S_4$	6.54	2.5	tet., metallic grey

Chief ores: thallium is generally obtained as the by-product of zinc and lead smelting.
World production/tonnes y^{-1}**:** 30
Main mining areas: see zinc and lead.
Reserves/tonnes**:** n.a.
Specimen: available as granules. *Danger!*

Abundances

Sun (relative to H = 1×10^{12})**:** 8.0
Earth's crust/p.p.m.**:** 0.6
Seawater/p.p.m.**:** 1.4×10^{-5}
Residence time/years**:** 10 000
Classification: accumulating
Oxidation state: I

Th

Atomic number: 90
Relative atomic mass ($^{12}C = 12.0000$)**:** 232.0381

CAS: [7440-29-1]

• CHEMICAL DATA

Description: Thorium is a radioactive, silvery metal that is soft and ductile, but alloys can be very strong. The bulk metal is protected by an oxide coating, but is attacked by steam, and slowly dissolved by acids. Thorium is used in refractory materials, nuclear fuel elements, and incandescent gas mantles.

Radii/pm: Th^{4+} 99; Th^{3+} 101; atomic 180
Electronegativity: 1.3 (Pauling); 1.11 (Allred); n.a. (absolute)
Effective nuclear charge: 1.95 (Slater)

Standard reduction potentials $E^{\ominus}/V$

	IV		0
acid	Th^{4+}	$\xrightarrow{-1.83}$	Th
base	ThO_2	$\xrightarrow{-2.56}$	Th

Oxidation states

Th^{II}	d^2	ThO, ThH_2
Th^{III}	d^1	ThI_3, $[Th(C_5H_5)_3]$
Th^{IV}	[Rn]	ThO_2, $[Th(OH_2)_x]^{4+}$ (aq), ThF_4, $ThCl_4$ etc., $[ThF_7]^{3-}$, Th^{4+} salts, complexes, $[Th(C_5H_5)_4]$

• PHYSICAL DATA

Melting point/K: 2023
Boiling point/K: *c.* 5060

ΔH_{fusion}/kJ mol^{-1}: < 19.2
ΔH_{vap}/kJ mol^{-1}: 543.9

Thermodynamic properties (298.15 K, 0.1 MPa)

State	$\Delta_f H^{\ominus}$/kJ mol^{-1}	$\Delta_f G^{\ominus}$/kJ mol^{-1}	$S^{\ominus}$/J K^{-1} mol^{-1}	C_p/J K^{-1} mol^{-1}
Solid	0	0	53.39	27.32
Gas	598.3	557.53	190.15	20.79

Density/kg m^{-3}: 11 720 [293 K]
Molar volume/cm^3: 19.80
Thermal conductivity/W m^{-1} K^{-1}: 54.0 [300 K]
Coefficient of linear thermal expansion/K^{-1}: 12.5×10^{-6}
Electrical resistivity/Ω m: 13.0×10^{-8} [273 K]
Mass magnetic susceptibility/kg^{-1} m^3: $+7.2 \times 10^{-9}$ (s)

Young's modulus/GPa: 78.3
Rigidity modulus/GPa: 30.8
Bulk modulus/GPa: 54.0
Poisson's ratio/GPa: 0.26

• BIOLOGICAL DATA

Biological role

None.

Toxicity

The radiotoxicity of thorium generally overrides other toxicity considerations, but it does have some industrial applications.

Lethal intake: LD_{50} (thorium (IV) nitrate, oral, mouse) = 1760 mg kg^{-1}

Hazards

Thorium compounds are moderately toxic; acute exposure can lead to dermatitis and chronic exposure causes cancer.

Levels in humans

Blood/mg dm^{-3}: 0.00016
Bone/p.p.m.: 0.002 – 0.012
Liver/p.p.m.: n.a.
Muscle/p.p.m.: n.a.
Daily dietary intake: 0.00005 – 0.003 mg
Total mass of element in average (70 kg) person: *c.* 0.1 mg

Discovered in 1829 by J.J. Berzelius at Stockholm, Sweden.
[Named after Thor, the Scandinavian god of war]
French, *thorium*; German, *Thorium*; Italian, *torio*; Spanish, *torio*

Thorium

[thor-iuhm]

• NUCLEAR DATA

Number of isotopes (including nuclear isomers): 25 **Isotope mass range:** 212 → 236

Key isotopes

Nuclide	Atomic mass	Natural abundance (%)	Nuclear spin I	Nuclear magnetic moment μ	Uses
^{232}Th*	232.038 054	100	0+		E

*^{232}Th is radioactive with a half-life of 1.4×10^{10} y and decay mode α (4.081 MeV); γ.
A table of radioactive isotopes is given in Appendix 1, on p251.

NMR [Reference: not recorded]

	^{229}Th*
Relative sensitivity (^1H = 1.00)	–
Receptivity (^{13}C = 1.00)	–
Magnetogyric ratio/rad $T^{-1} s^{-1}$	0.40×10^7
Nuclear quadrupole moment/m^2	4.300×10^{-28}
Frequency (^1H = 100 Hz; 2.3488T)/MHz	1.5

* Details of this isotope are given on p251.

• ELECTRON SHELL DATA

Ground state electron configuration: $[Rn]6d^2 7s^2$
Term symbol: $^3F_2$
Electron affinity ($M \rightarrow M^-$)/kJ mol^{-1}**:** n.a.

Ionization energies/kJ mol^{-1}**:**

1.	$M \rightarrow M^+$	587
2.	$M^+ \rightarrow M^{2+}$	1110
3.	$M^{2+} \rightarrow M^{3+}$	1978
4.	$M^{3+} \rightarrow M^{4+}$	2789

Electron binding energies/eV

K	1s	109 651
L_I	2s	20 472
L_{II}	$2p_{1/2}$	19 693
L_{III}	$2p_{3/2}$	16 300
M_I	3s	5182
M_{II}	$3p_{1/2}$	4830
M_{III}	$3p_{3/2}$	4046
M_{IV}	$3d_{3/2}$	3491
M_V	$3d_{5/2}$	3332

continued in Appendix 2, p258

Main lines in atomic spectrum
[Wavelength/nm(species)]

339.204 (II)
346.992 (II)
374.118 (II)
401.914 (II)
438.186 (II)

• CRYSTAL DATA

Crystal structure (cell dimensions/pm), space group
α-Th f.c.c. (a = 508.42), Fm3m
β-Th b.c.c. (a = 411), Im3m
$T(\alpha \rightarrow \beta)$ = 1673 K

X-ray diffraction: mass absorption coefficients (μ/ρ)/$cm^2 g^{-1}$**:** CuK_α 327 MoK_α 143
Neutron scattering length, $b/10^{-12}$ cm**:** 1.052
Thermal neutron capture cross-section, σ_a/barns**:** 7.37

• GEOLOGICAL DATA

Minerals

Mineral	Formula	Density	Hardness	Crystal appearance
Monazite	(Ce, La, Nd, Th, etc.)PO_4	5.20	5 – 5.5	mon., waxy/vit. yellow brown
Thorianite	ThO_2	9.7	6.5	cub., horny/sub-metallic dark

Chief ores: thorianite; monazite; dispersed on pegmatite rocks; also present in significant amounts in zircon (see Zr), titanite (see Ti), gadolinite (see Y) and betafite (see U).

World production/tonnes y^{-1}**:** 31 000

Main mining areas: USA, Brazil, India, Sri Lanka, Madagascar, Russia, Australia

Reserves/tonnes**:** 3.3×10^6

Specimen: not available as thorium metal.

Abundances

Sun (relative to H = 1×10^{12})**:** *c.* 2
Earth's crust/p.p.m.**:** 12
Seawater/p.p.m.**:** 9.2×10^{-6}
Residence time/years**:** 50
Classification: scavenged
Oxidation state: IV

Tm

Atomic number: 69
Relative atomic mass ($^{12}C = 12.0000$)**:** 168.93421

CAS: [7440-30-4]

• CHEMICAL DATA

Description: Thulium is a silvery metal, and rarest of all the so-called rare earth group (more correctly termed the lanthanides). It tarnishes in air and reacts with water. Thulium has few uses but some is employed as a radiation source in portable X-ray equipment.

Radii/pm: Tm^{4+} 87; Tm^{3+} 94; atomic 175; covalent 156
Electronegativity: 1.25 (Pauling); 1.11 (Allred); ≤ 3.4 eV (absolute)
Effective nuclear charge: 2.85 (Slater); 8.58 (Clementi); 11.80 (Froese-Fischer)

Standard reduction potentials $E^{\ominus}$/V

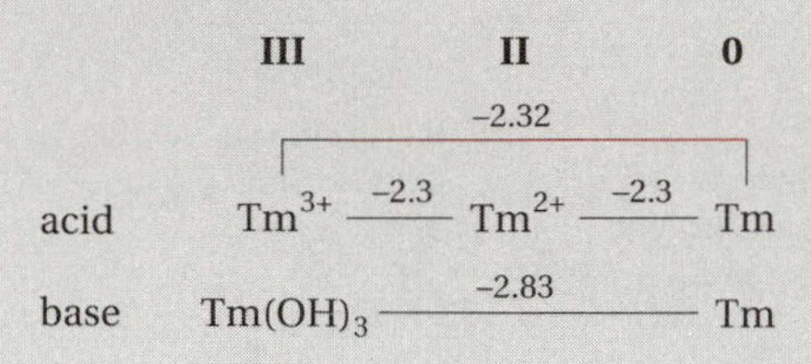

Oxidation states

Tm^{II}	f^{13}	$TmCl_2$, $TmBr_2$, TmI_2
Tm^{III}	f^{12}	Tm_2O_3, $Tm(OH)_3$, $[Tm(OH_2)_x]^{3+}$ (aq), Tm^{3+} salts, TmF_3, $TmCl_3$ etc., complexes

• PHYSICAL DATA

Melting point/K: 1818
Boiling point/K: 2220

ΔH_{fusion}/kJ mol^{-1}**:** 18.4
ΔH_{vap}/kJ mol^{-1}**:** 247

Thermodynamic properties (298.15 K, 0.1 MPa)

State	$\Delta_f H^{\ominus}$/kJ mol^{-1}	$\Delta_f G^{\ominus}$/kJ mol^{-1}	$S^{\ominus}$/J K^{-1} mol^{-1}	C_p/J K^{-1} mol^{-1}
Solid	0	0	74.01	27.03
Gas	232.3	197.5	190.113	20.786

Density/kg m^{-3}**:** 9321 [293 K]
Molar volume/cm^3**:** 18.12
Thermal conductivity/W m^{-1} K^{-1}**:** 16.8 [300 K]
Coefficient of linear thermal expansion/K^{-1}**:** 13.3×10^{-6}
Electrical resistivity/Ω m: 79.0×10^{-8} [298 K]
Mass magnetic susceptibility/kg^{-1} m^3**:** $+1.90 \times 10^{-6}$ (s)

Young's modulus/GPa**:** 74.0
Rigidity modulus/GPa**:** 30.5
Bulk modulus/GPa**:** 44.5
Poisson's ratio/GPa**:** 0.213

• BIOLOGICAL DATA

Biological role

None, but acts to stimulate metabolism.

Toxicity

Toxic intake: n.a.
Lethal intake: LD_{50} (chloride, oral, mouse) = 4290 mg kg^{-1}

Hazards

Thulium is mildly toxic by ingestion.

Levels in humans

Organs: n.a., but low
Daily dietary intake: n.a.
Total mass of element in average (70 kg) person: n.a.

Discovered in 1879 by P.T. Cleve at Uppsala, Sweden.
[Called after Thule, an ancient name for Scandinavia]
French, *thulium*; German, *Thulium*; Italian, *tulio*; Spanish, *tulio*

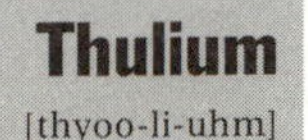

[thyoo-li-uhm]

• NUCLEAR DATA

Number of isotopes (including nuclear isomers): 28 **Isotope mass range:** $152 \rightarrow 176$

Key isotopes

Nuclide	Atomic mass	Natural abundance (%)	Nuclear spin *I*	Nuclear magnetic moment μ	Uses
^{169}Tm	168.934 212	100	1/2+	–0.231 6	NMR

A table of radioactive isotopes is given in Appendix 1, on p251.

NMR [Reference: not recorded]	^{169}Tm
Relative sensitivity (^{1}H = 1.00)	5.66×10^{-4}
Receptivity (^{13}C = 1.00)	3.21
Magnetogyric ratio/rad $T^{-1}s^{-1}$	-2.21×10^7
Nuclear quadrupole moment/m^2	–
Frequency (^{1}H = 100 Hz; 2.3488T)/MHz	8.271

• ELECTRON SHELL DATA

Ground state electron configuration: [Xe]$4f^{13}6s^2$
Term symbol: $^2F_{7/2}$
Electron affinity ($M \rightarrow M^-$)/kJ mol^{-1}**:** ≤ 50

Ionization energies/kJ mol^{-1}**:**

1.	$M \rightarrow M^+$	596.7
2.	$M^+ \rightarrow M^{2+}$	1163
3.	$M^{2+} \rightarrow M^{3+}$	2285
4.	$M^{3+} \rightarrow M^{4+}$	4119
5.	$M^{4+} \rightarrow M^{5+}$	6313

Electron binding energies/eV

K	1s	59 390
L_I	2s	10 116
L_{II}	$2p_{1/2}$	9617
L_{III}	$2p_{3/2}$	8648
M_I	3s	2307
M_{II}	$3p_{1/2}$	2090
M_{III}	$3p_{3/2}$	1885
M_{IV}	$3d_{3/2}$	1515
M_V	$3d_{5/2}$	1468

continued in Appendix 2, p258

Main lines in atomic spectrum
[Wavelength/nm(species)]

346.220 (II)
371.791 (I) (AA)
384.802 (II)
409.419 (I)
410.584 (I)
418.762 (I)

• CRYSTAL DATA

Crystal structure (cell dimensions/pm), space group
h.c.p. (*a* = 353.75, *c* = 555.46), $P6_3/mmc$

X-ray diffraction: mass absorption coefficients (μ/ρ)/cm^2 g^{-1}**:** CuK_α 140 MoK_α 80.8
Neutron scattering length, $b/10^{-12}$ cm**:** 0.707
Thermal neutron capture cross-section, σ_a/barns**:** 100

• GEOLOGICAL DATA

Minerals

Mineral	Formula	Density	Hardness	Crystal appearance
Bastnäsite*	(Ce,La, etc)CO_3F	4.9	4 – 4.5	hex., vit./greasy yellow
Monazite*	(Ce, La, Nd, Th, etc)PO_4	5.20	5 – 5.5	mon., waxy/vit. yellow-brown

*although not a major constituent, thulium is present in extractable amounts.

Chief ores: monazite, bastnäsite
World production/tonnes y^{-1}**:** *c.* 50
Main mining areas: USA, Brazil, India, Sri Lanka, Australia
Reserves/tonnes: *c.* 1×10^5
Specimen: available as chips, ingots or powder. Safe.

Abundances

Sun (relative to H = 1×10^{12})**:** 1.8
Earth's crust/p.p.m.**:** 0.48
Seawater/p.p.m.**:**
Atlantic surface: 1.3×10^{-7}
Atlantic deep: 1.6×10^{-7}
Pacific surface: 0.7×10^{-7}
Pacific deep: 3.3×10^{-7}
Residence time/years**:** n.a.
Classification: recycled
Oxidation state: III

Sn

Atomic number: 50
Relative atomic mass ($^{12}C = 12.0000$): **118.710**

CAS: [7440-31-5]

• CHEMICAL DATA

Description: Tin is a soft, pliable, silvery-white metal that is unreactive to oxygen (protected by an oxide film on the surface) and water. It dissolves in acids and bases. Tin is used in solder, alloys, tin plate, polymer additives and some anti-fouling paints.

Radii/pm: Sn^{4+} 74; Sn^{2+} 93; Sn^{4-} 294; atomic 141; covalent 140; van der Waals 200
Electronegativity: 1.96 (Pauling); 1.72 (Allred); 4.30 eV (absolute)
Effective nuclear charge: 5.65 (Slater); 9.10 (Clementi); 11.11 (Froese-Fischer)

Standard reduction potentials $E^{\ominus}$/V

	IV		II		0		–IV
acid	SnO_2	–0.088	SnO	–0.104	Sn	–1.071	SnH_4
	Sn^{4+}	0.15	Sn^{2+}	–0.137	(to Sn)		

[basic solutions contain many different forms]

Oxidation states

Sn^{II}	s^2	SnO, SnF_2, $SnCl_2$ etc., $[Sn(OH)]^+$ (aq), $[Sn_3(OH)_4]^{2+}$ (aq), Sn^{2+} salts
Sn^{IV}	d^{10}	SnO_2, SnF_4, $SnCl_4$ etc., $[SnCl_6]^{2-}$ (aq HCl), $[Sn(OH)_6]^{2-}$ (aq base), organotin compounds

Covalent bonds

Bond	r/ pm	E/ kJ mol^{-1}
Sn—H	170	< 314
Sn—C	217	225
Sn^{II}—O	195	557
Sn^{IV}—F	188	414
Sn^{IV}—Cl	233	323
Sn^{IV}—Br	246	273
Sn^{IV}—I	269	205
Sn—Sn (α)	281	146

• PHYSICAL DATA

Melting point/K: 505.118
Boiling point/K: 2543
ΔH_{fusion}/kJ mol^{-1}: 7.20
ΔH_{vap}/kJ mol^{-1}: 290.4

Thermodynamic properties (298.15 K, 0.1 MPa)

State	$\Delta_f H^{\ominus}$/kJ mol^{-1}	$\Delta_f G^{\ominus}$/kJ mol^{-1}	$S^{\ominus}$/J K^{-1} mol^{-1}	C_p/J K^{-1} mol^{-1}
Solid (α)	–2.09	0.13	44.14	25.77
Solid (β)	0	0	51.55	26.99
Gas	302.1	267.3	168.486	21.259

Density/kg m^{-3}: 5750 (α); 7310 (β) [293 K]; 6973 [liquid at m.p.]
Molar volume/cm^3: 16.24 (β)
Thermal conductivity/W m^{-1} K^{-1}: 66.6(α) [300 K]
Coefficient of linear thermal expansion/K^{-1}: 5.3×10^{-6} (α); 21.2×10^{-6} (β)
Electrical resistivity/Ω m: 11.0×10^{-8} (α) [273 K]
Mass magnetic susceptibility/kg^{-1} m^3: -4.0×10^{-9} (α); $+3.3 \times 10^{-10}$ (β)

Young's modulus/GPa: 49.9
Rigidity modulus/GPa: 18.4
Bulk modulus/GPa: 58.2
Poisson's ratio/GPa: 0.357

• BIOLOGICAL DATA

Biological role

May be essential to some organisms, including humans.

Toxicity

Toxic intake: low toxicity as metal and some inorganic tin (II) salts

Lethal intake: LD_{50} ($SnCl_2$, oral, rat) = 700 mg kg^{-1}

Hazards

Tin(II) salts can be poisonous by ingestion and other routes, and there is evidence that tin can have experimental carcinogenic and human mutagenic effects. Some organotin compounds are very toxic.

Levels in humans

Blood/mg dm^{-3}: *c.* 0.38
Bone/p.p.m.: 1.4
Liver/p.p.m.: 0.23 – 2.3
Muscle/p.p.m.: 0.33 – 2.4
Daily dietary intake: 0.2 – 3.5 mg
Total mass of element in average (70 kg) person: 20 mg

Known to ancient civilizations.
[Anglo-Saxon, *tin*; Latin, *stannum*]
French, *etain*; German, *Zinn*; Italian, *stagno*; Spanish, *estaño*

Tin

[tin]

• NUCLEAR DATA

Number of isotopes (including nuclear isomers): 37 **Isotope mass range:** 106 → 132

Key isotopes

Nuclide	Atomic mass	Natural abundance (%)	Nuclear spin I	Nuclear magnetic moment μ	Uses
^{112}Sn	111.904 826	0.97	0+		E
^{114}Sn	113.902 784	0.65	0+		E
^{115}Sn	114.903 348	0.36	1/2+	–0.918 84	E, NMR
^{116}Sn	115.901 747	14.53	0+		E
^{117}Sn	116.902 956	7.68	1/2+	–1.001 05	E, NMR
^{118}Sn	117.901 609	24.22	0+		E
^{119}Sn	118.903 310	8.58	1/2+	–1.047 29	E, NMR
^{120}Sn	119.902 220	32.59	0+		E
^{122}Sn	121.903 440	4.63	0+		E
^{124}Sn	123.905 274	5.79	0+		E

A table of radioactive isotopes is given in Appendix 1, on p251.

NMR [Reference: $Sn(CH_3)_4$]	^{115}Sn	^{117}Sn	^{119}Sn
Relative sensitivity (^1H = 1.00)	0.035	0.0452	0.3518
Receptivity (^{13}C = 1.00)	0.693	19.54	25.2
Magnetogyric ratio/rad $T^{-1}s^{-1}$	-8.7475×10^7	-9.5319×10^7	-9.9756×10^7
Nuclear quadrupole moment/m^2	–	–	–
Frequency (^1H = 100 Hz; 2.3488T)/MHz	32.699	35.625	37.272

• ELECTRON SHELL DATA

Ground state electron configuration: $[Kr]4d^{10}5s^25p^2$
Term symbol: $^3P_0$
Electron affinity ($M \to M^-$)/kJ mol^{-1}**:** 116

Ionization energies/kJ mol^{-1}**:**

1.	$M \to M^+$	708.6
2.	$M^+ \to M^{2+}$	1411.8
3.	$M^{2+} \to M^{3+}$	2943.0
4.	$M^{3+} \to M^{4+}$	3930.2
5.	$M^{4+} \to M^{5+}$	6974
6.	$M^{5+} \to M^{6+}$	(9900)
7.	$M^{6+} \to M^{7+}$	(12 200)
8.	$M^{7+} \to M^{8+}$	(14 600)
9.	$M^{8+} \to M^{9+}$	(17 000)
10.	$M^{9+} \to M^{10+}$	(20 600)

Electron binding energies/eV

K	1s	29 200
L_I	2s	4465
L_{II}	$2p_{1/2}$	4156
L_{III}	$2p_{3/2}$	3929
M_I	3s	884.7
M_{II}	$3p_{1/2}$	756.5
M_{III}	$3p_{3/2}$	714.6
M_{IV}	$3d_{3/2}$	493.2
M_V	$3d_{5/2}$	484.9

continued in Appendix 2, p258

Main lines in atomic spectrum
[Wavelength/nm(species)]

224.605 (I) (AA)
235.484 (I)
242.949 (I)
283.999 (I)
286.333 (I)
303.412 (I)

• CRYSTAL DATA

Crystal structure (cell dimensions/pm), space group

α-Sn (grey) cubic (a = 648.92), Fd3m
β-Sn (white) tetragonal (a = 583.16, c = 318.13), $I4_1$/amd
$T(\alpha \to \beta)$ = 286.4 K [β form at room temperature]

X-ray diffraction: mass absorption coefficients (μ/ρ)/cm^2 g^{-1}**:** CuK_α 256 MoK_α 31.1
Neutron scattering length, $b/10^{-12}$ cm: 0.6225
Thermal neutron capture cross-section, σ_a/barns**:** 0.626

• GEOLOGICAL DATA

Minerals

There are a few rare tin-containing minerals but only one ore of significance, cassiterite.

Mineral	Formula	Density	Hardness	Crystal appearance
Cassiterite	SnO_2	6.99	6 – 7	tet., adam./met. brown

Chief ores: cassiterite
World production/tonnes y^{-1}**:** 165 000
Main mining areas: Malaysia, Sumatra, Russia, China, Bolivia, Zaire
Reserves/tonnes: 4.5×10^6
Specimen: available as bars, beads, foil, granules, rod, shot and wire. Safe.

Abundances

Sun (relative to H = 1×10^{12})**:** 100
Earth's crust/p.p.m.**:** 2.2
Seawater/p.p.m.**:**
Atlantic surface: 2.3×10^{-6}
Atlantic deep: 5.8×10^{-6}
Residence time/years**:** n.a.
Classification: scavenged
Oxidation state: IV

Ti

Atomic number: 22
Relative atomic mass (^{12}C = 12.0000): **47.867**

CAS: [7440-32-6]

• CHEMICAL DATA

Discovery: Titanium was discovered in 1791 by Rev. W. Gregor at Creed, Cornwall, England, and independently in 1795 by M.H. Klaproth at Berlin, Germany.
Description: Titanium is hard, lustrous, silvery metal which resists corrosion due to an oxide layer on its surface. However, the powdered metal will burn if ignited. Titanium us unaffected by many acids (except HF, H_3PO_4 and concentrated H_2SO_4), and alkalis. White titanium dioxide is used in paints because of its covering power. The metal itself is used in chemical plants, lightweight alloys, hip replacement joints, etc.

Radii/pm: Ti^{4+} 69; Ti^{2+} 80; atomic 145; covalent 132
Electronegativity: 1.54 (Pauling); 1.32 (Allred); 3.45 eV (absolute)
Effective nuclear charge: 3.15 (Slater); 4.82 (Clementi); 6.37 (Froese-Fischer)

Standard reduction potentials $E^\ominus$/V

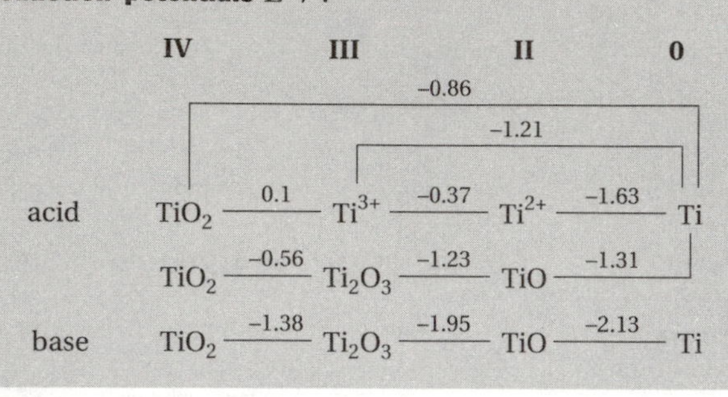

Oxidation states

Ti^{-II}	d^6	$[Ti(CO)_6]^{2-}$
Ti^{-I}	d^5	rare $[Ti(bipyridyl)_3]^-$
Ti^0	d^4	rare $[Ti(bipyridyl)_3]$, $[Ti(CO)_4\{CH_3C(CH_2PMe_2)_3\}]$
Ti^{II}	d^2	TiO, $TiCl_2$, $TiBr_2$, TiI_2, no solution chemistry (it reduces H_2O); complexes
Ti^{III}	d^1	Ti_2O_3, $[Ti(OH_2)_6]^{3+}$ (aq), TiF_3, $TiCl_3$ etc., complexes
Ti^{IV}	[Ar]	TiO_2, TiO^{2+} (aq), $[Ti(OH)_3]^{2+}$ (aq), TiF_4, $TiCl_4$ etc., titanates (TiO_3^{4-}, TiO_3^{2-}), complexes

• PHYSICAL DATA

Melting point/K: 1933
Boiling point/K: 3560
ΔH_{fusion}/kJ mol^{-1}: 20.9
ΔH_{vap}/kJ mol^{-1}: 428.9

Thermodynamic properties (298.15 K, 0.1 MPa)

State	$\Delta_f H^\ominus$/kJ mol^{-1}	$\Delta_f G^\ominus$/kJ mol^{-1}	$S^\ominus$/J K^{-1} mol^{-1}	C_p/J K^{-1} mol^{-1}
Solid	0	0	30.63	25.02
Gas	469.9	425.1	180.298	24.430

Density/kg m^{-3}: 4540 [293 K]; 4110 [liquid at m.p.]
Molar volume/cm^3: 10.55
Thermal conductivity/W m^{-1} K^{-1}: 21.9 [300 K]
Coefficient of linear thermal expansion/K^{-1}: 8.35×10^{-6}
Electrical resistivity/Ω m: 42.0×10^{-8} [293 K]
Mass magnetic susceptibility/kg^{-1} m^3: $+4.01 \times 10^{-8}$ (s)

Young's modulus/GPa: 120.2
Rigidity modulus/GPa: 45.6
Bulk modulus/GPa: 108.4
Poisson's ratio/GPa: 0.361

• BIOLOGICAL DATA

Biological role
None.

Toxicity
Low.
Toxic intake: low toxicity as metal, oxide (TiO_2) and inorganic titanium(IV) salts.
Lethal intake: non-lethal

Hazards
Some titanium compounds are dangerous to handle, such as $TiCl_3$, which is corrosive. Titanium has a stimulatory effect, and is a suspected carcinogen.

Levels in humans
Blood/mg dm^{-3}: 0.054
Bone/p.p.m.: n.a.
Liver/p.p.m.: 1.2 – 4.7
Muscle/p.p.m.: 0.9 – 2.2
Daily dietary intake: 0.8 mg
Total mass of element in average (70 kg) person: 20 mg

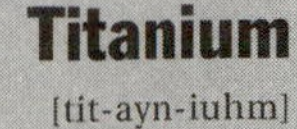

[tit-ayn-iuhm]

Discovery: see Chemical Data section.
[Named after the Titans, the sons of the Earth goddess of Greek mythology]
French, *titane*; German, *Titan*; Italian, *titanio*; Spanish, *titanio*

• NUCLEAR DATA

Number of isotopes (including nuclear isomers): 13 **Isotope mass range:** 41 → 53

Key isotopes

Nuclide	Atomic mass	Natural abundance (%)	Nuclear spin *I*	Nuclear magnetic moment μ	Uses
^{46}Ti	45.952 629	8.0	0+		E
^{47}Ti	46.951 764	7.3	5/2–	–0.78848	E, NMR
^{48}Ti	47.947 947	73.8	0+		E
^{49}Ti	48.947 871	5.5	7/2–	–1.10417	E, NMR
^{50}Ti	49.944 792	5.4	0+		

A table of radioactive isotopes is given in Appendix 1, on p251.

NMR [Reference: $[TiF_6]^{2-}$]	^{47}Ti	^{49}Ti
Relative sensitivity (^{1}H = 1.00)	2.09×10^{-3}	3.76×10^{-3}
Receptivity (^{13}C = 1.00)	0.864	1.18
Magnetogyric ratio/rad $T^{-1}s^{-1}$	1.5084×10^7	1.5080×10^7
Nuclear quadrupole moment/m^2	$+0.290 \times 10^{-28}$	$+0.240 \times 10^{-28}$
Frequency (^{1}H = 100 Hz; 2.3488T)/MHz	5.637	5.638

• ELECTRON SHELL DATA

Ground state electron configuration: $[Ar]3d^24s^2$
Term symbol: $^3F_2$
Electron affinity ($M \rightarrow M^-$)/kJ mol^{-1}**:** 7.6

Ionization energies/kJ mol^{-1}**:**

1.	$M \rightarrow M^+$	658
2.	$M^+ \rightarrow M^{2+}$	1310
3.	$M^{2+} \rightarrow M^{3+}$	2652
4.	$M^{3+} \rightarrow M^{4+}$	4175
5.	$M^{4+} \rightarrow M^{5+}$	9573
6.	$M^{5+} \rightarrow M^{6+}$	11 516
7.	$M^{6+} \rightarrow M^{7+}$	13 590
8.	$M^{7+} \rightarrow M^{8+}$	16 260
9.	$M^{8+} \rightarrow M^{9+}$	18 640
10.	$M^{9+} \rightarrow M^{10+}$	20 830

Electron binding energies/eV

K	1s	4966
L_I	2s	560.9
L_{II}	$2p_{1/2}$	460.2
L_{III}	$2p_{3/2}$	453.8
M_I	3s	58.7
M_{II}	$3p_{1/2}$	32.6
M_{III}	$3p_{3/2}$	32.6

Main lines in atomic spectrum
[Wavelength/nm(species)]
323.452 (II)
334.941 (II)
336.121 (II)
364.268 (I)
365.350 (I) (AA)
399.864 (I)

• CRYSTAL DATA

Crystal structure (cell dimensions/pm), space group
α-Ti h.c.p. (a = 295.11, c = 468.43), $P6_3/mmc$
β-Ti b.c.c. (a = 330.65), Im3m
$T(\alpha \rightarrow \beta)$ = 1155 K
High pressure form: (a = 462.5; c = 281.3), $P\bar{3}m1$

X-ray diffraction: mass absorption coefficients $(\mu/\rho)/cm^2\,g^{-1}$**:** CuK_α 208 MoK_α 24.2
Neutron scattering length, $b/10^{-12}$ cm**:** –0.3438
Thermal neutron capture cross-section, σ_a/barns**:** 6.09

• GEOLOGICAL DATA

Minerals

Mineral	Formula	Density	Hardness	Crystal appearance
Anatase	β-TiO_2	3.90	5.5 – 6	tet., adam./brown, green, etc.
Brookite	γ-TiO_2	4.14	5.5 – 6	orth., met. adam./brown, black
Ilmenite	$FeTiO_3$	4.72	5 – 6	rhom. met. black
Perovskite*	$CaTiO_2$	4.01	5.5	orth., adam./met. black
Rutile	α-TiO_2	4.23	6 – 6.5	tet., met. lustre brown/yellowish
Titanite	$CaTiSiO_5$	3.50	5 – 5.5	mon., adam./res. yellow/brown

*Varieties of perovskite can be rich in niobium, cerium and other rare earth elements and may be a source of these metals.

Chief ores: ilmenite; sometimes anatase is mined
World production/tonnes y^{-1}**:** 99 000 (titanium metal); 3×10^6 (TiO_2)
Main mining areas: Norway, India, Brazil, Canada, USA, Russia
Reserves/tonnes**:** 440×10^6
Specimen: available as crystals, foil, granules, powder, rod or wire. Safe.

Abundances

Sun (relative to H = 1×10^{12})**:** 1.12×10^5
Earth's crust/p.p.m.**:** 5600
Seawater/p.p.m.**:** 4.8×10^{-4}
Residence time/years**:** 50
Classification: n.a.
Oxidation state: IV

W

Atomic number: 74
Relative atomic mass ($^{12}C = 12.0000$): **183.84**

CAS: [7440-33-7]

• CHEMICAL DATA

Description: Tungsten is generally obtained as a dull grey powder, which is difficult to melt. The bulk metal is lustrous and silvery white, and resists attack by oxygen, acids and alkalis. Tungsten is used in alloys, to which it imparts great strength, in light bulb filaments and cutting tools.

Radii/pm: W^{6+} 62; W^{4+} 68; atomic 137; covalent 130
Electronegativity: 2.36 (Pauling); 1.40 (Allred); 4.40 eV (absolute)
Effective nuclear charge: 4.35 (Slater); 9.85 (Clementi); 14.22 (Froese-Fischer)

Standard reduction potentials $E^{\ominus}$/V

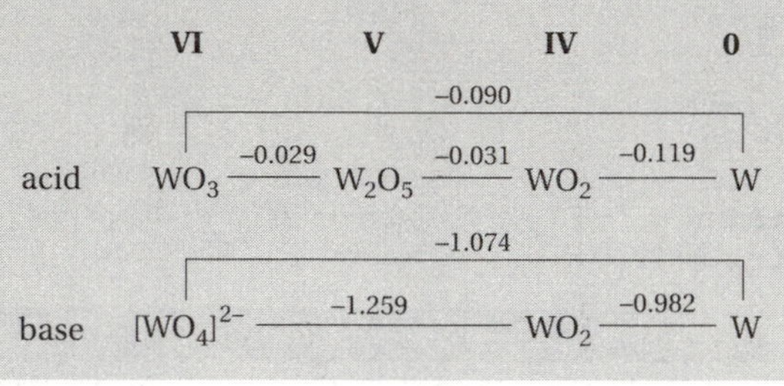

Oxidation states

W^{-IV}	d^{10}	$[W(CO)_4]^{4-}$
W^{-II}	d^8	$[W(CO)_5]^{2-}$
W^{-I}	d^7	$[W_2(CO)_{10}]^{2-}$
W^0	d^6	$[W(CO)_6]$
W^I	d^5	$[W(CO)_3(\eta\text{-}C_5H_5)]_2$
W^{II}	d^4	WCl_2, WBr_2, WI_2, complexes
W^{III}	d^3	WCl_3, WBr_3, WI_3, complexes
W^{IV}	d^2	WO_2, WF_4, WCl_4 etc., WS_2, complexes
W^V	d^1	W_2O_5, WF_5, WCl_5, $[WF_6]^-$, complexes
W^{VI}	$d^0[f^{14}]$	WO_3, $[WO_4]^{2-}$, WF_6, WCl_6, WCl_4O, polytungstates, complexes

[There are no aqua ions of W in any oxidation state.]

• PHYSICAL DATA

Melting point/K: 3680 ± 20
Boiling point/K: 5930
ΔH_{fusion}/kJ mol^{-1}: 35.2
ΔH_{vap}/kJ mol^{-1}: 799.1

Thermodynamic properties (298.15 K, 0.1 MPa)

State	$\Delta_f H^{\ominus}$/kJ mol^{-1}	$\Delta_f G^{\ominus}$/kJ mol^{-1}	$S^{\ominus}$/J K^{-1} mol^{-1}	C_p/J K^{-1} mol^{-1}
Solid	0	0	32.64	24.27
Gas	849.4	807.1	173.950	21.309

Density/kg m^{-3}: 19 300 [293 K]; 17 700 [liquid at m.p.]
Molar volume/cm^3: 9.53
Thermal conductivity/W m^{-1} K^{-1}: 174 [300 K]
Coefficient of linear thermal expansion/K^{-1}: 4.59×10^{-6}
Electrical resistivity/Ω m: 5.65×10^{-8} [300 K]
Mass magnetic susceptibility/kg^{-1} m^3: $+4.0 \times 10^{-9}$ (s)

Young's modulus/GPa: 411
Rigidity modulus/GPa: 160.6
Bulk modulus/GPa: 311
Poisson's ratio/GPa: 0.28

• BIOLOGICAL DATA

Biological role

None.

Toxicity

Toxic intake: mildly toxic
Lethal intake: LD_{50} (metal, rat) = 2000 mg kg^{-1}

Hazards

Tungsten dust is a skin and eye irritant, and an experimental teratogen.

Levels in humans

Blood/mg dm^{-3}: 0.001
Bone/p.p.m.: 0.00025
Liver/p.p.m.: n.a.
Muscle/p.p.m.: n.a.
Daily dietary intake: 0.001 – 0.015 mg
Total mass of element in average (70 kg) person: *c.* 0.02 mg

Discovered in 1783 by J.J. and F. Elhuijar at Vergara, Spain.
[Swedish, *tung sten* = heavy stone; wolfram is named after wolframite]
French, *tungstène*; German, *Wolfram*; Italian, *wolframio* (*tungsteno*); Spanish, *wolframio*

Tungsten

[tung-sten]

• NUCLEAR DATA

Number of isotopes (including nuclear isomers): 29 **Isotope mass range:** 160 → 190

Key isotopes

Nuclide	Atomic mass	Natural abundance (%)	Nuclear spin I	Nuclear magnetic moment μ	Uses
^{180}W	179.946 701	0.12	0+		E
^{182}W	181.948 202	26.3	0+		E
^{183}W	182.950 220	14.3	1/2–	+0.117 784 7	E, NMR
^{184}W	183.950 928	30.7	0+		E
^{186}W	185.954 357	28.6	0+		E

A table of radioactive isotopes is given in Appendix 1, on p252.

NMR [Reference: WF_6]	^{183}W
Relative sensitivity (^1H = 1.00)	7.20×10^{-4}
Receptivity (^{13}C = 1.00)	0.0589
Magnetogyric ratio/rad $T^{-1}s^{-1}$	1.1154×10^7
Nuclear quadrupole moment/m^2	–
Frequency (^1H = 100 Hz; 2.3488T)/MHz	4.161

• ELECTRON SHELL DATA

Ground state electron configuration: $[Xe]4f^{14}5d^4 6s^2$
Term symbol: $^5D_0$
Electron affinity ($M \rightarrow M^-$)/kJ mol^{-1}**:** 78.6

Ionization energies/kJ mol^{-1}**:**

1.	$M \rightarrow M^+$	770
2.	$M^+ \rightarrow M^{2+}$	(1700)
3.	$M^{2+} \rightarrow M^{3+}$	(2300)
4.	$M^{3+} \rightarrow M^{4+}$	(3400)
5.	$M^{4+} \rightarrow M^{5+}$	(4600)
6.	$M^{5+} \rightarrow M^{6+}$	(5900)

Electron binding energies/eV

K	1s	69 525
L_I	2s	12 100
L_{II}	$2p_{1/2}$	11 544
L_{III}	$2p_{3/2}$	10 207
M_I	3s	2820
M_{II}	$3p_{1/2}$	2575
M_{III}	$3p_{3/2}$	2281
M_{IV}	$3d_{3/2}$	1949
M_V	$3d_{5/2}$	1809

continued in Appendix 2, p258

Main lines in atomic spectrum
[Wavelength/nm(species)]

202.998 (II)
207.911 (II)
255.135 (I) (AA)
400.875 (I)
407.436 (I)
429.461 (I)

• CRYSTAL DATA

Crystal structure (cell dimensions/pm), space group
b.c.c. (a = 316.522), Im3m

X-ray diffraction: mass absorption coefficients (μ/ρ)/$cm^2\ g^{-1}$**:** CuK_α 172 MoK_α 99.1
Neutron scattering length, $b/10^{-12}$ cm**:** 0.486
Thermal neutron capture cross-section, σ_a/barns**:** 18.3

• GEOLOGICAL DATA

Minerals

Mineral	Formula	Density	Hardness	Crystal appearance
Ferberite	$FeWO_4$	7.40	4 – 4.5	mon., met. black
Scheelite	$CaWO_4$	6.10	4.5 – 5	tet., vit./adam. colourless
Wolframite	$(Fe,Mn)WO_4$	7.3	4 – 4.5	mon., sub-met./adam. greyish-black

Chief ores: scheelite and wolframite
World production/tonnes y^{-1}**:** 45 100
Main mining areas: China, Malaysia, Burma, Bolivia, Canada, Australia, Japan, USA
Reserves/tonnes: 1.5×10^6
Specimen: available as foil, powder, rod or wire. Safe.

Abundances

Sun (relative to H = 1×10^{12})**:** 50
Earth's crust/p.p.m.**:** 1
Seawater/p.p.m.**:** 9.2×10^{-5}
Residence time/years**:** n.a.
Oxidation state: VI

U

Atomic number: 92
Relative atomic mass ($^{12}C = 12.0000$)**:** 238.0289

CAS: [7440-61-1]

• CHEMICAL DATA

Discovery: Uranium was discovered in 1789 by M.J. Klaproth at Berlin, Germany. It was first isolated as the metal in 1841 by W. M. Peligot at Paris, France.

Description: Uranium is a silvery, ductile, malleable, radioactive metal. It tarnishes in air and is attacked by steam and acids, but not by alkalis. Uranium is used as nuclear fuel and in nuclear weapons.

Radii/pm: U^{6+} 80; U^{5+} 89; U^{4+} 97; U^{3+} 103; atomic 154;
Electronegativity: 1.38 (Pauling); 1.22 (Allred); n.a. eV (absolute)
Effective nuclear charge: 1.80 (Slater)

Standard reduction potentials $E^{\ominus}$/V

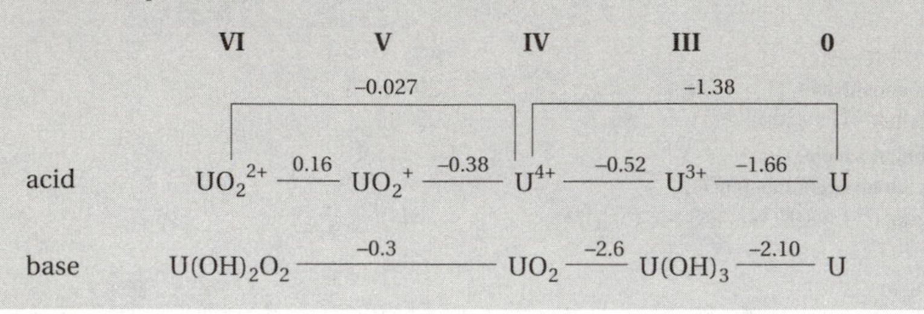

Oxidation states

U^{II}	f^3d^1	UO?
U^{III}	f^3	$[U(OH_2)_x]^{3+}$ (aq) unstable, UF_3, UCl_3 etc., $[U(C_5H_5)_3]$
U^{IV}	f^2	UO_2, $[U(OH_2)_x]^{4+}$ (aq) and salts, UF_4, UCl_4 etc., $[UCl_6]^{2-}$, $[U(C_5H_5)_4]$
U^{V}	f^1	U_2O_5, UO_2^{+} (aq) unstable, UF_5, UCl_5, UBr_5, $[UF_6]^{-}$, $[UF_7]^{2-}$, $[UF_8]^{3-}$
U^{VI}	[Rn]	UO_3, UO_2^{2+} (aq) and salts, UF_6, UCl_6, complexes

Mixed valence oxides: U_4O_9, U_3O_7, U_3O_8

• PHYSICAL DATA

Melting point/K: 1405.5
Boiling point/K: 4018
ΔH_{fusion}/kJ mol^{-1}: 15.5
ΔH_{vap}/kJ mol^{-1}: 422.6

Thermodynamic properties (298.15 K, 0.1 MPa)

State	$\Delta_f H^{\ominus}$/kJ mol^{-1}	$\Delta_f G^{\ominus}$/kJ mol^{-1}	$S^{\ominus}$/J K^{-1} mol^{-1}	C_p/J K^{-1} mol^{-1}
Solid	0	0	50.21	27.665
Gas	535.6	491.2	199.77	23.694

Density/kg m^{-3}: 18 950 [293 K]; 17 907 [liquid at m.p.]
Molar volume/cm^3: 12.56
Thermal conductivity/W m^{-1} K^{-1}: 27.6 [300 K]
Coefficient of linear thermal expansion/K^{-1}: 12.6×10^{-6}
Electrical resistivity/Ω m: 30.8×10^{-8} [273 K]
Mass magnetic susceptibility/kg^{-1} m^3: $+2.16 \times 10^{-8}$ (s)
Young's modulus/GPa: 175.8
Rigidity modulus/GPa: 73.1
Bulk modulus/GPa: 97.9
Poisson's ratio/GPa: 0.20

• BIOLOGICAL DATA

Biological role

None.

Toxicity

The radiotoxicity of uranium generally overrides other toxicity considerations, but it also has high chemical toxicity.

Lethal intake: LD_{50} (rats) = 36 mg kg^{-1}

Hazards

Uranium compounds are very poisonous, and may cause irreversible kidney damage. Soluble compounds, which can be absorbed through the skin, pass quickly through the body causing less damage than insoluble ones. The high toxicity of insoluble oxides is mainly due to their lodging as dust in the lungs. Uranium is recognized as a human carcinogen.

Levels in humans

Blood/mg dm^{-3}: 5×10^{-4}
Bone/p.p.m.: $(0.16 - 70) \times 10^{-3}$
Liver/p.p.m.: 0.003
Muscle/p.p.m.: 0.0009
Daily dietary intake: 0.001 – 0.002 mg
Total mass of element in average (70 kg) person: 0.1 mg (range 0.01 – 0.4 mg)

Discovery: see Chemical Data section.
[Named after the planet Uranus]
French, *uranium*; German, *Uran*; Italian, *uranio*; Spanish, *uranio*

Uranium

[yoo-rayn-iuhm]

• NUCLEAR DATA

Number of isotopes (including nuclear isomers): 17 **Isotope mass range:** 226 → 242

Key isotopes

Nuclide	Atomic mass	Natural abundance (%)	Nuclear spin *I*	Nuclear magnetic moment μ	Uses
^{234}U*	234.040 946	0.005	0+	0	
^{235}U**	235.043 924	0.720	7/2–	–0.35	NMR
^{238}U***	238.050 784	99.275	0+	0	

*^{234}U is radioactive with a half-life of 2.45×10^5 y and decay mode α (4.856 MeV) with γ.
**^{235}U is radioactive with a half-life of 7.04×10^8 y and decay mode α (4.6793 MeV) with γ.
***^{238}U is radioactive with a half-life of 4.46×10^9 y and decay mode α (4.039 MeV) with γ.
A table of other radioactive isotopes is given in Appendix 1, on p252.

NMR [Reference: UF_6]

	^{233}U	^{235}U
Relative sensitivity (^{1}H = 1.00)		1.21×10^{-4}
Receptivity (^{13}C = 1.00)		5.4×10^{-3}
Magnetogyric ratio/rad $T^{-1}s^{-1}$		-0.4926×10^7
Nuclear quadrupole moment/m^2	3.663×10^{-28}	4.936×10^{-28}
Frequency (^{1}H = 100 Hz; 2.3488T)/MHz		1.790

• ELECTRON SHELL DATA

Ground state electron configuration: [Rn]$5f^3 6d^1 7s^2$
Term symbol: $^5L_6$
Electron affinity ($M \rightarrow M^-$)/kJ mol^{-1}**:** n.a.

Ionization energies/kJ mol^{-1}**:**

1. $M \rightarrow M^+$		584
2. $M^+ \rightarrow M^{2+}$		1420

Electron binding energies/eV

K	1s	115 606
L_I	2s	21 757
L_{II}	$2p_{1/2}$	20 948
L_{III}	$2p_{3/2}$	17 166
M_I	3s	5548
M_{II}	$3p_{1/2}$	5182
M_{III}	$3p_{3/2}$	4303
M_{IV}	$3d_{3/2}$	3728
M_V	$3d_{5/2}$	3552

continued in Appendix 2, p258

Main lines in atomic spectrum
[Wavelength/nm(species)]
356.659 (I) (AA)
358.488 (I)
367.007 (II)
385.958 (II)
389.036 (II)
409.013 (II)

• CRYSTAL DATA

Crystal structure (cell dimensions/pm), space group
α-U orthorhombic (a = 284.785, b = 585.801, c = 494.553), Cmcm
β-U tetragonal (a = 1076.0, b = 565.2), $P4_2$/mnm or $P4_2$/nm
γ-U b.c.c. (a = 352.4), Im3m
$T(\alpha \rightarrow \beta)$ = 941 K; $T(\beta \rightarrow \gamma)$ = 1047 K

X-ray diffraction: mass absorption coefficients (μ/ρ)/$cm^2 g^{-1}$**:** CuK_α 352 MoK_α 153
Neutron scattering length, $b/10^{-12}$ cm**:** 0.8417
Thermal neutron capture cross-section, σ_a/barns**:** 7.57

• GEOLOGICAL DATA

Minerals

Many minerals are known.

Mineral	Formula	Density	Hardness	Crystal appearance
Autunite	$Ca(UO_2)_2(PO_2)_2.10H_2O$	3.1	2 – 2.5	tet., vitreous pearly yellow-green
Betafite	$(Ca,U)_2(Ti,Nb,Ta)_2O_6(OH)$	4.5	4 – 5.5	cub., waxy/vitr., pitch black
Carnotite	$K_2(UO_2)_2(VO_4)_2.3H_2O$	4.95	n.a./soft	mon., dull/earthy/pearly
Samarskite	$(Y,Ce,U,Fe)_3(Nb,Ta,Ti)_5O_{16}$	5.69	5 – 6	orth., vitreous/resinous black
Torbernite	$Cu(UO_2)_2(PO_4)_2.8\text{-}12H_2O$	3.22	2 – 2.5	tet., vitreous emerald green
Uraninite	UO_2	10.6	5 – 6	cub., sub-metallic pitch black

Chief ores: uraninite; autunite; carnotite; some varieties of samarskite contain up to 23% uranium (see tantalum); torbernite is a minor ore.
World production/tonnes y^{-1}**:** 35 000
Main mining areas: Canada, Zaire, Czech Republic, USA.
Reserves/tonnes**:** 3.5×10^6, plus 6.3×10^6 in phosphate ores.
Specimen: not commercially available as uranium metal, but compounds can be bought.
Radioactive!

Abundances

Sun (relative to H = 1×10^{12})**:** < 4
Earth's crust/p.p.m.**:** 2.4
Seawater/p.p.m.**:** 3.13×10^{-3}
Residence time/years**:** 300 000
Classification: accumulating
Oxidation state: VI

V

Atomic number: 23
Relative atomic mass ($^{12}C = 12.0000$): **50.9415**

CAS: [7440-62-2]

• CHEMICAL DATA

Discovery: Vanadium was discovered in 1801 by A.M. del Rio at Mexico City, Mexico. Rediscovered in 1831 by N.G. Selfström at Falun, Sweden.
Description: Vanadium is a shiny, silvery metal, which is soft when pure. It resists corrosion due to a protective film of oxide on the surface. Vanadium is attacked by concentrated acids, but not by alkalis, not even when these are molten. The metal is used mainly as alloys, especially in steels.

Radii/pm: V^{5+} 59; V^{4+} 61; V^{3+} 65; V^{2+} 72; atomic 132
Electronegativity: 1.63 (Pauling); 1.45 (Allred); 3.6 eV (absolute)
Effective nuclear charge: 3.30 (Slater); 4.98 (Clementi); 6.65 (Froese-Fischer)

Standard reduction potentials $E^{\ominus}$/V

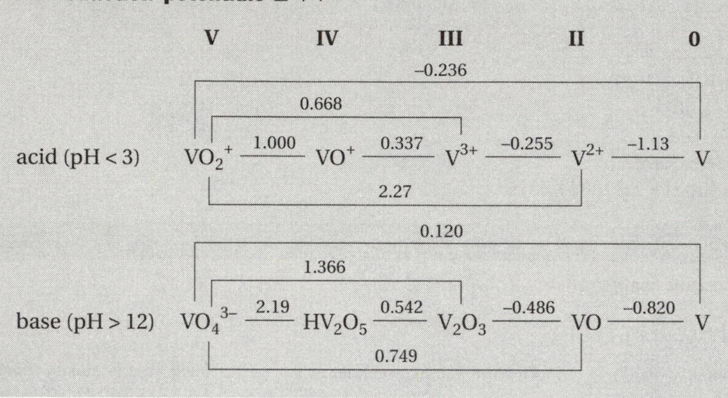

Oxidation states

V^{-III}	d^8	rare $[V(CO)_5]^{3-}$
V^{-I}	d^6	$[V(CO)_6]^-$
V^0	d^5	$[V(CO)_6]$
V^I	d^4	$[V(dipyridyl)_3]^+$
V^{II}	d^3	VO, $[V(OH_2)_6]^{2+}$ (aq), VF_2, VCl_2, complexes
V^{III}	d^2	V_2O_3, $[V(OH_2)_6]^{3+}$ (aq), VF_3, VCl_3, $[VCl_4]^-$
V^{IV}	d^1	VO_2, VO^{2+} (aq), VF_4, VCl_4, complexes
V^V	d^0 [Ar]	V_2O_5, VO_2^+ (aq), VO_4^{3-} (aq alkali), VF_5, $[VF_6]$, complexes

• PHYSICAL DATA

Melting point/K: 2160
Boiling point/K: 3650
ΔH_{fusion}/kJ mol^{-1}: 17.6
ΔH_{vap}/kJ mol^{-1}: 458.6

Thermodynamic properties (298.15 K, 0.1 MPa)

State	$\Delta_f H^{\ominus}$/kJ mol^{-1}	$\Delta_f G^{\ominus}$/kJ mol^{-1}	$S^{\ominus}$/J K^{-1} mol^{-1}	C_p/J K^{-1} mol^{-1}
Solid	0	0	28.91	24.89
Gas	514.21	754.43	182.298	26.012

Density/kg m^{-3}: 6110 [292 K]; 5550 [liquid at m.p.]
Molar volume/cm^3: 8.34
Thermal conductivity/W m^{-1} K^{-1}: 30.7 [300 K]
Coefficient of linear thermal expansion/K^{-1}: 8.3×10^{-6}
Electrical resistivity/Ω m: 24.8×10^{-8} [293 K]
Mass magnetic susceptibility/kg^{-1} m^3: $+6.28 \times 10^{-8}$ (s)
Young's modulus/GPa: 127.6
Rigidity modulus/GPa: 46.7
Bulk modulus/GPa: 158
Poisson's ratio/GPa: 0.365

• BIOLOGICAL DATA

Biological role

Essential to some species including humans; it also acts to stimulate metabolism.

Toxicity

Toxic intake: toxicity varies
Lethal intake: LD_{50} (V_2O_5, oral, rat) = 10 mg kg^{-1}

Hazards

Vanadium and its compounds irritate the eyes and lungs; the fumes of volatile compounds are highly toxic. Some vanadium compounds have experimental mutagenic effects.

Levels in humans

Blood/mg dm^{-3}: < 0.0002
Bone/p.p.m.: 0.0035
Liver/p.p.m.: 0.006
Muscle/p.p.m.: 0.02
Daily dietary intake: 0.04 mg
Total mass of element in average (70 kg) person: 0.11 mg

Discovery: see Chemical Data section.
[Named after Vanadis, a Scandinavian goddess]
French, *vanadium*; German, *Vanadium*; Italian, *vanadio*; Spanish, *vanadio*

Vanadium

[van-ay-di-uhm]

• NUCLEAR DATA

Number of isotopes (including nuclear isomers): 11 **Isotope mass range:** 44 → 55

Key isotopes

Nuclide	Atomic mass	Natural abundance (%)	Nuclear spin I	Nuclear magnetic moment μ	Uses
^{50}V*	49.947 161	0.25	6+	+3.347 45	E, NMR
^{51}V	50.943 962	99.75	7/2–	+5.157 4	E, NMR

*May be radioactive with a half-life of >3.9 × 10^{17} y.
A table of radioactive isotopes is given in Appendix 1, on p252.

NMR [Reference: $VOCl_3$]

	^{50}V	^{51}V
Relative sensitivity (^1H = 1.00)	0.0555	0.38
Receptivity (^{13}C = 1.00)	0.755	2150
Magnetogyric ratio/rad $T^{-1}s^{-1}$	2.6491×10^7	7.0362×10^7
Nuclear quadrupole moment/m^2	0.210×10^{-28}	-0.052×10^{-28}
Frequency (^1H = 100 Hz; 2.3488T)/MHz	9.970	26.289

• ELECTRON SHELL DATA

Ground state electron configuration: [Ar]$3d^34s^2$
Term symbol: $^4F_{3/2}$
Electron affinity ($M \rightarrow M^-$)/kJ mol^{-1}**:** 50.7

Ionization energies/kJ mol^{-1}**:**

1.	M	→ M^+	650
2.	M^+	→ M^{2+}	1414
3.	M^{2+}	→ M^{3+}	2828
4.	M^{3+}	→ M^{4+}	4507
5.	M^{4+}	→ M^{5+}	6294
6.	M^{5+}	→ M^{6+}	12 326
7.	M^{6+}	→ M^{7+}	14 489
8.	M^{7+}	→ M^{8+}	16 760
9.	M^{8+}	→ M^{9+}	19 860
10.	M^{9+}	→ M^{10+}	22 240

Electron binding energies/eV

K	1s	5465
L_I	2s	626.7
L_{II}	$2p_{1/2}$	519.8
L_{III}	$2p_{3/2}$	512.1
M_I	3s	66.3
M_{II}	$3p_{1/2}$	37.2
M_{III}	$3p_{3/2}$	37.2

Main lines in atomic spectrum
[Wavelength/nm(species)]

318.398 (I)
318.540 (I) (AA)
411.178 (I)
437.924 (I)
438.472 (I)
399.864 (I)

• CRYSTAL DATA

Crystal structure (cell dimensions/pm), space group
b.c.c. (a = 302.40), Im3m

X-ray diffraction: mass absorption coefficients (μ/ρ)/cm^2 g^{-1}**:** CuK_α 233 MoK_α 27.5
Neutron scattering length, $b/10^{-12}$ cm**:** –0.0382
Thermal neutron capture cross-section, σ_a/barns**:** 5.08

• GEOLOGICAL DATA

Minerals

Vanadium occurs in many minerals.

Mineral	Formula	Density	Hardness	Crystal appearance
Carnotite	$K_2(UO_2)_2(VO_4)_2.3H_2O$	4.95	n.a.*	mon., dull/earthy/pearly yellow
Descloizite	$PbZn(VO_4)(OH)$	6.2	3 – 3.5	orth., greasy brown/red
Patrónite	VS_4	2.81	2	mon., grey-black
Vanadinite	$Pb_5(VO_4)_3Cl$	6.86	2.7 – 3	hex., sub-res./sub-adam. orange/red

*soft

Chief ores: descloizite, patronite, vanadite, carnotite.

World production/tonnes y^{-1}**:** 7000

Main mining areas: not mined as such, but generally obtained as a by-product of other ores, and from Venezuelan oils.

Reserves/tonnes: n.a.

Specimen: available as foil, granules, powder, rod or turnings. *Care!*

Abundances

Sun (relative to H = 1×10^{12})**:** 1.05×10^4
Earth's crust/p.p.m.**:** 160
Seawater/p.p.m.**:**
Atlantic surface: 1.1×10^{-3}
Atlantic deep: n.a.
Pacific surface: 1.6×10^{-3}
Pacific deep: 1.8×10^{-3}
Residence time/years**:** 50 000
Classification: recycled
Oxidation state: V

Atomic number: 54
Relative atomic mass ($^{12}C = 12.0000$): 131.29

CAS: [7440-63-3]

• CHEMICAL DATA

Discovery: Xenon was discovered in 1898 by Sir William Ramsay and M.W. Travers at London, England.
Description: Xenon is a colourless, odourless gas obtained from liquid air. It is inert towards all other chemicals except fluorine gas, with which it reacts to form xenon fluorides. From these a range of other compounds, such as oxides, acids and salts can be made. Xenon has little commercial use, but in research it is employed as a supercritical fluid in various ways.

Radii / pm: Xe^+ 190; atomic 218; covalent 209; van der Waals 216
Electronegativity: 2.6 (Pauling); 2.40 (Allred); 5.85 eV (absolute)
Effective nuclear charge: 8.25 (Slater); 12.42 (Clementi); 15.61 (Froese-Fischer)

Standard reduction potentials $E^{\ominus}$/V

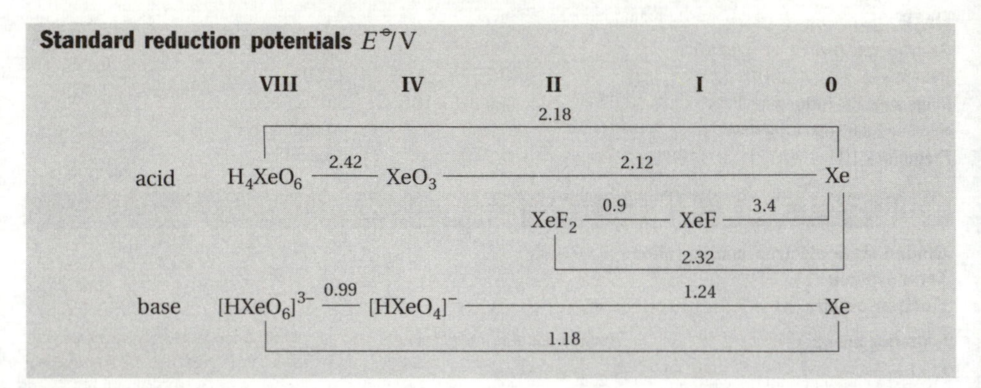

Oxidation states

Xe^0	[Xe]	clathrates: $Xe_8(OH_2)_{46}$, $Xe(quinol)_3$ (see argon)
Xe^{II}	s^2p^4	XeF_2, $[XeF]^+[AsF_6]^-$
Xe^{IV}	s^2p^2	XeF_4
Xe^{VI}	s^2	XeO_3, XeF_4O, XeF_2O_2, XeF_6, $[XeF_7]^-$, $[XeF_8]^{2-}$, $[XeF_5]^+[AsF_6]^-$
Xe^{VIII}	d^{10}	XeO_4, XeF_2O_3, Ba_2XeO_6, $[XeO_6]^{4-}$ (aq)

Covalent bonds

Bond	r/ pm	E/ kJ mol^{-1}
Xe^{VI}—O	176	84
Xe^{VI}—F	190	126

• PHYSICAL DATA

Melting point / K: 161.3
Boiling point / K: 166.1
Critical temperature / K: 289.75
Critical pressure / kPa: 5895

ΔH_{fusion} /kJ mol^{-1}: 3.10
ΔH_{vap} /kJ mol^{-1}: 12.65

Thermodynamic properties (298.15 K, 0.1 MPa)

State	$\Delta_f H^{\ominus}$/kJ mol^{-1}	$\Delta_f G^{\ominus}$/kJ mol^{-1}	$S^{\ominus}$/J K^{-1} mol^{-1}	C_p/J K^{-1} mol^{-1}
Gas	0	0	169.683	20.786

Density / kg m^{-3}: 3540 [s., m.p.]; 2939 [liq, b.p.]; 5.8971 [gas, 273 K]
Molar volume / cm^3: 37.09 [161 K]
Thermal conductivity / W m^{-1} K^{-1}: 0.00569 [300 K] (g)
Mass magnetic susceptibility / kg^{-1} m^3: -4.20×10^{-9} (g)

• BIOLOGICAL DATA

Biological role
None.

Toxicity
Non-toxic.

Hazards
Xenon is a harmless gas, although it could asphyxiate if it excluded oxygen from the lungs.

Levels in humans

Blood / mg dm^{-3}: trace
Bone / p.p.m.: nil
Liver / p.p.m.: nil
Muscle / p.p.m.: nil
Daily dietary intake: n.a. but low
Total mass of element in average (70 kg) person: n.a., but small

Discovery: see Chemical Data section.
[Greek, *xenos* = stranger]
French, *xénon*; German, *Xenon*; Italian, *xeno*; Spanish, *xenón*

Xenon

[zee-non]

• NUCLEAR DATA

Number of isotopes (including nuclear isomers): 35 **Isotope mass range:** 114 → 142

Key isotopes

Nuclide	Atomic mass	Natural abundance (%)	Nuclear spin *I*	Nuclear magnetic moment μ	Uses
^{124}Xe	123.905 894	0.10	0+		
^{126}Xe	125.904 281	0.09	0+		
^{128}Xe	127.903 531	1.91	0+		
^{129}Xe	128.904 780	26.4	1/2+	–0.777 977	NMR
^{130}Xe	129.903 509	4.1	0+		
^{131}Xe	130.905 072	21.2	3/2+	+0.6868	NMR
^{132}Xe	131.904 144	26.9	0+		
^{134}Xe	133.905 395	10.4	0+		
^{136}Xe	135.907 214	8.9	0+		

A table of radioactive isotopes is given in Appendix 1, on p252.

NMR [Reference: $XeOF_4$]	^{129}Xe	^{131}Xe
Relative sensitivity ($^1H = 1.00$)	0.0212	2.76×10^{-3}
Receptivity ($^{13}C = 1.00$)	31.8	3.31
Magnetogyric ratio/rad $T^{-1}s^{-1}$	-7.4003×10^7	2.1939×10^7
Nuclear quadrupole moment/m^2	–	-0.120×10^{-28}
Frequency ($^1H = 100$ Hz; 2.3488T)/MHz	27.660	8.199

• ELECTRON SHELL DATA

Ground state electron configuration: $[Kr]4d^{10}5s^25p^6 = [Xe]$
Term symbol: $^1S_0$
Electron affinity ($M \rightarrow M^-$)/kJ mol^{-1}**:** –41 (calc.)

Ionization energies/kJ mol^{-1}**:**

1.	$M \rightarrow M^+$	1170.4
2.	$M^+ \rightarrow M^{2+}$	2046
3.	$M^{2+} \rightarrow M^{3+}$	3097
4.	$M^{3+} \rightarrow M^{4+}$	(4300)
5.	$M^{4+} \rightarrow M^{5+}$	(5500)
6.	$M^{5+} \rightarrow M^{6+}$	(6600)
7.	$M^{6+} \rightarrow M^{7+}$	(9300)
8.	$M^{7+} \rightarrow M^{8+}$	(10 600)
9.	$M^{8+} \rightarrow M^{9+}$	(19 800)
10.	$M^{9+} \rightarrow M^{10+}$	(23 000)

Electron binding energies/eV

K	1s	34 561
L_I	2s	5453
L_{II}	$2p_{1/2}$	5107
L_{III}	$2p_{3/2}$	4786
M_I	3s	1148.7
M_{II}	$3p_{1/2}$	1002.1
M_{III}	$3p_{3/2}$	940.6
M_{IV}	$3d_{3/2}$	689.0
M_V	$3d_{5/2}$	676.4

continued in Appendix 2, p258

Main lines in atomic spectrum
[Wavelength/nm(species)]

823.164 (I)
828.012 (II)
881.941 (II)
3106.923 (I)
3507.025 (I)

• CRYSTAL DATA

Crystal structure (cell dimensions/pm), space group
f.c.c. (88 K) ($a = 619.7$), Fm3m

X-ray diffraction: mass absorption coefficients (μ/ρ)/cm^2 g^{-1}**:** CuK_α 306 MoK_α 39.2
Neutron scattering length, $b/10^{-12}$ cm**:** 0.492
Thermal neutron capture cross-section, σ_a/barns**:** 23.9

• GEOLOGICAL DATA

Minerals

None - only exists as a gas in the atmosphere.

Chief source: liquid air
World production/tonnes y^{-1}**:** *c.* 0.6
Reserves/tonnes: 2×10^9
Specimen: available in small pressurized canisters. Safe.

Abundances

Sun (relative to $H = 1 \times 10^{12}$)**:** n.a.
Earth's crust/p.p.m.: 2×10^{-6}
Atmosphere/p.p.m. (volume)**:** 0.086
Seawater/p.p.m.: 1×10^{-4}
Residence time/years**:** n.a.
Oxidation state: 0

Yb

Atomic number: 70
Relative atomic mass ($^{12}C = 12.0000$): 173.04

CAS: [7440-64-4]

• CHEMICAL DATA

Description: Ytterbium is a soft, silvery-white metal of the so-called rare earth group (more correctly termed the lanthanides). It is slowly oxidised by air and reacts with water. It has few uses but some is employed in stress gauges.

Radii/pm: Yb^{3+} 86; Yb^{2+} 113; atomic 194; covalent 170
Electronegativity: n.a. (Pauling); 1.06 (Allred); ≤ 3.5 eV (absolute)
Effective nuclear charge: 2.85 (Slater); 8.59 (Clementi); 11.90 (Froese-Fischer)

Standard reduction potentials E°/V

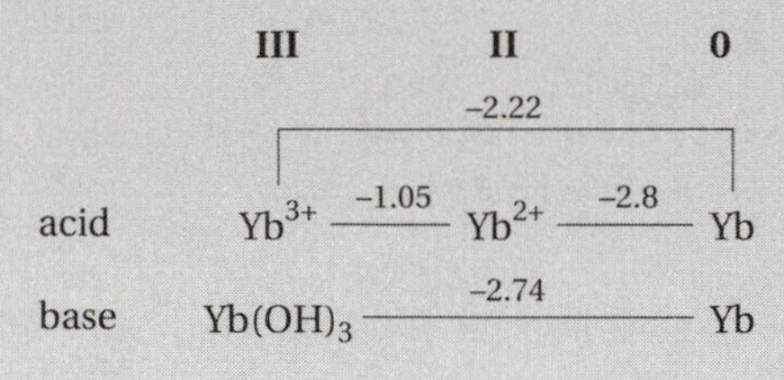

Oxidation states

Yb^{II}	f^{14}	YbO, YbS, YbF_2, $YbCl_2$ etc.
Yb^{III}	f^{13}	Yb_2O_3, $Yb(OH)_3$, $[Yb(OH_2)_x]^{3+}$ (aq), Yb^{3+} salts, YbF_3, $YbCl_3$ etc., $[YbCl_6]^{3-}$, complexes

• PHYSICAL DATA

Melting point/K: 1097
Boiling point/K: 1466
ΔH_{fusion}/kJ mol^{-1}: 9.20
ΔH_{vap}/kJ mol^{-1}: 159

Thermodynamic properties (298.15 K, 0.1 MPa)

State	$\Delta_f H^{\circ}$/kJ mol^{-1}	$\Delta_f G^{\circ}$/kJ mol^{-1}	S°/J K^{-1} mol^{-1}	C_p/J K^{-1} mol^{-1}
Solid	0	0	59.87	26.74
Gas	152.3	118.4	173.126	20.786

Density/kg m^{-3}: 6965 [293 K]
Molar volume/cm^3: 24.84
Thermal conductivity/W m^{-1} K^{-1}: 34.9 [300 K]
Coefficient of linear thermal expansion/K^{-1}: 25.0×10^{-6}
Electrical resistivity/Ω m: 29.0×10^{-8} [293 K]
Mass magnetic susceptibility/kg^{-1} m^3: $+1.81 \times 10^{-8}$ (s)

Young's modulus/GPa: 23.9
Rigidity modulus/GPa: 9.9
Bulk modulus/GPa: 30.5
Poisson's ratio/GPa: 0.207

• BIOLOGICAL DATA

Biological role

None, but acts to stimulate metabolism.

Toxicity

Toxic intake: n.a.
Lethal intake: LD_{50} (chloride, oral, mouse) = 6700 mg kg^{-1}

Hazards

Ytterbium is mildly toxic by ingestion and is a skin and eye irritant. It is also a suspected teratogen.

Levels in humans

Organs: n.a., but low
Daily dietary intake: n.a.
Total mass of element in average (70 kg) person: n.a.

Discovered in 1878 by J.C. Galissard de Marignac at Geneva, Switzerland.
[Named after Ytterby, Sweden]
French, *ytterbium*; German, *Ytterbium;* Italian, *itterbio*; Spanish, *yterbio*

Ytterbium

[i-terb-iuhm]

• NUCLEAR DATA

Number of isotopes (including nuclear isomers): 29 **Isotope mass range:** 154 → 179

Key isotopes

Nuclide	Atomic mass	Natural abundance (%)	Nuclear spin *I*	Nuclear magnetic moment *μ*	Uses
^{168}Yb	167.933 894	0.13	0+		E
^{170}Yb	169.934 759	3.05	0+		E
^{171}Yb	170.936 323	14.3	1/2–	+0.491 9	E, NMR
^{172}Yb	171.936 378	21.9	0+		E
^{173}Yb	172.938 208	16.12	5/2–	–0.677 6	E, NMR
^{174}Yb	173.938 859	31.8	0+		E
^{176}Yb	175.942 564	12.7	0+		E

A table of radioactive isotopes is given in Appendix 1, on p252.

NMR [Rarely studied]	^{171}Yb	^{173}Yb
Relative sensitivity (^{1}H = 1.00)	5.46×10^{-3}	1.33×10^{-3}
Receptivity (^{13}C = 1.00)	4.05	1.14
Magnetogyric ratio/rad $T^{-1}s^{-1}$	4.718×10^7	1.310×10^7
Nuclear quadrupole moment/m^2	–	$+2.800 \times 10^{-28}$
Frequency (^{1}H = 100 Hz; 2.3488T)/MHz	17.631	4.852

• ELECTRON SHELL DATA

Ground state electron configuration: [Xe]$4f^{14}6s^2$
Term symbol: $^1S_0$
Electron affinity ($M \rightarrow M^-$)/kJ mol^{-1}: ≤ 50

Ionization energies/kJ mol^{-1}:

1. M	→	M^+	603.4
2. M^+	→	M^{2+}	1176
3. M^{2+}	→	M^{3+}	2415
4. M^{3+}	→	M^{4+}	4220
5. M^{4+}	→	M^{5+}	6328

Electron binding energies/eV

K	1s	61 332
L_I	2s	10 486
L_{II}	$2p_{1/2}$	9978
L_{III}	$2p_{3/2}$	8944
M_I	3s	2398
M_{II}	$3p_{1/2}$	2173
M_{III}	$3p_{3/2}$	1950
M_{IV}	$3d_{3/2}$	1576
M_V	$3d_{5/2}$	1528

continued in Appendix 2, p258

Main lines in atomic spectrum
[Wavelength/nm(species)]

289.138 (II)
328.937 (II)
346.437 (I)
369.420 (II)
398.799 (I) (AA)
555.647 (I)

• CRYSTAL DATA

Crystal structure (cell dimensions/pm), space group

α-Yb f.c.c. (*a* = 548.62), Fm3m
β-Yb b.c.c. (*a* = 444), Im3m
$T(\alpha \rightarrow \beta)$ = 1 073 K

X-ray diffraction: mass absorption coefficients $(\mu/\rho)/cm^2\ g^{-1}$: CuK_α 146 MoK_α 84.5
Neutron scattering length, $b/10^{-12}$ cm: 1.243
Thermal neutron capture cross-section, σ_a/barns: 34.8

• GEOLOGICAL DATA

Minerals

Mineral	Formula	Density	Hardness	Crystal appearance
Euxenite*	$(Y,Ca,Ce,U,Th)(Nb,Ta,Ti)_2O_6$	5.0	5.5 – 6.5	orth., met. greasy black
Xenotime*	YPO_4	4.8	4 – 5	tet., vit./res. yellow-brown

*These ores contain extractable amounts of ytterbium.

Chief ores: euxenite, xenotime
World production/tonnes y^{-1}: *c.* 50
Main mining areas: USA, Canada. Euxenite is also found in Greenland and Brazil.
Reserves/tonnes: *c.* 1×10^6
Specimen: available as chips or ingots. Safe. Yb powder is a skin and eye irritant. *Care!*

Abundances

Sun (relative to H = 1×10^{12}): 8
Earth's crust/p.p.m.: 3.3
Seawater/p.p.m.:
Atlantic surface: 5×10^{-7}
Atlantic deep: 7.5×10^{-7}
Pacific surface: 3.7×10^{-7}
Pacific deep: 22×10^{-7}
Residence time/years: 400
Classification: recycled
Oxidation state: III

Y

Atomic number: 39
Relative atomic mass (^{12}C = 12.0000)**:** 88.90585

CAS: [7440-65-6]

• CHEMICAL DATA

Description: Yttrium is a soft, silvery-white metal that is stable in air due to the formation of an oxide film on its surface. It burns if ignited, and is attacked by water and forms hydrogen. Yttrium is used in various ways: to give red colours in TV screens; in X-ray filters; superconductors; and for superalloys.

Radii/pm: Y^{3+} 106; atomic 181; covalent 162
Electronegativity: 1.22 (Pauling); 1.11 (Allred); 3.19 eV (absolute)
Effective nuclear charge: 3.00 (Slater); 6.26 (Clementi); 8.72 (Froese-Fischer)

Standard reduction potentials $E^{\ominus}$/V

	III		0
acid	Y^{3+}	−2.37	Y
base	$Y(OH)_3$	−2.85	Y

Oxidation states

Y^{III} [Kr] Y_2O_3, $Y(OH)_3$, $[Y(OH_2)_x]^{3+}$ (aq), Y^{3+} salts, $Y_2(CO_3)_3$, YF_3, YCl_3 etc., YOCl, some complexes. [YH_2-YH_3 consists of $Y^{3+}H^-$, with complex bonding]

• PHYSICAL DATA

Melting point/K: 1795
Boiling point/K: 3611
ΔH_{fusion}/kJ mol^{-1}: 17.2
ΔH_{vap}/kJ mol^{-1}: 393.3

Thermodynamic properties (298.15 K, 0.1 MPa)

State	$\Delta_f H^{\ominus}$/kJ mol^{-1}	$\Delta_f G^{\ominus}$/kJ mol^{-1}	$S^{\ominus}$/J K^{-1} mol^{-1}	C_p/J K^{-1} mol^{-1}
Solid	0	0	44.43	26.53
Gas	421.3	381.1	179.48	25.86

Density/kg m^{-3}: 4469 [293 K]
Molar volume/cm^3: 19.89
Thermal conductivity/W m^{-1} K^{-1}: 17.2 [300 K]
Coefficient of linear thermal expansion/K^{-1}: 10.6×10^{-6}
Electrical resistivity/Ω m: 57.0×10^{-8} [298 K]
Mass magnetic susceptibility/kg^{-1} m^3: $+2.70 \times 10^{-8}$ (s)

Young's modulus/GPa: 66.3
Rigidity modulus/GPa: 25.5
Bulk modulus/GPa: n.a.
Poisson's ratio/GPa: 0.265

• BIOLOGICAL DATA

Biological role

None.

Toxicity

Toxic intake: n.a.
Lethal intake: LD_{50} (chloride, intraperitoneal, mouse) = 88 mg kg^{-1}

Hazards

Ytrrium is thought to be mildly toxic by ingestion, and yttrium salts are suspected of being carcinogenic.

Levels in humans

Blood/mg dm^{-3}: 0.0047
Bone/p.p.m.: 0.07
Liver/p.p.m.: < 0.01
Muscle/p.p.m.: 0.02
Daily dietary intake: 0.016 mg
Total mass of element in average (70 kg) person: 0.6 mg

Discovered in 1794 by J. Gadolin at Åbo, Finland.
[Named after Ytterby, Sweden]
French, *yttrium*; German, *Yttrium*; Italian, *ittrio*; Spanish, *ytrio*

Yttrium

[it-ree-uhm]

• NUCLEAR DATA

Number of isotopes (including nuclear isomers): 32 **Isotope mass range:** 80 → 99

Key isotopes

Nuclide	Atomic mass	Natural abundance (%)	Nuclear spin *I*	Nuclear magnetic moment μ	Uses
^{89}Y	89.905 849	100	1/2–	–0.1368	NMR

A table of radioactive isotopes is given in Appendix 1, on p253.

NMR [Reference: $Y(NO_3)_3$]	^{89}Y
Relative sensitivity (^1H = 1.00)	1.18×10^{-4}
Receptivity (^{13}C = 1.00)	0.668
Magnetogyric ratio/rad $T^{-1} s^{-1}$	-1.3108×10^7
Nuclear quadrupole moment/m^2	–
Frequency (^1H = 100 Hz; 2.3488T)/MHz	4.899

• ELECTRON SHELL DATA

Ground state electron configuration: $[Kr]4d^15s^2$
Term symbol: $^2D_{3/2}$
Electron affinity ($M \rightarrow M^-$)/kJ mol^{-1}**:** 29.6

Ionization energies/kJ mol^{-1}:

		kJ mol⁻¹
1.	$M \rightarrow M^+$	616
2.	$M^+ \rightarrow M^{2+}$	1181
3.	$M^{2+} \rightarrow M^{3+}$	1980
4.	$M^{3+} \rightarrow M^{4+}$	5963
5.	$M^{4+} \rightarrow M^{5+}$	7430
6.	$M^{5+} \rightarrow M^{6+}$	8970
7.	$M^{6+} \rightarrow M^{7+}$	11 200
8.	$M^{7+} \rightarrow M^{8+}$	12 400
9.	$M^{8+} \rightarrow M^{9+}$	14 137
10.	$M^{9+} \rightarrow M^{10+}$	18 400

Electron binding energies/eV

		eV
K	1s	17 038
L_I	2s	2373
L_{II}	$2p_{1/2}$	2156
L_{III}	$2p_{3/2}$	2080
M_I	3s	392.0
M_{II}	$3p_{1/2}$	310.6
M_{III}	$3p_{3/2}$	298.8
M_{IV}	$3d_{3/2}$	157.7
M_V	$3d_{5/2}$	155.8

continued in Appendix 2, p258

Main lines in atomic spectrum
[Wavelength/nm(species)]
360.073 (II)
371.030 (I)
377.433 (II)
407.738 (I)
410.238 (I) (AA)
437.494 (I)

• CRYSTAL DATA

Crystal structure (cell dimensions/pm), space group
α-Y h.c.p. (a = 364.74, c = 573.06), $P6_3/mmc$
β-Y b.c.c. (a = 411), Im3m
$T(\alpha \rightarrow \beta)$ = 1763 K

X-ray diffraction: mass absorption coefficients $(\mu/\rho)/cm^2\,g^{-1}$**:** CuK_α 134 MoK_α 100
Neutron scattering length, $b/10^{-12}$ cm**:** 0.775
Thermal neutron capture cross-section, σ_a/barns**:** 1.28

• GEOLOGICAL DATA

Minerals

Mineral	Formula	Density	Hardness	Crystal appearance
Bastnäsite-Y*	$(Y,Ce)CO_3F$	4.0	4 – 4.5	tri. translucent brick-red
Fergusonite	$YNbO_4$	5.7	5.5 – 6.5	tet., vitreous, sub-metallic black
Gadolinite	$Be_2FeY_2Si_2O_{10}$	4.4	6.5 – 7	mon., vit./greasy greenish-black
Polycrase-Y	$Y(Ti,Nb)_2(O,OH)_6$	5.0	5.5 – 6	orth., sub-metallic black
Samarskite	$(Y,Ce,U,Fe)_3(Nb,Ta,Ti)_5O_{16}$	5.69	5 – 6	orth., vit./resinous black
Xenotime	YPO_4	4.8	4 – 5	tet., vit./resinous yellow-brown

*A variety of this mineral that is particularly rich in yttrium.

Chief ores: xenotime, bastnäsite, fergusonite, samarskite

World production/tonnes y^{-1}**:** 400

Main mining areas: xenotine in USA, bastnäsite in Russia, fergusonite in Norway, Russia, Madagascar

Reserves/tonnes**:** *c.* 9×10^6

Specimen: available as chips, ingots or powder. Safe.

Abundances

Sun (relative to H = 1×10^{12})**:** 125
Earth's crust/p.p.m.**:** 30
Seawater/p.p.m.**:** 9×10^{-6}
Residence time/years**:** n.a.
Oxidation state: III

Zn

Atomic number: 30
Relative atomic mass ($^{12}C = 12.0000$)**:** 65.39

CAS: [7440-66-6]

• CHEMICAL DATA

Description: Zinc is a bluish-white metal that is brittle when cast. It tarnishes in air and reacts with acids and alkalis. Zinc is used for galvanizing iron, in alloys such as brass, and in batteries. Zinc oxide is used in rubber and as a polymer stabiliser.

Radii/pm: Zn^{2+} 83; atomic 133; covalent 125
Electronegativity: 1.65 (Pauling); 1.66 (Allred); 4.45 eV (absolute)
Effective nuclear charge: 4.35 (Slater); 5.97 (Clementi); 8.28 (Froese-Fischer)

Standard reduction potentials $E^{\ominus}$/V

	II		0
acid	Zn^{2+}	−0.7626	Zn
base	$[Zn(OH)_4]^{2-}$	−1.285	Zn
	$Zn(OH)_2$	−1.246	Zn

Oxidation states

Zn^{I}	$d^{10}s^{1}$	rare Zn_2^{2+} in $Zn/ZnCl_2$ glass
Zn^{II}	d^{10}	ZnO, ZnS, $Zn(OH)_2$, $[Zn(OH_2)_6]^{2+}$ (aq), $[Zn(OH)_4]^{2-}$ (aq alkali), Zn^{2+} salts, ZnF_2, $ZnCl_2$ etc., many complexes

• PHYSICAL DATA

Melting point/K: 692.73
Boiling point/K: 1180
ΔH_{fusion}/kJ mol^{-1}: 6.67
ΔH_{vap}/kJ mol^{-1}: 115.3

Thermodynamic properties (298.15 K, 0.1 MPa)

State	$\Delta_f H^{\ominus}$/kJ mol^{-1}	$\Delta_f G^{\ominus}$/kJ mol^{-1}	$S^{\ominus}$/J K^{-1} mol^{-1}	C_p/J K^{-1} mol^{-1}
Solid	0	0	41.63	25.40
Gas	130.729	95.145	160.984	20.786

Density/kg m^{-3}: 7133 [293 K]; 6577 [liquid at m.p.]
Molar volume/cm^3: 9.17
Thermal conductivity/W m^{-1} K^{-1}: 116 [300 K]
Coefficient of linear thermal expansion/K^{-1}: 25.0×10^{-6}
Electrical resistivity/Ω m: 5.916×10^{-8} [293 K]
Mass magnetic susceptibility/kg^{-1} m^3: -2.20×10^{-9} (s)
Young's modulus/GPa: 104.5
Rigidity modulus/GPa: 41.9
Bulk modulus/GPa: 69.4
Poisson's ratio/GPa: 0.249

• BIOLOGICAL DATA

Biological role

Essential for all species.

Toxicity

Toxic intake: generally regarded as of low toxicity.
Lethal intake: LD_{50} (chloride, oral, rat) = 350 mg kg^{-1}

Hazards

Zinc metal is a human skin irritant but otherwise is non-toxic, and so are most common compounds; however, zinc salts have been shown to be experimental carcinogens.

Levels in humans

Blood/mg dm^{-3}: 7.0
Bone/p.p.m.: 75 – 170
Liver/p.p.m.: 240
Muscle/p.p.m.: 240
Daily dietary intake: 5 – 40 mg
Total mass of element in average (70 kg) person: 2.3 g

Known in India and China before 1500.
[German, *zink*]
French, *zinc*; German, *Zink*; Italian, *zinco*; Spanish, *cinc*

Zinc

[zink]

• NUCLEAR DATA

Number of isotopes (including nuclear isomers): 23 **Isotope mass range:** 57 → 78

Key isotopes

Nuclide	Atomic mass	Natural abundance (%)	Nuclear spin I	Nuclear magnetic moment μ	Uses
^{64}Zn	63.929 145	48.6	0+		E
^{66}Zn	65.926 034	27.9	0+		E
^{67}Zn	66.927 129	4.1	5/2–	+0.875 15	E, NMR
^{68}Zn	67.924 846	18.8	0+		E
^{70}Zn	69.925 325	0.6	0+		E

A table of radioactive isotopes is given in Appendix 1, on p253.

NMR [Reference: $Zn(ClO_4)_2$]	^{67}Zn
Relative sensitivity ($^1H = 1.00$)	2.85×10^{-3}
Receptivity ($^{13}C = 1.00$)	0.665
Magnetogyric ratio/rad $T^{-1}s^{-1}$	1.6737×10^7
Nuclear quadrupole moment/m^2	$+0.150 \times 10^{-28}$
Frequency (^1H = 100 Hz; 2.3488T)/MHz	6.254

• ELECTRON SHELL DATA

Ground state electron configuration: $[Ar]3d^{10}4s^2$
Term symbol: $^1S_0$
Electron affinity ($M \rightarrow M^-$)/kJ mol^{-1}: 9

Ionization energies/kJ mol^{-1}:

1.	$M \rightarrow M^+$	906.4
2.	$M^+ \rightarrow M^{2+}$	1733.3
3.	$M^{2+} \rightarrow M^{3+}$	3832.6
4.	$M^{3+} \rightarrow M^{4+}$	5730
5.	$M^{4+} \rightarrow M^{5+}$	7970
6.	$M^{5+} \rightarrow M^{6+}$	10 400
7.	$M^{6+} \rightarrow M^{7+}$	12 900
8.	$M^{7+} \rightarrow M^{8+}$	16 800
9.	$M^{8+} \rightarrow M^{9+}$	19 600
10.	$M^{9+} \rightarrow M^{10+}$	23 000

Electron binding energies/eV

K	1s	9659
L_I	2s	1196.2
L_{II}	$2p_{1/2}$	1044.9
L_{III}	$2p_{3/2}$	1021.8
M_I	3s	139.8
M_{II}	$3p_{1/2}$	91.4
M_{III}	$3p_{3/2}$	88.6
M_{IV}	$3d_{3/2}$	10.2
M_V	$3d_{5/2}$	10.1

Main lines in atomic spectrum
[Wavelength/nm(species)]

213.856 (I) (AA)
250.199 (II)
255.795 (II)
330.259 (I)
334.502 (I)
491.162 (II)
636.234 (I)

• CRYSTAL DATA

Crystal structure (cell dimensions/pm), space group
h.c.p. (a = 266.47, c = 494.69), $P6_3/mmc$

X-ray diffraction: mass absorption coefficients $(\mu/\rho)/cm^2\ g^{-1}$: CuK_α 60.3 MoK_α 55.4
Neutron scattering length, $b/10^{-12}$ cm: 0.5680
Thermal neutron capture cross-section, σ_a/barns: 1.11

• GEOLOGICAL DATA

Minerals

Mineral	Formula	Density	Hardness	Crystal appearance
Hemimorphite*	$Zn_4Si_2O_7(OH)_2.H_2O$	3.475	4.5 – 5	orth., vit./pearly white
Hydrozincite	$Zn_5(CO_3)_2(OH)_6$	4.00	2 – 2.5	mon., earthy/silky white etc.
Sphalerite**	ZnS	4.08	3.5	cub., yellow to brown
Smithsonite	$ZnCO_3$	4.2	4 – 4.5	rhom., vit. grey-white
Willemite	Zn_2SiO_4	4.1	5.5	rhom., vitreous-resinous white/pale green

*Also known as calamine; ** also known as zinc blende.

Chief ores: sphalerite (with cadmium, gallium and indium as by-products), smithsonite, hemimorphite. Hydrozincite is also mined for zinc when found in economic amounts.

World production/tonnes y^{-1}: 5.02×10^6

Main mining areas: USA, Canada, Australia, Austria, Russia, Turkey.

Reserves/tonnes: 110×10^6

Specimen: available as dust, foil, granules, pieces, powder, shot, sticks and wire. Safe.

Abundances

Sun (relative to H = 1×10^{12}): 2.82×10^4
Earth's crust/p.p.m.: 75
Seawater/p.p.m.:
Atlantic surface: 0.5×10^{-4}
Atlantic deep: 1.0×10^{-4}
Pacific surface: 0.5×10^{-4}
Pacific deep: 5.2×10^{-4}
Residence time/years: 5000
Classification: recycled
Oxidation state: II

Zr

Atomic number: 40
Relative atomic mass ($^{12}C = 12.0000$)**:** 91.224

CAS: [7440-67-7]

• CHEMICAL DATA

Discovery: Zirconium was discovered in 1789 by M.H. Klaproth at University of Berlin, Germany. First isolated in 1824 by J.J. Berzelius at Stockholm, Sweden.
Description: Zirconium is a hard, lustrous, silvery metal which is very resistant towards corrosion due to an oxide layer on the surface. However, it will burn in air if ignited. Zirconium is unaffected by acids (except HF) and alkalis. It is used in alloys, coloured glazes, and nuclear reactors. Its oxides are used in foundry crucibles, bricks, ceramics and abrasives.

Radii/pm: Zr^{4+} 87; Zr^{2+} 109; atomic 160; covalent 145
Electronegativity: 1.33 (Pauling); 1.22 (Allred); 3.64 eV (absolute)
Effective nuclear charge: 3.15 (Slater); 6.45 (Clementi); 9.20 (Froese-Fischer)

Standard reduction potentials $E^\ominus/V$

IV		0
Zr^{4+}	$\xrightarrow{-1.55}$	Zr

Oxidation states

Zr^{-II}	d^6	$[Zr(CO)_6]^{2-}$
Zr^0	d^4	$[Zr(bipyridyl)_3]$, $[Zr(CO)_4\{MeC(CH_2PMe_2)_3\}]$
Zr^I	d^3	ZrCl ?
Zr^{II}	d^2	ZrO ?, $ZrCl_2$
Zr^{III}	d^1	$ZrCl_3$, $ZrBr_3$, ZrI_3, Zr^{3+} reduces H_2O
Zr^{IV}	d^0 [Kr]	ZrO_2, $[Zr(OH)]^{3+}$ (aq), ZrF_4, $ZrCl_4$ etc., $[ZrF_6]^{2-}$, $[ZrF_7]^{3-}$, $[ZrF_8]^{4-}$, zirconates, complexes

• PHYSICAL DATA

Melting point/K: 2125
Boiling point/K: 4650
ΔH_{fusion}/kJ mol^{-1}: 23.0
ΔH_{vap}/kJ mol^{-1}: 581.6

Thermodynamic properties (298.15 K, 0.1 MPa)

State	$\Delta_f H^\ominus$/kJ mol^{-1}	$\Delta_f G^\ominus$/kJ mol^{-1}	$S^\ominus$/J K^{-1} mol^{-1}	C_p/J K^{-1} mol^{-1}
Solid	0	0	38.99	25.36
Gas	608.8	566.5	181.36	26.65

Density/kg m^{-3}: 6506 [293 K]; 5800 [liquid at m.p.]
Molar volume/cm^3: 14.02
Thermal conductivity/W m^{-1} K^{-1}: 22.7 [300 K]
Coefficient of linear thermal expansion/K^{-1}: 5.78×10^{-6}
Electrical resistivity/Ω m: 42.1×10^{-8} [293 K]
Mass magnetic susceptibility/kg^{-1} m^3: $+1.68 \times 10^{-8}$ (s)

Young's modulus/GPa: 98
Rigidity modulus/GPa: 35
Bulk modulus/GPa: 89.8
Poisson's ratio/GPa: 0.38

• BIOLOGICAL DATA

Biological role

None.

Toxicity

Generally regarded as of low toxicity.
Lethal intake: LD_{50} (chloride, oral, rat) = 1688 mg kg^{-1}

Hazards

Zirconium and its compounds are not regarded as particularly poisonous except by excessive ingestion of zirconium salts. Zirconium dust is a very dangerous fire hazard.

Levels in humans

Blood/mg dm^{-3}: 0.011
Bone/p.p.m.: < 0.1
Liver/p.p.m.: 0.11
Muscle/p.p.m.: 0.08
Daily dietary intake: *c.* 0.05 mg
Total mass of element in average (70 kg) person: 1 mg

Discovery: see Chemical Data section.
[Arabic, *zargun* = gold colour]
French, *zirconium*; German, *Zirconium*; Italian, *zirconio*; Spanish, *circonio*

Zirconium

[zer-koh-ni-uhm]

• NUCLEAR DATA

Number of isotopes (including nuclear isomers): 25 **Isotope mass range:** 82 → 101

Key isotopes

Nuclide	Atomic mass	Natural abundance (%)	Nuclear spin I	Nuclear magnetic moment μ	Uses
^{90}Zr	89.904 703	51.45			E
^{91}Zr	90.905 644	11.22	5/2+	–1.303	E, NMR
^{92}Zr	91.905 039	17.15			E
^{94}Zr	93.906 314	17.38			E
^{96}Zr*	95.908 275	2.80			E

*May be radioactive with a half-life of $>3.6 \times 10^{17}$ y.
A table of radioactive isotopes is given in Appendix 1, on p253.

NMR [Reference: no agreed standard]	^{91}Zr
Relative sensitivity ($^1H = 1.00$)	9.48×10^{-3}
Receptivity ($^{13}C = 1.00$)	6.04
Magnetogyric ratio/rad $T^{-1}s^{-1}$	-2.4868×10^7
Nuclear quadrupole moment/m^2	-0.206×10^{-28}
Frequency ($^1H = 100$ Hz; 2.3488T)/MHz	9.330

• ELECTRON SHELL DATA

Ground state electron configuration: $[Kr]4d^25s^2$
Term symbol: $^3F_2$
Electron affinity ($M \rightarrow M^-$)/kJ mol^{-1}**:** 41.1

Ionization energies/kJ mol^{-1}**:**

1.	$M \rightarrow M^+$	660
2.	$M^+ \rightarrow M^{2+}$	1267
3.	$M^{2+} \rightarrow M^{3+}$	2218
4.	$M^{3+} \rightarrow M^{4+}$	3313
5.	$M^{4+} \rightarrow M^{5+}$	7860
6.	$M^{5+} \rightarrow M^{6+}$	(9500)
7.	$M^{6+} \rightarrow M^{7+}$	(11200)
8.	$M^{7+} \rightarrow M^{8+}$	(13800)
9.	$M^{8+} \rightarrow M^{9+}$	(15700)
10.	$M^{9+} \rightarrow M^{10+}$	(17500)

Electron binding energies/eV

K	1s	17998
L_I	2s	2532
L_{II}	$2p_{1/2}$	2307
L_{III}	$2p_{3/2}$	2223
M_I	3s	430.3
M_{II}	$3p_{1/2}$	343.5
M_{III}	$3p_{3/2}$	329.8
M_{IV}	$3d_{3/2}$	181.1
M_V	$3d_{5/2}$	178.8

continued in Appendix 2, p258

Main lines in atomic spectrum
[Wavelength/nm(species)]
339.198 (II)
343.823 (II)
349.621 (II)
360.119 (I) (AA)
389.032 (I)

• CRYSTAL DATA

Crystal structure (cell dimensions/pm), space group
α-Zr h.c.p. ($a = 323.21$, $c = 514.77$), $P6_3/mmc$
β-Zr b.c.c. ($a = 361.6$), Im3m
$T(\alpha \rightarrow \beta) = 1135$ K
High pressure form: ($a = 503.6$, $c = 310.9$), $P\bar{3}m1$

X-ray diffraction: mass absorption coefficients (μ/ρ)/cm^2 g^{-1}**:** CuK_α 143 MoK_α 15.9
Neutron scattering length, $b/10^{-12}$ cm**:** 0.716
Thermal neutron capture cross-section, σ_a/barns**:** 0.184

• GEOLOGICAL DATA

Minerals

Mineral	Formula	Density	Hardness	Crystal appearance
Baddeleyite	ZrO_2	5.82	6.5	mon., greasy/vit. col./ yellow/ green
Zircon	$ZrSiO_4$	4.7	7.5	tet., stubby crystals, colourless/yellow/grey

Chief ores: zircon, baddeleyite
World production/tonnes y^{-1}**:** 7000 (Zr metal); 7×10^5 (zircon)
Main mining areas: Australia, Brazil, USA, Sri Lanka
Reserves/tonnes**:** $> 1 \times 10^9$
Specimen: available as foil, powder, rod or wire. Safe.

Abundances

Sun (relative to $H = 1 \times 10^{12}$)**:** 560
Earth's crust/p.p.m.**:** 190
Seawater/p.p.m.**:** 9×10^{-6}
Residence time/years**:** n.a.
Oxidation state: IV

APPENDIX 1
Radioactive isotopes of the elements

There is an entry for every element in this appendix although not all the information about radioactive isotopes is to be found herein. Elements with naturally–occurring radioactive isotopes have these included in the table in the main section of the book. Elements for which all the isotopes are radioactive also have them listed in the main section – these are mainly the transuranium elements.

The tables in the main section, and in this appendix, give only the longer lived isotopes of the other elements, and a footnote to either table indicates the maximum half life of other short–lived isotopes. The sources of the data are given in the Key.

Uses: the abbreviations are R = research, D = medical diagnosis,T = medical therapy.

Actinium

Isotopes of actinium with half-lives longer than 10 minutes are given in the main table on page 17.

Aluminium/Aluminum (US)

Nuclide	Atomic mass	Half life ($T_{1/2}$)	Decay mode and energy (MeV)	Nuclear spin I	Nuclear magnetic moment μ	Uses
^{26}Al	25.986 892	7.2×10^5 y	β^+ (4.005) 82%; EC, 18%; γ	5+		R
^{28}Al	27.981 910	2.25 m	β^- (4.642); γ	3+	3.24	
^{29}Al	28.980 446	6.5 m	β^- (3.68); γ	5/2+		

Other radioisotopes of aluminium have half-lives shorter than 10 seconds.

Americium

Isotopes of americium with half-lives longer than 1 hour are given in the main table on page 21.

Antimony

Nuclide	Atomic mass	Half life ($T_{1/2}$)	Decay mode and energy (MeV)	Nuclear spin I	Nuclear magnetic moment μ	Uses
^{119}Sb	118.903 948	38.1 h	EC (0.59); γ	5/2+	+3.45	
^{120}Sb	119.903 821	15.89 m	β^+ (2.68) 41%; EC 59%; γ	1+	+2.3	
^{120m}Sb		5.76 d	EC; γ	8–	2.34	
^{122}Sb	121.905 179	2.73 d	β^- (1.9820) 98%; β^+ (0.619) 2%; γ	2–	–1.90	
^{124}Sb	123.905 038	60.30 d	β^- (2.905); γ	3–	1.2	R
^{125}Sb	124.905 252	2.758 y	β^- (0.767); γ	7/2+	+2.63	
^{126}Sb	125.907 250	12.4 d	β^- (3.67); γ	8–	1.3	
^{127}Sb	126.906 919	3.84 d	β^- (1.58); γ	7/2+	2.6	

Other radioisotopes of antimony have half-lives shorter than 1 day.

Argon

Nuclide	Atomic mass	Half life ($T_{1/2}$)	Decay mode and energy (MeV)	Nuclear spin I	Nuclear magnetic moment μ	Uses
^{37}Ar	36.966 776	35.0 d	EC (0.814); no γ	3/2+	+1.15	R
^{39}Ar	38.962 314	268 y	β^- (0.565); no γ	7/2–	–1.3	
^{41}Ar	40.964 501	1.82 h	β^- (2.492); γ	7/2–		
^{42}Ar	41.963 050	33 y	β^- (0.60); no γ	0+		
^{44}Ar	43.963 650	11.87 m	β^- (3.54); γ	0+		

Other radioisotopes of argon have half-lives of less than 10 minutes.

Arsenic

Nuclide	Atomic mass	Half life ($T_{1/2}$)	Decay mode and energy (MeV)	Nuclear spin *I*	Nuclear magnetic moment μ	Uses
^{71}As	70.927 114	2.72 d	β^+ (2.013) 32%; EC 68%; γ	5/2–	+1.6735	
^{72}As	71.926 755	26.0 h	β^+ (4.355) 77%; γ	2–	–2.1566	
^{73}As	72.923 827	80.3 d	EC (0.346); γ	3/2–		R
^{74}As	73.923 827	17.78 d	β^- (2.562) 31%; EC 37%; β^- (1.354); γ	2–	–1.597	
^{76}As	75.922 393	26.3 h	β^- (2.97); γ	2–	–0.906	
^{77}As	76.920 646	38.8 h	β^- (0.6904); γ	3/2–		

Other radioisotopes of arsenic have half-lives of less than 2 hours.

Astatine

Isotopes of astatine with half-lives longer than 10 minutes are given in the main table on page 29.

Barium

Nuclide	Atomic mass	Half life ($T_{1/2}$)	Decay mode and energy (MeV)	Nuclear spin *I*	Nuclear magnetic moment μ	Uses
^{128}Ba	127.908 237	2.43 d	EC (0.45); γ	0+		
^{131}Ba	130.906 902	11.7 d	EC (1.36); γ	1/2+	0.7081	
^{133}Ba	132.905 988	10.53 y	EC (0.516); γ	1/2+	0.7717	
^{133m}Ba		1.621 d	IT (0.288); γ	11/2–	–0.91	
^{135m}Ba	134.905 665	1.20 d	IT (0.2682); γ	11/2–	–1.00	R
^{140}Ba	139.910 581	12.75 d	β^- (1.05); γ	0+		

Other radioisotopes of barium have half-lives shorter than 1 day.

Berkelium

Isotopes of berkelium with half-lives longer than 1 hour are given in the main table on page 33.

Beryllium

Nuclide	Atomic mass	Half life ($T_{1/2}$)	Decay mode and energy (MeV)	Nuclear spin *I*	Nuclear magnetic moment μ	Uses
^{6}Be	6.019 725	5.9×10^{-21} s	2p.α	0+		
^{7}Be	7.016 928	53.82 d	EC (0.862); γ	3/2+		R
^{8}Be	8.005 305	*c.* 7×10^{-17} s	2α (0.046); no γ	0+		
^{10}Be	10.013 534	1.52×10^{6} y	β^- (0.5561); no γ	0+		R
^{11}Be	11.021 658	13.8 s	β^-,α(11.48); γ	1/2+		
^{12}Be	12.026 92	2.4×10^{-2} s	β^- (11.71); n	0+		
^{14}Be	13.037 5	4×10^{-3} s	β^- (16.4)	0+		

Bismuth

Nuclide	Atomic mass	Half life ($T_{1/2}$)	Decay mode and energy (MeV)	Nuclear spin *I*	Nuclear magnetic moment μ	Uses
^{205}Bi	204.977 365	15.31 d	EC (2.71); γ	9/2–	+4.16	
^{206}Bi	205.978 478	6.243 d	EC (3.76); γ	6+	+4.60	
^{207}Bi	206.978 446	35 y	EC (2.397); γ	9/2–	4.08	R
^{208}Bi	207.979 717	3.68×10^{5} y	EC (2.878); γ	5+		
^{210}Bi	209.984 095	5.01 d	β^- (1.162); γ	1–	–0.0445	
^{210m}Bi		3×10^{6} y	α (4.96); β^-; γ	9–		

Other radioisotopes of bismuth have half-lives shorter than 1 day.

Bohrium

All isotopes of bohrium are given in the main table on page 39.

Boron

Nuclide	Atomic mass	Half life ($T_{1/2}$)	Decay mode and energy (MeV)	Nuclear spin *I*	Nuclear magnetic moment μ	Uses
^{8}B	8.024 605	0.770 s	β^+ (17.979); 2 α	2+	1.0355	
^{9}B	9.013 328	8×10^{-19} s	p2 α	3/2–	+1.8007	
^{12}B	12.014 352	0.0202 s	β^- (13.369); $\beta^-\alpha$ 1.6%; γ	1+	+1.0031	
^{13}B	13.017 78	0.0174 s	β^- (13.436); β^-n 0.25%; γ	3/2–	+3.17778	

Other radioisotopes of boron have half-lives shorter than 15 ms.

Bromine

Nuclide	Atomic mass	Half life ($T_{1/2}$)	Decay mode and energy (MeV)	Nuclear spin *I*	Nuclear magnetic moment μ	Uses
^{75}Br	74.925 753	1.62 h	β^+ (3.0) 76%; EC 24%; γ	3/2–	0.75	
^{76}Br	75.924 528	16.0 h	β^+ (4.956) 57%; EC 43%; γ	1–	0.54821	
^{77}Br	76.921 378	2.376 d	EC (1.365) 99%; β^+ 1%; γ	3/2–		R
^{78}Br	77.921 144	6.46 m	β^+ (3.3574) 92%; EC 8%; γ	1+		
^{80}Br	79.918 528	17.66 m	β^- (2.00) 92%; EC (1.870) 6%; β^+ 2%; γ	1+	0.5140	
^{80m}Br		4.42 h	IT (0.04885); γ	5–	+1.3177	
^{82}Br	81.916 802	1.471 d	β^- (3.093); γ	5–	+1.6270	
^{83}Br	82.915 179	2.40 h	β^- (0.98); γ	3/2–		

Other radioisotopes of bromine have half-lives shorter than 1 hour.

Cadmium

Nuclide	Atomic mass	Half life ($T_{1/2}$)	Decay mode and energy (MeV)	Nuclear spin *I*	Nuclear magnetic moment μ	Uses
^{107}Cd	106.906 613	6.52 h	EC (1.42) 99%; β^+; γ	5/2+	–0.615 055	
^{109}Cd	108.904 953	462 d	EC (0.214); γ	5/2+	–0.827 846	R
^{113m}Cd	112.904 400	14.1 y	β^- (0.59) 99.9%; γ	11/2–	–1.087	
^{115}Cd	114.905 430	2.228 h	β^- (1.448); γ	1/2+	–0.648 426	
^{115m}Cd		44.6 d	β^- (1.629); γ	11/2–	–1.042	R
^{117}Cd	116.907 228	2.49 h	β^- (2.53); γ	1/2+		
^{117m}Cd		3.4 h	β^- (2.66); γ	11/2–		

Other radioisotopes of cadmium have half-lives shorter than 1 hour.

Caesium/Cesium (US)

Nuclide	Atomic mass	Half life ($T_{1/2}$)	Decay mode and energy (MeV)	Nuclear spin *I*	Nuclear magnetic moment μ	Uses
^{129}Cs	128.906 027	1.336 d	EC (1.192); γ	1/2+	+1.49	
^{131}Cs	130.905 444	9.69 d	EC (0.35); γ	5/2+	+3.54	
^{132}Cs	131.906 431	6.48 d	EC 98%; β^+ (2.121) 0.3%; β^- (1.28) 2%; γ	2–	+2.22	
^{134}Cs	133.906 696	2.065 y	β^- (2.06); γ	4+	+2.990	R
^{135}Cs	134.905 885	2.3×10^{6} y	β^- (0.205); no γ	7/2+	+2.732	
^{136}Cs	135.907 289	13.16 d	β^- (2.548); γ	5+	+3.71	
^{137}Cs	136.907 073	30.3 y	β^- (1.176); γ	7/2+	+2.84	R, T

Other radioisotopes of caesium have half-lives shorter than 1 day.

Calcium

Nuclide	Atomic mass	Half life ($T_{1/2}$)	Decay mode and energy (MeV)	Nuclear spin I	Nuclear magnetic moment μ	Uses
^{41}Ca	40.962 278	1.03×10^5 y	EC (0.421); no γ	7/2–	–1.595	R
^{45}Ca	44.956 185	162.7 d	β^- (0.257); no γ	7/2–	–1.327	R
^{47}Ca	46.954 543	4.536 d	β^- (1.988); γ	7/2–	–1.38	R, D
^{49}Ca	48.955 672	8.72 m	β^- (5.263); γ	3/2–		

Other radioisotopes of calcium have half-lives shorter than 1 minute.

Californium

Isotopes of californium with half-lives longer than 10 minutes are given in the main table on page 51.

Carbon

Nuclide	Atomic mass	Half life ($T_{1/2}$)	Decay mode and energy (MeV)	Nuclear spin I	Nuclear magnetic moment μ	Uses
^{10}C	10.016 86	19.3 s	β^+ (3.650); γ	0+		
^{11}C	11.011 43	20.3 m	β^+EC (1.982); γ	3/2–	–0.964	
^{14}C*	14.003 241	5715 y	β^- (0.156 48); no γ	0+		R

*Traces of this isotope occur naturally.
Other radioisotopes of carbon have half-lives shorter than 3 seconds.

Cerium

Nuclide	Atomic mass	Half life ($T_{1/2}$)	Decay mode and energy (MeV)	Nuclear spin I	Nuclear magnetic moment μ	Uses
^{134}Ce	133.908 890	3.16 d	EC (0.5); γ	0+		
^{135}Ce	134.909 117	17.7 h	β^+ (2.03) 1%; EC 99%; γ	1/2+		
^{137m}Ce	136.907 780	1.43 d	IT (0.254) 99%; EC 0.8%; γ	11/2–	0.70	
^{139}Ce	138.906 631	137.6 d	EC (0.26); γ	3/2+	0.9	
^{141}Ce	140.908 271	32.5 d	β^- (0.581); γ	7/2–	1.1	R
^{143}Ce	142.912 383	1.38 d	β^- (1.462); γ	3/2–	*c.* 1	
^{144}Ce	143.913 643	284.6 d	β^- (0.319); γ	0+		

Other radioisotopes of cerium have half-lives shorter than 10 hours.

Chlorine

Nuclide	Atomic mass	Half life ($T_{1/2}$)	Decay mode and energy (MeV)	Nuclear spin I	Nuclear magnetic moment μ	Uses
^{34m}Cl	33.973 763	32.2 m	β^+; IT; γ	3+		
^{36}Cl	35.968 306	3.01×10^5 y	β^- (0.709); β^+; EC (1.142); no γ	2+	+1.28547	R
^{38}Cl	37.968 010	37.2 m	β^- (4.917); γ	2–	2.05	
^{39}Cl	38.968 005	55.6 m	β^- (3.44); γ	3/2–		
^{40}Cl	39.970 440	1.38 m	β^- (7.5); γ	2–		

Other radioisotopes of chlorine have half-lives shorter than 1 minute.

Chromium

Nuclide	Atomic mass	Half life ($T_{1/2}$)	Decay mode and energy (MeV)	Nuclear spin I	Nuclear magnetic moment μ	Uses
^{48}Cr	47.954 033	21.6 h	EC (1.65); γ			
^{49}Cr	48.951 338	42.3 m	β^+, EC (2.627); γ	5/2–	0.476	
^{51}Cr	50.944 768	27.70 d	EC (0.751); γ	7/2–	–0.934	R, D
^{55}Cr	54.940 842	3.497 m	β^- (2.603); γ	3/2–		
^{56}Cr	55.940 643	5.9 m	β^- (1.62); γ	0+		

Other radioisotopes of copper have half-lives shorter than 1 minute.

Cobalt

Nuclide	Atomic mass	Half life ($T_{1/2}$)	Decay mode and energy (MeV)	Nuclear spin I	Nuclear magnetic moment μ	Uses
^{55}Co	54.942 001	17.53 h	β^+ (3.452); EC; γ	7/2–	+4.822	
^{56}Co	55.939 841	77.3 d	β^+ (4.566); EC; γ	4+	+3.85	
^{57}Co	56.936 294	271.8 d	EC (0.836); γ	7/2–	+4.72	R, D
^{58}Co	57.935 755	70.88 d	β^+ (2.30); EC; γ	2+	+4.04	R, D
^{60}Co	59.933 819	5.27 y	β^- (2.824); γ	5+	+3.799	R, D, T
^{61}Co	60.932 478	1.650 h	β^- (1.322); γ	7/2–		

Other radioisotopes of cobalt have half-lives shorter than 1 hour.

Copper

Nuclide	Atomic mass	Half life ($T_{1/2}$)	Decay mode and energy (MeV)	Nuclear spin I	Nuclear magnetic moment μ	Uses
^{60}Cu	59.937 366	23.7 m	β^+ (6.127); EC; γ	2+	+1.219	
^{61}Cu	60.933 461	3.35 h	β^+ (2.239); γ	3/2–	+2.14	
^{64}Cu	63.929 765	12.701 h	EC 41%; β^- (0.578) 39%; β^+ (1.675) 19%; γ	1+	–0.217	R, D
^{67}Cu	66.927 747	2.580 d	β^- (0.58); γ	3/2–		R

Other radioisotopes of copper have half-lives shorter than 10 minutes.

Curium

Isotopes of curium with half-lives longer than 1 hour are in the main table on page 65.

Dubnium

All isotopes of dubnium are given in the main table on page 67.

Dysprosium

Nuclide	Atomic mass	Half life ($T_{1/2}$)	Decay mode and energy (MeV)	Nuclear spin I	Nuclear magnetic moment μ	Uses
^{154}Dy	153.924 429	3×10^6 y	α	0+		
^{155}Dy	154.925 747	9.9 h	β^+ (2.10) 2%; EC 98%; γ	3/2–	–0.385	
^{157}Dy	156.925 460	8.1 h	EC (1.34); γ	3/2–	–0.301	
^{159}Dy	158.925 735	144 d	EC (0.366); γ	3/2–	–0.354	
^{166}Dy	165.932 803	3.40 d	β^- (0.486); γ	0+		

Other radioisotopes of dysprosium have half-lives shorter than 3 hours.

Einsteinium

Isotopes of einsteinium with half-lives longer than 1 hour are given in the main table on page 71.

Erbium

Nuclide	Atomic mass	Half life ($T_{1/2}$)	Decay mode and energy (MeV)	Nuclear spin I	Nuclear magnetic moment μ	Uses
^{160}Er	159.929 080	1.19 d	EC (0.33); no γ	0+		
^{165}Er	164.930 723	10.36 h	EC (0.377); no γ	5/2–	+0.643	
^{169}Er	168.934 588	9.40 d	β^- (0.350); γ	1/2–	+0.515	
^{171}Er	170.938 027	7.52 h	β^- (1.490); γ	5/2–	0.66	
^{172}Er	171.939 353	2.05 d	β^- (0.89); γ	0+		

Other radioisotopes of erbium have half-lives shorter than 4 hours.

Europium

Nuclide	Atomic mass	Half life ($T_{1/2}$)	Decay mode and energy (MeV)	Nuclear spin I	Nuclear magnetic moment μ	Uses
^{148}Eu	147.918 125	54.5 d	EC (3.12); γ	5–	+2.34	
^{149}Eu	148.917 926	93.1 d	EC (0.691); γ	5/2+	+3.57	
^{150m}Eu	149.919 702	36 y	EC (2.26); γ	5–	+2.71	
^{152}Eu	151.921 742	13.48 y	EC (1.874) 72%; β^- (1.819) 28%; γ	3–	–1.91	R
$^{152m_1}Eu$		9.30 h	β^- 72%; EC 28%; γ	0–		R
^{154}Eu	153.922 975	8.59 y	β^- (1.969) 99.9%; EC (0.728) 0.02%; γ	3–	+2.000	
^{155}Eu	154.922 889	4.71 y	β^- (0.252); γ	5/2+	+1.6	

Other radioisotopes of europium have half-lives shorter than 6 days.

Fermium

Isotopes of fermium with half-lives longer than 1 hour are in the main table on page 77.

Fluorine

Nuclide	Atomic mass	Half life ($T_{1/2}$)	Decay mode and energy (MeV)	Nuclear spin I	Nuclear magnetic moment μ	Uses
^{17}F	17.002 095	64.5 s	β^+ (2.761); no γ	5/2+	+4.722	
^{18}F	18.000 937	1.83 h	β^+ EC (1.655) 97%; no γ	1+		R, D
^{20}F	19.999 981	11.00 s	β^- (7.029); γ	2+	+2.094	

Other radioisotopes of fluorine have half-lives shorter than 5 seconds.

Francium

Isotopes of francium with half-lives longer than 1 minute are in the main table on page 81.

Gadolinium

Nuclide	Atomic mass	Half life ($T_{1/2}$)	Decay mode and energy (MeV)	Nuclear spin I	Nuclear magnetic moment μ	Uses
^{146}Gd	145.918 304	48.3 d	EC (1.03) 99.8%; β^+ 0.2%; γ	0+		
^{147}Gd	146.918 943	1.588 d	EC (2.19); γ	7/2–	1.0	
^{148}Gd	147.918 113	75 y	α; no γ	0+		
^{149}Gd	148.919 344	9.3 d	EC (1.32); γ	7/2–	0.9	
^{150}Gd	149.918 662	1.8×10^6 y	α (2.73); no γ	0+		
^{151}Gd	150.920 346	124 d	EC (0.464); γ	7/2–	0.8	
^{153}Gd	152.921 745	241.6 d	EC (0.485); γ	3/2–	0.4	R, D

Other radioisotopes of aluminium have half-lives shorter than 1 day.

Gallium

Nuclide	Atomic mass	Half life ($T_{1/2}$)	Decay mode and energy (MeV)	Nuclear spin I	Nuclear magnetic moment μ	Uses
^{66}Ga	65.931 590	9.5 h	β^+ (5.175) 56%; EC 43%; γ	0+		
^{67}Ga	66.928 204	3.260 d	EC (1.001); γ	3/2–	+1.8507	R, D
^{68}Ga	67.927 981	1.103 h	β^+ (2.921) 90%; EC 10%; γ	1+	0.01175	R
^{72}Ga	71.926 365	14.10 h	β^- (3.99); γ	3–	–0.13224	
^{73}Ga	72.925 169	4.87 h	β^- (1.59); γ	3/2–		

Other radioisotopes of gallium have half-lives shorter than 1 hour.

Germanium

Nuclide	Atomic mass	Half life ($T_{1/2}$)	Decay mode and energy (MeV)	Nuclear spin I	Nuclear magnetic moment μ	Uses
^{66}Ge	65.933 847	2.26 h	β^+ (2.10) 27%; EC 73%; γ	0+		
^{68}Ge	67.928 096	270.8 d	EC (0.11); γ	0+		R
^{69}Ge	68.927 969	1.63 d	β^+ (2.225) 36%; EC 64%; γ	5/2+	0.735	
^{71}Ge	70.924 953	11.2 d	EC (0.236); γ	1/2–	+0.547	
^{75}Ge	74.922 858	1.380 h	β^- (1.178); γ	1/2–	+0.510	
^{77}Ge	76.923 548	11.30 h	β^- (2.70); γ	7/2+		
^{78}Ge	77.922 853	1.45 h	β^- (0.95); γ	0+		

Other radioisotopes of germanium have half-lives shorter than 1 hour.

Gold

Nuclide	Atomic mass	Half life ($T_{1/2}$)	Decay mode and energy (MeV)	Nuclear spin I	Nuclear magnetic moment μ	Uses
^{194}Au	193.965 348	1.64 d	β^+ (2.49) 3%; EC 97%; γ	1–	0.075	
^{195}Au	194.965 013	186.12 d	EC (0.227); γ	3/2+	+0.148	
^{196}Au	195.966 544	6.18 d	EC (1.505) 92%; γ	2–	+0.591	
^{198}Au	197.968 217	2.694 d	β^- (1.373); γ	2–	+0.5934	R, D, T
^{198m}Au		2.30 d	IT (0.812); γ	12–		
^{199}Au	198.968 740	3.14 d	β^- (0.453); γ	3/2+	+0.2715	

Other radioisotopes of gold have half-lives shorter than 1 day.

Hafnium

Nuclide	Atomic mass	Half life ($T_{1/2}$)	Decay mode and energy (MeV)	Nuclear spin I	Nuclear magnetic moment μ	Uses
^{172}Hf	171.939 460	1.87 y	EC (0.35); γ	0+		
^{175}Hf	174.941 507	70 d	EC (0.686); γ	5/2–	0.54	
^{178m_2}Hf	177.943 696	31 y	IT; γ	16+	7.3	
^{179m_2}Hf	178.945 812	25.1 d	IT (1.1057); γ	25/2–	7.4	
^{181}Hf	180.949 096	42.4 d	β^- (1.028); γ	1/2–		R
^{182}Hf	181.950 550	9×10^6 y	β^- (0.37); γ	0+		

Other radioisotopes of hafnium have half-lives shorter than 1 day.

Hassium

All isotopes of hassium are given in the main table on page 93.

Helium

Nuclide	Atomic mass	Half life ($T_{1/2}$)	Decay mode and energy (MeV)	Nuclear spin I	Nuclear magnetic moment μ	Uses
^{6}He	6.018 886	0.807 s	β^- (3.5010); γ	0+		
^{8}He	8.033 92	0.119 s	β^- (14); n; γ	0+		

Other radioisotopes of helium have half-lives shorter than 1 hour.

Holmium

Nuclide	Atomic mass	Half life ($T_{1/2}$)	Decay mode and energy (MeV)	Nuclear spin I	Nuclear magnetic moment μ	Uses
^{160m}Ho	159.928 720	5.0 h	IT (0.060) 67%; EC (3.35) 33%; γ	2–	+2.52	
^{161}Ho	160.927 849	2.48 h	EC (0.857); γ	7/2–	+4.25	
^{163}Ho	162.928 731	4570 y	EC (0.004); no γ	7/2–	+4.23	
^{166}Ho	165.932 281	1117 d	β^- (1.854); γ	0–		
^{166m}Ho		1200 y	β^-; γ	7–	3.6	
^{167}Ho	166.933 127	3.1 h	β^- (1.01); γ	7/2–		

Other radioisotopes of holmium have half-lives shorter than 1 hour.

Hydrogen

Isotopes of hydrogen are given in the main table on page 99.

Indium

Nuclide	Atomic mass	Half life ($T_{1/2}$)	Decay mode and energy (MeV)	Nuclear spin *I*	Nuclear magnetic moment μ	Uses
^{109}In	108.907 133	4.2 h	β^+ (2.02) 8%; EC 92%; γ	9/2+	+5.54	
^{110m}In	109.907 230	4.9 h	EC; γ	7+	+4.72	
^{111}In	110.905 109	2.805 d	EC (0.86); γ	9/2+	+5.50	R, D
^{113m}In	112.904 061	1.658 h	IT (0.3917)	1/2–	–0.210	R, D
^{114m}In	113.904 916	49.51 d	IT (0.190) 97%; EC (1.6) 3%; γ	5+	+4.7	R
^{115m}In	114.903 880	4.486 h	IT (0.336) 95%; β^- 5%; γ	1/2–	–0.255	
^{117m}In	116.904 517	1.94 h	β^- (1.769) 53%; IT 47%; γ	1/2–	0.25	

Other radioisotopes of indium have half-lives shorter than 1 hour.

Iodine

Nuclide	Atomic mass	Half life ($T_{1/2}$)	Decay mode and energy (MeV)	Nuclear spin *I*	Nuclear magnetic moment μ	Uses
^{123}I	122.905 594	13.2 h	EC (1.234); γ	5/2+	2.82	R, D
^{124}I	123.906 207	4.18 d	β^+ (3.157) 23%; EC 77%; γ	2–	1.1	
^{125}I	124.904 620	59.4 d	EC (0.179); γ	5/2+	2.82	R, D
^{126}I	125.905 624	13.0 d	EC; β^+ (2.151); β^- (1.26); γ	2–		
^{129}I	128.904 986	1.7×10^7 y	β^- (0.191); γ	7/2+	+2.621	R
^{130}I	129.906 713	12.36 h	β^- (2.98); γ	5+		
^{131}I	130.906 114	8.040 d	β^- (0.971); γ	7/2+	+2.742	R, D, T
^{133}I	132.907 780	20.8 h	β^- (1.77); γ	7/2+	+2.86	

Other radioisotopes of iodine have half-lives shorter than 10 hours.

Iridium

Nuclide	Atomic mass	Half life ($T_{1/2}$)	Decay mode and energy (MeV)	Nuclear spin *I*	Nuclear magnetic moment μ	Uses
^{189}Ir	188.958 712	13.2 d	EC (0.53); γ	3/2+	0.13	
^{190}Ir		11.8 d	EC (2.0); γ	4+	0.04	
^{192}Ir	191.962 580	73.83 d	β^- (1.460); γ	4–	+1.92	R, T
^{192m_2}Ir		241 y	IT (0.161); γ	9+		
^{193m}Ir	192.962 917	10.53 d	IT (0.0802); γ	11/2–		
^{194m}Ir	193.965 069	170 d	β^-; γ	11		

Other radioisotopes of iridium have half-lives shorter than 1 day.

Iron

Nuclide	Atomic mass	Half life (T...)	Decay mode and energy (MeV)	Nuclear spin	Nuclear magnetic	Uses
^{52}Fe	51.948 114	8.28 h	β^+ (2.37); 57%; EC 43%; γ	0+		R
^{55}Fe	54.938 296	2.73 y	EC (0.2314); no γ	3/2–		R, D
^{59}Fe	58.934 877	44.51 d	β^- (1.565); γ	3/2–	0.29	R, D
^{60}Fe	59.934 080	1.5×10^6 y	β^- (0.243)	0+		

Other radioisotopes of iron have half-lives shorter than 10 minutes.

Krypton

Nuclide	Atomic mass	Half life ($T_{1/2}$)	Decay mode and energy (MeV)	Nuclear spin *I*	Nuclear magnetic moment μ	Uses
^{76}Kr	75.925 959	14.8 h	EC (1.33); γ	0+		
^{79}Kr	78.920 084	1.455 d	β^+ (1.63) 7%; EC 93%; γ	1/2–		
^{81}Kr	80.916 590	2.1×10^5 y	EC (0.28); γ	7/2+		
^{81m}Kr		13.1 s	IT (0.1903); γ	1/2		R, D
^{85}Kr	84.912 531	10.73 y	β^- (0.687); γ	9/2+	1.005	R

Other radioisotopes of krypton have half-lives shorter than 5 hours.

Lanthanum

Nuclide	Atomic mass	Half life ($T_{1/2}$)	Decay mode and energy (MeV)	Nuclear spin *I*	Nuclear magnetic moment μ	Uses
^{132}La	131.910 100	4.8 h	β^+ (4.71) 40%; EC 60%; γ	2–		
^{133}La	132.908 140	3.91 h	β^+ (2.0) 4%; EC 96%; γ	5/2+		
^{135}La	134.906 953	19.5 h	EC (1.20); γ	5/2+		
^{137}La	136.906 460	60 000 y	EC (0.60); no γ	7/2+	+2.70	
^{140}La	139.909 471	1.678 d	β^- (3.761); γ	3–		
^{141}La	140.910 91	3.90 h	β^- (2.46); γ	7/2+		
^{142}La	141.914 09	1.54 h	β^- (4.52); γ	2–		

Other radioisotopes of lanthanum have half-lives shorter than 1 hour.

Lawrencium

All isotopes of lawrencium are given in the main table on page 113.

Lead

Nuclide	Atomic mass	Half life ($T_{1/2}$)	Decay mode and energy (MeV)	Nuclear spin *I*	Nuclear magnetic moment μ	Uses
^{200}Pb	199.971 790	21.5 h	EC (0.8); γ	0+		
^{202}Pb	201.972 134	53 000 y	EC (0.05); γ	0+		
^{203}Pb	202.973 265	2.162 d	EC (0.97); γ	5/2–	+0.686	
^{205}Pb	204.974 458	1.51×10^7 y	EC (0.053); no γ	5/2–	+0.712	R
^{210}Pb*	209.984 163	22.6 y	β^- (0.063); γ	0+		R, T
^{212}Pb	211.991 871	10.64 h	β^- (0.57); γ	0+		
^{214}Pb*	213.999 798	27 m	β^- (1.03); γ	0+		

*Traces of these isotopes occur naturally.
Other radioisotopes of lead have half-lives shorter than 10 hours.

Lithium

Nuclide	Atomic mass	Half life ($T_{1/2}$)	Decay mode and energy (MeV)	Nuclear spin *I*	Nuclear magnetic moment μ	Uses
^{8}Li	8.022 485	0.84 s	$\beta^-\alpha$(16.005); γ	2+	+1.6536	
^{9}Li	9.026 789	0.177 s	β^- (13.6068) 65%, n2α 35%; γ	3/2–	+3.439	

Other radioisotopes of lithium have half-lives shorter than 10 milliseconds.

Lutetium

Nuclide	Atomic mass	Half life ($T_{1/2}$)	Decay mode and energy (MeV)	Nuclear spin *I*	Nuclear magnetic moment μ	Uses
^{170}Lu	169.938 452	2.01 d	EC (3.46); γ	0+		
^{171}Lu	170.937 911	8.24 d	EC (1.480); γ	7/2+	2.0	
^{172}Lu	171.939 085	6.70 d	EC (2.520); γ	4–	2.25	
^{173}Lu	172.938 929	1.37 y	EC (0.672); γ	7/2+	2.3	
^{174}Lu	173.940 336	3.3 y	EC (1.376); γ	1–	1.9	
^{174m}Lu		142 d	IT (0.17086) 99.3%; EC 0.7%; γ	6–	2.3	
^{177}Lu	176.943 752	6.75 d	β^- (0.498); γ	7/2+	+2.239	
^{177m}Lu		160.7 d	IT (0.9702) 22%; β^- 78%; γ	23/2–	2.9	

Other radioisotopes of lutetium have half-lives shorter than 2 days.

Magnesium

Nuclide	Atomic mass	Half life ($T_{1/2}$)	Decay mode and energy (MeV)	Nuclear spin *I*	Nuclear magnetic moment μ	Uses
^{23}Mg	22.994 124	11.32 s	β^+ (4.058); γ	3/2+		
^{27}Mg	26.984 341	9.45 m	β^- (2.610); γ	0+		R
^{28}Mg	27.983 876	21.0 h	β^- (1.832); γ	1/2+		R

Other radioisotopes of magnesium have half-lives shorter than 10 seconds.

Manganese

Nuclide	Atomic mass	Half life ($T_{1/2}$)	Decay mode and energy (MeV)	Nuclear spin I	Nuclear magnetic moment μ	Uses
^{52}Mn	51.945 568	5.591 d	β^+ (4.712); EC; γ	6+	+3.063	
^{53}Mn	52.941 291	3.7×10^6 y	EC (0.596); no γ	7/2–	5.024	
^{54}Mn	53.940 361	312.2 d	EC (1.377); γ	3+	+3.282	R
^{56}Mn	55.938 906	2.578 h	β^- (3.696); γ	3+	+3.2266	

Other radioisotopes of manganese have half-lives shorter than 30 minutes.

Meitnerium

All isotopes of meitnerium are given in the main table on page 125.

Mendelevium

All isotopes of mendelevium are given in the main table on page 127.

Mercury

Nuclide	Atomic mass	Half life ($T_{1/2}$)	Decay mode and energy (MeV)	Nuclear spin I	Nuclear magnetic moment μ	Uses
^{194}Hg	193.965 391	520 y	EC (0.04); no γ	0+		
^{195m}Hg	194.966 640	1.67 d	IT (0.3186) 54%; EC 46%; γ	13/2+	–1.044 65	
^{197}Hg	196.967 187	64.1 h	EC (0.599); γ	1/2–	+0.527 374	R, D
^{197m}Hg		23.8 h	IT (0.2989) 93%; γ	13/2+	–1.027 68	
^{203}Hg	202.972 848	46.61 d	β^- (0.492); γ	5/2–	+0.8489	R, D

Other radioisotopes of mercury have half-lives shorter than 24 hours.

Molybdenum

Nuclide	Atomic mass	Half life ($T_{1/2}$)	Decay mode and energy (MeV)	Nuclear spin I	Nuclear magnetic moment μ	Uses
^{90}Mo	89.913 933	5.7 h	β^+ (2.49) 25%; EC 75%; γ	0+		
^{93}Mo	92.906 813	3 500 y	EC (0.406); γ	5/2+		
^{93m}Mo		6.9 h	IT (2.425) >99%; γ	21/2+	+9.21	
^{99}Mo	98.907 711	2.748 d	β^- (1.357); γ	1/2+	0.375	

Other radioisotopes of molybdenum have half-lives shorter than 20 minutes.

Neodymium

Nuclide	Atomic mass	Half life ($T_{1/2}$)	Decay mode and energy (MeV)	Nuclear spin I	Nuclear magnetic moment μ	Uses
^{138}Nd	137.911 820	5.1 h	EC (*c.* 1.1); γ	0+		
^{139m}Nd	138.911 920	5.5 h	IT (0.231) 12%; β^+ 88%; γ	11/2–		
^{140}Nd	139.909 306	3.37 d	EC (0.22); no γ	0+		
^{141}Nd	140.909 594	2.49 h	EC (1.823) 98%; β^+ 2%; γ	3/2+	1.01	
^{147}Nd	146.916 097	10.98 h	β^- (0.896); γ	5/2–	0.58	R
^{149}Nd	148.920 145	1.73 h	β^- (1.688); γ	5/2–	0.35	

Other radioisotopes of neodymium have half-lives shorter than 1 hour.

Neon

Nuclide	Atomic mass	Half life ($T_{1/2}$)	Decay mode and energy (MeV)	Nuclear spin *I*	Nuclear magnetic moment μ	Uses
^{23}Ne	22.994 465	37.2 s	β^- (4.376); γ	3/2+	–1.08	
^{24}Ne	23.993 613	3.38 m	β^- (2.468); γ	0+		

Other radioisotopes of neon have half-lives shorter than 20 seconds.

Neptunium

Isotopes of neptunium with half-lives longer than 1 hour are in the main table on page 137.

Nickel

Nuclide	Atomic mass	Half life ($T_{1/2}$)	Decay mode and energy (MeV)	Nuclear spin *I*	Nuclear magnetic moment μ	Uses
^{56}Ni	55.943 124	6.10 d	EC (2.14); no γ	0+		
^{57}Ni	56.937 99	35.6 h	β^+ (3.265); EC; γ	3/2–	0.88	
^{59}Ni	58.934 349	76 000 y	EC; no γ	3/2–		
^{63}Ni	62.929 669	100 y	β^- (0.065); no γ	1/2–		R
^{65}Ni	64.930 086	2.57 h	β^- (2.134); γ	5/2–	0.69	
^{66}Ni	65.929 116	54.6 h	β^- (0.24); no γ	0+		

Other radioisotopes of nickel have half-lives shorter than 1 minute.

Niobium

Nuclide	Atomic mass	Half life ($T_{1/2}$)	Decay mode and energy (MeV)	Nuclear spin *I*	Nuclear magnetic moment μ	Uses
^{91m}Nb	90.906 991	62 d	IT 97%; γ	1/2–		
^{92}Nb	91.907 192	3×10^7 y	EC (2.006); γ	7+		
^{92m}Nb		10.13 d	EC >99%; γ	2+	6.114	
^{93m}Nb	93.906 377	16.1 y	IT (0.0304); γ	1/2–		
^{94}Nb	93.907 280	2.4×10^4 y	β^- (2.045); γ	6+		
^{95}Nb	94.906 835	34.97 d	β^- (0.926); γ	9/2+	6.141	

Other radioisotopes of niobium have half-lives shorter than 100 hours.

Nitrogen

Nuclide	Atomic mass	Half life ($T_{1/2}$)	Decay mode and energy (MeV)	Nuclear spin *I*	Nuclear magnetic moment μ	Uses
^{13}N	13.005 738	9.97 m	β^+ (2.2205); no γ	1/2	–0.3222	
^{16}N	16.006 099	7.13 s	$\beta^-\alpha$(10.4187); β^-; γ	2–		
^{17}N	17.008 450	4.17 s	β^+, β^-n(8.680); γ	1/2–		

Other radioisotopes of nitrogen have half-lives shorter than 1 second.

Nobelium

All isotopes of nobelium are given in the main table on page 145.

Osmium

Nuclide	Atomic mass	Half life ($T_{1/2}$)	Decay mode and energy (MeV)	Nuclear spin *I*	Nuclear magnetic moment μ	Uses
^{182}Os	181.952 120	21.5 h	EC (0.9); γ	0+		
^{185}Os	184.954 041	93.6 d	EC (1.013); γ	1/2–		
^{191}Os	190.960 920	15.4 d	β^- (0.314); γ	9/2–		
^{193}Os	192.964 138	30.5 h	β^- (1.139); γ	3/2–	+0.75	
^{194}Os	193.965 173	6.0 y	β^- (0.097); γ	0+		

Other radioisotopes of osmium have half-lives shorter than 20 hours.

Oxygen

Nuclide	Atomic mass	Half life ($T_{1/2}$)	Decay mode and energy (MeV)	Nuclear spin *I*	Nuclear magnetic moment μ	Uses
^{14}O	14.008 595	70.60 s	β^+ (5.1430); γ	0+		
^{15}O	15.003 065	122.2 s	β^+ (2.754); no γ	1/2–	0.719	

Other radioisotopes of oxygen have half-lives shorter than 1 minute.

Palladium

Nuclide	Atomic mass	Half life ($T_{1/2}$)	Decay mode and energy (MeV)	Nuclear spin *I*	Nuclear magnetic moment μ	Uses
^{100}Pd	99.908 527	3.7 d	EC (037); γ	0+		
^{103}Pd	102.906 114	16.99 d	EC (0.5); γ	5/2+		
^{107}Pd	106.905 127	6.5×10^6 y	β^- (0.035); γ	5/2+		
^{109}Pd	108.905 954	13.5 h	β^- (1.116); γ	5/2+		

Other radioisotopes of palladium have half-lives shorter than 10 hours.

Phosphorus

Nuclide	Atomic mass	Half life ($T_{1/2}$)	Decay mode and energy (MeV)	Nuclear spin *I*	Nuclear magnetic moment μ	Uses
^{30}P	29.978 307	2.50 m	β^+ (4.226); no γ	1+		
^{32}P	31.973 907	14.28 d	β^- (1.710); no γ	1+	–0.2524	R, D, T
^{33}P	32.971 725	25.3 d	β^- (0.249); no γ	1/2+		R

Other radioisotopes of phosphorus have half-lives shorter than 20 seconds.

Platinum

Nuclide	Atomic mass	Half life ($T_{1/2}$)	Decay mode and energy (MeV)	Nuclear spin *I*	Nuclear magnetic moment μ	Uses
^{193}Pt	192.962 977	60 y	EC (0.057); no γ	1/2–		
^{193m}Pt		4.33 d	IT (0.1498); γ	13/2+	–0.75	
^{195m}Pt	194.964 766	4.02 d	IT (0.2592); γ	13/2+	0.61	
^{197}Pt	196.967 315	18.3 h	β^- (0.719); γ	1/2–	0.51	
^{200}Pt	199.971 417	12.5 h	β^- (0.65); γ	0+		

Other radioisotopes of platinum have half-lives shorter than 11 hours.

Plutonium

Isotopes of plutonium with half-lives longer than 1 hour are in the main table on page 157.

Potassium

Nuclide	Atomic mass	Half life ($T_{1/2}$)	Decay mode and energy (MeV)	Nuclear spin *I*	Nuclear magnetic moment μ	Uses
^{38}K	37.969 080	7.63 m	β^+ (5.913); γ	3+	–1.37	
^{42}K	41.962 402	12.36 h	β^- (3.523); γ	2–	–1.1425	R, D
^{43}K	42.960 717	22.3 h	β^- (1.82); γ	3/2+	+0.163	D
^{44}K	43.961 56	22.1 m	β^- (5.66); γ	2–	–0.856	
^{45}K	44.960 696	17.8 m	β^- (4.20); γ	3/2+	+0.1734	

Other radioisotopes of potassium have half-lives shorter than 2 minutes.

Praseodymium

Nuclide	Atomic mass	Half life ($T_{1/2}$)	Decay mode and energy (MeV)	Nuclear spin *I*	Nuclear magnetic moment μ	Uses
^{137}Pr	136.910 680	1.28 h	β^+ (2.71) 26%; EC 74%; γ	5/2+		
^{138m}Pr	137.910 748	2.1 h	β^+ 24%; EC 76% γ	7–		
^{139}Pr	138.908 917	4.41 h	β^+ (2.13) 8%; EC 92%; γ	5/2+		
^{142}Pr	141.910 039	19.12 h	β^- (2.162); γ	2–	+0.223	R
^{143}Pr	142.910 814	13.57 d	β^- (0.934); IT (0.059) 99+%; γ	7/2+		R
^{145}Pr	144.914 501	5.98 h	β^- (1.81); γ	7/2+		

Other radioisotopes of praseodymium have half-lives shorter than 1 hour.

Promethium

Isotopes of promethium with half-lives longer than 1 minute are in the main table on page 165.

Protactinium

Isotopes of protactinium with half-lives longer than 1 hour are in the main table on page 167.

Radium

Isotopes of radium with half-lives longer than 5 minutes are in the main table on page 169.

Radon

Isotopes of radon with half-lives longer than 10 minutes are in the main table on page 171.

Rhenium

Nuclide	Atomic mass	Half life ($T_{1/2}$)	Decay mode and energy (MeV)	Nuclear spin *I*	Nuclear magnetic moment μ	Uses
^{181}Re	180.950 020	20 h	EC (1.8); γ	5/2+	3.19	
^{182}Re	181.951 210	2.67 d	EC (2.80); γ	7+	2.8	
^{183}Re	182.950 817	70 d	EC (0.556); γ	5/2+		
^{184}Re	183.952 530	38 d	EC (1.49); 92%	3–	2.53	
^{184m}Re		165 d	EC 25%; IT (0.188) 75%; γ	8+	+2.9	
^{186}Re	185.954 984	3.78 d	β^- (1.070) 92%; EC (0.585); 8%; γ	1–	+1.739	
^{186m}Re		20 000 y	IT (0.150); γ			
^{188}Re	187.958 106	16.94 h	β^- (2.120); γ	1–	+1.788	
^{189}Re	188.959 219	24 h	β^- (1.01); γ	5/2+		

Other radioisotopes of rhenium have half-lives shorter than 15 hours.

Rhodium

Nuclide	Atomic mass	Half life ($T_{1/2}$)	Decay mode and energy (MeV)	Nuclear spin *I*	Nuclear magnetic moment μ	Uses
^{99}Rh	98.908 192	16 d	β^+ (2.10) 4%; EC 96%; γ	1/2–		
^{100}Rh	99.908 116	20.8 h	β^+ (3.63); EC; γ	1–		
^{101}Rh	100.906 159	3.3 y	EC (0.54); γ	1/2–		
^{101m}Rh		4.35 d	EC 92%; IT (0.1573); 8%; γ	9/2+	+5.51	
^{102}Rh	101.906 814	2.9 y	EC (2.28); γ	6+	4.04	
^{102m}Rh		207 d	IT 5%; β^- 19%; β^+ 14%; EC 62%; γ			
^{105}Rh	104.905 686	35.4 h	β^- (0.567); γ	7/2+	+4.45	

Other radioisotopes of rhodium have half-lives shorter than 5 hours.

Rubidium

Nuclide	Atomic mass	Half life ($T_{1/2}$)	Decay mode and energy (MeV)	Nuclear spin *I*	Nuclear magnetic moment μ	Uses
^{81}Rb	80.918 990	4.57 h	β^+ (2.24) 27%; EC 73%; γ	3/2–	+2.060	
^{82m}Rb	81.918 195	6.47 h	β^+ 26%; EC 74%; γ	5–	+1.5	
^{83}Rb	82.915 144	86.2 d	EC (0.93); γ	5/2–	+1.425	R
^{84}Rb	83.914 39	32.9 d	β^+ (2.682) 22%; EC 75%; β^- (0.893) 3%; γ	2–	–1.324 12	
^{86}Rb	85.911 172	18.65 d	β^- (1.775); γ	2–	–1.6920	R

Other radioisotopes of rubidium have half-lives shorter than 1 hour.

Ruthenium

Nuclide	Atomic mass	Half life ($T_{1/2}$)	Decay mode and energy (MeV)	Nuclear spin *I*	Nuclear magnetic moment μ	Uses
^{95}Ru	94.910 414	1.64 h	EC (2.57) 85%; β^+ 15%; γ	5/2+		
^{97}Ru	96.907 556	2.89 d	EC (1.11); γ	5/2+	–0.78	
^{103}Ru	102.906 323	39.27 d	β^- (0.764); γ	3/2+	0.20	R
^{105}Ru	104.907 744	4.44 h	β^- (1.92); γ	3/2+	–0.3	
^{106}Ru	105.907 321	1.020 y	β^- (0.039); no γ	0+		R, T

Other radioisotopes of ruthenium have half-lives shorter than 1 hour.

Rutherfordium

All isotopes of rutherfordium are given in the main table on page 181.

Samarium

Nuclide	Atomic mass	Half life ($T_{1/2}$)	Decay mode and energy (MeV)	Nuclear spin *I*	Nuclear magnetic moment μ	Uses
^{145}Sm	144.913 409	340 d	EC (0.618); γ	7/2–	–1.1	
^{146}Sm	145.913 053	1.03×10^8 y	α; no γ	0+		
^{151}Sm	150.919 929	90 y	β^- (0.076); γ	5/2–	–0.363	R
^{153}Sm	152.922 094	1.929 d	β^- (0.809); γ	3/2+	–0.0216	
^{156}Sm	155.925 518	9.4 h	β^- (0.72); γ	0+		

Other radioisotopes of samarium have half-lives shorter than 1 hour.

Scandium

Nuclide	Atomic mass	Half life ($T_{1/2}$)	Decay mode and energy (MeV)	Nuclear spin *I*	Nuclear magnetic moment μ	Uses
^{43}Sc	42.961 150	3.89 h	β^+ (2.221); EC; γ	7/2–	+4.62	
^{44}Sc	43.959 404	3.93 h	β^- (3.655); EC; γ	2+	+2.56	
^{44m}Sc		2.44 d	IT (0.27); EC (3.926); γ	6+	+3.88	
^{46}Sc	45.955 170	83.81 d	β^- (2.367); γ	4+	+3.03	R
^{47}Sc	46.952 408	3.349 d	β^- (0.601); γ	7/2–	+5.34	
^{48}Sc	47.952 235	43.7 h	β^- (3.99); γ	6+		

Other radioisotopes of scandium have half-lives shorter than 1 hour.

Seaborgium

All isotopes of seaborgium are given in the main table on page 187.

Selenium

Nuclide	Atomic mass	Half life ($T_{1/2}$)	Decay mode and energy (MeV)	Nuclear spin I	Nuclear magnetic moment μ	Uses
^{72}Se	71.927 110	8.5 d	EC (0.33); γ	0+		
^{73}Se	72.926 768	7.1 h	β^+ (2.74) 65%; EC 35%; γ	9/2+		
^{75}Se	74.922 521	119.8 d	β^- (0.864); γ	5/2+	0.67	R, D
^{79}Se	78.918 498	65 000 y	β^- (0.149); γ	7/2+	–1.018	

Other radioisotopes of selenium have half-lives shorter than 1 hour.

Silicon

Nuclide	Atomic mass	Half life ($T_{1/2}$)	Decay mode and energy (MeV)	Nuclear spin I	Nuclear magnetic moment μ	Uses
^{31}Si	30.975 362	2.62 h	β^- (1.49); γ	3/2+		
^{32}Si	31.974 148	160 y	β^- (0.227); no γ	0+		R

Other radioisotopes of silicon have half-lives shorter than 10 seconds.

Silver

Nuclide	Atomic mass	Half life ($T_{1/2}$)	Decay mode and energy (MeV)	Nuclear spin I	Nuclear magnetic moment μ	Uses
^{104}Ag	103.908 623	69 m	β^+ (4.28); 16%; EC 84%; γ	5+	3.92	
^{105}Ag	104.906 520	41.3 d	EC (1.34); γ	1/2–	0.1014	
^{106m}Ag	105.906 662	8.4 d	EC; γ	6+	3.71	
^{108m}Ag	107.905 952	130 y	EC 92%; IT (0.079) 8%; γ	6+	+3.580	
^{110m}Ag	109.906 111	249.8 d	β 99%; IT (0.1164) 1%; γ	6+	+3.607	R
^{111}Ag	110.905 295	7.47 d	β^- (1.037); γ	1/2–	–0.146	
^{112}Ag	111.907 010	3.13 h	β^- (3.96); γ	2–	0.0547	

Other radioisotopes of silver have half-lives shorter than 1 hour.

Sodium

Nuclide	Atomic mass	Half life ($T_{1/2}$)	Decay mode and energy (MeV)	Nuclear spin I	Nuclear magnetic moment μ	Uses
^{22}Na	21.994 434	2.605 y	β^+ (2.842); 90%; EC 10%; γ	3+	+1.746	R, D
^{24}Na	23.990 961	14.96 y	β^- (5.514); γ	4+	+1.690	R, D

Other radioisotopes of sodium have half-lives shorter than 1 minute.

Strontium

Nuclide	Atomic mass	Half life ($T_{1/2}$)	Decay mode and energy (MeV)	Nuclear spin I	Nuclear magnetic moment μ	Uses
^{82}Sr	81.918 414	25.36 d	EC (0.21); no γ	1/2–		R
^{83}Sr	82.917 566	1.350 d	β^+ (2.27) 24%; EC 76%; γ	7/2+	–0.898	
^{85}Sr	84.912 937	64.84 d	EC (1.085); γ	9/2+	–1.001	R, D
^{87m}Sr	86.908 884	2.80 h	IT (0.3885); γ	1/2–		R, D
^{89}Sr	88.907 450	50.52 d	β^- (1.492); γ	5/2+	–1.149	
^{90}Sr	89.907 738	29.1 y	β^- (0.546); no γ	0+		R, T
^{91}Sr	90.910 187	9.5 h	β^- (2.70); no γ	5/2+	–0.887	

Other radioisotopes of strontium have half-lives shorter than 3 hours.

Sulfur

Nuclide	Atomic mass	Half life ($T_{1/2}$)	Decay mode and energy (MeV)	Nuclear spin *I*	Nuclear magnetic moment μ	Uses
^{35}S	34.969 031	87.2 d	β^- (0.1674); no γ	3/2+	+1.00	R
^{37}S	36.971 125	5.05 m	β^- (4.865); γ			
^{38}S	37.971 162	2.84 h	β^- (2.94); γ	0+	+1.00	

Other radioisotopes of sulfur have half-lives shorter than 1 minute.

Tantalum

Nuclide	Atomic mass	Half life ($T_{1/2}$)	Decay mode and energy (MeV)	Nuclear spin *I*	Nuclear magnetic moment μ	Uses
^{173}Ta	172.943 650	3.6 h	β^+ (2.8); 24%; EC 76%; γ	5/2–		
^{174}Ta	173.944 340	1.12 h	β^+ (4.00); 27%; EC 73%; γ	3+		
^{175}Ta	174.943 650	10.5 h	EC (2.00); γ	7/2+	2.27	
^{176}Ta	175.944 730	8.1 h	EC (3.1); γ	1–		
^{177}Ta	176.944 460	2.356 d	EC (1.166); γ	7/2+	2.25	
^{178m}Ta	177.945 750	2.4 h	EC; γ	7–		
^{179}Ta	178.945 930	1.8 y	EC (0.11); no γ	7/2+		
^{182}Ta	181.950 149	114.4 d	β^- (1.814); γ	3–	+3.02	
^{183}Ta	182.951 369	5.1 d	β^- (1.07); γ	7/2+	+2.36	
^{184}Ta	183.954 005	8.7 d	β^- (2.87); γ	5–		

Other radioisotopes of tantalum have half-lives shorter than 1 hour.

Technetium

Isotopes of technetium with half-lives longer than 5 hours are given in the main table on page 203.

Tellurium

Nuclide	Atomic mass	Half life ($T_{1/2}$)	Decay mode and energy (MeV)	Nuclear spin *I*	Nuclear magnetic moment μ	Uses
^{118}Te	117.905 908	6.00 d	EC (0.34); no γ	0+		
^{121}Te	120.904 947	16.8 d	EC (1.04); γ	1/2+		
^{121m}Te		154 d	IT 89%; EC 11%; γ	11/2–		
^{123m}Te	122.904 271	119.7 d	IT (0.247); γ	11/2–	–0.93	
^{125m}Te	124.904 433	58 d	IT (0.145); γ	11/2–	–0.99	
^{127}Te	126.905 277	9.4 h	β^- (0.697); γ	3/2+	0.64	
^{127m}Te		109 d	IT (0.088) 98%; β^- (0.77) 2%; γ	11/2–	–1.04	
^{129m}Te	128.906 594	33.5 d	IT (0.105) 63%; β^- 37%; γ	11/2–		

Other radioisotopes of tellurium have half-lives shorter than 5 days.

Terbium

Nuclide	Atomic mass	Half life ($T_{1/2}$)	Decay mode and energy (MeV)	Nuclear spin *I*	Nuclear magnetic moment μ	Uses
^{153}Tb	152.923 440	2.34 d	EC (1.571); γ	5/2+	3.5	
^{155}Tb	154.923 499	5.3 d	EC (0.82); γ	3/2+	2.0	
^{156}Tb	155.924 742	5.3 d	EC (2.444); γ	3–	1.4	
^{157}Tb	156.924 023	150 y	EC (0.058); γ	3/2+		
^{158}Tb	157.925 411	180 y	EC (1.221) 80%; β^- (0.936) 20%; γ	3–	+1.76	
^{160}Tb	159.927 163	72.3 d	β^- (1.834); γ	3–	+1.79	
^{161}Tb	160.927 566	6.91 d	β^- (0.593); γ	3/2+	2.2	

Other radioisotopes of terbium have half-lives shorter than 1 day.

Thallium

Nuclide	Atomic mass	Half life ($T_{1/2}$)	Decay mode and energy (MeV)	Nuclear spin *I*	Nuclear magnetic moment μ	Uses
^{198}Tl	197.940 460	5.3 h	β^+ (3.5) 1%; EC 99%; γ	2–	0.00	
^{199}Tl	198.969 870	7.4 h	EC (1.54); γ	1/2–	+1.60	
^{200}Tl	199.970 934	1.087 d	EC (2.46); γ	2–	0.04	
^{201}Tl	200.970 794	3.038 d	EC (0.48); γ	1/2+	+1.66	D
^{202}Tl	201.972 085	12.23 d	EC (1.36); γ	2–	0.06	
^{204}Tl		3.78 y	β^- (0.763) 97%; EC (0.345) 3%; γ	2–	0.09	

Other radioisotopes of thallium have half-lives shorter than 5 hours.

Thorium

Nuclide	Atomic mass	Half life ($T_{1/2}$)	Decay mode and energy (MeV)	Nuclear spin *I*	Nuclear magnetic moment μ	Uses
^{227}Th*	227.027 703	18.72 d	α (6.146); γ	3/2+		
^{228}Th*	228.028 715	1.913 y	α (5.520); γ	0+		
^{229}Th	229.031 755	7900 y	α (5.168); γ	5/2+	+0.46	R, NMR
^{230}Th*	230.033 127	75 400 y	α (4.771); γ	0+		R
^{231}Th*	231.036 298	1.063 d	β^- (0.389); γ	5/2+		
^{234}Th*	234.036 593	24.1 d	β^- (0.270); γ	0+		

*Traces of these isotopes are found in some minerals.
Other radioisotopes of thorium have half-lives shorter than 1 hour.

Thulium

Nuclide	Atomic mass	Half life ($T_{1/2}$)	Decay mode and energy (MeV)	Nuclear spin *I*	Nuclear magnetic moment μ	Uses
^{165}Tm	164.932 432	1.253 d	EC (1.592); γ	1/2+	0.139	
^{167}Tm	166.932 848	9.24 d	EC (0.748); γ	1/2+	–0.197	
^{168}Tm	167.934 170	93.1 d	EC (1.679); γ	3+	0.23	
^{170}Tm	169.935 198	128.6 d	β^- (0.968) 99.8%; EC (0.315) 0.2%; γ	1–	0.2476	
^{171}Tm	170.936 427	1.92 y	β^- (0.097); γ	1/2+	–0.2303	
^{172}Tm	171.938 397	2.65 d	β^- (1.88); γ	2–		

Other radioisotopes of thulium have half-lives shorter than 10 hours.

Tin

Nuclide	Atomic mass	Half life ($T_{1/2}$)	Decay mode and energy (MeV)	Nuclear spin *I*	Nuclear magnetic moment μ	Uses
^{110}Sn	109.907 858	4.1 h	EC (0.58); γ	0+		
^{111}Sn	110.907 741	35 m	β^+ (2.45); 31%; EC 69%; γ	7/2+	+0.61	
^{113}Sn	112.905 176	115.1 d	EC (1.038); γ	1/2+	–0.879	R
^{117m}Sn	116.902 956	13.6 d	IT (0.3146); γ	11/2–	–1.40	
^{119m}Sn	118.903 310	293 d	IT (0.0895); γ	11/2–	–1.4	R
^{121}Sn	120.904 238	1.128 d	β^- (0.388); no γ	3/2+	0.698	
^{123}Sn	122.905 722	129.2 d	β^- (1.403); γ	11/2–	–1.370	
^{125}Sn	124.907 785	9.63 d	β^- (2.360); γ	11/2–	–1.35	
^{126}Sn	125.907 654	10^5 y	β^- (0.38); γ	0+		
^{127}Sn	126.910 355	2.12 h	β^- (3.20); γ	11/2–		

Other radioisotopes of tin have half-lives shorter than 1 hour.

Titanium

Nuclide	Atomic mass	Half life ($T_{1/2}$)	Decay mode and energy (MeV)	Nuclear spin *I*	Nuclear magnetic moment μ	Uses
^{44}Ti	43.959 689	67 y	EC (0.265); γ	0+		R
^{45}Ti	44.958 124	3.078 h	β^+ (2.063) 86%; EC 14%; no γ	7/2–	0.095	

Other radioisotopes of titanium have half-lives shorter than 10 minutes.

Tungsten

Nuclide	Atomic mass	Half life ($T_{1/2}$)	Decay mode and energy (MeV)	Nuclear spin *I*	Nuclear magnetic moment μ	Uses
^{178}W	177.945 840	21.6 d	EC (0.09); no γ	0+		
^{181}W	180.948 192	121.2 d	EC (0.19); no γ	9/2+		
^{185}W	184.953 416	74.8 d	β^- (0.433); γ	3/2–		
^{187}W	186.957 153	23.9 h	β^- (1.312); γ	3/2–	0.62	
^{188}W	187.958 480	69.4 d	β^- (0.349); γ	0+		

Other radioisotopes of tungsten have half-lives shorter than 3 hours.

Uranium

Nuclide	Atomic mass	Half life ($T_{1/2}$)	Decay mode and energy (MeV)	Nuclear spin *I*	Nuclear magnetic moment μ	Uses
^{230}U	230.033 921	20.8 d	α (5.992); γ	0+		
^{231}U	231.036 270	4.2 d	EC (0.33); γ	5/2–		
^{232}U	232.037 130	68.9 y	α (5.414); no γ	0+		
^{233}U	233.039 628	1.59×10^5 y	α (4.909); γ	5/2+	+0.59	
^{236}U	236.045 562	2.34×10^7 y	α (4.569); γ	0+		
^{237}U	237.048 724	6.75 d	β^- (0.519); γ	1/2+		

Other radioisotopes of uranium have half-lives shorter than 1 day.

Vanadium

Nuclide	Atomic mass	Half life ($T_{1/2}$)	Decay mode and energy (MeV)	Nuclear spin *I*	Nuclear magnetic moment μ	Uses
^{47}V	46.954 906	32.6 m	β^+ (2.927); EC; no γ	3/2–		
^{48}V	47.952 257	15.98 d	β^+ (4.015); γ	4+	2.01	
^{49}V	48.948 517	337 d	EC (0.601); no γ	7/2–	4.47	

Other radioisotopes of vanadium have half-lives shorter than 5 minutes.

Xenon

Nuclide	Atomic mass	Half life ($T_{1/2}$)	Decay mode and energy (MeV)	Nuclear spin *I*	Nuclear magnetic moment μ	Uses
^{122}Xe	121.908 170	20.1 h	EC (1.0); γ	0+		
^{123}Xe	122.908 469	2.00 h	EC 77%; β^+ (2.68) 23%; γ	1/2+		
^{125}Xe	124.906 397	17.1 h	EC (1.654); γ	1/2+		
^{127}Xe	126.905 182	36.4 d	EC (0.663); γ	1/2+	–0.504	R, D
^{129m}Xe	128.904 780	8.89 d	IT (0.236); γ	11/2–		
^{131m}Xe	130.905 072	11.92 d	IT (0.164); γ	11/2+	+0.6908	
^{133}Xe	132.905 888	5.243 d	β^- (0.427); γ	3/2+	+0.813	D
^{133m}Xe?		2.19 d	IT (0.233); γ	11/2–	–1.082	
^{135}Xe	134.907 130	9.10 h	β^- (1.16); γ	3/2+	0.903	

Other radioisotopes of xenon have half-lives shorter than 1 hour.

Ytterbium

Nuclide	Atomic mass	Half life ($T_{1/2}$)	Decay mode and energy (MeV)	Nuclear spin *I*	Nuclear magnetic moment μ	Uses
^{164}Yb	163.934 530	1.26 h	EC (1.0); no γ	0+		
^{166}Yb	165.933 875	2.363 d	EC (0.30); γ	0+		
^{169}Yb	168.935 186	32.03 d	EC (0.908); γ	7/2+	–0.63	R, D
^{175}Yb	174.941 273	4.19 d	β^- (0.469); γ	7/2	0.6	
^{177}Yb	176.945 253	1.9 h	β^- (1.398); γ	9/2+		
^{178}Yb	177.946 639	1.23 h	β^- (0.63); γ	0+		

Other radioisotopes of ytterbium have half-lives shorter than 1 hour.

Yttrium

Nuclide	Atomic mass	Half life ($T_{1/2}$)	Decay mode and energy (MeV)	Nuclear spin *I*	Nuclear magnetic moment μ	Uses
^{86}Y	85.914 893	14.74 h	EC (5.24); β^+; γ	4	<0.6	
^{87}Y	86.910 882	3.35 d	EC (1.861) > 99%; γ	1/2–		
^{87m}Y		13 h	IT 98%; β^+ 0.7%; EC; γ	9/2+		
^{88}Y	87.909 508	106. 6d	EC (3.623) 99.8%; β^+ 0.2%; γ	4–		R
^{90}Y	89.907 152	2.67 d	β^- (2.283); no γ	2–	–1.630	R
^{91}Y	90.907 303	58.5 d	β^- (1.545); γ	1/2–	0.1461	
^{93}Y	92.909 571	10.2 h	β^- (2.88); γ	1/2–		

Other radioisotopes of yttrium have half-lives shorter than 5 hours.

Zinc

Nuclide	Atomic mass	Half life ($T_{1/2}$)	Decay mode and energy (MeV)	Nuclear spin *I*	Nuclear magnetic moment μ	Uses
^{62}Zn	61.934 332	9.22 h	β^+ (1.63) 3%; EC 97%; γ	0+		
^{65}Zn	64.929 243	243.8 d	β^+ (1.352) 98%; EC 1.5%; γ	5/2–	+0.7690	R
^{69m}Zn	68.926 552	13.76 h	IT (0.439); γ	9/2+		R
^{71m}Zn	70.927 727	3.97 h	β^-; γ	9/2+		

Other radioisotopes of zinc have half-lives shorter than 1 hour.

Zirconium

Nuclide	Atomic mass	Half life ($T_{1/2}$)	Decay mode and energy (MeV)	Nuclear spin *I*	Nuclear magnetic moment μ	Uses
^{86}Zr	85.916 290	16.5 h	EC (1.3); γ	0+		
^{88}Zr	87.910 225	83.4 d	EC (0.67); γ	0+		
^{89}Zr	88.908 890	3.27 d	EC 77%; β^+ (2.83) 23%; γ	9/2+		
^{93}Zr	92.906 474	1.5×10^6 y	β^- (0.09); γ	5/2+		
^{95}Zr	94.908 042	64.02 d	β^- (1.125); γ	5/2+		
^{97}Zr	96.910 950	16.8 h	β^- (2.658); γ	1/2–		

Other radioisotopes of zirconium have half-lives shorter than 2 hours.

APPENDIX 2
Electron binding energies

Actinium

Electron binding energies /eV

N_{I}	4s	1269
N_{II}	$4p_{1/2}$	1080
N_{III}	$4p_{3/2}$	890
N_{IV}	$4d_{3/2}$	675
N_{V}	$4d_{5/2}$	639
N_{VI}	$4f_{5/2}$	319
N_{VII}	$4f_{7/2}$	319
O_{I}	5s	272
O_{II}	$5p_{1/2}$	215
O_{III}	$5p_{3/2}$	167
O_{IV}	$5d_{3/2}$	80
O_{V}	$5d_{5/2}$	80
P_{I}	6s	–
P_{II}	$6p_{1/2}$	–
P_{III}	$6p_{3/2}$	–

Antimony

Electron binding energies /eV

N_{I}	4s	153.2
N_{II}	$4p_{1/2}$	95.6
N_{III}	$4p_{3/2}$	95.6
N_{IV}	$4d_{3/2}$	33.3
N_{V}	$4d_{5/2}$	32.1

Astatine

Electron binding energies /eV

N_{I}	4s	1042
N_{II}	$4p_{1/2}$	886
N_{III}	$4p_{3/2}$	740
N_{IV}	$4d_{3/2}$	533
N_{V}	$4d_{5/2}$	507
N_{VI}	$4f_{5/2}$	210
N_{VII}	$4f_{7/2}$	210
O_{I}	5s	195
O_{II}	$5p_{1/2}$	148
O_{III}	$5p_{3/2}$	115
O_{IV}	$5d_{3/2}$	40
O_{V}	$5d_{5/2}$	40

Barium

Electron binding energies /eV

N_{I}	4s	253.5
N_{II}	$4p_{1/2}$	192
N_{III}	$4p_{3/2}$	178.6
N_{IV}	$4d_{3/2}$	92.6
N_{V}	$4d_{5/2}$	89.9
N_{VI}	$4f_{5/2}$	–
N_{VII}	$4f_{7/2}$	–
O_{I}	5s	30.3
O_{II}	$5p_{1/2}$	17.0
O_{III}	$5p_{3/2}$	14.8

Bismuth

Electron binding energies /eV

N_{I}	4s	939
N_{II}	$4p_{1/2}$	805.2
N_{III}	$4p_{3/2}$	678.8
N_{IV}	$4d_{3/2}$	464.0
N_{V}	$4d_{5/2}$	440.1
N_{VI}	$4f_{5/2}$	162.3
N_{VII}	$4f_{7/2}$	157.0
O_{I}	5s	159.3
O_{II}	$5p_{1/2}$	119.0
O_{III}	$5p_{3/2}$	92.6
O_{IV}	$5d_{3/2}$	26.9
O_{V}	$5d_{5/2}$	23.8

Cadmium

Electron binding energies /eV

N_{I}	4s	109.8
N_{II}	$4p_{1/2}$	63.9
N_{III}	$4p_{3/2}$	63.9
N_{IV}	$4d_{3/2}$	11.7
N_{V}	$4d_{5/2}$	10.7

Caesium/Cesium (US)

Electron binding energies /eV

N_{I}	4s	232.3
N_{II}	$4p_{1/2}$	172.4
N_{III}	$4p_{3/2}$	161.3
N_{IV}	$4d_{3/2}$	79.8
N_{V}	$4d_{5/2}$	77.5
N_{VI}	$4f_{5/2}$	–
N_{VII}	$4f_{7/2}$	–
O_{I}	5s	22.7
O_{II}	$5p_{1/2}$	14.2
O_{III}	$5p_{3/2}$	12.1

Cerium

Electron binding energies /cV

N_{I}	4s	291.0
N_{II}	$4p_{1/2}$	223.3
N_{III}	$4p_{3/2}$	206.5
N_{IV}	$4d_{3/2}$	109
N_{V}	$4d_{5/2}$	–
N_{VI}	$4f_{5/2}$	0.1
N_{VII}	$4f_{7/2}$	0.1
O_{I}	5s	37.8
O_{II}	$5p_{1/2}$	19.8
O_{III}	$5p_{3/2}$	17.0

Dysprosium

Electron binding energies /eV

N_{I}	4s	414.2
N_{II}	$4p_{1/2}$	333.5
N_{III}	$4p_{3/2}$	293.2
N_{IV}	$4d_{3/2}$	153.6
N_{V}	$4d_{5/2}$	153.6
N_{VI}	$4f_{5/2}$	8.0
N_{VII}	$4f_{7/2}$	4.3
O_{I}	5s	49.9
O_{II}	$5p_{1/2}$	26.3
O_{III}	$5p_{3/2}$	26.3

Erbium

Electron binding energies /eV

N_{I}	4s	449.8
N_{II}	$4p_{1/2}$	366.2
N_{III}	$4p_{3/2}$	320.2
N_{IV}	$4d_{3/2}$	167.6
N_{V}	$4d_{5/2}$	167.6
N_{VI}	$4f_{5/2}$	–
N_{VII}	$4f_{7/2}$	4.7
O_{I}	5s	50.6
O_{II}	$5p_{1/2}$	31.4
O_{III}	$5p_{3/2}$	24.7

Europium

Electron binding energies /eV

N_{I}	4s	360
N_{II}	$4p_{1/2}$	284
N_{III}	$4p_{3/2}$	257
N_{IV}	$4d_{3/2}$	133
N_{V}	$4d_{5/2}$	127.7
N_{VI}	$4f_{5/2}$	0
N_{VII}	$4f_{7/2}$	0
O_{I}	5s	32
O_{II}	$5p_{1/2}$	22
O_{III}	$5p_{3/2}$	22

Francium

Electron binding energies /eV

N_{I}	4s	1153
N_{II}	$4p_{1/2}$	980
N_{III}	$4p_{3/2}$	810
N_{IV}	$4d_{3/2}$	603
N_{V}	$4d_{5/2}$	577
N_{VI}	$4f_{5/2}$	268
N_{VII}	$4f_{7/2}$	268
O_{I}	5s	234
O_{II}	$5p_{1/2}$	182
O_{III}	$5p_{3/2}$	140
O_{IV}	$5d_{3/2}$	58
O_{V}	$5d_{5/2}$	58
P_{I}	6s	34
P_{II}	$6p_{1/2}$	15
P_{III}	$6p_{3/2}$	15

Gadolinium

Electron binding energies /eV

N_{I}	4s	378.6
N_{II}	$4p_{1/2}$	286
N_{III}	$4p_{3/2}$	271
N_{IV}	$4d_{3/2}$	–
N_{V}	$4d_{5/2}$	142.6
N_{VI}	$4f_{5/2}$	8.6
N_{VII}	$4f_{7/2}$	8.6
O_{I}	5s	36
O_{II}	$5p_{1/2}$	20
O_{III}	$5p_{3/2}$	20

Gold

Electron binding energies /eV

N_I	4s	762.1
N_{II}	$4p_{1/2}$	642.7
N_{III}	$4p_{3/2}$	546.3
N_{IV}	$4d_{3/2}$	353.2
N_V	$4d_{5/2}$	335.1
N_{VI}	$4f_{5/2}$	87.6
N_{VII}	$4f_{7/2}$	83.9
O_I	5s	107.2
O_{II}	$5p_{1/2}$	74.2
O_{III}	$5p_{3/2}$	57.2

Hafnium

Electron binding energies /eV

N_I	4s	538
N_{II}	$4p_{1/2}$	438.2
N_{III}	$4p_{3/2}$	380.7
N_{IV}	$4d_{3/2}$	220.0
N_V	$4d_{5/2}$	211.5
N_{VI}	$4f_{5/2}$	15.9
N_{VII}	$4f_{7/2}$	14.2
O_I	5s	64.2
O_{II}	$5p_{1/2}$	38
O_{III}	$5p_{3/2}$	29.9

Holmium

Electron binding energies /eV

N_I	4s	432.4
N_{II}	$4p_{1/2}$	343.5
N_{III}	$4p_{3/2}$	308.2
N_{IV}	$4d_{3/2}$	160
N_V	$4d_{5/2}$	160
N_{VI}	$4f_{5/2}$	8.6
N_{VII}	$4f_{7/2}$	5.2
O_I	5s	49.3
O_{II}	$5p_{1/2}$	30.8
O_{III}	$5p_{3/2}$	24.1

Indium

Electron binding energies /eV

N_I	4s	122.9
N_{II}	$4p_{1/2}$	73.5
N_{III}	$4p_{3/2}$	73.5
N_{IV}	$4d_{3/2}$	17.7
N_V	$4d_{5/2}$	16.9

Iodine

Electron binding energies /eV

N_I	4s	186
N_{II}	$4p_{1/2}$	123
N_{III}	$4p_{3/2}$	123
N_{IV}	$4d_{3/2}$	50
N_V	$4d_{5/2}$	50

Iridium

Electron binding energies /eV

N_I	4s	691.1
N_{II}	$4p_{1/2}$	577.8
N_{III}	$4p_{3/2}$	495.8
N_{IV}	$4d_{3/2}$	311.9
N_V	$4d_{5/2}$	296.3
N_{VI}	$4f_{5/2}$	63.8
N_{VII}	$4f_{7/2}$	60.8
O_I	5s	95.2
O_{II}	$5p_{1/2}$	63.0
O_{III}	$5p_{3/2}$	48.0

Krypton

Electron binding energies /eV

N_I	4s	27.5
N_{II}	$4p_{1/2}$	14.1
N_{III}	$4p_{3/2}$	14.1

Lanthanum

Electron binding energies /eV

N_I	4s	274.7
N_{II}	$4p_{1/2}$	205.8
N_{III}	$4p_{3/2}$	196.0
N_{IV}	$4d_{3/2}$	105.3
N_V	$4d_{5/2}$	102.5
N_{VI}	$4f_{5/2}$	–
N_{VII}	$4f_{7/2}$	–
O_I	5s	34.3
O_{II}	$5p_{1/2}$	19.3
O_{III}	$5p_{3/2}$	16.8

Lead

Electron binding energies /eV

N_I	4s	891.8
N_{II}	$4p_{1/2}$	761.9
N_{III}	$4p_{3/2}$	634.5
N_{IV}	$4d_{3/2}$	434.3
N_V	$4d_{5/2}$	412.2
N_{VI}	$4f_{5/2}$	141.7
N_{VII}	$4f_{7/2}$	136.9
O_I	5s	147
O_{II}	$5p_{1/2}$	106.4
O_{III}	$5p_{3/2}$	83.3
O_{IV}	$5d_{3/2}$	20.7
O_V	$5d_{5/2}$	18.1

Lutetium

Electron binding energies /eV

N_I	4s	506.8
N_{II}	$4p_{1/2}$	412.4
N_{III}	$4p_{3/2}$	359.2
N_{IV}	$4d_{3/2}$	206.1
N_V	$4d_{5/2}$	196.3
N_{VI}	$4f_{5/2}$	8.9
N_{VII}	$4f_{7/2}$	7.5
O_I	5s	57.3
O_{II}	$5p_{1/2}$	33.6
O_{III}	$5p_{3/2}$	26.7

Mercury

Electron binding energies /eV

N_I	4s	802.2
N_{II}	$4p_{1/2}$	680.2
N_{III}	$4p_{3/2}$	576.6
N_{IV}	$4d_{3/2}$	378.2
N_V	$4d_{5/2}$	358.8
N_{VI}	$4f_{5/2}$	104.0
N_{VII}	$4f_{7/2}$	99.9
O_I	5s	127
O_{II}	$5p_{1/2}$	83.1
O_{III}	$5p_{3/2}$	64.5
O_{IV}	$5d_{3/2}$	9.6
O_V	$5d_{5/2}$	7.8

Molybdenum

Electron binding energies /eV

N_I	4s	63.2
N_{II}	$4p_{1/2}$	37.6
N_{III}	$4p_{3/2}$	35.5

Neodymium

Electron binding energies /eV

N_I	4s	319.2
N_{II}	$4p_{1/2}$	243.3
N_{III}	$4p_{3/2}$	224.6
N_{IV}	$4d_{3/2}$	120.5
N_V	$4d_{5/2}$	120.5
N_{VI}	$4f_{5/2}$	1.5
N_{VII}	$4f_{7/2}$	1.5
O_I	5s	37.5
O_{II}	$5p_{1/2}$	21.1
O_{III}	$5p_{3/2}$	21.1

Niobium

Electron binding energies /eV

N_I	4s	56.4
N_{II}	$4p_{1/2}$	32.6
N_{III}	$4p_{3/2}$	30.8

Osmium

Electron binding energies /eV

N_I	4s	658.2
N_{II}	$4p_{1/2}$	549.1
N_{III}	$4p_{3/2}$	470.7
N_{IV}	$4d_{3/2}$	293.1
N_V	$4d_{5/2}$	278.5
N_{VI}	$4f_{5/2}$	53.4
N_{VII}	$4f_{7/2}$	50.7
O_I	5s	84
O_{II}	$5p_{1/2}$	58
O_{III}	$5p_{3/2}$	44.5

Palladium

Electron binding energies /eV

N_I	4s	87.1
N_{II}	$4p_{1/2}$	55.7
N_{III}	$4p_{3/2}$	50.9

Platinum

Electron binding energies /eV

N_I	4s	725.4
N_{II}	$4p_{1/2}$	609.1
N_{III}	$4p_{3/2}$	519.4
N_{IV}	$4d_{3/2}$	331.6
N_V	$4d_{5/2}$	314.6
N_{VI}	$4f_{5/2}$	74.5
N_{VII}	$4f_{7/2}$	71.2
O_I	5s	101.7
O_{II}	$5p_{1/2}$	65.3
O_{III}	$5p_{3/2}$	51.7

Polonium

Electron binding energies /eV

N_I	4s	995
N_{II}	$4p_{1/2}$	851
N_{III}	$4p_{3/2}$	705
N_{IV}	$4d_{3/2}$	500
N_V	$4d_{5/2}$	473
N_{VI}	$4f_{5/2}$	184
N_{VII}	$4f_{7/2}$	184
O_I	5s	177
O_{II}	$5p_{1/2}$	132
O_{III}	$5p_{3/2}$	104
O_{IV}	$5d_{3/2}$	31
O_V	$5d_{5/2}$	31

Praseodymium

Electron binding energies / eV

N_I	4s	304.5
N_{II}	$4p_{1/2}$	236.3
N_{III}	$4p_{3/2}$	217.6
N_{IV}	$4d_{3/2}$	115.1
N_V	$4d_{5/2}$	115.1
N_{VI}	$4f_{5/2}$	2.0
N_{VII}	$4f_{7/2}$	2.0
O_I	5s	37.4
O_{II}	$5p_{1/2}$	22.3
O_{III}	$5p_{3/2}$	22.3

Promethium

Electron binding energies / eV

N_I	4s	–
N_{II}	$4p_{1/2}$	242
N_{III}	$4p_{3/2}$	242
N_{IV}	$4d_{3/2}$	120
N_V	$4d_{5/2}$	120
N_{VI}	$4f_{5/2}$	n.a.
N_{VII}	$4f_{7/2}$	n.a.
O_I	5s	n.a.
O_{II}	$5p_{1/2}$	n.a.
O_{III}	$5p_{3/2}$	n.a.

Protactinium

Electron binding energies / eV

N_I	4s	1387
N_{II}	$4p_{1/2}$	1224
N_{III}	$4p_{3/2}$	1007
N_{IV}	$4d_{3/2}$	743
N_V	$4d_{5/2}$	708
N_{VI}	$4f_{5/2}$	371
N_{VII}	$4f_{7/2}$	360
O_I	5s	310
O_{II}	$5p_{1/2}$	232
O_{III}	$5p_{3/2}$	232
O_{IV}	$5d_{3/2}$	94
O_V	$5d_{5/2}$	94
P_I	6s	–
P_{II}	$6p_{1/2}$	–
P_{III}	$6p_{3/2}$	–

Radium

Electron binding energies / eV

N_I	4s	1208
N_{II}	$4p_{1/2}$	1058
N_{III}	$4p_{3/2}$	879
N_{IV}	$4d_{3/2}$	636
N_V	$4d_{5/2}$	603
N_{VI}	$4f_{5/2}$	299
N_{VII}	$4f_{7/2}$	299
O_I	5s	254
O_{II}	$5p_{1/2}$	200
O_{III}	$5p_{3/2}$	153
O_{IV}	$5d_{3/2}$	68
O_V	$5d_{5/2}$	68
P_I	6s	44
P_{II}	$6p_{1/2}$	19
P_{III}	$6p_{3/2}$	19

Radon

Electron binding energies / eV

N_I	4s	1097
N_{II}	$4p_{1/2}$	929
N_{III}	$4p_{3/2}$	768
N_{IV}	$4d_{3/2}$	567
N_V	$4d_{5/2}$	541
N_{VI}	$4f_{5/2}$	238
N_{VII}	$4f_{7/2}$	238
O_I	5s	214
O_{II}	$5p_{1/2}$	164
O_{III}	$5p_{3/2}$	127
O_{IV}	$5d_{3/2}$	48
O_V	$5d_{5/2}$	48
P_I	6s	26

Rhenium

Electron binding energies / eV

N_I	4s	625.4
N_{II}	$4p_{1/2}$	518.7
N_{III}	$4p_{3/2}$	446.8
N_{IV}	$4d_{3/2}$	273.9
N_V	$4d_{5/2}$	260.5
N_{VI}	$4f_{5/2}$	42.9
N_{VII}	$4f_{7/2}$	40.5
O_I	5s	83
O_{II}	$5p_{1/2}$	45.6
O_{III}	$5p_{3/2}$	34.6

Rhodium

Electron binding energies / eV

N_I	4s	81.4
N_{II}	$4p_{1/2}$	50.5
N_{III}	$4p_{3/2}$	47.3

Rubidium

Electron binding energies / eV

N_I	4s	30.5
N_{II}	$4p_{1/2}$	16.3
N_{III}	$4p_{3/2}$	15.3

Ruthenium

Electron binding energies / eV

N_I	4s	75.0
N_{II}	$4p_{1/2}$	46.5
N_{III}	$4p_{3/2}$	43.2

Samarium

Electron binding energies / eV

N_I	4s	347.2
N_{II}	$4p_{1/2}$	265.6
N_{III}	$4p_{3/2}$	247.4
N_{IV}	$4d_{3/2}$	129.0
N_V	$4d_{5/2}$	129.0
N_{VI}	$4f_{5/2}$	5.2
N_{VII}	$4f_{7/2}$	5.2
O_I	5s	37.4
O_{II}	$5p_{1/2}$	21.3
O_{III}	$5p_{3/2}$	21.3

Silver

Electron binding energies / eV

N_I	4s	97.0
N_{II}	$4p_{1/2}$	63.7
N_{III}	$4p_{3/2}$	58.3

Strontium

Electron binding energies / eV

N_I	4s	38.9
N_{II}	$4p_{1/2}$	21.6
N_{III}	$4p_{3/2}$	20.1

Tantalum

Electron binding energies / eV

N_I	4s	563.4
N_{II}	$4p_{1/2}$	463.4
N_{III}	$4p_{3/2}$	400.9
N_{IV}	$4d_{3/2}$	237.9
N_V	$4d_{5/2}$	226.4
N_{VI}	$4f_{5/2}$	23.5
N_{VII}	$4f_{7/2}$	21.6
O_I	5s	69.7
O_{II}	$5p_{1/2}$	42.2
O_{III}	$5p_{3/2}$	32.7

Technetium

Electron binding energies / eV

N_I	4s	69.5
N_{II}	$4p_{1/2}$	42.3
N_{III}	$4p_{3/2}$	39.9

Tellurium

Electron binding energies / eV

N_I	4s	169.4
N_{II}	$4p_{1/2}$	103.3
N_{III}	$4p_{3/2}$	103.3
N_{IV}	$4d_{3/2}$	41.9
N_V	$4d_{5/2}$	40.4

Terbium

Electron binding energies / eV

N_I	4s	396.0
N_{II}	$4p_{1/2}$	322.4
N_{III}	$4p_{3/2}$	284.1
N_{IV}	$4d_{3/2}$	150.5
N_V	$4d_{5/2}$	150.5
N_{VI}	$4f_{5/2}$	7.7
N_{VII}	$4f_{7/2}$	2.4
O_I	5s	45.6
O_{II}	$5p_{1/2}$	28.7
O_{III}	$5p_{3/2}$	22.6

Thallium

Electron binding energies / eV

N_I	4s	846.2
N_{II}	$4p_{1/2}$	720.5
N_{III}	$4p_{3/2}$	609.5
N_{IV}	$4d_{3/2}$	405.7
N_V	$4d_{5/2}$	385.0
N_{VI}	$4f_{5/2}$	122.2
N_{VII}	$4f_{7/2}$	117.8
O_I	5s	136
O_{II}	$5p_{1/2}$	94.6
O_{III}	$5p_{3/2}$	73.5
O_{IV}	$5d_{3/2}$	14.7
O_V	$5d_{5/2}$	12.5

Thorium

Electron binding energies / eV

N_I	4s	1330
N_{II}	$4p_{1/2}$	1168
N_{III}	$4p_{3/2}$	966.4
N_{IV}	$4d_{3/2}$	712.1
N_V	$4d_{5/2}$	675.2
N_{VI}	$4f_{5/2}$	342.4
N_{VII}	$4f_{7/2}$	333.1
O_I	5s	290
O_{II}	$5p_{1/2}$	229
O_{III}	$5p_{3/2}$	182
O_{IV}	$5d_{3/2}$	92.5
O_V	$5d_{5/2}$	85.4
P_I	6s	41.4
P_{II}	$6p_{1/2}$	24.5
P_{III}	$6p_{3/2}$	16.6

Thulium

Electron binding energies / eV

N_I	4s	470.9
N_{II}	$4p_{1/2}$	385.9
N_{III}	$4p_{3/2}$	332.6
N_{IV}	$4d_{3/2}$	175.5
N_V	$4d_{5/2}$	175.5
N_{VI}	$4f_{5/2}$	–
N_{VII}	$4f_{7/2}$	4.6
O_I	5s	54.7
O_{II}	$5p_{1/2}$	31.8
O_{III}	$5p_{3/2}$	25.0

Tin

Electron binding energies / eV

N_I	4s	137.1
N_{II}	$4p_{1/2}$	83.6
N_{III}	$4p_{3/2}$	83.6
N_{IV}	$4d_{3/2}$	24.9
N_V	$4d_{5/2}$	23.9

Tungsten

Electron binding energies / eV

N_I	4s	594.1
N_{II}	$4p_{1/2}$	490.4
N_{III}	$4p_{3/2}$	423.6
N_{IV}	$4d_{3/2}$	255.9
N_V	$4d_{5/2}$	243.5
N_{VI}	$4f_{5/2}$	33.6
N_{VII}	$4f_{7/2}$	31.4
O_I	5s	75.6
O_{II}	$5p_{1/2}$	45.3
O_{III}	$5p_{3/2}$	36.8

Uranium

Electron binding energies / eV

N_I	4s	1439
N_{II}	$4p_{1/2}$	1271
N_{III}	$4p_{3/2}$	1043
N_{IV}	$4d_{3/2}$	778.3
N_V	$4d_{5/2}$	736.2
N_{VI}	$4f_{5/2}$	388.2
N_{VII}	$4f_{7/2}$	377.4
O_I	5s	321
O_{II}	$5p_{1/2}$	257
O_{III}	$5p_{3/2}$	192
O_{IV}	$5d_{3/2}$	102.8
O_V	$5d_{5/2}$	94.2
P_I	6s	43.9
P_{II}	$6p_{1/2}$	26.8
P_{III}	$6p_{3/2}$	16.8

Xenon

Electron binding energies / eV

N_I	4s	213.2
N_{II}	$4p_{1/2}$	146.7
N_{III}	$4p_{3/2}$	145.5
N_{IV}	$4d_{3/2}$	69.5
N_V	$4d_{5/2}$	67.5
N_{VI}	$4f_{5/2}$	–
N_{VII}	$4f_{7/2}$	–
O_I	5s	23.3
O_{II}	$5p_{1/2}$	13.4
O_{III}	$5p_{3/2}$	12.1

Ytterbium

Electron binding energies / eV

N_I	4s	480.5
N_{II}	$4p_{1/2}$	388.7
N_{III}	$4p_{3/2}$	339.7
N_{IV}	$4d_{3/2}$	191.2
N_V	$4d_{5/2}$	182.4
N_{VI}	$4f_{5/2}$	2.5
N_{VII}	$4f_{7/2}$	1.3
O_I	5s	52.0
O_{II}	$5p_{1/2}$	30.3
O_{III}	$5p_{3/2}$	24.1

Yttrium

Electron binding energies / eV

N_I	4s	43.8
N_{II}	$4p_{1/2}$	24.4
N_{III}	$4p_{3/2}$	23.1

Zirconium

Electron binding energies / eV

N_I	4s	50.6
N_{II}	$4p_{1/2}$	28.5
N_{III}	$4p_{3/2}$	27.1

APPENDIX 3
Minerals additional data

This appendix contains additional data to the minerals tables in the Geological Data sections of some of the elements

Aluminium

Mineral	Formula	Density	Hardness	Crystal appearance
Andalusite	Al_2SiO_5	3.14	6.5–7.5	orth., trans., pink, red etc.
Corundum	Al_2O_3	4.0	9	rhom., vit. colourless/brown (gem)
Sillimanite	Al_2SiO_5	3.25	6.5–7.5	orth., trans., colourless, white, etc.
Topaz	$Al_2SiO_4(F,OH)_2$	3.5	9	orth., trans., colourless, yellow, etc.

This does not complete the list, there are many other aluminium silicates than those given here. See Key for reference works to minerals.

Calcium

Mineral	Formula	Density	Hardness	Crystal appearance
Gypsum	$CaSO_4.2H_2O$	2.317	2	mon., vit. colourless
Shortite	$Na_2Ca_2(CO_3)_3$	2.63	3	orth., vit. colourless
Vaterite	$CaCO_3$	2.54	n.a.	hex., translucent/colourless

There are many other minerals in which calcium is present

Copper

Mineral	Formula	Density	Hardness	Crystal appearance
Chrysocolla	$(Cu,Al)_2H_2Si_20_5(OH)_4$	2	2–4	orth., vit. earthy green/blue
Covellite	Cu_2S	4.7	1.5–2	hex., sub-metallic blue
Cuprite	Cu_2O	6.14	3.5–4	cub., adamantine sub-metallic red
Dioptase*	$CuSiO_3.H_2O$	3.3	5	rhom., vit. green
Enargite	Cu_3AsS_4	4.45	3	orth., metallic grey-black
Malachite	$Cu_2(CO_3)(OH)_2$	4.05	3.5–4	mon., vit./silky earthy green
Rosasite	$(Cu,Zn)_2(CO_3)(OH)_2$	4.1	4.5	mon., green-blue
Tetrahedrite	$(Cu,Fe)_{12}Sb_4S_{13}$	4.97	3–4.5	cub., metalic grey/black

* Used in jewelry

Magnesium

Mineral	Formula	Density	Hardness	Crystal appearance
Epsomite	$MgSO_4.7H_2O$	1.677	2–2.5	orth., vit./silky/earthy colourless
Kiersite	$MgSO_4.H_2O$	2.571	3.5	mon., vit., colourless-white
Magnesite	$MgCO_3$	3.00	4	rhom., vit. colourless compact
Pyrope	$Mg_3Al_2(SiO_4)_3$	3.51	6.5–7.5	cub., vit./resinous deep red
Spinel	$MgAl_2O_4$	3.55	7.5–8	cub., vit. pink (gem)

Silicon

Mineral	Formula	Density	Hardness	Crystal appearance
Muscovite*	$KAl_2(Si_3Al)O_{10}(OH)_2$	2.8	2.5–3	mon., trans. colourless-pale green
Talc	$Mg_3Si_4O_{10}(OH)_2$	2.7	1	tric.,trans./colourless-white
Tremolite†	$Ca_2Mg_5Si_8O_{22}(OH)_2$	3.0	5–6	mon., vit. colourless-grey

*mica; †asbestos-like mineral

Other groups of silicates include the following:

Potassium magnesium silicates: Biotite; Chrysotile; Phlogopite

Calcium aluminium silicates: Margarite; Montmorillonite; Phrenite

Magnesium aluminium silicates: Clinochlore; Vermiculite

Sodium potassium calcium aluminium silicates: Albite; Analcime; Apophyllite; Kaolinite; Laumonite; Microcline; Muscovite; Nepheline; Oligoclase; Phlogopite; Pyrophyllite; Sodalite.

The periodic table

ON the inside front cover of this book you will find a periodic table of all the elements that have been reported up to 1995. The periodic table has been around for so long, adorning classrooms, lecture halls and chemical laboratories, that you might imagine this is the only possible version. It may come as a surprise to learn that more than 600 periodic tables have been published in the past 125 years. According to E. G. Mazurs' book, *Graphic representations of the periodic system during one hundred years*, there are about 150 basically different formats of the table, but hundreds of variations of these.

One of the first was drawn up by the Russian chemist Dimitri Mendeleyev on 1 March 1869. He is rightly seen as the discoverer of the periodic table because he based his on strictly scientific principles. Chemists had been groping towards a classification of the elements before this date, as they found more and more of them, and some came close to discovering the periodic table itself, as we shall see. Since the time of Mendeleyev, other chemists have redesigned his periodic table many times, sometimes in response to new elements being discovered or made artificially, sometimes in response to advances in our knowledge about the nature of atoms.

Looking back from our vantage point in time, we might perhaps see it as inevitable that a periodic table would be devised sooner or later. Once chemists knew about the subatomic particles which make up atoms, it would have been inevitable. The number of positive protons in the nucleus of an atom, the so-called atomic number, determines the element: hydrogen has one proton, helium two, lithium three, beryllium four, boron five, and so on up to element 100 (fermium) and beyond. The electrons in the orbits which surround the nucleus, and especially those in the outermost orbit are responsible for the *chemistry* of the element, and when the outer electron configuration is the same, then those elements should resemble one another. This explains the periodic repetition of chemical properties and behaviour that is observed in the groups of the periodic table.

The remarkable achievement of Mendeleyev was that he produced his periodic table 27 years before Joseph Thomson (1856–1940) discovered the electron in 1897,[1] and 42 years before Ernest Rutherford (1871–1937) discovered the positively charged atomic nucleus in 1911. The neutrons in the nucleus are also important—they are the key to explaining anomalous atomic weights and isotopes—and were discovered in 1930 by James Chadwick (1891–1974) who earned a Nobel prize in 1935 for his work.

Yet when Mendeleyev heard of the discovery of electrons in 1897, he rejected the idea that they came from atoms. He believed that atoms were indivisible. Indeed he predicted that the electron would disappear from science in the way that phlogiston had. This may strike us as mildly eccentric now, but in those days some eminent scientists were still not convinced of the existence of atoms, never mind electrons! One such sceptic was the electrochemist Wilhelm Ostwald (1853–1932).

A brief history of atoms and elements

The idea of atoms can be traced back to the ancient Greek philosophers who were the first to argue that matter must be composed of them. One argument they used to prove this went as follows: consider what happens when you slice horizontally through a cone. If matter is continuous then both faces of the slice must be exactly the same size, but since they are part of a cone then they must be different. The lower face must be larger, but that can only be so if we have sliced between two

[1] Joseph John Thomson won the Nobel prize for physics in 1906 for this work. Curiously the name for the electron preceded its discovery. Michael Faraday first suggested that there were particles of electricity in 1834, and the name electrine was suggested by George J. Stoney for the unit of electrical charge in 1874. He changed this to electron in 1891.

planes of atoms. The chief atomists were Leucippus and his pupil, Democritus, and their ideas were developed by Epicurus (341–270 BC). We know of their theories because of the book *De rerum natura (On Natural Things)* which was published around 55 BC by the Roman poet Lucretius, and is still in print.

Another philosophical debate concerned the idea of elements. To begin with the Greek philosophers thought in terms of *the* element and debated what it might be. For example, Heraclitus thought it was water, while others suggested air, or fire. Empedocles (*ca.* 400 BC) proposed there were four elements, an idea taken up by the great Aristotle (384–322 BC) who named them as earth, air, fire, and water.[2] This seemed very reasonable and was supported by commonsense observation, such as what happens when a stick burns in a fire. It can be seen to break down into the four elements: flames, steam, gases, and ash.

For 2,000 years Aristotle's ideas were accepted in Europe, almost without question, until the dawn of modern science in the seventeenth century. In the eighteenth century the concept of *chemical* elements emerged and it became clear that ten of these had indeed been known for thousands of years, but not recognized as such, see Table 1.

Table 1 The chemical elements known to the Ancients

Carbon as charcoal came with the discovery of fire.

Sulfur was to be found near volcanoes.

Copper was the first metal to be worked, around 5000 BC.

Gold and **silver** objects were first produced about 3000 BC.

Iron smelting led to the Iron Age which began around 2500 BC.

Tin was used earlier to forge bronze, an alloy of copper that was much stronger, but the element itself was not smelted as such until around 2100 BC.

Mercury was reported about 1500 BC. This forms when the ore cinnabar (HgS) is heated, and it also occurs naturally in cinnabar deposits.

Lead appeared around 1000 BC and was extensively mined. It became the most important metal in the Roman Empire.

Antimony objects date from around 1600 BC although this metal was not much used.

The Dark Ages in Europe traditionally began with the sack of Rome in AD 410. Science and technology declined, later to be revived by the Arabs and spread out with them during their years of expansion. Alchemy had it adherents and some elements were discovered during this period. Arsenic appears to have been isolated by the German alchemist, Magnus, in the middle of the thirteenth century; bismuth appeared towards the end of the fifteenth century; phosphorus was made by Hennig Brandt of Hamburg in 1669. This last element had amazing properties: it burst into flames spontaneously when exposed to air and it glowed in the dark. The discovery of phosphorus is seen as a turning point between alchemy and chemistry.

By 1700 there were 15 known elements, although these were still not recognized as *chemical* elements. To those above can be added zinc, which was smelted as such in the fifteenth century, and platinum, which was brought back to Europe from the New World where it was regarded it as a superior form of silver because it did not tarnish.

However, the eighteenth century saw the emergence of chemistry in Europe and the discovery of several new metals: cobalt (1735), nickel (1751), magnesium (1755), manganese (1774), chromium (1780), molybdenum (1781), tellurium (1783), tungsten (1783), zirconium (1789), uranium (1789) and a few gases: hydrogen (1766), nitrogen (1772), oxygen (*c.*1772), and chlorine (1774).

It was left to the great French chemist, Antoine Laurent de Lavoisier (1743–94),[3]

[2] In China scientific philosophy was based on there being five elements: air, earth, fire, water, and wood.

[3] Lavoisier was executed on 8 May 1794 during the Reign of Terror following the French Revolution of 1789. Although officially charged with adulterating tobacco, it was his links with the tax gathering organization, the *Fermier Générale*, that brought him to the guillotine. After his death his widow, Marie-Anne, continued to publish his work.

to bring order to the fledgling science, which he did with his remarkable book *Traité elémentaire de chemie* in 1789. In this he defined a chemical element as something which could not be further broken down (decomposed) and he listed 33 substances that came within the terms of his definition. He classified them into four categories: gases, non–metals, metals, and earths (see Table 2). Under gases he included heat and light. Another of Lavoisier's brilliant contributions to chemistry was to advocate that chemicals should be named after the elements of which they were composed. This de–mystified the language of chemistry of its alchemical names, and the result was that chemicals like 'butter of antimony' became antimony chloride, and 'lunar caustic' became silver nitrate.

Table 2 Lavoisier's elements

Gases	Non–metals	Metals		Earths†
[Light]	Sulfur	Antimony	Mercury	Lime
[Heat]	Phosphorus	Arsenic	Molybdenum	Magnesia
Oxygen	Carbon	Bismuth	Nickel	Barytes
Nitrogen	Chloride*	Cobalt	Platinum	Alumina
Hydrogen	Fluoride*	Copper	Silver	Silica
	Borate*	Gold	Tin	
		Iron	Tungsten	
		Lead	Zinc	
		Manganese		

* Lavoisier called these radicals, because he knew that although they were elements, they were always accompanied by another element.

† These are the inert oxides of the elements calcium, magnesium, barium, aluminium and silicon, which in Lavoisier's time could not be reduced to simpler substances.

A milestone in the development of chemistry came with the discovery of oxygen, the gas given off when calx of mercury (mercuric oxide) was heated. The discovery is attributed to Joseph Priestley (1733–1804) in 1774, but oxygen had already been isolated by Carl Wilhelm Scheele (1742–86) two years earlier, although not reported until 1777. Oxygen delivered the greatest blow to Aristotle's four elements when in 1781 Joseph Priestly (1733–1804) demonstrated that water was not an element because it could be formed from hydrogen and oxygen.

Another intellectual revolution was initiated by John Dalton (1766–1844) a science teacher living in Manchester, England. He pondered on the atomic nature of matter and realized that the Law of Fixed Proportions, which recognized that elements combined with one another in fixed ratios of weights, could be explained if they were composed of atoms. In 1803 Dalton presented a paper to the Manchester Literary and Philosophical Society entitled 'On the absorption of gases by water and other liquids'. In this he attempted to explain the different solubilities of gases by comparing the relative weights of them. This led him to propose a scale of atomic weights and when he published his talk in the newly launched *Proceedings* of the Society in 1805, he explored the idea further and appended a list of 20 elements with their atomic weights, and suggested atomic symbols—see Fig. 1. At a stroke Dalton's theory not only proposed the existence of atoms, but suggested that atoms of each element had individual weights and that these could be calculated relative to one another.

By 1810, chemistry was an established science, producing remarkable discoveries almost every year, not least of which was a constant supply of new elements. In the years between the publication of Lavoisier's *Traité* and Dalton's talk, another ten or so had been discovered: titanium (1791), yttrium (1794), beryllium (1797), vanadium (1801), niobium (1801), tantalum (1802), rhodium (1803), palladium (1803), osmium (1803), iridium (1803), and cerium (1803). This pace was to continue, especially when Humphry Davy discovered that Lavoisier's 'earths' could be decomposed by electrolysis into new elements. This led to the isolation of potassium (1807), sodium (1807), calcium (1808), and barium (1808). Strontium was isolated by Crawford in the same year.

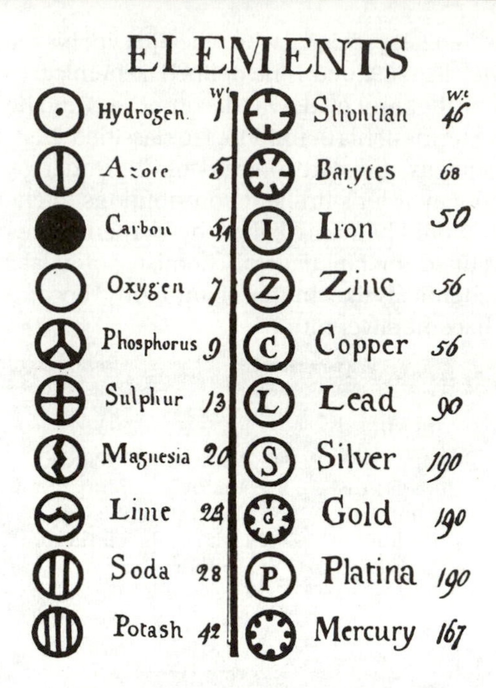

Fig. 1 Daltons table of elements

Boron was also discovered in 1808, then came iodine in 1811. The year 1817 was particularly fruitful with thorium, lithium, selenium, and cadmium being announced. By now the chemical similarity of groups of elements began to be commented on. In 1829 Johann Wolfgang Döbereiner (1780–1849) announced his Law of Triads, in which he noted that many elements came in groups of three ('triads'), and of which the weight of the middle element was the average of the lighter and heavier members. Lithium/sodium/potassium were one such triad and others were chlorine/bromine[4]/iodine and sulfur/selenium/tellurium. By the time Leopold Gmelin published the first edition of his compilation of essential chemical data, *Handbuch der Chemie*, in 1843 he noted there were ten triads, three tetrads, and even a pentad: nitrogen/phosphorus/arsenic/antimony/bismuth. Today we recognize this pentad as group 15 of the modern periodic table.

Occasionally there were curious insights into the underlying nature of matter—and one that was well in advance of its time was the hypothesis put forward by Prout in 1815. He proposed that all the elements were multiples of the atomic weight of hydrogen, which is the lightest. If this is taken as 1 then it explained why all the other elements had weights that were whole numbers.[5]

Although absolute atomic weights were not known as such, it was possible to compare the weights of the elements to one another, and this was done with respect to oxygen which forms compounds with almost all of them. Oxygen was taken to be 8, which followed from a comparison with hydrogen, and the 1 : 8 ratio of these elements in water. At the time the composition of water was assumed to be HO, not H_2O.

In fact the true formula could have been deduced as early as 1800 when William Nicholson and Anthony Carlisle electrolysed water and showed it decomposed into two volumes of hydrogen and one of oxygen. The significance of this was not realized, even though in 1811 Amadeo Avagadro (1776–1856) suggested that equal volumes of gases contained equal numbers of particles (molecules). Eventually it was deduced that the formula for water was H_2O, and the weight of oxygen was correspondingly adjusted to 16.

Most elements had relative weights that were whole numbers, and this continued to be true as more were discovered. Prout's hypothesis was tantalizingly almost correct in assuming that hydrogen was the element of elements. In a way he had

[4] Bromine had been discovered in 1826.

[5] Not quite all, a notable exception being chlorine (35.5).

been right. Most hydrogen atoms consist of a single proton, and it is the number of protons which determines the nature of all the elements. The nucleus of an atom is where 99.98 per cent of its weight resides. Because this is composed of protons and neutrons which have exactly the same mass, and because most elements have one dominant isotope, then it is not surprising that their weights will tend to whole numbers. A glance at the periodic table on the inside cover will show that this is so, revealing that most elements have atomic weights that fall within the limits ± 0.1 of an integer, with few having fractional numbers. Apart from chlorine, already mentioned, nickel (58.7) copper (63.5) and zinc (65.4) are notable exceptions.

From 1830 to 1860 only three new elements appeared: lanthanum (1839), erbium (1842), and terbium (1843). These metals were later to yield other metals because they were part of the group known as the rare earths, all of which have very similar properties, making them difficult to separate from one another.

The next era of element discovery came with the atomic spectroscope, which revealed that each element had a characteristic fingerprint pattern of lines in the visible spectrum. Merely submitting a new mineral to this technique immediately showed that a new element was present if new lines appeared in the spectrum. As a result rubidium, caesium, thallium and indium were announced in the years 1860–3.

The total of known elements was now in excess of 60 and chemists were beginning to ask whether there was a limit to the number. To answer this question required a theory that could explain the known and predict the unknown. The 'triads', 'tetrads' and 'pentads', were governed by regular increases in relative weight.[6] Clearly chemists might use this universal property to rank all the elements, if only they could talk in terms of relative atomic weights. The Italian chemist Stanislao Cannizzaro (1826–1910) was to be instrumental in bringing this about.

Atomic weights and the first periodic tables

In 1858 Cannizzaro published his 'Outline of a course of chemical philosophy' in which he showed how atomic weights could be deduced if Avogadro's Law was accepted, and he presented a paper at the First International Chemical Congress which was held at Karlsruhe in 1860. Cannizzaro argued that the law led directly to the atomic weights of gaseous elements, and thence to other elements. A given volume of oxygen gas is 16 times heavier than the same volume of hydrogen, so if the latter has an atomic weight of 1.0 then oxygen must be 16.0.[7] Cannizzaro's ideas were quickly accepted and copies of his table of atomic weights were eagerly sought by conference members. One fell into the hands of a young Russian student, Dimitri Mendeleyev, who was then engaged in postgraduate research in Germany.

The table of atomic weights triggered a new debate about the elements. The first attempt to arrange all of them in a regular pattern was made in 1862 by a French geologist Alexandre Emile Becuyer de Chancourtois (1820–86). He wrote down the elements in order of increasing atomic weight on a piece of tape and then wound this spiral–like around a cylinder—Fig. 2. The cylinder surface was divided into 16 parts, based on the atomic weight of oxygen. Chancourtois noted that certain triads came together down the cylinder, such as the alkali metals, lithium, sodium and potassium whose atomic weights are 7, 23 (7+16), and 39 (23 + 16). This coincidence was also true of the tetrad oxygen/sulfur/selenium/tellurium. He called his model the *Vis Tellurique* (Telluric Screw) and published it in 1862. He concluded: 'the properties of substances are the properties of numbers.' His paper had little impact among chemists, but this might have been because some of the alignments did not make chemical sense; for example one of his groups consisted of boron, aluminium, nickel, arsenic, lanthanum and palladium.

What Chancourtois had discovered was the *periodic* nature of the elements. In other words, elements with the same properties occurred at regular periodic inter-

[6] Rubidium and caesium extended the lithium/sodium/potassium triad to a pentad.
[7] This also proved that the correct chemical formula for water was H_2O, not HO.

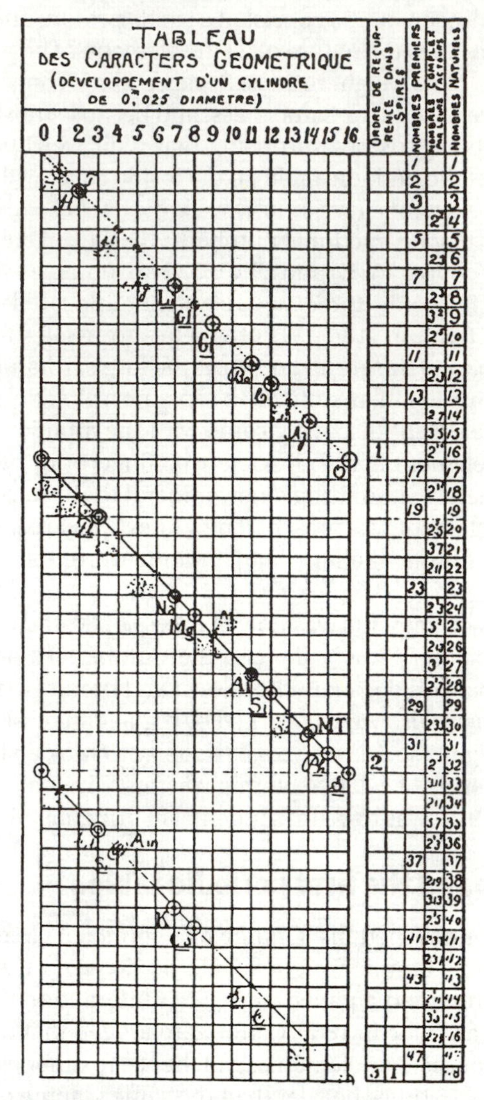

Fig. 2 A table of atomic weights by Alexandre Emile Becuyer de Chancourtois

vals, in this case when the atomic weights differed by 16, as with the alkali metals, or if they were multiples of 16, or nearly so, such as we find with oxygen = 16, sulfur = 32, selenium = 79 (5 × 16 = 80) and tellurium = 128 (8 × 16). Clearly there appeared to be an underlying *numerical* rhythm to the elements, even though we now know that it is more by luck than by logic that the elements of the oxygen–tellurium tetrad are such close multiples of 16.[8]

Another attempt to classify the elements was made by an Englishman, 27–year old John Alexander Reina Newlands (1837–98). He wrote articles for the London weekly *Chemical News*, in one of which (1865, volume 12, page 83) he arranged 56 elements into groups and noted that there seemed to be a repetition of properties with every eighth element. In 1864 he had read a paper entitled 'The law of octaves' to a meeting of the London Chemical Society, and the title was chosen by analogy with octaves in music. This was an ill–judged choice, and it is said that one member of his audience[9] mockingly asked Newlands whether he had ever thought of arranging the elements alphabetically instead. The society's journal refused to publish his paper. However, the Royal Society of London award him the Davy Medal in 1887 in belated recognition of his achievement.

[8] Although polonium, the next element in this group, was not to be discovered for another 36 years, it too would have fallen into line with its partners in the Telluric Screw because its most stable isotope has an atomic weight of 209 (13×16 = 208). Marie Curie isolated this radioactive element in 1898.

[9] Believed to be a Professor Carey Foster.

What was curious about Newlands approach was that, although he placed elements in groups that we would now recognize as part of the periodic table, he was clearly not aware of any underlying imperative for so doing. For example, he noticed that silicon and tin should be part of a triad, but he left no vacant place in the table for the missing element between them. Nor had he any qualms about putting two elements together in some of the boxes in his table.

William Odling, successor to Michael Faraday at the Royal Institution in London, was another chemist to speculate about relationships among the elements, and he published a paper entitled: 'On the proportional numbers of the elements' in the first volume of the *Quarterly Journal of Science* (1864, page 642). His arrangement of the elements came surprisingly close to that of Mendeleyev's first attempt, and he left gaps where there were missing elements. In fact he even left gaps that were later to be filled by helium and neon, although he thought these would be lighter elements of a pentad which expanded the triad of zinc/cadmium/mercury. He also suggested, wrongly, that lithium and thallium were the top and bottom of another pentad.

Another scientist to come near to discovering the periodic table was the German, Julius Lothar Meyer (1830–95). He published a table of 49 elements with their valencies in 1864, and he drew a graph of atomic volumes versus atomic weight in 1868 which shows a periodic rise and fall, and from which he deduced a periodic table.[10] He passed this paper to a colleague, Professor Remelé , for his comments. Unfortunately these were slow in coming, and before he could submit it for publication Mendeleyev's definitive paper had appeared.

Dimitri Mendeleyev

The first genuine periodic table of the elements was produced by a relatively unknown Russian professor of chemistry, Dimitri Ivanovich Mendeleyev (1834–1907). He was born in Tobolsk in western Siberia on 8 February 1834, the fourteenth child of a local school master. However, it was his mother, Maria, who raised the children after her husband became blind. She came from a family with interests in glassworks and paper mills and these she ran while struggling to educate her favourite son, Dimitri. Eventually she took him to Moscow, but failed to secure him a place at the university there, so instead went to St Petersburg, where he was allowed to enroll at the Central Pedagogic Institute to study physics and mathematics. He graduated with a gold medal for excellence in scholarship, and in 1859 went to study for his doctorate, first to Paris, working under Henri–Victor Regnault, and then to Heidelberg, supervised by Robert Bunsen.

In 1861 he returned to St Petersburg, that great city on the Baltic which prided itself on its European architecture and culture. In the 1860s it was at the leading edge of reform as Russia began the painful road from feudalism to a modern industrial society. A liberated outlook promoted education, science and the arts, in ways that were well ahead of their time. For example women were encouraged to become educated and take up professions. In this atmosphere of reform chemistry flourished. Even today we are reminded of this period when we use the Markovnikoff rule to predict the way an olefin will react, or listen to the music of Borodin, a chemist who is better known as the composer of symphonies.[11] But its most enduring chemical monument is the periodic table.

Mendeleyev became a professor of chemistry in 1867, and began to write his textbook *The principles of chemistry*, which was to run into many editions and be translated into French, German, and English. It was while preparing the second volume that he made his momentous discovery. In trying to find a format for the chapters of the book, he grouped the elements according to their valencies.

[10] Atomic volume is defined as atomic weight divided by density. It is the volume occupied by a mole (6.022×10^{23} atoms) of an element—see physical data section.

[11] The best known of which is the opera *Prince Igor* on which the musical *Scheherazade* is based.

On a winter's day in 1869 the breakthrough happened.[12] Mendeleyev's movements that day are well documented. He had planned to visit a cheese factory in the morning but this was called off, and instead he worked on his book. He wrote the name of each element on individual pieces of card, together with its atomic weight, a few physical properties and the formulae of any hydrides and oxides it formed. These cards he arranged in sets of increasing atomic weight with elements having the same valency, as shown by their hydrides or oxides, in the same row (see Fig. 3.) He produced one arrangement that particularly impressed him and wrote it down on to the back of an envelope. This can be considered his first periodic table, and the envelope still exists.[13]

After his midday meal he took a nap, and on waking he decided that a vertical arrangement of groups was a better way of depicting the periodic table. This event was to give rise to the romantic notion that the periodic table came to Mendeleyev in a dream. In any case he redrew his table and this has remained the standard format to this day.

What makes Mendeleyev's table so important was that he realized that he had stumbled on an underlying order to the elements and this gave him the confidence to make a few predictions, some of which were soon to be proved right, and some of which were later to be proved wrong.

If Mendeleyev's periodic table were correct, it should show gaps where elements, yet unknown, should come. Below aluminium and silicon were such vacant spaces. These elements he named eka–aluminium and eka–silicon. He predicted their atomic weights, physical properties such as melting points and densities, and said what the chemical composition of their oxides and other compounds would be. In another part of the table Mendeleyev was able to resolve a long–standing dispute over the formula for the oxide of beryllium. Some claimed it was Be_2O_3 while others said BeO. The former was inconsistent with beryllium's position in group two of his table, so Mendeleyev rightly said that the formula must be BeO.

He made another prediction with equal certainty, but in this Mendeleyev was wrong. He said that the atomic weight of tellurium or iodine must be incorrect. To the nearest whole numbers these were 128 (Te) and 127 (I). Clearly a periodic table

			Ti=50	Zr=90	?=180.
			V=51	Nb=94	Ta=182.
			Cr=52	Mo=96	W=186.
			Mn=55	Rh=104,4	Pt=197,4
			Fe=56	Ru=104,4	Ir=198.
			Ni–Co=59	Pl=106,6,	Os=199.
H=1			Cu=63,4	Ag=108	Hg=200.
	Be=9,4	Mg=24	Zn=65,2	Cd=112	
	B=11	Al=27,4	?=68	Ur=116	Au=197?
	C=12	Si=28	?=70	Sn=118	
	N=14	P=31	As=75	Sb=122	Bi=210
	O=16	S=32	Se=79,4	Te=128?	
	F=19	Cl=35,5	Br=80	I=127	
Li=7	Na=23	K=39	Rb=85,4	Cs=133	Tl=204
		Ca=40	Sr=87,6	Ba=137	Pb=207.
		?=45	Ce=92		
		?Er=56	La=94		
		?Yt=60	Di=95		
		?In=75,6	Th=118?		

Fig. 3 Mendeleyev's first periodic table

[12] In St Petersburg the date was 17 February, in the rest of Europe it was 1 March. In Russia the Julian calendar of 365 days was still in use, rather than the Gregorian calendar of 365.25 days.

[13] Mendeleyev liked playing cards and these might have given him the idea for his element cards. He also liked playing patience which might have suggested various ways of arranging the cards.

based on increasing atomic weights must have iodine before tellurium, and yet when Mendeleyev placed their cards in this order it resulted in iodine coming below selenium, and tellurium below bromine. This was quite at odds with the chemistry, so he suggested that the atomic weight of tellurium or iodine had been wrongly measured. Much effort was to be wasted on trying to correct the 'error' but without success. The true explanation of this anomaly requires an understanding of isotopes, and these were not to be discovered for another 40 years, as we shall see.

The most convincing demonstration of the correctness of Mendeleyev's periodic table was the discovery of two of the missing elements he had predicted. Eka–aluminium was discovered by Paul–Émile Lecoq de Boisbaudran (1838–1912) in 1875 at Paris, and he called it gallium after the Latin name for his native France. He measured its properties, including the density which he said was 4.7 g cm^{-3}. He was told that the new element was Mendeleyev's eka–aluminium and that most of its properties had already been foretold. Boisbaudran was then aware of the discrepancy between his density and that predicted by Mendeleyev, which was 5.9 g cm^{-3}, so he checked his measurements, discovered he had made an error, and reported a corrected value of 5.956 g cm^{-3}.[14]

In 1879 Lars Fredrik Nilson (1840–99) discovered an element which Mendeleyev called eka–boron at Uppsala, Sweden, and this too was named after the region of its discovery, scandium, from the Latin for Scandinavia. It too had the properties Mendeleyev predicted for the missing element after calcium. For example its atomic weight was 44 (predicted 44) and density 3.86 g cm^{-3} (predicted 3.5).

A few years later, in 1886, Lemens Alexander Winkler (1838–1904) discovered eka–silicon at Freiberg in Germany, and he also named it after his native land. Germanium was almost exactly as Mendeleyev had predicted, as Table 3 shows.

Table 3 The predicted and observed* properties of germanium

	Predicted for ekasilicon (Mendeleyev)	Found for germanium (Winkler)
Atomic weight	72	72.32
Density/g cm^{-3}	5.5	5.47
Atomic volume/cm^3	13	13.22
Valency	4	4
Specific heat/cal g^{-1}	0.073	0.076
Density of oxide/g cm^{-3}	4.7	4.703
Boiling point of tetrachloride/°C	<100	86

*These are slightly different from today's accepted values.

Mendeleyev also noted that the titanium group was lacking a heavy member and urged that this be sought among titanium ores. In fact it came to light as late as 1923 when the D. Coster and George von Hevsey found it in a zirconium ore and named it hafnium, from the Latin name for Copenhagen, the city where it was discovered. Mendeleyev had guessed its atomic weight would be 180, it was measured as 178.5.

Closer inspection of Mendeleyev's first table (Fig. 3) reveals that he was sometimes way off target with his groups of elements, and with his predictions. He got the group headed by manganese quite wrong for reasons we shall come to. He placed mercury in the same group as copper and silver, even though it meant placing it in the wrong order of atomic weights. He therefore queried the atomic weight of gold, which he put in the boron group under uranium, whose atomic weight was believed to be 116, half what it should be.

None of these errors detracts from the intellectual achievement which is the periodic table, and its discovery, which came long before chemists were even aware of the entities that shape it. These entities were the subatomic particles, protons, electrons, and neutrons. The protons in the nucleus were to lead to the concept of atomic number, but it is the surrounding electrons upon which the periodic table is

[14] It is actually 5.907 g cm^{-3} at 25 °C.

really based; they were not to be discovered for another generation. Mendeleyev based his table on the two properties, atomic weight and valency, which closely mirror atomic number and electron configurations.

Others came close to making the same discovery, but perhaps what distinguishes Mendeleyev was his ability to grasp that the known elements fitted into a scheme that was predetermined. The elements were not being arranged to *make* a periodic table, but to *fit* the periodic table. The difference may seem trivial to us today, but for its time the mental jump was truly one of genius. The certainty that he was right was surely what gave Mendeleyev the confidence not only to forecast the discovery of several new elements, but also to forecast their properties. He even predicted that a whole row of his periodic table awaited discovery because of the gap in atomic weights between cerium (140) and tantalum (182). These we now recognize as the so–called rare earths. Two of these, erbium and terbium, were already known to Mendeleyev.[15]

The development of the periodic table

Considering his isolation from the main centres of chemistry, Mendeleyev's discovery might easily have gone unnoticed. However, he was determined that it should be publicized and he immediately wrote a paper, entitled 'The relationship of properties and the atomic weights of the elements', which was read at a meeting of the Russian Chemical Society on 18 March.[16] Mendeleyev was away on an official inspection tour of factories that day and it was delivered by a colleague Nikolai Menshutkin. It was published in the very first edition of a new journal, *The Russian Journal of General Chemistry.* When copies of this reached Germany later that year, Mendeleyev's article was summarized in a German journal *Zeitschrift für Chemie* which thereby gave it wide circulation among the chemical community. The importance of the discovery was immediately recognized, and a German translation of Mendeleyev's paper was published that same year. The periodic table that this contained is shown in Fig. 4 and it remained the standard version for almost a century.

Mendeleyev's table had eight columns labelled with the Roman numerals I to VIII, corresponding to the chemical valencies, or oxidation states, of the elements. This property is revealed by the chemical formula of the highest oxide. Yet it brings together elements that may be quite dissimilar in other ways, such as metals and non–metals. For example in group V we find vanadium pentoxide, V_2O_5, and phosphorus pentoxide, P_2O_5, which have little in common. Mendeleyev consequently split the columns of his periodic table into two sub–groups which he labelled A and

Reihen	Gruppe I. — R^2O	Gruppe II. — RO	Gruppe III. — R^2O^3	Gruppe IV. RH^4 RO^2	Gruppe V. RH^3 R^2O^5	Gruppe VI. RH^2 RO^3	Gruppe VII. RH R^2O^7	Gruppe VIII. — RO^4
1	H=1							
2	Li=7	Be=9,4	B=11	C=12	N=14	O=16	F=19	
3	Na=23	Mg=24	Al=27,3	Si=28	P=31	S=32	Cl=35,5	
4	K=39	Ca=40	—=44	Ti=48	V=51	Cr=52	Mn=55	Fe=56, Co=59, Ni=59, Cu=63.
5	(Cu=63)	Zn=65	—=68	—=72	As=75	Se=78	Br=80	
6	Rb=85	Sr=87	?Yt=88	Zr=90	Nb=94	Mo=96	—=100	Ru=104, Rh=104, Pd=106, Ag=108.
7	(Ag=108)	Cd=112	In=113	Sn=118	Sb=122	Te=125	J=127	
8	Cs=133	Ba=137	?Di=138	?Ce=140	—	—	—	— — — —
9	(—)	—	—	—	—	—	—	
10	—	—	?Er=178	?La=180	Ta=182	W=184	—	Os=195, Ir=197, Pt=198, Au=199.
11	(Au=199)	Hg=200	Tl=204	Pb=207	Bi=208	—	—	
12	—	—	—	Th=231	—	U=240	—	— — — —

Fig. 4 Mendeleyev's periodic table as published in *Zeitschrift für Chemie*

[15] He did not recognize erbium and terbium as part of a missing group because in his first table he assumed their atomic weights were 56 and 60, which are a third of their real values of 167 and 173.

[16] This was 6 March 1869 by the Julian calendar.

B. Vanadium was in VA and phosphorus in VB. In group IA, for example, there were the so–called alkali metals, lithium, sodium, potassium, rubidium and caesium and in group IB were copper, silver and gold. The same pattern was repeated in the other columns with the exception of group VIII which contained metals which were very similar and which occurred in three sets of three. These were iron/cobalt/nickel, ruthenium/rhodium/palladium, and osmium/iridium/platinum.

Examination of Fig. 4 reveals other problems that Mendeleyev wrestled with. In group III he has ?Yt, ?Di and ?Er referring to the elements yttrium, didymium, and erbium, but questioning whether these were correctly placed. They were, in terms of their chemical preference for oxidation state III, but were misplaced because they are part of the group of 14 rare earths. In any case didymium turned out not to be an element but was a mixture of two rare earth elements praseodymium and neodymium.

Mendeleyev's group VIIA has gaps below manganese, and understandably so, because the heavier members of this group, i.e. technetium, rhenium and bohrium, are either radioactive or rare. All the isotopes of technetium are radioactive and none is stable enough to have existed throughout geological time. Some technetium isotopes have half–lives of several million of years, but this still means that they have long since vanished from the Earth's crust. Technetium was finally made by Emilio Segrè in 1937 by bombarding molybdenum with heavy hydrogen nuclei. Below technetium comes rhenium, one of the rarest elements on the planet, and one which was not isolated until 1925, and below rhenium is bohrium of which only a few atoms have ever been made.

Isotopes

The American Theodore W. Richards (1868–1928) won the Nobel Prize in chemistry in 1914 for his accurate work in measuring the atomic weights of elements. The year previous to winning the award he had uncovered something rather disturbing about the atomic weight of lead. It varied according to where the lead ore was mined. Admittedly the variations were small, but they could not be ignored, and they proved that lead must be composed of atoms with different masses.[17] This finding undermined the concept of fixed atomic weights, which was one of the pillars on which the periodic table rested.

Yet the periodic table persisted, and the reason was that the *chemical* behaviour of lead was not affected by its atomic weight. We now know that atoms of the same element can vary in weight because there are different isotopes. These have the same number of protons in the nucleus but differing numbers of neutrons.

The word isotope was coined by Frederick Soddy (1877–1956), and his work in this area was recognized with a Nobel Prize in 1921. His theory explained why tellurium (element 52, atomic weight 127.6) could be heavier than iodine (element 53, atomic weight 126.9). Tellurium has eight isotopes of which the heavier ones are ^{129}Te comprising 32 per cent and ^{130}Te 34 per cent. Iodine, one the other hand, has only one isotope, ^{127}I. It is not surprising, therefore, that the atomic weight of tellurium should be higher than that of iodine.

Today we know of other pairs of adjacent atoms in the periodic table where the atom with the larger atomic number has the smaller atomic weight. They are:

- argon (atomic number 18; atomic weight 39.948)/potassium (19; 39.098)
- cobalt (27; 58.933)/nickel (28; 58.69)
- thorium (90; 232.038)/protactinium (91; 231.036)
- uranium (92; 238.029)/neptunium (93; 237.048)

We also know that isotopes can have slight differences in chemical behaviour, and this is most marked with hydrogen. We can see by comparing the properties of ordinary water with those of so–called 'heavy water' in which ^{1}H, the common isotope,

[17] Lead varies because it is produced in Nature as the end-product of the radioactive decay of other elements, and these produce different isotopes of lead.

has been replaced by ^{2}H, often referred to as deuterium. For example, the melting points are 273.15 and 276.96 K, the boiling points are 373.15 and 374.57 K, and the densities (at 300 K) are 997 and 1104 kg m^{-3} respectively.

Even with a heavy element like uranium the isotopic composition can lead to detectable differences in behaviour, and this was essential to separating the two isotopes of uranium needed for nuclear weapons and nuclear fuel. In this case the isotopic difference affected the rate at which the vapour of uranium hexafluoride, UF_6, diffused through a porous barrier. Although $^{235}UF_6$ is only 0.84 per cent lighter than $^{238}UF_6$, this is enough to cause the former to diffuse at a measurably faster rate than the latter. Today isotopes play an important part in the study of chemistry, and especially in the analysis of materials and the monitoring of chemical reactions. Important isotopes are indicated in the nuclear data sections of this book.

The discovery of isotopes highlighted another problem regarding atomic weights. It was no longer good enough to measure all atomic weights relative to oxygen, taken to be exactly 16.0000 because this did not reflect the isotopic composition which is 99.75 per cent ^{16}O, 0.05 per cent ^{17}O and 0.20 per cent ^{18}O. Physicists redefined their scale of atomic weights basing them on the isotope ^{16}O taken as exactly 16.0000. This then meant that the atomic weight of naturally occurring oxygen was 16.0044. Chemists clung to the older scale of atomic weights for many years, but the two scales were brought into line in 1961 when it was agreed to re–base atomic weights on ^{12}C which was defined as 12.0000. This did not change the physicists atomic weight tables and chemists suffered only a 0.033 per cent decrease in the values they used, not enough to cause serious inconvenience.

A multiplicity of periodic tables

Within two years of Mendeleyev's table appearing, other chemists were proposing alternative versions and these have continued to appear year after year. Often they repeat arrangements that have already been proposed. Most tables are two dimensional representations although there have been several three dimensional versions in the shapes of cylinders, pyramids, spirals and even trees. While these make excellent displays for museums and chemistry departments, they lack the convenience of a flat format. Computer generated versions offer exciting interactive possibilities and we can expect these to use explorative shapes as teaching aids.

Basically there have been two approaches to devising a periodic table: the first lists all the elements in a continuous line, rather like the numbers on a tape measure, and this is then looped in such a way that like elements come together. The second version chops the tape into segments and stacks these in rows, again bringing like elements together.

The former approach is what Chancourtois used in 1862, and what many others have done since. The most whimsical one of this type is that produced in Russia by Romanoff in 1934 to celebrate the centenary of Mendeleyev's birth (see Fig 5).

Circular versions of the continuous type of table have also been proposed, but despite their elegance and the tantalizing analogy with electrons in shells around a nucleus (see Fig. 6) they all suffer the drawback of being difficult to read and of abstracting the information they contain. This is because they tend to crowd together the more important elements at their centre while giving the less important elements more room at the periphery.

Noble gases

The preferred way of arranging the elements is still the matrix format that Mendeleyev used, but now there is more purpose to it. The modern table denotes each row as the filling of an electron shell, which finally results in a noble gas element when the shell is full. These noble gases have a special place in the development of the periodic table. These were discovered in the 1890s when Lord Rayleigh (1842–1919), William Ramsay (1852–1916), and Morris Travers (1872–1961) isolated them at University College, London. They had actually been discov-

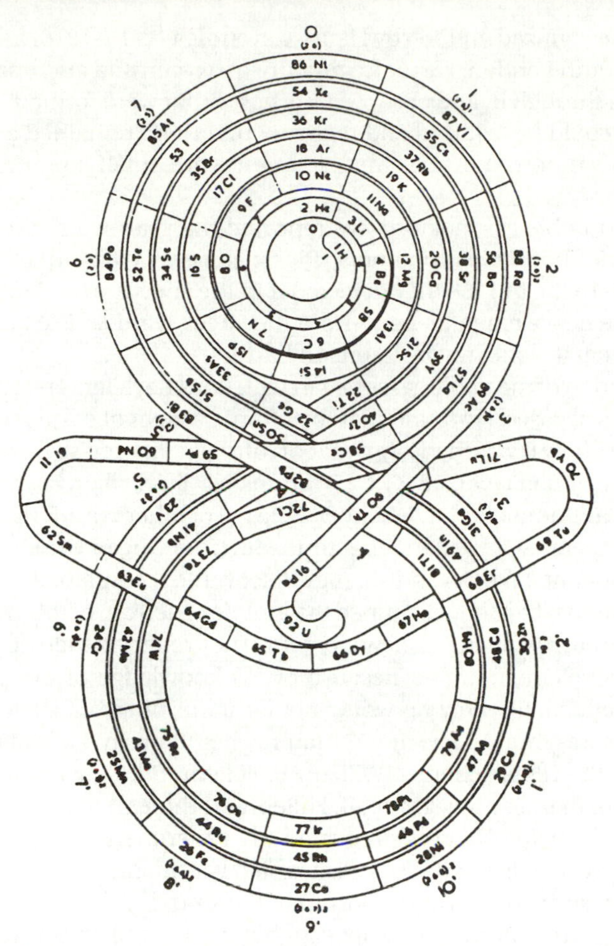

Fig. 5 A whimsical periodic table produced in Russia by Romanoff in 1934

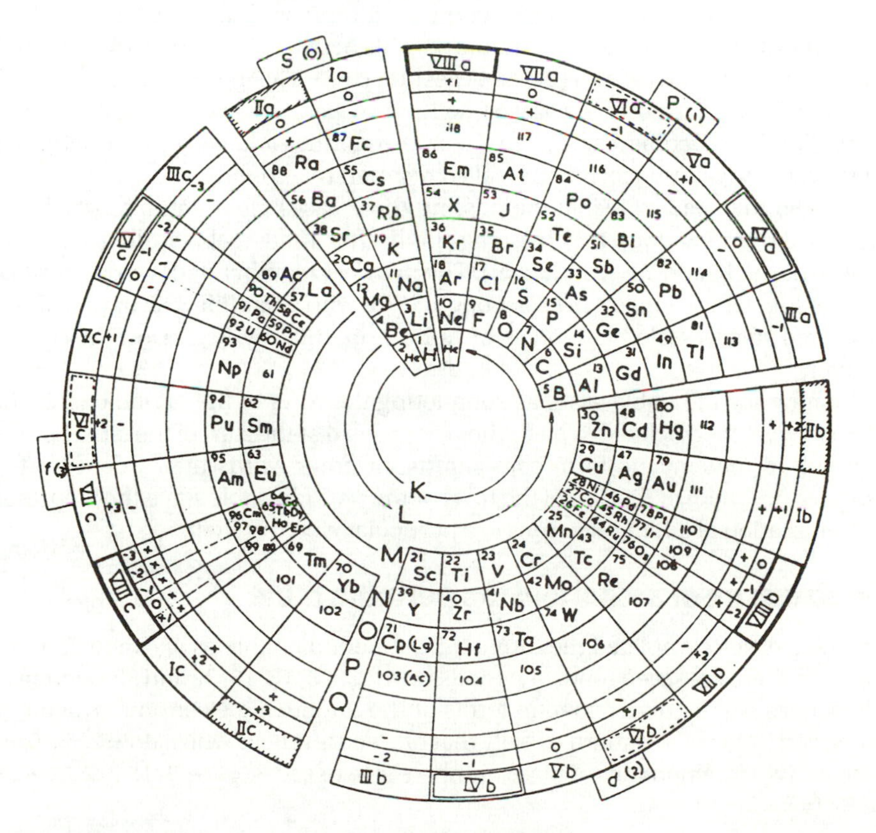

Fig. 6 A circular periodic table

ered, but not recognized, in 1766 by Henry Cavendish (1713–1810) at his laboratory in Clapham, South London. He took a mixture of oxygen and air, repeatedly passed electric sparks through it, and absorbed the gases that were formed. He found that not all the air could be oxidized and reported that there remained a residual 1 per cent. This gas we now know was mainly argon with small amounts of the other noble gases.

The lightest noble gas, helium, was reported the year before Mendeleyev produced his table, by two astronomers, Pierre Janssen (1824–1907) and Norman Lockyer (1836–1920). In 1868 they detected it in the spectrum of light from the sun. They called the new element helium (from the Greek word *helios*) but it was generally assumed that it would not be found on Earth.

The separation of the noble gases began with Lord Rayleigh's researches into the atomic weights of gases. He found that the atomic weight of oxygen was always the same no matter how it was made, but that of nitrogen differed slightly. The nitrogen extracted from air had a density of 1.257 grams per litre, whereas that obtained by decomposing ammonia had a density of 1.251. Together with Ramsay he investigated the discrepancy, by examining atmospheric nitrogen to see if it contained another component. Like Cavendish, they discovered it contained about 1 per cent of another gas and when they examined the atomic spectrum of this they found lines that could only be explained by a new element. They reported their discovery in 1894 and named the element argon. Because of his knowledge of the periodic table, Ramsay realized that this new gas would not be alone, but would be part of a group.

Helium gas was found a few months later, when Ramsay's attention was drawn to a report by the US geochemist William F. Hillebrand, that when uranium ores were heated, an inert gas was given off. Hillebrand thought this might be nitrogen. Ramsay, with the help of Travers, repeated the experiment, collected the gas and found it had a yellow line in its spectrum which was identical with one discovered discovered 30 years earlier in the spectrum of the sun.

Having measured the atomic weights of helium (4) and argon (40) Ramsay was able to deduce that there would be another gaseous element of atomic number about 20, and others heavier than argon with atomic weights around 82 and 132. These would fill gaps in the list of atomic weights between bromine (80) and rubidium (85), and between iodine (127) and caesium (133). The missing gases were likely to be found in the atmosphere as well, so Ramsay and Travers, careful distilled liquid air. This had become available in sizeable quantities following the development of a special refrigeration technique, called regenerative cooling.

By 1899 they had extracted the missing three gases: neon (atomic weight 20), krypton (84) and xenon (131), naming them after the Greek words *neon* (new), *kryptos* (hidden), and *xenos* (stranger). The heaviest member of the group is radioactive radon, whose longest lived isotope is ^{222}Rn with a half-life 3.82 days. This was discovered inside sealed ampoules of radium in 1900 by F. E. Dorn at Halle in Germany.

We now see the noble gases as completing the rows of the periodic table, but chemists were first inclined to place them as group 0 at the start of the table because they had no tendency to form compounds, in other words their valency was 0. Today we put them at the end of the rows of the periodic table since they represent the culmination of adding electrons to a particular electron shell.

The structure of atoms and the periodic table

The noble gases completed the format of the periodic table as we know it. In the long form there are seven rows, or periods, with 2, 8, 8, 18, 18, 32 and 32 elements.[18] This pattern can be understood in terms of the underlying electronic structure of atoms, and indeed the pattern is really that of electron shells which hold 2, 6, 10 and 14 electrons. Combinations of these numbers give rise to 8 (= 2 + 6), 18 (= 2 + 6 + 10), and 32 (= 2 + 6 + 10 + 14).

[18] This last row of the table currently extends to 24 although should eventually reach 32.

This underlying structure could only be understood once the nature of the atom was revealed. This puzzle began to unravel with the discovery of the electron in 1896 by J. J. Thompson. A few year after this Lord Rutherford bombarded thin gold foil with α particles, and discovered that atoms of gold must consist of a tiny, positively charged nucleus in which almost all the mass was concentrated. In 1913 H. G. J. Moseley (1887–1915) formulated the property of atomic number, which is the number of protons of positive charge in the nucleus. He showed that the sequence of elements in the periodic table was really the order of their atomic numbers. Neils Bohr (1885–1962), that same year, linked the form of the periodic table to the atomic structure of atoms. He published a periodic table based on electron energy levels in 1923.

If the elements are arranged in rows in order of increasing atomic number, and in columns having the same electron outer shell, then we arrive at the long form of the periodic table as shown inside the front cover. Across a row of the periodic table we are adding electrons to a particular shell until that shell is full when we arrive at one of the noble gases. Consequently these represent a natural break in the table.

Despite these advances in atomic theory, Mendeleyev's eight-column periodic table remained the common type until after World War II. Then the extended or long form slowly displaced it as the man–made elements were announced, and interest centred on the final row. Although the long form of the table can be traced back to Mendeleyev, he championed the eight-column version. The long form was revived by Lang in 1893, and then developed by the great Swiss inorganic chemist Alfred Werner (1866–1919) in 1905 (Fig. 7). The long form of the table was first used in a textbook by H. G. Deming in 1923, and it became popular in America in the 1930s when it was distributed free by the firm Merck & Company. With the resurgence of inorganic chemistry in the 1960s, the new long form eventually triumphed and today is the accepted format for textbooks, posters and educational displays.

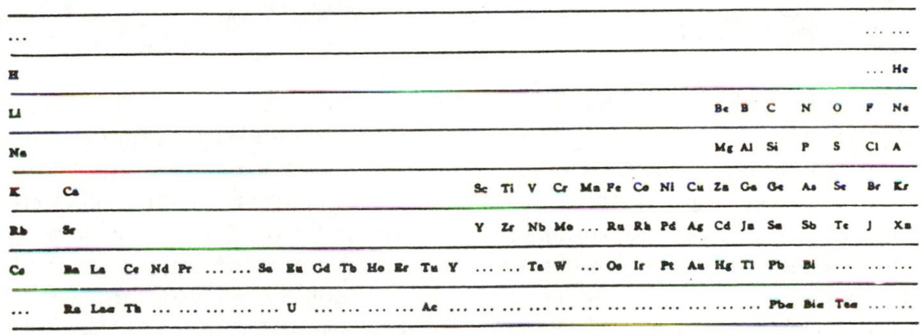

... ...

H ... He

Li Be B C N O F Ne

Na Mg Al Si P S Cl A

K Ca Sc Ti V Cr Mn Fe Co Ni Cu Zn Ga Ge As Se Br Kr

Rb Sr Y Zr Nb Mo ... Ru Rh Pd Ag Cd In Sn Sb Te J Xn

Cs Ba La Ce Nd Pr ... ... Sm Eu Gd Tb Ho Er Tu Y ... ... Ta W ... Os Ir Pt Au Hg Tl Pb Bi ... ...

... Ra Laα Th ... ... U ... ... Ac ... ... Pbα Biα Teα ... ...

Fig. 7 Alfred Werner's periodic table

To understand the periodic table, it is necessary to realize that atoms have shells and sub–shells of electrons around the nucleus and we can imagine these being filled one–by–one with electrons, starting at the levels of lowest energy nearest the nucleus. Studies of the spectra of atoms showed there to be lines corresponding to the energies of electromagnetic radiation (ultraviolet, visible, and infrared) which represented electrons jumping between these various levels. The spectral lines gave rise to a system of quantum numbers that are related to the electron shells.

First there are the principal electron shells and these are numbered according to the principal quantum number, n, which has values 1–7. Each shell has sub–shells and for principal quantum number n there are n of these. They are given a secondary quantum number, l, the so–called orbital quantum number, and this can have values of 0, 1, 2 . . . $(n-1)$. In other words l is always numerically less than n. This means that for the first shell, where $n = 1$, there is only the one shell for which $l = 0$. The values of n and l for the first four shells and sub–shells are as in Table 4.

Table 4 The electron shells and sub-shells which surround the nucleus of an atom

Principal shell n	Sub-shells l	Notation
1	0	1.0 (1s)
2	0, 1	2.0 (2s), 2.1 (2p)
3	0, 1, 2	3.0 (3s), 3.1 (3p), 3.2 (3d)
4	0, 1, 2, 3	4.0 (4s), 4.1 (4p), 4.2 (4d), 4.3 (4f)

Although l has the mathematical values 0, 1, 2, 3 the common notation for these is s, p, d, f which are based on their spectral lines. Thus the first orbital sub-shell is called 1s rather than 1.0, and the second sub-shells are 2s and 2p rather than 2.0 and 2.1.

The shells are filled up with electrons in the order of increasing $n + l$ starting with 1s, then 2s, 2p, 3s, and so on. If two shells have the same numerical value for $n + l$ then the one with the smaller n is filled first. For example if $n + l = 4$, then of the two combinations 4 + 0 or 3 + 1 the latter comes first. The order of filling the electron shells is reflected in the periodic table (see Table 5).

Table 5 The order of filling electrons shells and the periodic table*

		s groups	f groups	d groups	p groups
row 1		1s (1+0=1)			
row 2	→	2s (2+0=2)	→	→	2p (2+1=3)
row 3	→	3s (3+0=3)	→	→	3p (3+1=4)
row 4	→	4s (4+0=4)	→	3d (3+2=5)	4p (4+1=5)
row 5	→	5s (5+0=5)	→	4d (4+2=6)	5p (5+1=6)
row 6	→	6s (6+0=6)	4f (4+3=7)	5d (5+2=7)	6p (6+1=7)
row 7	→	7s (7+0=7)	5f (5+3=8)	6d (6+2=8)	

*The numbers in brackets are $n + l$.

The sub-shells can hold increasing numbers of electrons. The s shells can hold 2, the p shells 6, the d shells 10, and the f shells 14. We recognized these are the basis of the various blocks of the periodic table. The s-block elements consist of two groups (numbered 1, the alkali metals, and 2, the alkaline earths), the p-block elements consist of six groups (numbered 13 to 18), the d-block elements have ten groups (numbered 3 to 12) and the f-block elements consist of two rows 4f and 5f which are not given group numbers. The format of Table 5 reflects the format of the periodic table in the front of *The elements* where the elements are grouped into s, p, d, and f blocks.

There are versions of the periodic table in which the groups are arranged in order of l, the secondary quantum number, i.e.

s group → p group → d group → f group

and while this may seem more logical, it is not possible to place the elements in such tables, based on electron shells, strictly according to increasing atomic numbers. R. Gardner was the first to suggest such a table in 1930, but they did not catch on because they lack the ordering of elements that chemists expect. For instance the third row of such a table would include the elements of the 4s, 4p, 4d, and 4f groups and reads as follows:

4s	/	4p	/	4d	/	4f
K (19), Ca(20)	/	Ga(31), Ge(32) . . . Kr(36)	/	Y(39), Zr(40) . . . Cd (48)	/	La(57), Ce(58) . . . Yb(70)

The dark side

Illuminating as it appears, the periodic table has its dark side. It has led chemists astray, as it did with the tellurium/iodine paradox; it still has its unresolved tensions, and more recently it has led to bitter disputes. One of the latter occurred in the 1970s and 1980s as attempts were made to resolve the conflict between the European and American system of numbering the groups of the periodic table. When the short periodic table of eight groups was turned into the long form of 18 groups, the Europeans numbered the groups on the left half IA to VIII and on the right hand half IB to VIIB with the noble gases being group 0. The Americans on the other hand kept the IA/IB system of Mendeleyev so that their long form of the table was numbered IA, IIA, etc. for the s- and p-block elements, and 1B, IIB, etc. for the d block. Both systems numbered the alkali metals groups IA, and the alkaline earth metals IIA but after that they diverged.

Various suggestions were made to resolve the difficulty but finally the International Union of Pure and Applied Chemistry (IUPAC) suggested the groups simply be numbered 1 through to 18 across the periodic table, a scheme that had first been proposed by a Swedish chemist Arne Ölander in the 1950s. After much heart searching the American Chemical Society (ACS) finally agreed. The f-block does not fit into this system but this poses no problem since the 4f and 5f periods of elements are best dealt with as rows of the periodic table rather than groups.

Renumbering the groups was a storm in a tea cup compared to the controversy of naming new elements. These are mainly the radioactive elements that come beyond uranium, plus the missing ones from the body of the periodic table, elements 43 and 61. All these isotopes are radioactive and have long gone from the Earth's crust. As we have seen, technetium (43) was first made in 1937.

Despite several claims to have discovered element 61 in the early years of this century, the first promethium atoms were only made in 1945 by nuclear processes. In discovering these elements the periodic table was useful in that it indicated that they were yet to be found, while suggesting what properties they would have when finally they were discovered.

The periodic table was less helpful when attempts were first made to make the elements beyond uranium. The first of these were produced by bombarding this with neutrons. The transuranium elements were expected to be like the ones above them in the periodic tables of the day. For example, uranium came below tungsten in the table[19] so the next element should come below rhenium, and the one after that would be below osmium. This misconception led several eminent scientists astray.

In the 1930s research in this area was being undertaken by Enrico Fermi and his team in Rome, by Otto Hahn, Lise Meitner, and Fritz Strassmann in Berlin, and by Irene Curie and Paul Savitch in Paris. It was observed that bombarding uranium with neutrons gave a product with a half-life of 2.3 days. However, the new isotope had chemical properties more like uranium than the rhenium it should have resembled so it was not recognized as a new element. The sequence of neutron bombardment really had produced a new element. It was correctly identified as such by Edwin MacMillan and Philip Abelson at Berkeley in May 1940 and they named it neptunium.

$$^{238}U \rightarrow {}^{239}U\ (t_{1/2} = 23\ \text{min}) \rightarrow {}^{239}Np\ (t_{1/2} = 2.3\ \text{days})$$

The transuranium elements

The element with the highest atomic number that can be found as mineral deposits is uranium (92). Elements beyond this need to be synthesized and are called the transuranium elements. There are alternative ways of doing this. Some new elements have been made by bombarding an existing element with neutrons. These are absorbed by the nucleus relatively easily because they are not charged and so are not repelled. However, absorbing a neutron does not in itself create a new element, it merely makes another isotope of the existing element. What it may

[19] There are several similarities between the chemistry of these two elements, for example both display a wide range of oxidation states, and they have similar oxides UO_2 / U_2O_5 / UO_3 and WO_2 / W_2O_5 / WO_3.

do is to make the nucleus unstable, and this may result in the nucleus ejecting an electron, called a β particle, which has a negative charge. Consequently the nuclear charge increases by +1, and a new element is formed. This is how neptunium (93), americium (95), einsteinium (99) and fermium (100) were made.

Another method of making a new element is to bombard a target of a heavy element with nuclei such as hydrogen (atomic number 1), helium (2), carbon (6), nitrogen (7), or oxygen (8), hoping thereby to fuse them into the nucleus and so create a new heavier element. The heaviest elements are synthesized this way. The difficulty here is that both target and missile are positively charged and so repel each other. If a new, merged, nucleus is to form then the missile nuclei must have a very high energy. This can be done by accelerating them in machines called cyclotrons running at high voltages.

Because the transuranium elements are synthetic, it does not mean that they may not once have existed on Earth. It may simply mean that their half–lives are short compared to the age of our planet, which is about 4.5 billion years old (4.5×10^9 y). For example, even if there had been a million tonnes of neptunium when the Earth was formed there would still have been time for this to undergo over 2000 half lives. In fact it would require only 91 half lives, taking 195 000 000 years, to reduce a million tonnes of neptunium to a single atom.

The longest lived isotopes of the transuranium elements are listed in Table 6.

Table 6 The half–lives of the longest lived isotopes of the transuranium elements

Atomic number	Element	Longest lived isotope	Half life
93	neptunium	^{237}Np	2 140 000 y
94	plutonium	^{244}Pu	82 000 000 y
95	americium	^{243}Am	7370 y
96	curium	^{247}Cm	15 600 000 y
97	berkelium	^{247}Bk	1400 y
98	californium	^{251}Cf	890 y
99	einsteinium	^{254}Es	275 d
100	fermium	^{257}Fm	101 d
101	mendelevium	^{258}Md	56 d
102	nobelium	^{259}No	58 m
103	lawrencium	^{260}Lr	3 m
104	rutherfordium	^{261}Rf	65 s
105	dubnium	^{262}Db	34 s
106	seaborgium	^{266}Sg	27.3 s
107	bohrium	^{262}Bh	0.1 s
108	hassium	^{265}Hs	2×10^{-3} s
109	meitnerium	^{266}Mt	3×10^{-3} s
110	un–named	269	1.7×10^{-4} s
111	un–named	?	?
112	un–named	277	2.8×10^{-4} s

The transfermium controversy

As Table 6 shows, elements beyond fermium have very unstable nuclei, and it becomes progressively harder to make and detect them. As a result the claims to have discovered them are often disputed. Curiously, although these elements have no use, and little chemistry, each one possesses a trivial feature to which an undue amount of importance is attributed—its name. The first person to make a single atom of one of these heavy elements had the indisputable right to name it. When the discovery was disputed then an element could have more than one name, and this is what happened with elements 104, 105, and 106.

The first claim for element 104 came in 1964 from a group of of scientists at Dubna, near Moscow, who reported isotope 260. They named the element kurchatovium after Igor Kurchatov, who had developed the Russian atomic bomb. The claim was disputed in 1969 by a group of scientists led by Albert Ghiorso at

Berkeley, California, who reported isotope 257 and named it rutherfordium after Lord Rutherford. By bombarding ^{249}Cf with ^{12}C nuclei they made several thousand atoms of element 104.

Element 105 was likewise to cause dissent. Two isotopes, 260 and 261, were reported in 1967 by the scientists at Dubna but the element was not given a name, although they later called it nielsbohrium in tribute to Neils Bohr. The claim was disputed in 1970 by the Berkeley group, who reported isotope 260, and named it hahnium after Otto Hahn, the German chemist who first observed uranium fission. Since then several atoms of element 105 have been made from ^{249}Cf by bombarding it with ^{15}N nuclei.

Which names should we use? The International Union of Pure and Applied Chemistry (IUPAC) and The International Union of Pure and Applied Physics (IUPAP) are the bodies which confirm names, and they suggested a compromise. Elements 104 and above were to be named according to a system based on the atomic number of the elements. This was derived from Greek and Latin terms in which the digit 0 would be called nil, 1 would be called un, 2 became bi, 3 tri, 4 quad, 5 pent, 6 hex, 7 sept, 8 oct, and 9 was enn. Since all the elements in this part of the periodic table are metals they all were given the ending –ium to their names. Thus element 104 became unnilquadium (un–nil–quad–ium), 105 became unnilpentium, and 106 unnilhexium.

This expedient solution was only a temporary one, while the claims and counter-claims of the rival groups were judged. In 1987 IUPAC and IUPAP set up a nine–member committee,[20] called the Transfermium Working Group (TWG) to look into the competing priority claims for the discovery of all the transfermium elements. They reported their deliberations in 1992 entitled 'Discovery of the transfermium elements', published in full in the journal *Progress in Particle and Nuclear Physics* **29**, 453 .

The TWG confirmed that there could be no dispute about the discovery and naming of mendelevium (101), which was discovered by the group at The Lawrence Berkeley Laboratory in Berkeley, California; nor about nobelium (102), which was first reported by the group at the Joint Institute for Nuclear Research at Dubna, near Moscow. Lawrencium (103) was reported with varying degrees of completeness and conviction by both Berkeley and Dubna over a period of years and credit should be shared by both. The TWG came to the same conclusion for elements 104 and 105.

Element 106 was also claimed both by the Americans and the Russians but the TWG confirmed the US claim that element 106 was conclusively proved in 1974 by teams at the Lawrence Berkeley Laboratory in Berkeley, California and the Lawrence Livermore National Laboratory led by Ghiorso. He announced the name of the element as seaborgium in March 1994, in honour of Glenn Seaborg who had been instrumental in producing several transuranium elements after World War II. Several atoms of seaborgium have been made by bombarding ^{249}Cf with ^{18}O nuclei using an 88–inch–diameter cyclotron which produces about a billion atoms per hour of which only one is seaborgium.

Element 107 was first made in 1981 at the nuclear research facility Gesellschaft für Schwerionenforschung (GSI) in Darmstadt, Germany and was named nielsbohrium. Peter Armbruster, Gottfried Münzenberg and their co–workers at the heavy ion research facility are credited with the discovery. The Dubna group had reported it earlier but the TWG found their evidence unconvincing. The Darmstadt group used the so–called cold fusion method in which a target of bismuth was bombarded with atoms of chromium and an atom of the element 107 was detected.

In 1984 Peter Armbruster, Gottfried Münzenberg, and their co–workers at GSI again were the first to make hassium, whose name they derived from *Hassia*, the Latin name for Hesse, the German state in which GSI is located. The Russian group at Dubna also produced element 108 in the same year, but the TWG decided that

[20] Its members were R. C. Barber, N. N. Greenwood, A.Z. Hrynkiewicz, Y. P. Jeannin, M. Lefort, S. Sakai, I. Ulehla, A. H. Wapstra, and D. H. Wilkinson.

the major credit should to to the Darmstadt group. They again used the so–called cold fusion method in which a target of lead was bombarded with atoms of iron to give an atom of hassium. The IUPAC committee thought the German state of Hesse did not merit naming an element in its honour and suggested hahnium instead, thereby adding to the confusion because the American Chemical Society had given this to element 105.

Element 109 was discovered in 1982, and in fact was found before element 108. It too was made by Peter Armbruster, Gottfried Münzenberg, and their co–workers at GSI. They named it meitnerium, after the Austrian physicist Lise Meitner, who was the first scientist to realize that spontaneous nuclear fission was possible.[21] A single atom of meitnerium was made by the cold fusion method in which a target of bismuth was bombarded with atoms of iron, and to date fewer than ten atoms of this element have been produced.

To resolve disputes over names, formulas, and symbols, chemists turn to IUPAC. They considered the problem and came up with an approved set of names for these new elements: 104 was to be dubnium; 105 jolontium; 106 rutherfordium; 107 bohrium; 108 hahnium, and 109 meitnerium. In choosing these names they ignored the wishes of the undisputed discoverers of elements 106 (seaborgium) and 108 (hassium). As might be imagined, few chemists were happy with the new names, especially as they conflicted with those already in use by the American Chemical Society.

IUPAC was urged to think again and in February 1997 it came up with a revised list of names which took into account the wishes of the discoverers and the sensibilities of national groups. The approved names of the transfermium elements are now:

- 101 mendelevium (Md)
- 102 nobelium (No)
- 103 lawrencium (Lr)
- 104 rutherfordium (Rf)
- 105 dubnium (Db)
- 106 seaborgium (Sg)
- 107 bohrium (Bh)
- 108 hassium (Hs)
- 109 meitnerium (Mt)

There is little doubt that these will be used, albeit rarely, because there will never be much chemistry to report for these elusive elements, nor for those which immediately follow them.

Late in 1994 elements 110 and 111 were reported by Armbruster. A single atom of the former was made by accelerating nickel atoms through 311 MeV and bombarding a spinning lead target. Its half–life was 0.17 milliseconds, and it decays by emitting an α particle rather than undergoing nuclear fission. A single atom of element 111 was made similarly. In 1996 Armbruster reported element 112, made by fusing zinc and lead nuclei. This had a half-life of 0.28 milliseconds and also decayed by α emission.[22]

The half-life and decay of element 112 hints at increased nuclear stability, although this is not immediately apparent from Table 6. Yet this 'stability' is not so strange as it appears. Several years ago it was suggested that as the atomic number increased we would reach an 'island of stability' for those elements that came at the bottom of the p block of elements, and especially for the one which comes below lead in the table. This would be element 114, and isotope–298 (114 protons and 184 neutrons) would be particularly stable. This isotope corresponds to energy levels within the nucleus being complete—rather like electron energy levels around the nucleus being filled when an atom is a noble gas.

Elements adjacent to 114 would also have enhanced stability of the nucleus so that element 112 (below mercury), element 113 (below thallium) and element 115 (below bismuth) might all have isotopes that would be stable enough for them to be collected, and from which chemical compounds might even be made. This island of

[21] This belated tribute to Lise Meitner (1878–1968) was well deserved. She and Otto Hahn had discovered protactinium in 1917. Lise was also instrumental in discovering nuclear fission, although this was not recognized when Hahn was awarded the Nobel Prize for chemistry in 1944. She had fled Germany in 1938, when her native Austria was annexed by the Nazis.

[22] These elements have yet to be named and are not included in the main tables of *The elements*.

stability is expected to extend to element 118, which would complete row 7p at the bottom of the p block of the periodic table.

Even though we may never reach the elements of the island of stability we can predict what element 114 will be like because it will come below lead in the table. Assuming the half–life of one of its isotopes was long lived, enough for there to be weighable quantities of the element produced, we would expect it to be a soft metal with a low melting point and a very high density. Its most stable oxide would be MO, and chloride MCl_2. There should also be a higher oxide MO_2, and chloride MCl_4. There would be salts of the ion M^{2+} and possibly complex ions of the higher oxidation state M^{4+}, such as MCl_6^{2-} and maybe even MCl_8^{4-} since its large size should enable more chloride ions to surround it than surround lead in $PbCl_6^{2-}$. It would of course be debatable whether these ions could withstand the intense radioactivity of the element, and so they might never be prepared.

Misplaced elements?

There are three elements which pose problems of location in any periodic table. These are hydrogen (1), lanthanum (57), and actinium (89).

Hydrogen has an electron configuration 1s which should place it in the s block of the table above lithium (2s) and this is where it is to be found in some tables. But hydrogen is not a metal like lithium. Indeed by the same logic the noble gas helium ($1s^2$) should therefore be placed above beryllium, but no form of the table places it so because it is clearly a noble gas and must go above neon even though this has the electron configuration of a filled p shell. Some tables place hydrogen by itself, or with helium, in the very centre of the table, floating free above the other elements. Others place hydrogen above fluorine, although it shares little in common with the halogen gases. Some authors give it double billing and place it above both lithium and fluorine. In The Elements I have placed it next to helium and above fluorine and neon, but put both of them in a separate block, labelled 1s, as a mark of their uniqueness.

The other problem concerns the f block of elements which comes between the s block and d block at the bottom of the table. For reasons of economy of space this block is generally written below the d block, as in the inside front cover. Which elements belong to the f block? Traditionally, and for historical reasons, lanthanum and actinium are placed in group 3, below scandium and yttrium. After lanthanum come the other so-called rare earths or lanthanides, which are in fact the upper row of the f block and this begins with cerium. After actinium comes the actinides, the lower row of the f block, which starts with thorium. This arrangement, common in chemistry textbooks, has been challenged as ill-judged. The weight of evidence is that lanthanum and actinium are the first members of the respective rows of the f block. The rows of the f block have 14 elements and so the final ones in each row are ytterbium and nobelium respectively. In 1982 William B. Jensen argued cogently for this change to be made to periodic tables in an article in the *Journal of Chemical Education*, volume 59, page 634. The reasons he gives seem indisputable and based entirely on the physical and chemical properties of these elements, and consequently I have chosen this arrangement for *The elements.*

The periodic table is the hallmark of inorganic chemistry. It summarizes the chemical elements in a simple yet logical table that can even be made into a work of art. As long as chemistry is studied there will be a periodic table. And even if someday we communicate with another part of the universe, we can be sure that one thing both cultures will have in common is an ordered system of the elements that will be instantly recognizable by both intelligent life forms.

Further reading

Abelson, P.H. (1992) 'Discovery of neptunium' *Transuranium elements* (ed. Morss, L.R. and Fuger, J.) American Chemical Society, Washington DC.

Bensaude–Vincent, B. (1984) 'La genèse du tableau de Mendeleev' in *La Recherche*, 15, 1206.

Brock, W.H. (1992) *The Fontana History of Chemistry*, Fontana Press, London.

Cassebaum, H. and Kauffman, G.B., (1971) *Isis*, **62**, 314.
Emsley, J. 'Mendeleyev's dream table' in *New Scientist*, 7 March 1984.
Emsley, J. (1987) 'The development of the periodic table of the chemical elements' in *Interdisciplinary Science Reviews*, **12**, 23.
Gilreath, E.S. (1958) *Fundamental concepts of inorganic chemistry*, McGraw-Hill, New York.
Holden, N.E. (1984) 'Mendeleyev and the periodic classification of the elements' *Chemistry International*, **6**, 18.
Mazurs, E.G. (1974) *Graphic representations of the periodic system during one hundred years*, University of Alabama Press, Tuscloosa.
Rouvray, D.H. (1994) 'Turning the tables on Mendeleev' in *Chemistry in Britain*, May, 373.
Sanderson, R.T. (1967) *Inorganic Chemistry*, Reinhold Publishing Corp., New York, .
Scerri, E.R. (1994) 'Plus ça change...' in *Chemistry in Britain*, May, 379.
Seaborg, G.T. (1990) *The Elements Beyond Uranium*, John Wiley & Sons Inc., New York.
Van Spronsen, J.W. (1969) *The periodic system of chemical elements: a history of the first hundred years*, Elsevier, Amsterdam.
Venables, F.P. (1896) *The development of the periodic law*, Chemical Publishing Co., Easton, PA.
Weeks, M.E. and Leicester, H.M. (1968) *Discovery of the elements* (7th edn), Chapter 14, Journal of Chemical Education, Easton, PA.

The discovery of the elements

There now follows two listings, the first in chronological order of discovery, the second with the elements in order of increasing atomic number.

Discovery of the elements in chronological order

Year	Element	Discoverer	Place
The ancient world			
pre–history	Carbon	–	–
pre–history	Sulfur	–	–
***c.*5000 BC**	Copper	–	–
***c.*3000 BC**	Silver	–	–
***c.*3000 BC**	Gold	–	–
***c.*2500 BC**	Iron	–	–
***c.*2100 BC**	Tin	–	–
***c.*1600 BC**	Antimony	–	–
***c.*1500 BC**	Mercury	–	–
***c.*1000 BC**	Lead	–	–
The Middle Ages			
***c.*1250**	Arsenic	Magnus	Germany
pre–1500*	Zinc	–	–
***c.*1500**	Bismuth	–	–
1669	Phosphorus	Brandt, H.	Hamburg
pre–1700	Platinum	–	–
The eighteenth century			
1735	Cobalt	Brandt, G.	Stockholm
1751	Nickel	Cronstedt	Stockholm
1755	Magnesium	Black	Edinburgh
1766	Hydrogen	Cavendish	London
1772	Nitrogen	Rutherford	Edinburgh
1772	Oxygen	Scheele	Uppsala
		Priestley	Leeds
1774	Chlorine	Scheele	Uppsala
1774	Manganese	Grahn	Stockholm
1780	Chromium	Vauquelin	Paris
1781	Molybdenum	Hjelm	Uppsala
1783	Tellurium	von Reichenstein	Sibiu, Romania
1783	Tungsten	Elhuijar and Elhuijar	Vergara, Spain
1789	Zirconium	Klaproth	Berlin
1789	Uranium	Klaproth	Berlin
1791	Titanium	Gregor	Creed, Cornwall
		Klaproth	Berlin
1794	Yttrium	Gadolin	Åbo, Finland
1797	Beryllium	Vauquelin	Paris
The nineteenth century			
1801	Vanadium	del Rio	Mexico
1801	Niobium	Hatchett	London

* Zinc was known as the copper–zinc alloy, brass, around 20 BC.

1802	Tantalum	Ekeberg	Uppsala
1803	Rhodium	Wollaston	London
1803	Palladium	Wollaston	London
1803	Osmium	Tennant	London
1803	Iridium	Tennant	London
1803	Cerium	Berzelius and Hisinger	Vestmanland, Sweden
1807	Potassium	Davy	London
1807	Sodium	Davy	London
1808	Boron	Lussac and Thenard	Paris
		Davy	London
1808	Calcium	Davy	London
1808	Strontium	Crawford	Edinburgh
1808	Ruthenium	Sniadecki	Vilno, Poland
1808	Barium	Davy	London
1811	Iodine	Courtois	Paris
1815	Thorium	Berzelius	Stockholm
1817	Lithium	Arfvedson	Stockholm
1817	Selenium	Berzelius	Stockholm
1817	Cadmium	Davy	London
1824	Silicon	Berzelius	Stockholm
1825	Aluminium	Oersted	Copenhagen
1826	Bromine	Balard	Montpellier
		Löwig	Heidelberg
1839	Lanthanum	Mosander	Stockholm
1842	Erbium	Mosander	Stockholm
1843	Terbium	Mosander	Stockholm
1860	Caesium	Bunsen and Kirchhoff	Heidelberg
1861	Rubidium	Bunsen and Kirchhoff	Heidelberg
1861	Thallium	Crookes	London
1863	Indium	Reich and Richter	Freiberg
1875	Gallium	de Boisbaudran	Paris
1878	Holmium	Cleve	Uppsala
		Delafontaine and Soret	Geneva
1878	Ytterbium	de Marignac	Geneva
1879	Scandium	Nilson	Uppsala
1879	Samarium	de Boisbaudran	Paris
1879	Thulium	Cleve	Uppsala
1880	Gadolinium	de Marignac	Geneva
1885	Praseodymium	von Welsbach	Vienna
1885	Neodymium	von Welsbach	Vienna
1886	Germanium	Winkler	Freiberg
1886	Fluorine	Moissan	Paris
1886	Dysprosium	de Boisbaudran	Paris
1894	Argon	Rayleigh and Ramsay	London and Bristol
1895	Helium	Ramsay	London
1898	Krypton	Ramsay and Travers	London
1898	Neon	Ramsay and Travers	London
1898	Xenon	Ramsay and Travers	London
1898	Polonium	Curie (Marie)	Paris
1898	Radium	Curie and Curie	Paris
1899	Actinium	Debierne	Paris

The twentieth century

1900	Radon	Dorn	Halle
1901	Europium	Demarçay	Paris

1907	Lutetium	Urbain	Paris
		James	New Hampshire, USA
1917	Protactinium	Hahn and Meitner	Berlin
		Fajans	Karlsruhe
		Soddy, Cranston, and Fleck	Glasgow
1923	Hafnium	Coster and Hevesey	Copenhagen
1925	Rhenium	Noddack, Tacke, and Berg	Berlin
1937	Technetium	Perrier, and Segré	Palermo
1939	Francium	Perey	Paris
1940	Neptunium	McMillan and Abelson	Berkeley, California
1940	Astatine	Corson, Mackenzie, and Segré	Berkeley, California
1940	Plutonium	Seaborg, Wahl and Kennedy	Berkeley, California
1944	Americium	Seaborg, James, Morgan and Ghiorso	Chicago
1944	Curium	Seaborg, James, and Ghiorso	Berkeley, California
1945	Promethium	Marinsky, Glendenin, and Coryell	Oak Ridge, USA
1949	Berkelium	Thompson, Ghiorso, and Seaborg	Berkeley, California
1950	Californium	Thompson, Street, Ghiorso, and Seaborg	Berkeley, California
1952	Einsteinium	Choppin, Thompson, Ghiorso, and Harvey	Berkeley, California
1952	Fermium	Choppin, Thompson, Ghiorso, and Harvey	Berkeley, California
1955	Mendelevium	Ghiorso, Harvey, Choppin, Thompson and Seaborg	Berkeley, California
1958	Nobelium	Various	Dubna, Moscow
1961	Lawrencium	Ghiorso, Sikkeland, Larsh	Berkeley, California
1964	Rutherfordium	Various	Dubna, Moscow and Berkeley, California
1967	Dubnium	Various	Dubna, Moscow and Berkeley, California
1974	Seaborgium	Ghiorso, and others	Berkeley, California
1981	Bohrium	Armbruster, Gottfried Münzenberg, and others	Darmstadt, Germany
1982	Meitnerium	Armbruster, Gottfried Münzenberg, and others	Darmstadt, Germany
1984	Hassium	Armbruster, Gottfried Münzenberg, and others	Darmstadt, Germany
1994	Element 110	Armbruster, Hofmann and others	Darmstadt, Germany
1994	Element 111	Armbruster, Hofmann and others	Darmstadt, Germany
1996	Element 112	Armbruster, Hofmann and others	Darmstadt, Germany

Discovery of the elements in order of atomic number

Element		Year	Discoverer	Place
1	Hydrogen	1766	Cavendish	London
2	Helium	1895	Ramsay	London
3	Lithium	1817	Arfvedson	Stockholm
4	Beryllium	1797	Vauquelin	Paris
5	Boron	1808	Lussac and Thenard	Paris
			Davy	London
6	Carbon	pre–history	–	–
7	Nitrogen	1772	Rutherford	Edinburgh
8	Oxygen	1772	Scheele	Uppsala
			Priestley	Leeds
9	Fluorine	1886	Moissan	Paris
10	Neon	1898	Ramsay and Travers	London
11	Sodium	1807	Davy	London
12	Magnesium	1755	Black	Edinburgh
13	Aluminium	1825	Oersted	Copenhagen
14	Silicon	1824	Berzelius	Stockholm
15	Phosphorus	1669	Brandt, H.	Hamburg
16	Sulfur	pre–history	–	–
17	Chlorine	1774	Scheele	Uppsala
18	Argon	1894	Rayleigh and Ramsay	London and Bristol
19	Potassium	1807	Davy	London
20	Calcium	1808	Davy	London
21	Scandium	1879	Nilson	Uppsala
22	Titanium	1791	Gregor	Creed, Cornwall
			Klaproth	Berlin
23	Vanadium	1801	del Rio	Mexico
24	Chromium	1780	Vauquelin	Paris
25	Manganese	1774	Grahn	Stockholm
26	Iron	*c.*2500 BC	–	–
27	Cobalt	1735	Brandt, G.	Stockholm
28	Nickel	1751	Cronstedt	Stockholm
29	Copper	*c.*5000 BC	–	–
30	Zinc	pre–1500*	–	–
31	Gallium	1875	de Boisbaudran	Paris
32	Germanium	1886	Winkler	Freiberg
33	Arsenic	*c.*1250	Magnus	Germany
34	Selenium	1817	Berzelius	Stockholm
35	Bromine	1826	Balard	Montpellier
			Löwig	Heidelberg
36	Krypton	1898	Ramsay and Travers	London
37	Rubidium	1861	Bunsen and Kirchhoff	Heidelberg
38	Strontium	1808	Crawford	Edinburgh
39	Yttrium	1794	Gadolin	Åbo, Finland
40	Zirconium	1789	Klaproth	Berlin
41	Niobium	1801	Hatchett	London
42	Molybdenum	1781	Hjelm	Uppsala
43	Technetium	1937	Perrier and Segré	Palermo
44	Ruthenium	1808	Sniadecki	Vilno, Poland
45	Rhodium	1803	Wollaston	London
46	Palladium	1803	Wollaston	London
47	Silver	*c.*3000 BC	–	–
48	Cadmium	1817	Davy	London
49	Indium	1863	Reich and Richter	Freiberg

* Zinc was known as the copper–zinc alloy, brass, around 20 BC.

50	Tin	*c.*2100 BC	–	–
51	Antimony	*c.*1600 BC	–	–
52	Tellurium	1783	von Reichenstein	Sibiu, Romania
53	Iodine	1811	Courtois	Paris
54	Xenon	1898	Ramsay and Travers	London
55	Caesium	1860	Bunsen and Kirchhoff	Heidelberg
56	Barium	1808	Davy	London
57	Lanthanum	1839	Mosander	Stockholm
58	Cerium	1803	Berzelius and Hisinger	Vestmanland, Sweden
59	Praseodymium	1885	von Welsbach	Vienna
60	Neodymium	1885	von Welsbach	Vienna
61	Promethium	1945	Marinsky, Glendenin and Coryell	Oak Ridge, USA
62	Samarium	1879	de Boisbaudran	Paris
63	Europium	1901	Demarçay	Paris
64	Gadolinium	1880	de Marignac	Geneva
65	Terbium	1843	Mosander	Stockholm
66	Dysprosium	1886	de Boisbaudran	Paris
67	Holmium	1878	Cleve	Uppsala
			Delafontaine and Soret	Geneva
68	Erbium	1842	Mosander	Stockholm
69	Thulium	1879	Cleve	Uppsala
70	Ytterbium	1878	de Marignac	Geneva
71	Lutetium	1907	Urbain	Paris
			James	New Hampshire, USA
72	Hafnium	1923	Coster and Hevesey	Copenhagen
73	Tantalum	1802	Ekeberg	Uppsala
74	Tungsten	1783	Elhuijar and Elhuijar	Vergara, Spain
75	Rhenium	1925	Noddack, Tacke and Berg	Berlin
76	Osmium	1803	Tennant	London
77	Iridium	1803	Tennant	London
78	Platinum	pre-1700	–	–
79	Gold	*c.*3000 BC	–	–
80	Mercury	*c.*1500 BC	–	–
81	Thallium	1861	Crookes	London
82	Lead	*c.*1000 BC	–	–
83	Bismuth	*c.*1500	–	–
84	Polonium	1898	Curie (Marie)	Paris
85	Astatine	1940	Corson, Mackenzie and Segrè	Berkeley, California
86	Radon	1900	Dorn	Halle
87	Francium	1939	Perey	Paris
88	Radium	1898	Curie and Curie	Paris
89	Actinium	1899	Debierne	Paris
90	Thorium	1815	Berzelius	Stockholm
91	Protactinium	1917	Hahn and Meitner	Berlin
			Fajans	Karlsruhe
			Soddy, Cranston and Fleck	Glasgow
92	Uranium	1789	Klaproth	Berlin
93	Neptunium	1940	McMillan and Abelson	Berkeley, California
94	Plutonium	1940	Seaborg, Wahl and Kennedy	Berkeley, California
95	Americium	1944	Seaborg, James, Morgan and Ghiorso	Chicago
96	Curium	1944	Seaborg, James and Ghiorso	Berkeley, California
97	Berkelium	1949	Thompson, Ghiorso and Seaborg	Berkeley, California

98	Californium	1950	Thompson, Street, Ghiorso and Seaborg	Berkeley, California
99	Einsteinium	1952	Choppin, Thompson Ghiorso and Harvey	Berkeley, California
100	Fermium	1952	Choppin, Thompson, Ghiorso and Harvey	Berkeley, California
101	Mendelevium	1955	Ghiorso, Harvey, Choppin, Thompson and Seaborg	Berkeley, California
102	Nobelium	1958	Various	Dubna, Moscow
103	Lawrencium	1961	Ghiorso, Sikkeland, Larsh	Berkeley, California
104	Rutherfordium	1964	Various	Dubna, Moscow and Berkeley, California
105	Dubnium	1967	Various	Dubna, Moscow and Berkeley, California
106	Seaborgium	1974	Ghiorso, and others	Berkeley, California
107	Bohrium	1981	Armbruster, Gottfried Münzenberg, and others	Darmstadt, Germany
108	Hassium	1984	Armbruster, Gottfried Münzenberg, and others	Darmstadt, Germany
109	Meitnerium	1982	Armbruster, Münzenberg and others	Darmstadt, Germany
110	Un–named	1994	Armbruster, Hofmann and others	Darmstadt, Germany
111	Un–named	1994	Armbruster, Hofmann and others	Darmstadt, Germany
112	Un–named	1996	Armbruster, Hofmann and others	Darmstadt, Germany

The abundance of elements in the Earth's crust

The following figures give the weight in parts per million, which is equivalent to grams per tonne, and they are arranged in order of decreasing abundance.

No.	Element	Abundance
8	Oxygen	474 000
14	Silicon	277 000
13	Aluminium	82 000
26	Iron	41 000
20	Calcium	41 000
11	Sodium	23 000
12	Magnesium	23 000
19	Potasium	21 000
22	Titanium	5600
1	Hydrogen	1520
15	Phosphorus	1000
25	Manganese	950
9	Fluorine	950
56	Barium	500
6	Carbon	480
38	Strontium	370
16	Sulfur	260
40	Zirconium	190
23	Vanadium	160
17	Chlorine	130
24	Chromium	100
37	Rubidium	90
28	Nickel	80
30	Zinc	75
58	Cerium	68
29	Copper	50
60	Neodymium	38
57	Lanthanum	32
39	Yttrium	30
7	Nitrogen	25
3	Lithium	20
27	Cobolt	20
41	Niobium	20
31	Gallium	18
21	Scandium	16
82	Lead	14
90	Thorium	12
5	Boron	10
59	Praseodymium	9.5
62	Samarium	7.9
64	Gadolinium	7.7
66	Dysprosium	6
70	Ytterbium	5.3
68	Erbium	3.8
72	Hafnium	3.3
55	Caesium	3
4	Beryllium	2.6
92	Uranium	2.4
50	Tin	2.2
63	Europium	2.1
73	Tantalum	2
32	Germainium	1.8
33	Arsenic	1.5
42	Molybdenum	1.5
67	Holmium	1.4
18	Argon	1.2
65	Terbium	1.1
74	Tungsten	1
81	Thallium	0.6
71	Lutetium	0.51
69	Thulium	0.48
35	Bromine	0.37
51	Antimony	0.2
53	Iodine	0.14
48	Cadmium	0.11
47	Silver	0.07
34	Selenium	0.05
80	Mercury	0.05
49	Indium	0.049
83	Bismuth	0.048
2	Helium	0.008
52	Tellurium	*c.*0.005
79	Gold	0.0011
44	Ruthenium	*c.*0.001
78	Platinum	*c.*0.001
46	Palladium	6×10^{-4}
75	Rhenium	4×10^{-4}
45	Rhodium	2×10^{-4}
76	Osmium	1×10^{-4}
10	Neon	7×10^{-5}
36	Krypton	1×10^{-5}
77	Iridium	3×10^{-6}
54	Xenon	2×10^{-6}
88	Radium	6×10^{-7}
84	Polonium	trace
85	Astatine	trace
86	Radon	trace
89	Actinium	trace
91	Protactinium	trace
94	Plutonium	trace
43	Technetium	nil
61	Promethium	trace
87	Francium	trace
93	Neptunium	trace
95	Americium	nil
96	Curium	nil
97	Berkelium	nil
98	Califorium	nil
99	Einsteinium	nil
100	Fermium	nil
101	Mendelevium	nil
102	Nobelium	nil
103	Lawrencium	nil
104	Rutherfordium	nil
105	Dubnium	nil
106	Seaborgium	nil
107	Bohrium	nil
108	Hassium	nil
109	Meitnerium	nil
110	Un-named	nil
111	Un-named	nil
112	Un-named	nil

Index

The following is an alphabetical listing of the information included under each element and the section where it is to be found; 'heading' means it is in the information boxes at the top of each page.

The elements in alphabetical order with formulae and atomic numbers

Actinium	Ac	89
Aluminium	Al	13
Americium	Am	95
Antimony	Sb	51
Argon	Ar	18
Arsenic	As	33
Astatine	At	85
Barium	Ba	56
Berkelium	Bk	97
Beryllium	Be	4
Bismuth	Bi	83
Bohrium	Bh	107
Boron	B	5
Bromine	Br	35
Cadmium	Cd	48
Caesium	Cs	55
Calcium	Ca	20
Californium	Cf	98
Carbon	C	6
Cerium	Ce	58
Chlorine	Cl	17
Chromium	Cr	24
Cobolt	Co	27
Copper	Cu	29
Curium	Cm	96
Dubnium	Db	105
Dysprosium	Dy	66
Einsteinium	Es	99
Erbium	Er	68
Europium	Eu	63
Fermium	Fm	100
Fluorine	F	9
Francium	Fr	87
Gadolinium	Gd	64
Gallium	Ga	31
Germanium	Ge	32
Gold	Au	79
Hafnium	Hf	72
Hassium	Hs	108
Helium	He	2
Holmium	Ho	67
Hydrogen	H	1
Indium	In	49
Iodine	I	53
Iridium	Ir	77
Iron	Fe	26
Krypton	Kr	36
Lanthanum	La	57
Lawrencium	Lr	103
Lead	Pb	82
Lithium	Li	3
Lutetium	Lu	71
Magnesium	Mg	12
Manganese	Mn	25
Meitnerium	Mt	109
Mendelevium	Md	101
Mercury	Hg	80
Molybdenum	Mo	42
Neodymium	Nd	60
Neon	Ne	10
Neptunium	Np	93
Nickel	Ni	28
Niobium	Nb	41
Nitrogen	N	7
Nobelium	No	102
Osmium	Os	76
Oxygen	O	8
Palladium	Pa	46
Phosphorus	P	15
Platinum	Pt	78
Plutonium	Pu	94
Polonium	Po	84
Potassium	K	19
Praseodymium	Pr	59
Promethium	Pm	61
Protactinium	Pa	91
Radium	Ra	88
Radon	Rn	86
Rhenium	Re	75
Rhodium	Rh	45
Rubidium	Rb	37
Ruthenium	Ru	44
Rutherfordium	Rf	104
Samarium	Sm	62
Scandium	Sc	21
Seaborgium	Sg	106
Selenium	Se	34
Silicon	Si	14
Silver	Ag	47
Sodium	Na	11
Strontium	Sr	38
Sulfur	S	16
Tantalum	Ta	73
Technetium	Tc	43
Tellurium	Te	52
Terbium	Tb	65
Thallium	Tl	81
Thorium	Th	90
Thulium	Tm	69
Tin	Sn	50
Titanium	Ti	22
Tungsten	W	74
Uranium	U	92
Vanadium	V	23
Xenon	Xe	54
Ytterbium	Yb	70
Yttrium	Y	39
Zinc	Zn	30
Zirconium	Zr	40